T0205349

Mou-Hsiung Chang

Stochastic Control of Hereditary Systems and Applications

 Springer

Mou-Hsiung Chang
4300 S. Miami Blvd.
U.S. Army Research Office
Durham, NC 27703-9142
USA
mouhsiung.chang@us.army.mil

Managing Editors
B. Rozovskiĭ
Division of Applied Mathematics
182 George St.
Providence, RI 02912
USA
rozovski@dam.brown.edu

G. Grimmett
Centre for Mathematical Sciences
Wilberforce Road
Cambridge CB3 0WB
UK
G.R.Grimmett@statslab.cam.ac.uk

ISBN 978-1-4419-2605-0   e-ISBN 978-0-387-75816-9
DOI: 10.1007/978-0-387-75816-9
ISSN: 0172-4568 Stochastic Modelling and Applied Probability

Mathematics Subject Classification (2000): 93E20, 34K50, 90C15

Printed on acid-free paper

9 8 7 6 5 4 3 2 1

springer.com

This book is dedicated to my wife, Yuen-Man Chang.

Preface

This research monograph develops the Hamilton-Jacobi-Bellman (HJB) theory via the dynamic programming principle for a class of optimal control problems for stochastic hereditary differential equations (SHDEs) driven by a standard Brownian motion and with a bounded or an unbounded but fading memory. These equations represent a class of infinite-dimensional stochastic systems that become increasingly important and have wide range of applications in physics, chemistry, biology, engineering, and economics/finance. The wide applicability of these systems is due to the fact that the reaction of real-world systems to exogenous effects/signals is never "instantaneous" and it needs some time, time that can be translated into a mathematical language by some delay terms. Therefore, to describe these delayed effects, the drift and diffusion coefficients of these stochastic equations depend not only on the current state but also explicitly on the past history of the state variable.

The theory developed herein extends the finite-dimensional HJB theory of controlled diffusion processes to its infinite-dimensional counterpart for controlled SHDEs in which a certain infinite-dimensional Banach space or Hilbert space is critically involved in order to account for the bounded or unbounded memory. Another type of infinite-dimensional HJB theory that is not treated in this monograph but arises from real-world application problems can often be modeled by controlled stochastic partial differential equations. Although they are both infinite dimensional in nature and are both in the infancy of their developments, the SHDE exhibits many characteristics that are not in common with stochastic partial differential equations. Consequently, the HJB theory for controlled SHDEs is parallel to and cannot be treated as a subset of the theory developed for controlled stochastic partial differential equations. Therefore, the effort for writing this monograph is well warranted.

The stochastic control problems treated herein include discounted optimal classical control and optimal stopping for SHDEs with a bounded memory over a finite time horizon. Applications of the dynamic programming principles developed specifically for control of stochastic hereditary equations yield an infinite-dimensional Hamilton-Jacobi-Bellman equation (HJBE) for finite time horizons discounted optimal classical control problem, a HJB variational inequality (HJBVI) for optimal stopping problems, and a HJB quasi-variational inequality (HJBQVI) for combined optimal classical-impulse control problems. As an application to its theoretical developments, characterizations of pricing functions in terms of the infinite-dimensional Black-Scholes equation and an infinite-dimensional HJBVI, are derived for European and American option pricing problems in a financial market that consists of a riskless bank account and a stock whose price dynamics depends explicitly on the past historical prices instead of just the current price alone. To further illustrate the roles that the theory of stochastic control of hereditary differential systems played in real-world applications, a chapter is devoted to the development of theory of combined optimal classical-impulse control that is specifically applicable to an infinite time horizon discounted optimal investment-consumption problem in which capital gains taxes and fixed plus proportional transaction costs are taken into consideration. To address some computational issues, a chapter is devoted to Markov chain approximations and finite difference approximations of the viscosity solution of infinite-dimensional HJBEs and HJBVIs. It is well known that the value functions for most of optimal control problems, deterministic or stochastic, are not smooth enough to be a classical solution of HJBEs, HJBVIs, or HJBQVIs. Therefore, the theme of this monograph is centered around development of the value function as the unique viscosity solution of these equations or inequalities.

This monograph can be used as an introduction and/or a research reference for researchers and advanced graduate students who have special interest in theory and applications of optimal control of SHDEs. The monograph is intended to be as much self-contained as possible. Some knowledge in measure theory, real analysis, and functional analysis will be helpful. However, no background material is assumed beyond knowledge of the basic theory of Itô integration and stochastic (ordinary) differential equations driven by a standard Brownian motion. Although the theory developed in this monograph can be extended with additional efforts to hereditary differential equations driven by semi martingales such as Levy processes, we restrain our treatments to systems driven by Brownian motion only for the sake of clarity in theory developments.

This monograph is largely based on the current account of relevant research results contributed by many researchers on controlled SHDEs and on some research done by the author during his tenure as a faculty member at the University of Alabama in Huntsville and more recently as a member of the scientific staff at the U.S. Army Research Office. Most of the material in this monograph is the product of some recently published or not-yet-published results obtained by the author and his collaborators, Roger Youree, Tao Pang, and Moustapha Pemy. The list of references is certainly not exhaustive and is likely to have omitted works done by other researchers. The author apologizes for any inadvertent omissions in this monograph of their works.

The author would like to thank Boris Rozovskiĭ for motivating the submission of the manuscript to Springer and for his encouragement. Sincere thanks also go to Achi Dosanjh and Donna Lukiw, Mathematics Editors of Springer, for their editorial assistance and to Frank Holzwarth, Frank Ganz, and Frank McGuckin for their help on matters that are related to svmono.cls and LaTex.

The author acknowledges partial support by a staff research grant (W911NF-04-D-0003) from the U.S. Army Research Office for the development of some of the recent research results published in journals and contained in this monograph. However, the views and conclusions contained herein are those of the author and should not be interpreted as necessarily representing the official policies or endorsements, either expressed or implied, of the U.S. Army.

Research Triangle Park, North Carolina, USA   *Mou-Hsiung Chang*
October 2007

Contents

Notation

- $\equiv$ denotes "is defined as".

- $\mathbb{Z}$, $\aleph_0$, and $\aleph$ denote the sets of integers, nonnegative integers, and positive integers, respectively.
- $\Re$ and $\Re_+$ denote the sets of real numbers and non-negative real numbers respectively.
- $\mathbf{Q}$ denotes the set of all rational numbers and $\mathbf{Q}_+$ the set of all nonnegative rational numbers.
- $\Re^n$ denotes the n-dimensional Euclidean space equipped with the inner product defined by $x \cdot y = \sum_{i=1}^{n} x_i y_i$ and the Euclidean norm $|\cdot|$ defined by $|x| = (x \cdot x)^{1/2}$ for $x = (x_1, x_2, \ldots, x_n)$, $y = (y_1, y_2, \ldots, y_n) \in \Re^n$.
- $\Re^n_+ = \{(x_1, x_2, \ldots, x_n) \in \Re^n \mid x_i \geq 0, i = 1, 2, \ldots, n\}$.
- $\Re^{n \times m}$- the space of all $n \times m$ real matrices $A = [a_{ij}]$ equipped with the norm $|A| = \{\sum_{i=1}^{n} \sum_{j=1}^{m} |a_{ij}|^2\}^{1/2}$.
- $A^\top$- the transpose of the matrix A.
- $\mathcal{S}^n$- the set of all $n \times n$ symmetric matrices.
- trace $A = \sum_{i=1}^{n} a_{ii}$, the trace of matrix $A = [a_{ij}] \in \mathcal{S}^n$.
- $a \vee b = max\{a, b\}$, $a \wedge b = min\{a, b\}$, $a^+ = a \vee 0$, and $a^- = -(a \wedge 0)$ for all real numbers a and b.
- $m : [0, T] \times C([t - r, T], \Re^n) \rightarrow \mathbf{C}$ denotes the memory map.
- Ξ- a generic real separable Banach or Hilbert space equipped with the norm $\|\cdot\|_\Xi$.
- $\mathcal{B}(\Xi)$- the Borel σ-algebra of subsets of Ξ, $i.e.$, the smallest σ-algebra of subsets of Ξ that contains all open (and hence all closed) subsets of Ξ.
- Ξ^* denotes the class of bounded linear functionals on Ξ equipped with the operator norm $\|\cdot\|_\Xi^*$.
- $\Xi^\dagger$ denotes the class of bounded bilinear functionals on Ξ and equipped with the operator norm $\|\cdot\|_\Xi^\dagger$.
- $\mathcal{L}(\Xi, \Theta)$ denotes the collection of bounded linear transformations (maps) $\Phi : \Xi \rightarrow \Theta$ and equipped with the operator norm $\|\Phi\|_{\mathcal{L}(\Xi, \Theta)}$.

- $\mathcal{L}(\Xi) = \mathcal{L}(\Xi,\Xi)$.
- $C_b(\Xi)$ denotes the class of bounded continuous functions $\Phi : \Xi \to \Re$ and equipped with the norm $\|\Phi\|_{C_b(\Xi)} = \sup_{\mathbf{x}\in\Xi} |\Phi(\mathbf{x})|$.
- $D\Phi(\phi) \in \Xi^*$ denotes the first order Fréchet derivative of $\Phi : \Xi \to \Re$ at $\phi \in \Xi$.
- $D^2\Phi(\phi) \in \Xi^\dagger$ denotes the second order Fréchet derivative of $\Phi : \Xi \to \Re$ at $\phi \in \Xi$.
- $C^2(\Xi)$ denotes the space of twice continuously Fréchet differentiable functions $\Phi : \Xi \to \Re$.
- $C^2_{lip}(\Xi)$ denotes the class of $\Phi \in C^2(\Xi)$ satisfying the following condition: there exists a constant $K > 0$ such that

$$\|D^2\Phi(\mathbf{x}) - D^2\Phi(\mathbf{y})\|^\dagger_\Xi \le K\|\mathbf{x} - \mathbf{y}\|_\Xi, \quad \forall \mathbf{x}, \mathbf{y} \in \Xi.$$

- $C([a,b];\Re^n)(-\infty < a < b < \infty)$ the separable Banach space of continuous functions $\phi : [a,b] \to \Re^n$ equipped with the sup-norm $\|\cdot\|_\infty$ defined by

$$\|\phi\|_\infty = \sup_{t\in[a,b]} |\phi(t)|.$$

- $C^{1,2}_{lip}([0,T] \times \Xi)$ denotes the class of functions $\Phi : [0,T] \times \Xi \to \Re$ that is continuously differentiable with respect to its first variable $t \in [0,T]$ and twice continuously Fréchet differentiable with respect to its second variable $\mathbf{x} \in \Xi$ and $\Phi(t,\cdot) \in C^2_{lip}(\Xi)$ uniformly for all $t \in [0,T]$.
- $\partial_t\Phi(t,x_1,\cdots,x_n) = \frac{\partial}{\partial t}\Phi(t,x_1,\cdots,x_n)$, $\partial_i\Phi(t,x_1,\cdots,x_n) = \frac{\partial}{\partial x_i}\Phi(t,x_1,\cdots,x_n)$, and $\partial_{ij}\Phi(t,x_1,\cdots,x_n) = \frac{\partial^2}{\partial x_i \partial x_j}\Phi(t,x_1,\cdots,x_n)$ for $i,j = 1,2,\cdots,n$.
- $L^2(a,b;\Re^n)(-\infty < a < b < \infty)$ the separable Hilbert space of Lebesque square integrable functions $\phi : [a,b] \to \Re^n$ equipped with the L^2-norm $\|\cdot\|_2$ defined by

$$\|\phi\|_2 = \int_a^b |\phi(t)|^2 dt.$$

- $D([a,b];\Re^n)(-\infty < a < b < \infty)$ the space of functions $\phi : [a,b] \to \Re^n$ that are continuous from the right on $[a,b)$ and have a finite left-hand-limits on $(a,b]$. The space $D([a,b];\Re^n)$ is a complete metric space equipped with Skorohod metric as defined in Definition ( ).
- $r > 0$ is the duration of the bounded memory or delay.
- $\mathbf{C} = C([-r,0];\Re^n)$.
- $\mathbf{M} \equiv \Re \times L^2_\rho((-\infty,0];\Re)$–the ρ-weighted separable Hilbert space equipped with the inner product

$$\langle (x,\phi),(y,\varphi)\rangle_{\mathbf{M}} = xy + \int_{-\infty}^0 \rho(\theta)\phi(\theta)\varphi(\theta)d\theta,$$

$$\forall (x,\phi),(y,\varphi) \in \mathbf{M},$$

and the Hilbertian norm $\|(x,\phi)\|_\rho = \langle (x,\phi),(x,\phi)\rangle_\rho^{1/2}$. In the above, $\rho :$
$(-\infty, 0] \to [0,\infty)$ is a certain given function.
- $(\Omega, \mathcal{F}, P, \mathbf{F})$ (where $\mathbf{F} = \{\mathcal{F}(s), s \geq 0\}$) denotes a complete filtered probability space that satisfies the usual conditions.
- $E[X|\mathcal{G}]$ denotes the conditional expectation of the Ξ-valued random vari-
able X (defined on $(\Omega, \mathcal{F}, P)$) given the sub-σ-algebra $\mathcal{G} \subset \mathcal{F}$. Denote
$E[X] = E[X|\mathcal{F}]$.
- $L^2(\Omega, \Xi; \mathcal{G})$ denotes the collection of $\mathcal{G}$-measurable ($\mathcal{G} \subset \mathcal{F}$) and Ξ-valued
random variables X such that

$$\|X\|_{L^2(\Omega,\Xi)} \equiv \int_\Omega \|X(\omega)\|_\Xi^2 dP(\omega) < \infty.$$

Denote $L^2(\Omega, \Xi) = L^2(\Omega, \Xi; \mathcal{F})$.
- $\mathcal{T}_a^b(\mathbf{F})$ denotes the collection of $\mathbf{F}$-stopping times τ such that $0 \leq a \leq \tau \leq b$,
P-a.s.. $\mathcal{T}_a^b(\mathbf{F}) = \mathcal{T}(\mathbf{F})$ when $a = 0$ and $b = \infty$.
- $(\Omega, \mathcal{F}, P, \mathbf{F}, W(\cdot))$ (or simply $W(\cdot)$ denotes the standard Brownian motion
of appropriate dimension.
- $\mathcal{L}_w\Phi(x_1, \cdots, x_n) = \frac{1}{2}\sum_{j=1}^m \partial_j^2 \Phi(x_1, \cdots, x_n)$.
- $\nabla_x \Phi(x)$ and $\nabla_x^2 \Phi(x)$ denote, respectively, the gradient and Hessian matrix
of $\Phi : \mathfrak{R}^n \to \mathfrak{R}$.
- $\mathbf{1}_{\{0\}} : [-r, 0] \to \mathfrak{R}$, where $\mathbf{1}_{\{0\}}(0) = 1$ and $\mathbf{1}_{\{0\}}(\theta) = 0$ for $-r \leq \theta < 0$.
- $\mathbf{C} \oplus \mathbf{B} = \{\phi + v\mathbf{1}_{\{0\}} \mid \phi \in \mathbf{C}, v \in \mathfrak{R}^n\}$ equipped with the norm
$\|\phi + v\mathbf{1}_{\{0\}}\|_{\mathbf{C}\oplus\mathbf{B}} = \|\phi\|_{\mathbf{C}} + |v|$.
- $\bar{\Gamma}$ is the continuous isometric extension of Γ from $\mathbf{C}^*$ (respectively, $\mathbf{C}^\dagger$) to
$(\mathbf{C} \oplus \mathbf{B})^*$ (respectively, $(\mathbf{C} \oplus \mathbf{B})^\dagger$).
- $\tilde{\phi} : [-r, T] \to \mathfrak{R}^n$ (respectively, $\tilde{\phi} : (-\infty, T] \to \mathfrak{R}^n$ is the extension of
$\phi : [-r, 0] \to \mathfrak{R}^n$ (respectively, $\phi : (-\infty, 0] \to \mathfrak{R}^n$), where $\tilde{\phi}(t) = \phi(0)$ if $t \geq 0$
and $\tilde{\phi}(t) = \phi(t)$ if $t \leq 0$.
- $\{T(t), t \geq 0\} \subset \mathcal{L}(\Xi)$ denotes $\mathcal{C}_0$-semigroup of bounded linear operators on
Ξ.
- $J(t, \psi; u(\cdot))$ denotes the discounted objective functional for the optimal
classical control problem.
- $\mathcal{U}[t, T]$ denotes the class of admissible controls the optimal classical control
problem.
- $\hat{\mathcal{U}}[t, T]$ denotes the class of admissible relaxed controls the optimal classical
control problem.
- $J(t, \psi; \tau)$ denotes the discounted objective functional for the optimal stopping problem.
- $V(t, \psi)$ denotes the value function for the optimal classical control problem,
the optimal stopping problem, and the pricing function, etc.
- $\mathbf{A}$ denotes the weak infinitesimal generator of a Ξ-valued Markov process
with domain $\mathcal{D}(\mathbf{A})$, where $\Xi = \mathbf{C}$ or $\Xi = \mathbf{M}$.
- $\mathcal{S}$ denotes the shift operator with domain $\mathcal{D}(\mathcal{S})$.
- $\lfloor a \rfloor$ denotes the integral part of $a \in \mathfrak{R}$.
- $h^{(N)} = \frac{r}{N}$ for $N \in \aleph$.

- $T^{(N)} = \lfloor \frac{T}{h^{(N)}} \rfloor$.
- $\lfloor t \rfloor_N = h^{(N)} \lfloor \frac{t}{h^{(N)}} \rfloor$.
- G_κ denotes the liquidating function for the hereditary portfolio optimization problem.
- $\mathbf{S}_\kappa$ denotes the solvency region for the hereditary portfolio optimization problem.
- $\mathcal{M}_\kappa$ denotes the intervention operator for the hereditary portfolio optimization problem.
- $\partial \mathbf{S}_\kappa$ denotes the boundary of $\mathbf{S}_\kappa$.
- $(C, \mathcal{T})$ denotes the class of admissible consumption-trading strategies.

Introduction and Summary

This monograph develops the Hamilton-Jacobi-Bellman theory via the dynamic programming principle for a class of optimal control problems for stochastic hereditary differential equations (SHDEs) driven by a standard Brownian motion and with a bounded memory of duration $r > 0$ or an unbounded $r = \infty$ but fading memory. One of the characteristics of these controlled stochastic equations is that their drift and diffusion coefficients at any time $s > 0$ depend not only on the present state $x(s)$ but also explicitly on the finitely past history of the state $x(s + \theta)$, $-r \leq \theta \leq 0$, over the time interval $[s - r, s]$, or on the infinitely past history of the state $x(s + \theta)$, $-\infty < \theta \leq 0$, over the time interval $(-\infty, s]$. These stochastic hereditary differential equations are often called stochastic delay differential equations (SDDEs), stochastic functional differential equations (SFDEs), or stochastic differential equations with aftereffects (especially among Russian and/or Eastern European mathematicians) in the literature. The mathematical models described by SHDEs are ubiquitous and represent a class of stochastic infinite-dimensional systems that have wide range of applications in physics (see, e.g., Frank [Fra02]), chemistry (see, e.g., Beta et al [BBMRE03], Singh and Fogler [SF98]), biology (see, e.g., Longtin et al [LMBM90], Vasilakos and Beuter [VB93], Frank and Beek [FB01], Peterka [Pet00]), engineering and communication (see, e.g., Kushner [Kus05], [Kus06], Ramezani et al [RBCY06], Yang et al [YBCR06]) and economics/finance (see, e.g., Chang and Youree [CY99], [CY07], Ivanov and Swishuk [IS04], Chang et al [CPP07d]), advertising (see, e.g., Gozzi and Marinelli [GM04], Gozzi et al. [GMS06]), and vintage capital (see Fabbri et al. [FFG06], Fabbri and Gozzi [FG06], Fabbi [Fab06]), just to mention a few. The wide applicability of these systems is mainly due to the fact that the reaction of real-world systems to exogenous signals/forces is never "instantaneous" and it needs some time, time which can be translated into a mathematical language by some delay terms. To illustrate some of these examples and to describe the class of optimal control problems for the SHDEs treated in this monograph, we first adopt the following conventional notation.

A. Basic Notation

Throughout the volume, we will adopt the following conventional notation:
1. Throughout the volume, all (time) intervals I will be interpreted as $I \cap \Re$. In particular, the interval $[a, b]$ will be interpreted as $[a, b] \cap \Re$ for all $-\infty \leq a < b \leq \infty$. Therefore, if $b = \infty$ and $a > -\infty$ (respectively, $b < \infty$ and $a = -\infty$), then $[a, b]$ will be interpreted as $[a, \infty)$ (respectively, $(-\infty, b]$). Other intervals such as $(a, b]$, $[a, b)$, and (a, b) shall be interpreted similarly.

2. For $0 < r < \infty$, let $\mathbf{C} \equiv C([-r, 0]; \Re^n)$ be the space of continuous functions $\phi : [-r, 0] \to \Re^n$. The space $\mathbf{C}$ is a real separable Banach space equipped with the uniform topology, that is,

$$\|\phi\| = \sup_{\theta \in [-r,0]} |\phi(\theta)|, \quad \phi \in \mathbf{C}.$$

3. For the unbounded memory case $r = \infty$, we will work with $\mathbf{M} \equiv \Re \times L_\rho^2(-\infty, 0; \Re)$, the separable ρ-weighted Hilbert space equipped with the inner product

$$\langle (x, \phi), (y, \varphi) \rangle_{\mathbf{M}} = xy + \langle \phi, \varphi \rangle_{\rho,2}$$

$$= xy + \int_{-\infty}^{0} \phi(\theta) \varphi(\theta) \rho(\theta) \, d\theta, \quad \forall (x, \phi), (y, \varphi) \in \mathbf{M},$$

and the Hilbertian norm $\|(x, \phi)\|_{\mathbf{M}} = \sqrt{\langle (x, \phi), (x, \phi) \rangle_{\mathbf{M}}}$. In the above, $\rho : (-\infty, 0] \to [0, \infty)$ is a given function that satisfies the following two conditions:

Condition (A1). ρ is summable on $(-\infty, 0]$, that is,

$$0 < \int_{-\infty}^{0} \rho(\theta) \, d\theta < \infty.$$

Condition (A2). For every $t \leq 0$ one has

$$\bar{K}(t) = ess \sup_{\theta \in (-\infty, 0]} \frac{\rho(t + \theta)}{\rho(\theta)} \leq \bar{K} < \infty,$$

$$\underline{K}(t) = ess \sup_{\theta \in (-\infty, 0]} \frac{\rho(\theta)}{\rho(t + \theta)} < \infty.$$

Note that ρ will be referred to as the fading function for the case of infinite memory $r = \infty$.

4. For both bounded ($0 < r < \infty$) and unbounded ($r = \infty$) memory, we adopt the following conventional notation commonly used for (deterministic) functional differential equations (see, e.g., Hale [Hal77], Hale and Lunel [HL93]):

If $0 < T \leq \infty$ and $\varphi : [-r, T] \to \Re^n$ ($r \leq \infty$) is a measurable function, we define, for each $s \in [0, T]$, the function $\psi_s : [-r, 0] \to \Re^n$ by

$$\psi_s(\theta) = \psi(s + \theta), \quad \theta \in [-r, 0].$$

In particular, if $\psi \in C([-r, T]; \Re^n)$ for $r < \infty$, then $\psi_s \in \mathbf{C}$, and if $\psi \in L^2(-r, T; \Re^n)$, then $\psi_s \in L^2(-r, 0; \Re^n)$. The same notation is also applied to any stochastic process $x(\cdot) = \{x(s), s \in [-r, T]\}$. Therefore, for each $s \in [0, T]$, $x_s(\theta) = x(s + \theta)$, and $\theta \in [-r, 0]$, represents the segment of the process $x(\cdot)$ over the finite time interval $[s - r, s]$ or the infinite time interval $(-\infty, s]$. The readers are cautioned to make the distinction between $x(s)$, the value of $x(\cdot)$ at time s, and x_s, the segment of $x(\cdot)$ over the time interval $[s - r, s]$ or $(-\infty, s]$, throughout the monograph.

5. Throughout, $K > 0$ denotes a generic positive constant whose values may change from line to line. We sometimes use the notation $K(p)$ if the constant depends explicitly on the parameter or expression p.

6. Explanations of numbering system used in this monograph are in order. Section x.y stands for Section y of Chapter x. Subsection x.y.z means Subsection z in Section y of Chapter x. Definitions, lemmas, theorems, and remarks are numbered exactly the same way.

Remarks.
1. In studying the finite time horizon control problems, we normally denote $T \geq 0$ as the (fixed) terminal time and $t \in [0, T]$ as a variable initial time. For infinite time horizon control problems (i.e., $T = \infty$), we normally work with autonomous controlled or uncontrolled SHDE and in this case, we can and will assume the initial time $t = 0$.

2. The state variable for each of the control problems will be either $x_s \in \mathbf{C}$, $s \in [t, T]$ (mainly for Chapters 3, 4, 5, and 6) or $(S(s), S_s) \in \mathbf{M}$, $s \geq 0$ (mainly for Chapter 7), depending on whether the initial data $x_t = \psi \in \mathbf{C}$ or $(S(0), S_0) = (\psi(0), \psi) \in \mathbf{M}$ as well as the dynamics of the controlled SHDEs involved.

B. Brief Descriptions of the Stochastic Control Problems and Summary of Results

This monograph treats the discounted optimal classical control, optimal stopping, and optimal classical-impulse control problems for the state systems described by a certain SHDE over a finite or an infinite time horizon. To illustrate the concepts and simplify the treatments, we consider only SHDEs driven by a standard Brownian motion of an appropriate dimension. Although they are increasingly important in real-world applications, the controlled SHDEs driven by Levy or fractional Brownian motions are not the subjects of investigation in this monograph. The omission of this type of SHDE is due to the fact that the treatments of SHDEs driven by Levy processes, although possible by using the concept of stochastic integration with respect to a semi-martingale, require additional technical background preparation and introductions of complicated notation that often unnecessarily obscure the presentation of the main ideas. On the other hand, a general and complete theory of SHDEs driven by fractional Brownian motion and its corresponding optimal control problems have yet to be developed. Interested readers are referred to Protter [Pro95] and Bensousen and Sulem [BS05] and the references contained therein for a theory of jumped SDEs (without delay) and its corresponding optimal control problems. The theory of SDEs without delay ($r = 0$) but driven by a fractional Brownian motion is currently a subject that draws a lot of research attention due to its applications in mathematical finance (see, e.g., Øksendal [Øks04]) and recently discovered phenomena of long-range dependence and self-similarity in modern data networks. On the other hand, the literature on optimal control of SDEs driven by fractional Brownian motion is rather scarce (see, e.g., Mazliak and Nourdin [MN06], Kleptsyna et al. [KLV04]). It is even scarcer in the area of SHDS driven by fractional Brownian motion. To the best of the author's knowledge, there are only two existing papers (see Ferrante and Rovina [FR05], Prakasa-Rao [Pra03]) that treat SHDE driven by a fractional Brownian motion and none addresses the corresponding optimal control problems.

Remark. In this monograph, we choose to treat the maximization of the expected objective functional instead of minimization of the expected cost functional for each of the optimal control problems described below. This is due to the fact that the maximization and minimization problems are related via the following simple relation:

$$\sup_{u\in U} J(u) = -\inf_{u\in U}[-J(u)]$$

for any real-valued function $J(u)$ that are dependent on the control variable u in a control set U. Therefore, all minimization problems of practical applications can be easily reformulated as a maximization problem through the above relation.

As an introduction and a summary, each of the aforementioned optimal control problems and their corresponding solutions will be briefly described as follows. The rigorous formulation of the problems and detailed description of the results obtained can be found in Chapter 3 through Chapter 7. The results on the existence and uniqueness of strong and weak solutions of the SHDE and the Markovian properties of their segmented solution processes are the subjects of Chapter 1. The stochastic calculus including the available Dynkin's and Itô's formulas of the segmented solution processes in (**C** or in **M**) are given in Chapter 2.

This introduction and summary chapter is intended to serve as a sneak preview of the serious stuffs to follow throughout the monograph. Readers who prefer having such a crashed course is not helpful are recommended to go directly to Chapter 1 without loss of reading continuity.

B1. The Optimal Classical Control Problem

The controlled SHDE for the discounted optimal control problem with the bounded memory $0 < r < \infty$ and over a finite time horizon is normally described as follows:

$$dx(s) = f(s, x_s, u(s))ds + g(s, x_s, u(s))dW(s), \quad s \in [t, T], \qquad (0.1)$$

where
(i) $0 < T < \infty$ is a fixed terminal time;

(ii) $W(\cdot) = \{W(s) = (W_1(s), W_2(s), \ldots, W_m(s)), s \geq 0\}$ is an m-dimensional standard Brownian motion defined on a certain complete filtered probability space $(\Omega, \mathcal{F}, P; \mathbf{F})$, with $\mathbf{F} = \{\mathcal{F}(s), s \geq 0\}$ being the P-augmented natural filtration generated by the Brownian motion $W(\cdot)$. Therefore,

$$\mathcal{F}(s) = \sigma\{W(t), 0 \leq t \leq s\} \vee \mathcal{N},$$

and $\mathcal{N}$ contains all P-null sets, that is,

$$\mathcal{N} = \{A \subset \Omega \mid \text{ there exists a } B \in \mathcal{F} \text{ such that } A \subset B \text{ and } P(B) = 0\}.$$

(iii) The controlled drift coefficient f and the controlled diffusion g are some appropriate deterministic functions from $[0, T] \times \mathbf{C} \times U$ into $\Re^n$ and $\Re^{n \times m}$, respectively.

(iv) The process $u(\cdot) = \{u(s), s \in [t, T]\}$ is a (classical) control process taking values in a control set U in a certain Euclidean space.

Given initial data $(t, \psi) \in [0, T] \times \mathbf{C}$, the main objective of the control problem is to find an admissible control $u^*(\cdot) \in \mathcal{U}[t, T]$ (see Section 3.1 for the definition of an admissible control and the class of admissible controls $\mathcal{U}[0, T]$) that maximizes the following discounted objective functional:

$$J(t, \psi; u(\cdot)) = E\left[\int_t^T e^{-\alpha(s-t)} L(s, x_s, u(s))\, ds + e^{-\alpha(T-t)}\Psi(x_T)\right], \qquad (0.2)$$

where $\alpha \geq 0$ denotes a discount factor and $L : [0, T] \times \mathbf{C} \times U \to \Re$ and $\Psi : \mathbf{C} \to \Re$ are respectively the instantaneous and terminal rewards functions that satisfy appropriate polynomial growth conditions in Section 3.1.

The value function (as a function of the initial datum $(t, \psi) \in [0, T] \times \mathbf{C}$) $V : [0, T] \times \mathbf{C} \to \Re$ of the control problem is defined by

$$V(t, \psi) = \sup_{u(\cdot) \in \mathcal{U}[t,T]} J(t, \psi; u(\cdot)). \tag{0.3}$$

Note that the optimal control problem over a finite time horizon briefly described above is referred to as an optimal classical control problem. This is because, although with a complication of having a bounded memory $(0 < r < \infty)$, it is among the first type of continuous time stochastic control problems treated in the literature (see, e.g., Fleming and Rishel [FR75] and Fleming and Soner [FS93] for controlled diffusion processes without delay) in which the effects of the control process $u(\cdot)$ on the dynamics of the state process $x(\cdot)$ are through the controlled drift f and diffusion g and, therefore, will not cause any jumped discontinuity of the controlled state trajectories. This is in contrast to recent studies in singular or impulse control problems for SDEs in the literature (see, e.g., Bensoussan and Lions [BL84], Zhu [Zhu91] and references contained therein).

The treatment and results of the optimal classical control problem are the main objectives of Chapter 3. In addition to the optimal classical control problem described above, we also consider the optimal relaxed control problem in which one wants to find an admissible relaxed control $\mu(\cdot, \cdot)$ that maximizes the expected objective functional

$$\hat{J}(t, \psi; \mu(\cdot, \cdot)) = E \left[\int_t^T \int_U e^{-\alpha(s-t)} L(s, x_s, u) \mu(s, du) \, ds + e^{-\alpha(T-t)} \Psi(x_T) \right] \tag{0.4}$$

among the class of admissible relaxed controls $\hat{\mathcal{U}}[t, T]$ and is subject to the following controlled equation for all $s \in [s, T]$:

$$dx(s) = \int_U f(s, x_s, u) \mu(s, du) \, ds + \int_U g(s, x_s, u) \mu(s, du) \, dW(s). \tag{0.5}$$

We again define the value function $\hat{V} : [0, T] \times \mathbf{C} \to \Re$ as

$$\hat{V}(t, \psi) = \sup_{\mu(\cdot, \cdot) \in \hat{\mathcal{U}}[t,T]} \hat{J}(t, \psi; \mu(\cdot, \cdot)).$$

It is known that any admissible classical control $u(\cdot) \in \mathcal{U}[t, T]$ can be written as an admissible relaxed control as

$$\mu_{u(\cdot)}(B) = \int_t^T \int_U \mathbf{1}_{\{(s,u) \in B\}} \delta_{u(s)} (du) ds, \quad B \in \mathcal{B}([t, T] \times U),$$

where $\boldsymbol{\delta}_u$ is the Dirac measure at $u \in U$. However, the converse is not true in general.

Since $\hat{\mathcal{U}}[t, T]$ is compact under weak convergence, by extending the classical admissible controls $\mathcal{U}[t, T]$ to the class of admissible relaxed controls $\hat{\mathcal{U}}[t, T]$, one can construct a weak convergent sequence of admissible relaxed controls $\{\mu^{(k)}(\cdot, \cdot)\}_{k=1}^{\infty}$ such that

$$\lim_{k \to \infty} \sup_{\mu^{(k)} \in \hat{\mathcal{U}}[t, T]} \hat{J}(t, \psi; \mu^{(k)}(\cdot, \cdot)) = \hat{V}(t, \psi) = V(t, \psi).$$

This establishes the existence of an optimal classical control. The approach described for the existence of optimal classical control is also a main ingredient for the semidiscretization scheme and Markov chain approximation of the optimal control problem in Chapter 5.

The main results obtained include the heuristic derivation via the dynamic programming principle (see Larssen [Lar02]), and under the smoothness conditions of the value function $V : [0, T] \times \mathbf{C} \to \Re$ of an infinite-dimensional Hamilton-Jacobi-Bellman equation (HJBE) over a finite time horizon. The HJBE can be briefly described as follows (see Section 3.4 for details):

$$\alpha V(t, \psi) - \partial_t V(t, \psi) - \max_{u \in U} [\mathbf{A}^u V(t, \psi) + L(t, \psi, u)] = 0, \qquad (0.6)$$

with the terminal condition $V(T, \cdot) = \Psi(\cdot)$ on $\mathbf{C}$, where the operator $\mathbf{A}^u$, $u \in U$, is given as

$$\mathbf{A}^u V(t, \psi) \equiv \mathcal{S}V(t, \psi) + \overline{DV(t, \psi)}(f(t, \psi, u)\mathbf{1}_{\{0\}})$$
$$+ \frac{1}{2} \sum_{j=1}^{m} \overline{D^2 V(t, \psi)}(g(t, \psi, u)\mathbf{e}_j \mathbf{1}_{\{0\}}, g(t, \psi, u)\mathbf{e}_j \mathbf{1}_{\{0\}}),$$

where $\mathbf{e}_j$ is the jth unit vector of the standard basis in $\Re^m$.

The above two expressions have terms that involve the infinitesimal generator, $\mathcal{S}V$, of a semigroup of shift operators, and the extensions, $\overline{DV(t, \psi)}$ and $\overline{D^2 V(t, \psi)}$, of the first- and second-order Freéchet derivatives of V with respect to its second variable that takes value in $\mathbf{C}$. Some brief explanations for these terms in the differential operator $\mathbf{A}^u$, $u \in U$, are given below. The precise definitions of these terms can be found in Sections 2.2 and 2.3.

First, $\mathcal{S}V(t, \psi)$ is defined as

$$\mathcal{S}V(t, \psi) = \lim_{\epsilon \downarrow 0} \frac{V(t, \tilde{\psi}_\epsilon) - V(t, \psi)}{\epsilon}, \qquad (0.7)$$

and $\tilde{\psi} : [-r, T] \to \Re^n$ is the extension of $\psi \in \mathbf{C}$ defined by

$$\tilde{\psi}(t) = \begin{cases} \psi(0) & \text{for } t \geq 0 \\ \psi(t) & \text{for } t \in [-r, 0). \end{cases} \qquad (0.8)$$

Second, $DV(t, \psi) \in \mathbf{C}^*$ and $D^2V(t, \psi) \in \mathbf{C}^\dagger$ are the first and second order Fréchet derivatives of V with respect to its second argument $\psi \in \mathbf{C}$, where $\mathbf{C}^*$ and $\mathbf{C}^\dagger$ are the spaces of bounded linear and bilinear functionals on $\mathbf{C}$, respectively. In addition, $\overline{DV(t, \psi)} \in (\mathbf{C} \oplus \mathbf{B})^*$ is the extension of of $DV(t, \psi)$ from $\mathbf{C}^*$ to $(\mathbf{C} \oplus \mathbf{B})^*$ (see Section 2.4) and $\overline{D^2V(t, \psi)} \in (\mathbf{C} \oplus \mathbf{B})^\dagger$ is the extension of $D^2V(t, \psi)$ from $\mathbf{C}^\dagger$ to $(\mathbf{C} \oplus \mathbf{B})^\dagger$ (also see Section 2.4).

Finally, the function $\mathbf{1}_{\{0\}} : [-r, 0] \to \Re$ is defined by

$$\mathbf{1}_{\{0\}}(\theta) = \begin{cases} 0 \text{ for } \theta \in [-r, 0) \\ 1 \quad \text{for } \theta = 0. \end{cases}$$

These terms are unique to the SHDEs and the HJBE described above and are nontrivial extensions of its counter parts for finite- dimensional controlled diffusion processes without delay ($r = 0$) (see, e.g., [FS93]) and/or controlled stochastic partial differential equations. For comparison purpose, the optimal control problem of finite-dimensional diffusion processes are described below.

The finite-dimensional optimal control problem consists of a controlled SDE described by

$$dx(s) = \bar{f}(s, x(s), u(s)) \, ds + \bar{g}(s, x(s), u(s)) \, dW(s), \quad s \in [t, T],$$

with the initial datum $(t, x) \in [0, T] \times \Re^n$ and the objective functional given by

$$\bar{J}(t, x; u(\cdot)) = E\left[\int_t^T e^{-\alpha(s-t)} \bar{L}(s, x(s), u(s)) \, ds + e^{-\alpha(T-t)} \bar{\Psi}(x(T)) \right],$$

where $\bar{f}, \bar{g}, \bar{L},$ and $\bar{\Psi}$ are appropriate functions defined on $[0, T] \times \Re^n \times U$ (respectively, $\Re^n$) instead of the infinite-dimensional spaces $[0, T] \times \mathbf{C} \times U$ (respectively, $\mathbf{C}$). In this case, the value function $\bar{V} : [0, T] \times \Re^n \to \Re$ is characterized by the following well-known finite- dimensional HJBE:

$$\alpha \bar{V}(t, x) - \partial_t \bar{V}(t, x) - \max_{u \in U} [\mathcal{L}^u \bar{V}(t, x) + L(t, x, u)] = 0,$$

with the terminal condition $\bar{V}(T, x) = \Psi(x)$ for all $x \in \Re^n$, where

$$\mathcal{L}^u \bar{V}(t, x) = \nabla_x \bar{V}(t, x) \cdot \bar{f}(t, x, u) + \frac{1}{2} \bar{g}(t, x, u) \cdot \nabla_x^2 \bar{V}(t, x) \bar{g}(t, x, u).$$

As in most of optimal control problems, deterministic or stochastic, finite or infinite dimensional, it is not known whether the value function is smooth enough to be a solution of the HJBE in the classical sense. Therefore, the concept of viscosity solution is developed for the infinite-dimensional HJBE (0.6). It is also shown in Sections 3.5 and 3.6 that the value function $V : [0, T] \times \mathbf{C} \to \Re$ is the unique viscosity solution of the HJBE (0.6). To characterize the optimal state-control pair, $(x^*(\cdot), u^*(\cdot))$, a generalized verification theorem in the viscosity solution framework is conjectured without proof in Section 3.7. Some special cases of the HJBE and a couple of application examples are illustrated in Section 3.8.

B2. The Optimal Stopping Problem

To describe the optimal stopping problem, we consider the following uncontrolled SHDE with bounded memory $0 < r < \infty$:

$$dx(s) = f(s, x_s)ds + g(s, x_s)\, dW(s), \quad s \in [t, T], \tag{0.9}$$

where, again,

$$W(\cdot) = \{W(s) = (W_1(s), W_2(s), \ldots, W_m(s)), s \geq t\}$$

is an m-dimensional standard Brownian motion as described in **B1**, and $f : [0, T] \times \mathbf{C} \to \Re^n$, $g : [0, T] \times \mathbf{C} \to \Re^{n \times m}$ are some functions that represent respectively the drift and diffusion component of the equation.

Let $\mathbf{G}(t) = \{\mathcal{G}(t, s), t \leq s \leq T\}$ be the filtration of the solution process $\{x(s), s \in [t - r, T]\}$ of (0.9), that is,

$$\mathcal{G}(t, s) = \sigma\{x(\lambda), t \leq \lambda \leq s \leq T\}.$$

For each initial time $t \in [0, T]$, let $\mathcal{T}_t^T(\mathbf{G})$ (or simply $\mathcal{T}_t^T$ when there is no danger of ambiguity) be the class of $\mathbf{G}(t)$-stopping times $\tau : \Omega \to \bar{\Re}$ such that $t \leq \tau \leq T$. Let $L : [0, T] \times \mathbf{C} \to \Re$ and $\Psi : \mathbf{C} \to \Re$ be functions that represent the instantaneous reward rate and the terminal reward of the optimal stopping problem, respectively.

Given an initial data $(t, \psi) \in [0, T] \times \mathbf{C}$ of the SHDE (0.9), the main objective of the optimal stopping problem is to find a stopping time $\tau^* \in \mathcal{T}_t^T$ that maximizes the following discounted objective functional:

$$J(t, \psi; \tau) = E\left[\int_t^\tau e^{-\alpha(s-t)} L(s, x_s)\, ds + e^{-\alpha(\tau-t)} \Psi(x_\tau) \right].$$

In this case, the value function $V : [0, T] \times \mathbf{C} \to \Re$ is defined to be

$$V(t, \psi) = \sup_{\tau \in \mathcal{T}_t^T} J(t, \psi; \tau). \tag{0.10}$$

In Chapter 4, the above optimal stopping problem is approached via two different methods: (1) the construction of the least superharmonic majorant of the terminal reward functional of an equivalent optimal stopping problem and (2) characterization of the value function of the optimal stopping problem in terms of an infinite dimensional HJB variational inequality (HJBVI).

Method (1) is presented in Section 4.2. The main result in that section provides the existence result of an optimal stopping.

It is shown in Section 4.3 that the value function $V : [0, T] \times \mathbf{C} \to \Re$ satisfies the following infinite- dimensional HJBVI if it is sufficiently smooth (see Smoothness Conditions defined in Section 2.6):

$$\max\{\Psi - V, \; \partial_t V + \mathbf{A} V + L - \alpha V\} = 0, \tag{0.11}$$

where

$$\mathbf{A} V(t, \psi) \equiv \mathcal{S} V(t, \psi) + \overline{D V(t, \psi)}(f(t, \psi) \mathbf{1}_{\{0\}})$$
$$+ \frac{1}{2} \sum_{j=1}^{m} \overline{D^2 V(t, \psi)}(g(t, \psi) e_j \mathbf{1}_{\{0\}}, g(t, \psi) e_j \mathbf{1}_{\{0\}}),$$

e_j is the jth unit vector of the standard basis in $\Re^m$, and $\mathcal{S} V(t, \psi)$, $\overline{D V(t, \psi)}$, $\overline{DV(t, \psi)}$, and $\mathbf{1}_{\{0\}}$ are as defined in **B1**.

The above inequality will be interpreted as follows

$$\partial_t V + \mathbf{A} V + L - \alpha V \leq 0 \text{ and } V \geq \Psi \tag{0.12}$$

and

$$(\partial_t V + \mathbf{A} V + L - \alpha V)(V - \Psi) = 0 \tag{0.13}$$

on $[0, T] \times \mathbf{C}$ and with the terminal condition $V(T, \psi) = \Psi(\psi)$ for all $\psi \in \mathbf{C}$.

Again, since the value function $V : [0, T] \times \mathbf{C} \to \Re$ is in general not smooth enough to satisfy the above equation in the classical sense, therefore the concept of viscosity solution is introduced in Section 4.4. It is shown that the value function V is a unique viscosity solution of the HJBVI (0.11). The detail of these results are the subject of discussion in Sections 4.5 and 4.6.

B3. Discrete Approximations

In **B1** we described the value function $V : [0, T] \times \mathbf{C} \to \Re$ of the finite time horizon optimal classical control problem as the unique viscosity solution of the following HJBE:

$$\alpha V(t, \psi) - \partial_t V(t, \psi) - \max_{u \in U} [\mathbf{A}^u V(t, \psi) + L(t, \psi, u)] = 0 \tag{0.14}$$

on $[0, T] \times \mathbf{C}$, and $V(T, \psi) = \Psi(\psi)$, $\forall \psi \in \mathbf{C}$.

The main objective in this subsection is to explore computational issues for the optimal classical control problem described in **B1** under appropriate assumptions. In particular, we present in Chapter 5 three different discrete approximations, namely (1) semidiscrete scheme, (2) Markov chain approximations, and (3) finite difference approximation for the problem.

Let N be a positive integer. We set $h^{(N)} \equiv \frac{r}{N}$ and define

$$\lfloor \cdot \rfloor_N : [0, T] \to \mathbf{I}^{(N)} \equiv \{ kh^{(N)} \mid k = 1, 2, \ldots \} \cap [0, T]$$

by $\lfloor t \rfloor_N = h^{(N)} \lfloor \frac{t}{h}^{(N)} \rfloor$ for $t \in [0, T]$, where $\lfloor a \rfloor$ is the integer part of the real number $a \in \Re$. As T is the terminal time horizon for the original control problem, $\lfloor T \rfloor_N$ will be the terminal time for the Nth approximating problem. It is clear that $\lfloor T \rfloor_N \to T$ and $\lfloor t \rfloor_N \to t$ for any $t \in [0, T]$. The set $\mathbf{I}^{(N)}$ is the time grid of discretization degree N.

Let $\pi^{(N)}$ be the partition of the interval $[-r, 0]$, that is,

$$\pi^{(N)} : r = -Nh^{(N)} < (-N+1)h^{(N)} < \cdots < -h^{(N)} < 0.$$

Define $\tilde{\pi}^{(N)} : \mathbf{C} \to (\Re^n)^{N+1}$ as the $(N+1)$-point mass projection of a continuous function $\phi \in \mathbf{C}$ based on the partition $\pi^{(N)}$, that is,

$$\tilde{\pi}^{(N)}\phi = (\phi(-Nh^{(N)}), \phi((-N+1)h^{(N)}), \ldots, \phi(-h^{(N)}), \phi(0)). \qquad (0.15)$$

Define $\Pi^{(N)} : (\Re^n)^{N+1} \to \mathbf{C}$ by $\Pi^{(N)}\mathbf{x} = \tilde{\mathbf{x}}$ for each

$$\mathbf{x} = (x(-Nh^{(N)}), x(-N+1)h^{(N)}), \ldots, x(-h^{(N)}), x(0)) \in (\Re^n)^{N+1},$$

and $\tilde{\mathbf{x}} \in \mathbf{C}$ by making the linear interpolation between the two (consecutive) time-space points $(kh, x(kh))$ and $((k+1)h, x((k+1)h)$. With a little abuse of notation, we also denote by $\Pi^{(N)} : \mathbf{C} \to \mathbf{C}$ the operator that maps a function $\varphi \in \mathbf{C}$ to its piece-wise linear interpolation $\Pi^{(N)}\varphi$ on the grid $\pi^{(N)}$.

B3.1. Semidiscrete Approximation Scheme

The semidiscrete approximation scheme presented in Section 5.2 of Chapter 5 consists of two steps. Step one is to temporally discretize the state process $x(\cdot) = \{x(s; t, \psi, u(\cdot)), s \in [t-r, T]\}$ but not the control process, and step two further temporally discretizes the control process $u(\cdot) = \{u(s), s \in [t, T]\}$ as well. The semidiscrete approximation scheme presented in that section is mainly due to Fischer and Nappo [FN07].

In the first step of the semi-discrete approximation, the controlled state process $x(\cdot)$, the objective functional $J(t, \psi; u(\cdot))$, and the value function $V(t, \psi)$ are approximated by its temporally discretized counterparts $z^{(N)}(\cdot) = \{z^{(N)}(s; t, \psi, u(\cdot)), s \in [t-r, T]\}$, $J^{(N)}(t, \psi, u(\cdot))$, and $V^{(N)}(t, \psi)$, respectively, where

$$z(s) = \psi(0) + \int_t^s f(\lambda, \Pi^{(N)} z_{\lfloor \lambda \rfloor_N}, u(\lambda)) \, d\lambda \qquad (0.16)$$

$$+ \int_t^s g(\lambda, \Pi^{(N)} z_{\lfloor \lambda \rfloor_N}, u(\lambda)) \, dW(\lambda), \quad s \in [t, \lfloor T \rfloor_N];$$

$$J^{(N)}(t, \psi, u(\cdot)) = E\left[\int_t^{\lfloor T \rfloor_N} e^{-\alpha(s-t)} L(s, \Pi^{(N)} z_{\lfloor s \rfloor_N}, u(s))\, ds \right. \tag{0.17}$$

$$\left. + e^{-\alpha(T-t)} \Psi^{(N)}(\Pi^{(N)} z_{T^{(N)}}) \right];$$

and

$$V^{(N)}(t, \psi) = \sup_{u(\cdot) \in \mathcal{U}[t,T]} J^{(N)}(t, \psi; u(\cdot)). \tag{0.18}$$

Assuming the global boundedness of $|f(t, \phi, u)|$, $|g(t, \phi, u)|$, $|L(t, \phi, u)|$, and $|\Psi(\phi)|$ by K_b and the Lipschitz continuity of these functions (with Lispschitz constant K_{lip}) in addition to assumptions made for **B1**, we have the following error bound for the approximation:

Theorem B3.1.1. Let the initial segment $\phi \in \mathbf{C}$ be γ-Hölder continuous with $0 < \gamma \leq K_H < \infty$ for some constant $K_H > 0$. Then there is a constant $\tilde{K}$ depending only on γ, K_H, K_{lip}, K_b, T, and the dimensions (n and m) such that for all $N = 1, 2, \ldots$ with $N \geq 2r$, and all initial time $t \in \mathbf{I}^{(N)}$, we have

$$|V(t, \phi) - V^{(N)}(t, \psi)| \leq \sup_{u(\cdot) \in \mathcal{U}[t,T]} |J(t, \phi; u(\cdot)) - J^{(N)}(t, \psi; u(\cdot))|$$

$$\leq \tilde{K}\left((h^{(N)})^\gamma \vee \sqrt{h^{(N)} \ln\left(\frac{1}{h^{(N)}}\right)}\right),$$

where $\psi \in \mathbf{C}(N)$ is such that $\psi|_{[-r,0]} = \phi$ and $\mathbf{C}(N)$ is the space of n-dimensional continuous functions defined on the extended interval $[-r - h^{(N)}, 0]$.

The second step of the semidiscretization scheme is to also discretize the temporal variable of the control process and consider the piecewise constant, right-continuous, and **F**-adapted control process $u(\cdot)$ as follows. For $M = 1, 2, \ldots$, set

$$\mathcal{U}^{(M)}[t, T] = \{u(\cdot) \in \mathcal{U}[t, T] \mid u(s) \text{ is } \mathcal{F}(\lfloor s \rfloor_M) \text{ measurable}$$

$$\text{and } u(s) = u(\lfloor s \rfloor_M) \text{ for each } s \in [t, T]\}. \tag{0.19}$$

For the purpose of approximating the control problem of degree N we will use strategies in $\mathcal{U}^{(NM)}[t, T]$. We write $\mathcal{U}^{(N,M)}[t, T]$ for $\mathcal{U}^{(NM)}[t, T]$. Note that $u(\cdot) \in \mathcal{U}^{(N,M)}[t, T]$ has M times finer discretization of that of the discretization of degree N.

With the same dynamics and the same discounted objective functional as in the previous subsection, for each $N = 1, 2, \ldots$ we introduce a family of value functions $\{V^{(N,M)}, M = 1, 2, \ldots\}$ defined on $[t, \lfloor T \rfloor_N] \times \mathbf{C}(N)$ by setting

$$V^{(N,M)}(t, \psi) := \sup_{u(\cdot) \in \mathcal{U}^{(N,M)}[t,T]} J^{(N)}(t, \psi; u(\cdot)). \tag{0.20}$$

We will refer to $V^{(N,M)}$ as the value function of degree (N, M). Note that by construction, we have $V^{(N,M)}(t, \psi) \leq V^{(N)}(t, \psi)$ for all $(t, \psi) \in [0, T^{(N)}] \times C(N)$, since $\mathcal{U}^{(M)}[t, T] \subset \mathcal{U}[t, T]$.

We have the following results for overall discretization error.

Theorem B3.1.2. Let $0 < \gamma \leq K_H < \infty$. Then there is a constant $\bar{K}$ depending on γ, K_H, K_{lip}, K_b, T, and dimensions n and m such that for all $\beta > 3$, $N = 1, 2, \ldots$ with $N \geq 2r$, and all initial datum $(t, \phi) \in \mathbf{I}^{(N)} \times \mathbf{C}$ with ϕ being γ-Hölder continuous, it holds that, with $h = \frac{r}{N^{1+\beta}}$,

$$|V^{(N)}(t, \phi) - V^{(N, \lceil N^\beta \rceil)}(t, \phi)|$$

$$\leq \bar{K} \left(r^{\frac{\gamma\beta}{1+\beta}} h^{\frac{\gamma}{1+\beta}} \vee r^{\frac{\beta}{2(1+\beta)}} \sqrt{\ln\left(\frac{1}{h}\right)} h^{\frac{1}{2(1+\beta)}} + r^{-\frac{\beta}{1+\beta}} h^{\frac{\beta-3}{4(1+\beta)}} \right).$$

In particular, with $\beta = 5$ and $h = \frac{r}{N^6}$, it holds that

$$|V^{(N)}(t, \phi) - V^{(N, N^5)}(t, \phi)| \leq \bar{K} \left(r^{\frac{5\gamma}{6}} h^{\frac{2\gamma-1}{12}} \vee r^{\frac{5}{12}} \sqrt{\ln\left(\frac{1}{h}\right)} + r^{-\frac{5}{6}} \right) h^{\frac{1}{12}}.$$

Theorem B3.1.3. Let $0 < \gamma \leq K_H$. Then there is a constant $\bar{K}(r)$ depending on γ, K_H, K_{lip}, K_b, T, dimensions n and m, and delay duration r such that for all $\beta > 3$, $N, M = 1, 2, \ldots$ with $N \geq 2r$ and $M \geq N^\beta$, and all initial datum $(t, \phi) \in \mathbf{I}^{(N)} \times \mathbf{C}$ with ϕ being γ-Hölder continuous, the following holds: If $\bar{u}(\cdot) \in \mathcal{U}^{(N,M)}[t, T]$ is such that

$$V^{(N,M)}(t, \phi) - J^{(N)}(t, \phi; \bar{u}(\cdot)) \leq \epsilon,$$

then with $h = \frac{r}{N^{1+\beta}}$,

$$V(t, \phi) - J(t, \phi; \bar{u}(\cdot)) \leq \bar{K}(r) \left(h^{\frac{\gamma}{1+\beta}} \vee \sqrt{\ln\left(\frac{1}{h}\right)} h^{\frac{1}{2(1+\beta)}} + h^{\frac{\beta-3}{4(1+\beta)}} \right) + \epsilon.$$

B3.2. Markov Chain Approximations

For notational simplicity, we assume $n = m = 1$ and consider the following autonomous one-dimensional controlled equation:

$$dx(t) = f(x_s, u(s)) \, ds + g(x_s) \, dW(s), \quad s \in [0, T], \tag{0.21}$$

with the initial segment $\psi \in \mathbf{C}$ (here $\mathbf{C} = C[-r, 0]$ throughout **B3.2**) at the initial time $t = 0$.

Using the spatial discretization $\mathbf{S}^{(N)} \equiv \{k\sqrt{h^{(N)}} \mid k = 0, \pm 1, \pm 2, \ldots, \}$ and letting $(\mathbf{S}^{(N)})^{N+1} = \mathbf{S}^{(N)} \times \cdots \times \mathbf{S}^{(N)}$ be the $(N+1)$-folds Cartesian product of $\mathbf{S}^{(N)}$, we call a one-dimensional discrete-time process,

$$\{\zeta(kh^{(N)}), k = -N, -N+1, \ldots, 0, 1, 2, \ldots, T^{(N)}\},$$

defined on a complete filtered probability space $(\Omega, \mathcal{F}, P, \mathbf{F})$, a *discrete chain of degree* N if it takes values in $\mathbf{S}^{(N)}$ and $\zeta(kh^{(N)})$ is $\mathcal{F}(kh^{(N)})$-measurable for all $k = 0, 1, 2, \ldots, T^{(N)}$, where $T^{(N)} = \lfloor \frac{T}{h^{(N)}} \rfloor$. We define the $(\mathbf{S}^{(N)})^{N+1}$-valued discrete process $\{\zeta_{kh^{(N)}}, k = 0, 1, 2, \ldots, T^{(N)}\}$ by setting for each $k = 0, 1, 2, \ldots, T^{(N)}$

$$\zeta_{kh^{(N)}} = \left(\zeta((k-N)h^{(N)}), \zeta((k-N+1)h^{(N)}), \ldots, \zeta((k-1)h^{(N)}), \zeta(kh^{(N)}) \right).$$

We note here that the $\mathbf{S}^{(N)}$-valued process $\zeta(\cdot)$ is not Markovian, but it is desirable under appropriate conditions that its corresponding $(\mathbf{S}^{(N)})^{N+1}$-valued segment process $\{\zeta_{kh^{(N)}}, k = 0, 1, \ldots, T^{(N)}\}$ is Markovian with respect to the discrete-time filtration $\{\mathcal{F}(kh^{(N)}), k = 0, 1, 2, \ldots\}$.

A sequence

$$u^{(N)}(\cdot) = \{u^{(N)}(kh^{(N)}), k = 0, 1, 2, \ldots, T^{(N)}\}$$

is said to be a discrete admissible control if $u^{(N)}(kh^{(N)})$ is $\mathcal{F}(kh^{(N)})$-measurable for each $k = 0, 1, 2, \ldots, T^{(N)}$, and

$$E\left[\sum_{k=0}^{T^{(N)}} \left| u^{(N)}(kh^{(N)}) \right|^2 \right] < \infty.$$

As described earlier in **B3.1**, we let $\mathcal{U}^{(N)}[0, T]$ be the class of continuous-time admissible control process $\bar{u}(\cdot) = \{\bar{u}(s), s \in [0, T]\}$, where for each $s \in [0, T]$, $\bar{u}(s) = \bar{u}(\lfloor s \rfloor_N)$ is $\mathcal{F}(\lfloor s \rfloor_N)$-measurable and takes only finite different values in U.

Given an one-step Markov transition functions $p^{(N)} : (\mathbf{S}^{(N)})^{N+1} \times U \times (\mathbf{S}^{(N)})^{N+1} \to [0, 1]$, where $p^{(N)}(\mathbf{x}, u; \mathbf{y})$ shall be interpreted as the probability that the $\zeta_{(k+1)h} = \mathbf{y} \in (\mathbf{S}^{(N)})^{N+1}$ given that $\zeta_{kh} = \mathbf{x}$ and $u(kh) = u$, where $h = h^{(N)}$. We define a sequence of *controlled Markov chains* associated with the initial segment ψ and $\bar{u}(\cdot) \in \mathcal{U}^{(N)}[0, T]$ as a sequence $\{\zeta^{(N)}(\cdot)\}_{N=1}^{\infty}$ of processes such that $\zeta^{(N)}(\cdot)$ is a $\mathbf{S}^{(N)}$-valued discrete chain of degree N defined on the same filtered probability space $(\Omega, \mathcal{F}, P, \mathbf{F})$ as $u^{(N)}(\cdot)$, provided the following conditions are satisfied:
(i) Initial Condition: $\zeta(-kh) = \psi(-kh) \in \mathbf{S}^{(N)}$ for $k = -N, \ldots, T^{(N)}$.
(ii) Extended Markov Property: For any $k = 1, 2, \ldots$, and

$$\mathbf{y} = (y(-Nh), y((-N+1)h), \ldots, y(0)) \in (\mathbf{S})^{N+1},$$

$$P\{\zeta_{(k+1)h} = \mathbf{y} \mid \zeta(ih), u(ih), i \leq k\}$$
$$= \begin{cases} p^{(N)}(\zeta_{kh}, u(kh); \mathbf{y}) & \text{if } y(-ih) = \zeta((-i+1)h) \text{ for } 1 \leq i \leq N \\ 0 & \text{otherwise} \end{cases}$$

(iii) Local Consistency with the Drift Coefficient:

$$b(kh) \equiv E_\psi^{(N,u)}[\zeta((k+1)h) - \zeta(kh)]$$
$$= hf(\Pi^{(N)}(\zeta_{kh}), u(kh)) + o(h)$$
$$\equiv hf^{(N)}(\zeta_{kh}, u(kh)),$$

where $f^{(N)} : \Re^{N+1} \times U \to \Re$ is defined by

$$f^{(N)}(\mathbf{x}, u) = f(\Pi^{(N)}(\mathbf{x}), u), \quad \forall(\mathbf{x}, u) \in \Re^{N+1} \times U,$$

$E_\psi^{N,u(\cdot)}$ is the conditional expectation given the discrete admissible control $u^{(N)}(\cdot) = \{u(kh), k = 0, 1, 2, \ldots, T^{(N)}\}$, and the initial function

$$\pi^{(N)}\psi \equiv (\psi(-Nh), \psi((-N+1)h), \ldots, \psi(-h), \psi(0)).$$

(iv) Local Consistency with the Diffusion Coefficient:

$$E_\psi^{N,u(\cdot)}[(\zeta((k+1)h) - \zeta(kh) - b(kh))^2] = hg^2(\Pi^{(N)}(\zeta_{kh}), u(kh)) + o(h)$$
$$\equiv h(g^{(N)})^2(\zeta_{kh}, u(kh)),$$

where $g^{(N)} : \Re^{N+1} \to \Re$ is defined by

$$g^{(N)}(\mathbf{x}) = g(\Pi^{(N)}(\mathbf{x})), \quad \forall \mathbf{x} \in \Re^{N+1}.$$

(v) Jump Heights: There is a positive number $\tilde{K}$ independent of N such that

$$\sup_k |\zeta((k+1)h) - \zeta(kh)| \leq \tilde{K}\sqrt{h} \quad \text{for some } \tilde{K} > 0.$$

Note that in the above and below, h and $\mathbf{S}$ are abbreviations for $h^{(N)}$ and $\mathbf{S}^{(N)}$, respectively.

Using the notation and concept developed in the previous subsection, we assume that the $(\mathbf{S}^{(N)})^{N+1}$-valued process $\{\zeta_{kh}, k = 0, 1, \ldots, T^{(N)}\}$ is a controlled $(\mathbf{S}^{(N)})^{N+1}$-valued Markov chain with initial datum $\zeta_0 = \pi^{(N)}\psi \in (\mathbf{S}^{(N)})^{N+1}$ and the Markov probability transition function $p^{(N)} : (\mathbf{S}^{(N)})^{N+1} \times U \times (\mathbf{S}^{(N)})^{N+1} \to [0, 1]$ that satisfies some appropriate local consistency conditions described in (iii) and (iv) above.

The objective functional $J^{(N)} : (\mathbf{S}^{(N)})^{N+1} \times \mathcal{U}^{(N)}[0, T] \to \Re$ and the value function $V^{(N)} : (\mathbf{S}^{(N)})^{N+1} \to \Re$ of the approximating optimal control problem are defined as follows.

Define the objective functional of degree N by

$$J^{(N)}(\pi^{(N)}; u^{(N)}(\cdot)) = E\left[\sum_{k=0}^{T^{(N)}-1} e^{-\alpha kh} L(\Pi^{(N)}(\zeta_{kh}), u(kh))h \right.$$
$$\left. + e^{-\alpha\lfloor T\rfloor_N} \Psi(\zeta_{\lfloor T\rfloor_N}^{(N)}) \right] \tag{0.22}$$

and the value function

$$V^{(N)}(\psi^{(N)}) = \sup_{u^{(N)}(\cdot)} J(\psi^{(N)}; u^{(N)}(\cdot)), \quad \psi \in \mathbf{C}, \tag{0.23}$$

with the terminal condition $V^{(N)}(\zeta^{(N)}_{\lfloor T \rfloor_N}) = \Psi(\zeta^{(N)}_{\lfloor T \rfloor_N})$.

For each $N = 1, 2, \ldots$, we have the following dynamic programming principle (DDP) (see Section 5.3.2 of Chapter 5) for the controlled Markov chain $\{\zeta^{(N)}_{kh}, k = 0, 1, \ldots, T^{(N)}\}$ and discrete admissible control process $u(\cdot) \in \mathcal{U}^{(N)}[0, T]$ as follows.

Proposition B3.2.1. Let $\psi \in \mathbf{C}$ and let $\{\zeta_{kh}, k = 0, 1, 2, \ldots, T^{(N)}\}$ be an $(\mathbf{S})^{N+1}$-valued Markov chain determined by the probability transition function $p^{(N)}$. Then for each $k = 0, 1, 2, \ldots, T^{(N)} - 1$,

$$V^{(N)}(\psi^{(N)})$$

$$= \sup_{u^{(N)}(\cdot)} E\left[e^{-\alpha(k+1)h} V^{(N)}(\zeta^{(N)}_{(k+1)h}) + \sum_{i=0}^{k} h e^{-\alpha ih} L(\Pi(\zeta_{ih}), u^{(N)}(ih)) \right].$$

The optimal control of the Markov chain can then be stated as follows: Given $N = 1, 2, \ldots$ and $\psi \in \mathbf{C}$, find an admissible discrete control process $u^{(N)}(\cdot)$ that maximizes the objective functional $J^{(N)}(\psi^{(N)}; u^{(N)}(\cdot))$ in (0.22).

The algorithm for computing the optimal discrete control $u^{(N)}(\cdot)$ and its corresponding value function $V^{(N)}(\psi^{(N)})$ is provided in Subsection 5.3.2 using a DDP for controlled Markov chain. Under some reasonable assumptions, we prove the following main convergence result:

$$\lim_{N \to \infty} V^{(N)}(\psi^{(N)}) = \hat{V}(\psi),$$

where $\psi \in \mathbf{C}$ is the initial segment, $\psi^{(N)} = \tilde{\pi}^{(N)}(\psi)$ is the point-mass projection of ψ into $(\mathbf{S}^{(N)})^{N+1}$, and $\hat{V}(\psi)$ is the value function of the optimal relaxed control problem mentioned earlier in **B1**. Since $\hat{V}(\psi) = V(\psi)$ for all initial segment ψ, the convergence result is valid for the optimal classical control problem.

We note here that a similar approach of the Markov chain approximation was investigated in Kushner [Kus05, Kus06] and Fischer and Reiss [FR06].

B3.3. Finite Difference Approximation

In Section 5.4 of Chapter 5, a finite difference approximation for the viscosity solution for the HJBE (0.14) is presented. The method detailed in that section and described below is based on the result obtained in Chang et al. [CPP07] and is an extension of results in Barles and Souganidis [BS91].

Given a positive integer M, we consider the following truncated optimal control problem with value function $V_M : [0, T] \times \mathbf{C} \to \Re$ satisfying the following truncated objective functional:

$$V_M(t, \psi) = \sup_{u(\cdot) \in \mathcal{U}[t,T]} E\left[\int_t^T e^{-\alpha(s-t)}(L(s, x_s, u(s)) \wedge M)\, ds \right.$$

$$\left. +e^{-\alpha(T-t)}(\Psi(x_T) \wedge M)\right], \tag{0.24}$$

where $a \wedge b$ is defined by $a \wedge b = \min\{a, b\}$ for all $a, b \in \Re$.

The corresponding HJBE for the truncated $V_M : [0, T] \times \mathbf{C} \to \Re$ defined in (0.24) is given by

$$\alpha V_M(t, \psi) - \partial_t V_M(t, \psi) - \max_{u \in U}\, [\mathbf{A}^u V_M(t, \psi) + L(t, \psi, u) \wedge M] = 0 \tag{0.25}$$

on $[0, T] \times \mathbf{C}$, and $V_M(T, \psi) = \Psi(\psi) \wedge M$, $\forall \psi \in \mathbf{C}$.

Similarly to the proof that $V : [0, T] \times \mathbf{C} \to \Re$ is the unique viscosity solution of the HJBE (0.14) (see Section 3.6 of Chapter 3 for detail), one can show that the value function V_M is the unique viscosity solution of (0.25). Moreover, it is easy to see that $V_M \to V$ as $M \to \infty$ pointwise on $[0, T] \times \mathbf{C}$.

To obtain a computational algorithm, we need only find the numerical solution for $V_M : [0, T] \times \mathbf{C} \to \Re$ for each M.

Let ϵ ($0 < \epsilon < 1$) be the stepsize for variable ψ and η ($0 < \eta < 1$) be the stepsize for t. We consider the finite difference operators Δ_ϵ, Δ_η and Δ_η^2 defined by

$$\Delta_\eta \Phi(t, \psi) = \frac{\Phi(t + \eta, \psi) - \Phi(t, \psi)}{\eta},$$

$$\Delta_\epsilon \Phi(t, \psi)(\phi + v\mathbf{1}_{\{0\}}) = \frac{\Phi(t, \psi + \epsilon(\phi + v\mathbf{1}_{\{0\}})) - \Phi(t, \psi)}{\epsilon},$$

$$\Delta_\epsilon^2 \Phi(t, \psi)(\phi + v\mathbf{1}_{\{0\}}, \varphi + w\mathbf{1}_{\{0\}}) = \frac{\Phi(t, \psi + \epsilon(\phi + v\mathbf{1}_{\{0\}})) - \Phi(t, \psi)}{\epsilon^2}$$
$$+\frac{\Phi(t, \psi - \epsilon(\varphi + w\mathbf{1}_{\{0\}})) - \Phi(t, \psi)}{\epsilon^2},$$

where $\phi, \varphi \in \mathbf{C}$ and $v, w \in \Re^n$.

Recall that

$$\mathcal{S}\Phi(t, \psi) = \lim_{\epsilon \to 0+} \frac{1}{\epsilon}\left[\Phi(t, \tilde{\psi}_\epsilon) - \Phi(t, \psi)\right].$$

Therefore, we define

$$\mathcal{S}_\epsilon \Phi(t, \psi) = \frac{1}{\epsilon}\left[\Phi(t, \tilde{\psi}_\epsilon) - \Phi(t, \psi)\right].$$

It is clear that $\mathcal{S}_\epsilon \Phi$ is an approximation of $\mathcal{S}\Phi$.

Let $C^{1,2}([0, T] \times \mathbf{C})$ be the space of continuous functions $\Phi : [0, T] \times \mathbf{C} \to \Re$ that are continuously differentiable with respect its first variable $t \in [0, T]$ and twice continuously Fréchet differentiable with respect to its second variable

$\psi \in \mathbf{C}$. The following preliminary results hold true (see Section 5.4 of Chapter 5): For any $\Phi : [0,T] \times \mathbf{C} \to \Re$, $\Phi \in C^{1,2}([0,T] \times \mathbf{C})$, such that Φ can be smoothly extended to $[0,T] \times (\mathbf{C} \oplus \mathbf{B})$, we have

$$\lim_{\epsilon \to 0} \Delta_\epsilon \Phi(t, \psi)(\phi + v\mathbf{1}_{\{0\}}) = \overline{D\Phi(t, \psi)}(\phi + v\mathbf{1}_{\{0\}}) \qquad (0.26)$$

and

$$\lim_{\epsilon \to 0} \Delta_\epsilon^2 \Phi(t, \psi)(\phi + v\mathbf{1}_{\{0\}}) = \overline{D^2\Phi(t, \psi)}(\phi + v\mathbf{1}_{\{0\}}, \varphi + w\mathbf{1}_{\{0\}}). \qquad (0.27)$$

With the discrete approximating scheme described earlier, the discretized version of (0.25) for the truncated V_M can be rewritten in the following form:

$$V_M(t, \psi) = \mathcal{T}_{\epsilon,\eta} V_M(t, \psi), \qquad (0.28)$$

where $\mathcal{T}_{\epsilon,\eta}$ is operator on $C_b([0,T] \times (\mathbf{C} \oplus \mathbf{B}))$ (the space of bounded continuous functions from $[0,T] \times (\mathbf{C} \oplus \mathbf{B})$ to $\Re$) defined by

$$\mathcal{T}_{\epsilon,\eta}\Phi(t, \psi) \equiv \max_{u \in U} \left[\frac{1}{\frac{2}{\epsilon} + \frac{1}{\eta} + \frac{m}{\epsilon^2} + \delta} \left(\frac{1}{\epsilon}\Phi(t, \tilde{\psi}_\epsilon) + \frac{\Phi(t, \psi + \epsilon(f(t, \psi, u)\mathbf{1}_{\{0\}}))}{\epsilon} \right. \right.$$

$$+ \frac{1}{2}\sum_{i=1}^m \frac{\Phi(t, \psi + \epsilon(g(t, \psi, u)\mathbf{e}_i\mathbf{1}_{\{0\}})) + \Phi(t, \psi - \epsilon(g(t, \psi, u)\mathbf{e}_i\mathbf{1}_{\{0\}}))}{\epsilon^2}$$

$$\left. \left. + \frac{\Phi(t + \eta, \psi)}{\eta} + L(t, \psi, u) \wedge M \right) \right]. \qquad (0.29)$$

By showing that $\mathcal{T}_{\epsilon,\eta}$ is a contraction map for each ϵ and η, one proves by the Banach fixed point that the strict contraction $\mathcal{T}_{\epsilon,\eta}$ has a unique fixed point denoted by $\Phi_{\epsilon,\eta}^M$. Given any function $\Phi_0 \in C_b([0,T] \times (\mathbf{C} \oplus \mathbf{B})$, we construct a sequence as follows: $\Phi_{n+1} = \mathcal{T}_{\epsilon,\eta}\Phi_n$ for $n \geq 0$. It is clear that

$$\lim_{n \to \infty} \Phi_n = \Phi_{\epsilon,\eta}^M. \qquad (0.30)$$

We have the following as one of the main theorems for Section 5.4 of Chapter 5.

Theorem B3.3.1. Let $\Phi_{\epsilon,\eta}^M$ denote the solution to (0.30). Then, as $(\epsilon, \eta) \to 0$, the sequence $\Phi_{\epsilon,\eta}^M$ converges uniformly on $[0,T] \times \mathbf{C}$ to the unique viscosity solution V_M of (0.25).

Based on the results described above and combining with the semidiscretization scheme summarized in **B3.1**, we can construct the computational algorithm for each $N = 1, 2, \ldots$ to obtain a numerical solution for the HJBE (0.14). For example, one algorithm can be like the following:

Step 0. Let $(t, \psi) \in [0,T] \times \mathbf{C}$ be the given initial datum.
(i) Compute $\lfloor t \rfloor_N$ and $\pi^{(N)}\psi$.

(ii) Choose any function $\Phi^{(0)} \in C_b([0,T] \times \mathbf{C} \oplus \mathbf{B})$.

(iii) Compute $\hat{\Phi}^{(0,N)}(\lfloor t \rfloor_N, \pi^{(N)}\psi) = \Phi^{(0)}(\lfloor t \rfloor_N, \pi^{(N)}\psi)$.

Step 1. Pick the starting values for $\epsilon(1), \eta(1)$. For example, we can choose $\epsilon(1) = 10^{-2}$ and $\eta(1) = 10^{-3}$.

Step 2. For the given $\epsilon, \eta > 0$, compute the function

$$\hat{\Phi}^{(1,N)}_{\epsilon(1),\eta(1)} \in C_b([0,T] \times (\mathbf{S}^{(N)})^{N+1})$$

by the following formula:

$$\hat{\Phi}^{(1,N)}_{\epsilon(1),\eta(1)} = \mathcal{T}^{(N)}_{\epsilon(1),\eta(1)} \Phi^{(0,N)},$$

where $\mathcal{T}^{(N)}_{\epsilon(1),\eta(1)}$, the semidiscretized version of $\mathcal{T}_{\epsilon,\eta}$, is defined on $C_b([0,T] \times (\mathbf{S}^{(N)})^{N+1})$ and can be found in Subsection 5.4.2 of Chapter 5.

Step 3. Repeat Step 2 for $i = 2, 3, \ldots$ using

$$\hat{\Phi}^{(i,N)}_{\epsilon(1),\eta(1)}(\lfloor t \rfloor_N, \pi^{(N)}\psi) = \mathcal{T}^{(N)}_{\epsilon(1),\eta(1)} \Phi^{(i-1,N)}_{\epsilon(1),\eta(1)}(\lfloor t \rfloor_N, \pi^{(N)}\psi).$$

Stop the iteration when

$$|\hat{\Phi}^{(i+1,N)}_{\epsilon(1),\eta(1)}(t,\psi) - \hat{\Phi}^{(i,N)}_{\epsilon(1),\eta(1)}(t,\psi)| \leq \delta_1,$$

where δ_1 is a preselected number that is small enough to achieve the accuracy we want. Denote the final solution by $\hat{\Phi}_{\epsilon(1),\eta(1)}(\lfloor t \rfloor_N, \pi^{(N)}\psi)$.

Step 4. Choose two sequences of $\epsilon(k)$ and $\eta(k)$ such that

$$\lim_{k\to\infty} \epsilon(k) = \lim_{k\to\infty} \eta(k) = 0.$$

For example, we may choose $\epsilon(k) = \eta(k) = 10^{-(2+k)}$. Now, repeat Step 2 and Step 3 for each $\epsilon(k)$ and $\eta(k)$ until

$$|\hat{\Phi}^{(i,N)}_{\epsilon(k+1),\eta(k+1)}(\lfloor t \rfloor_N, \pi^{(N)}\psi) - \hat{\Phi}^{(i,N)}_{\epsilon(k),\eta(k)}(\lfloor t \rfloor_N, \pi^{(N)}\psi)| \leq \delta_2,$$

where δ_2 is chosen to obtain the expected accuracy.

B4. Option Pricing

As an application of the optimal stopping problem outlined in **B2** and detailed in Chapter 4, we consider pricing problems of the American and European options briefly described below. The detail of this particular application is the subject of discussions in Chapter 6.

An *American option* is a contract conferred by the contract *writer* on the contract *holder* giving the *holder* the right (but not the obligation) to buy from or to sell to the *writer* a share of the stock at a prespecified price prior to or at the contract expiry time $0 < T < \infty$. The right for the *holder* to buy from the *writer* a share of the stock will be called a *call option* and the right to sell to the *writer* a share of the stock will be called a *put option*. If an *option* is purchased at time $t \in [0,T]$ and is exercised by the *holder* at an stopping time $\tau \in [t,T]$, then he will receive a discounted payoff of the amount $e^{-\alpha(\tau-t)}\Psi(S_\tau)$ from the *writer*, where $\Psi : \mathbf{C} \to [0,\infty)$ is the payoff function and $\{S(s), s \in [t,T]\}$ is the unit price of the underlying stock whose dynamics is described by the following one-dimensional SHDE:

$$\frac{dS(s)}{S(s)} = f(S_s)\,ds + g(S_s)\,dW(s), \quad s \in [t,T], \tag{0.31}$$

with initial data $S_t = \psi \in \mathbf{C}$ at time $t \in [0,T]$, where the mean growth rate $f(S_s)$ and the volatility rate $g(S_s)$ of the stock at time $s \in [t,T]$ depend explicitly on the stock prices S_s over the time interval $[s-r,s]$ instead of the stock price $S(s)$ at time s alone.

In order to secure such a contract, the contract *holder*, however, has to pay to the *writer* at contract purchase time t a fee that is mutually agreeable to both parties. The determination of such a fee is called the pricing of the *American option*. In determining a price for the *American option*, the *writer* of the option seeks to invest in the (B,S)-market the fee x received from the *holder* and trades over the time interval $[t,T]$ between the *bank* account and the underlined *stock* in an optimal and prudent manner so that his total wealth will replicate or exceed that of the discounted payoff $e^{-\alpha(\tau-t)}\Psi(S_\tau)$ he has to pay to the *holder* if and when the option is exercised at τ. The smallest such x is called the fair price of the *option*. Note that the fair price of the option x is a function of the initial data $(t,\psi) \in [0,T] \times \mathbf{C}$ and will be expressed as the rational pricing function $V : [0,T] \times \mathbf{C} \to [0,\infty)$.

Under the same scenario described above, the contract is called an European option and can be viewed as a special case of an American option if the option can only be exercised at the expiry time T instead of any time prior to or at the expiry time T as is stipulated in an American option.

In the (B,S)-market, it is assumed that the bank account $\{B(t), t \geq -r\}$ grows according to the following linear (deterministic) functional differential equation:

$$dB(t) = L(B_t)\,dt, \tag{0.32}$$

where

$$L(B_t) \equiv \int_{-r}^{0} B(t+\theta)d\eta(\theta)\,dt, \; t \geq 0,$$

and $\eta : [-r,0] \to \Re$ is a nondecreasing function (and therefore of bounded variation) such that $\eta(0) - \eta(-r) > 0$.

The following characterization of the pricing function of the American option $V : [0,T] \times \mathbf{C} \to \Re_+$ as an optimal stopping problem is obtained in Chang and Youree [CY99], [CY07], Chang et al. [CPP07a], [CPP07d] and will be treated in detail in Chapter 6.

Theorem B4.1. Given a (globally convex) payoff function $\Psi : \mathbf{C} \to \Re_+$ satisfying the polynomial growth condition

$$|\Psi(\psi)| \leq K(1 + \|\psi\|_2^k) \text{ for some constants } K > 0 \text{ and } k \geq 1,$$

then the pricing function $V : [0,T] \times \mathbf{C} \to \Re_+$ is given as follows:

$$V(t,\psi) = \sup_{\tau \in \mathcal{T}_t^T} \tilde{E}\left[e^{-\alpha(T-\tau)}\Psi(S_\tau)\Big| S_t = \psi\right],$$

where $\tilde{E}[\cdots]$ represents the expectation for a suitable probability measure $\tilde{P}$ defined in Section 6.3 and $\alpha > 0$ is the discount factor.

As a consequence of the optimal stopping problem, we also have the following result.

Theorem B4.2. Assume that the the payoff function $\Psi : \mathbf{C} \to \Re_+$ satisfies the polynomial growth condition described in Theorem (B4.1). Then the pricing function is the unique viscosity solution of the following HJBVI:

$$\max\{\partial_t V + \mathbf{A}V - \alpha V, \Psi - V\} = 0, \qquad (0.33)$$

with the terminal condition $V(T,\psi) = \Psi(\psi)$ for all $\psi \in \mathbf{C}$, where the operator $\mathbf{A}$ is defined by

$$\mathbf{A}\Phi(\psi) = \mathcal{S}\Phi(\psi) + \overline{D\Phi(\psi)}(\lambda\psi(0)\mathbf{1}_{\{0\}}) \qquad (0.34)$$
$$+ \frac{1}{2}\overline{D^2\Phi(\psi)}(\psi(0)g(\psi)\mathbf{1}_{\{0\}}, \psi(0)g(\psi)\mathbf{1}_{\{0\}}),$$

where $\lambda > 0$ is the effective interest rate of the Bank account that satisfies the following equation:

$$\lambda = \int_{-r}^0 e^{\lambda\theta}d\eta(\theta).$$

As a special case of the two results described in Theorem B4.1. and Theorem B4.2., we obtain the following infinite-dimensional Black-Scholes equation for the European option:

Theorem B4.3. Assume that the the payoff function $\Psi : \mathbf{C} \to \Re_+$ satisfies the polynomial growth condition as described in Theorem (B4.1). The the pricing function for the European option is the unique viscosity solution of the following infinite-dimensional Black-Scholes equation:

$$\partial_t V + \mathbf{A}V - \alpha V = 0. \qquad (0.35)$$

A computational algorithm is also developed for solving the above Black-Scholes equation in Section 6.7 of Chapter 6.

The reward function $\Psi : \mathbf{C} \to \Re_+$ considered here includes the standard call/put option that has been traded in option exchanges around the world.

Standard Options

The payoff function for the standard call option is given by $\Psi(S_s)=\max\{S(s)-q,0\}$, where $S(s)$ is the stock price at time s and $q > 0$ is the strike price of the standard call option. If the standard American option is exercised by the *holder* at the **G**-stopping time $\tau \in T_t^T$ (**G** $= \{\mathcal{G}(t), t \geq 0\}$ is the filtration generated by $\{S(t), t \geq 0\}$), the *holder* will receive the payoff of the amount $\Psi(S_\tau)$. Of course, the option will be called the standard European option if $\tau \equiv T$. In this case, the amount of the reward will be $\Psi(S_T) = \max\{S(T) - q, 0\}$. We offer a financial interpretation for the standard American (or European) call option as follows. If the option has not been exercised and the current stock price is higher than the strike price, then the *holder* can exercise the option and buy a share of the stock from the *writer* at the strike price q and immediately sell it at the open market and make an instant profit of the amount $S(\tau) - q > 0$. It the strike price is higher than the current stock price, then the option is worthless to the *holder*. In this case, the *holder* will not exercise the option and, therefore, the payoff will be zero.

The following path-dependent exotic options are special cases of our option pricing problem and gives justification for the formulation with bounded memory.

Modified Russian Option

The payoff process of a Russian call option can be expressed as

$$\max\left\{E\left[\sup_{s\in[\tau-r,\tau]} S(s)\right] - q, 0\right\} = \max\{E[\|S_\tau\| - q, 0\}$$

for some strike price $q > 0$. Note that the payoff of a Russian option depends on the highest price on the time interval $[T - r, T]$ if the option is exercised at $\tau \in T_t^T$.

Modified Asian Option

The payoff for the Asian call option is given by

$$\max\left\{E\left[\frac{1}{\tau}\int_{\tau-r}^{\tau} S(s)\,ds\right] - q, 0\right\}.$$

This is similar to a European call option with strike price $q > 0$ except that now the "averaged stock price" $\frac{1}{\tau}\int_{\tau-r}^{\tau} S(s)\,ds$, over the interval $[\tau - r, \tau]$, is used in place of the stock price $S(\tau)$ at option exercise time τ.

It is clear that when $r = 0$, (0.31) reduces to the following nonlinear stochastic ordinary differential equation:

$$\frac{dS(t)}{S(t)} = f(S(t)) \, dt + g(S(t)) \, dW(t), \ t \geq 0, \tag{0.36}$$

of which the Black-Scholes (B, S)-market (i.e., $f(S(t)) \equiv \mu \in \Re$ and $g(S(t)) \equiv \sigma > 0$) is a special case. When $r > 0$, (0.31) is general enough to include the following linear model considered by Chang and Youree [CY99] and pure discrete delay models considered by Arriojas et al. [AHMP07] and Kazmerchuk et al [KSW04a,KSW04b, and KSW04c]:

$$dS(t) = M(S_t) \, dt + N(S_t) \, dW(t) \quad ([CY99])$$
$$= \int_{-r}^{0} S(t + \theta) d\xi(\theta) \, dt + \int_{-r}^{0} S(t + \theta) \, d\zeta(\theta) dW(t), \ t \geq 0 \, ,$$

where $\xi, \zeta : [-r, 0] \to \Re$ are certain functions of bounded variation.

$$dS(t) = f(S_t) \, dt + g(S(t - b))S(t) \, dW(t), \quad t \geq 0 \ ([AHMP07]), \tag{0.37}$$

where $a, b > 0$, $f(S_t) = \mu S(t - a)S(t)$ or $f(S_t) = \mu S(t - a)$, and

$$\frac{dS(t)}{S(t)} = \mu S(t - a) \, dt + \sigma(S(t - b)) \, dW(t), \ t \geq 0 \ ([KSW04a]).$$

B5. Hereditary Portfolio Optimization Problem

As an illustration of a new class of combined optimal classical-impulse control problems that involve SHDEs with unbounded but fading memory ($r = \infty$), we briefly describe the hereditary portfolio optimization problem with capital gain taxes and fixed plus proportional transaction costs. This new optimal classical-impulse control problem is motivated by a realistic mathematical finance problem in consumption and investment and is not covered by any of the control problems outlined in **B1-B4**. The hereditary portfolio optimization problem and its solution will be detailed in Chapter 7 which is based on Chang [Cha07a], [Cha07b].

Consider a *small investor* who has assets in a financial market that consists of one *savings* account and one *stock* account. In this market, it is assumed that the *savings* account compounds continuously at a constant interest rate $\lambda > 0$ and $\{S(t), t \in (-\infty, \infty)\}$, the unit price process of the *stock*, satisfies the following nonlinear SHDE with infinite but fading memory:

$$dS(t) = S(t)[f(S_t) \, dt + g(S_t) \, dW(t)], \quad t \geq 0. \tag{0.38}$$

In the above equation, the process $\{W(t), t \geq 0\}$ is an one-dimensional standard Brownian motion. Note that $f(S_t)$ and $g(S_t)$ in (0.38) represent respectively the *mean growth rate* and the *volatility rate* of the *stock* price at time $t \geq 0$. The *stock* is said to have a hereditary price structure with infinite but fading memory because both the drift term $S(t)f(S_t)$ and the diffusion term $S(t)g(S_t)$ on the right-hand side of (0.38) explicitly depend on the entire past history prices $(S(t), S_t) \in \Re_+ \times L^2_{\rho,+}$ (where $L^2_{\rho,+} = L^2_\rho((-\infty, 0]; \Re_+))$ in a weighted fashion by the function ρ satisfying Conditions 2(i)-2(ii) of the Subsection A.

The main purpose of the *stock* account is to keep track of the inventories (i.e., the time instants and the base prices at which shares were purchased or short sold) of the underlying stock for the purpose of calculating the capital gains taxes and so forth. The space of stock inventories, **N**, will be the space of bounded functions $\xi : (-\infty, 0] \to \Re$ of the following form:

$$\xi(\theta) = \sum_{k=0}^{\infty} n(-k)\mathbf{1}_{\{\tau(-k)\}}(\theta), \quad \theta \in (-\infty, 0], \tag{0.39}$$

where $\{n(-k), k = 0, 1, 2, \ldots\}$ is a sequence in $\Re$ with $n(-k) = 0$ for all but finitely many k,

$$-\infty < \cdots < \tau(-k) < \cdots < \tau(-1) < \tau(0) = 0,$$

and $\mathbf{1}_{\{\tau(-k)\}}$ is the indicator function at $\tau(-k)$. Note that $n(-k) > 0$ (respectively, $n(-k) < 0$) represents the number of shares of the stock purchased (respectively, short sold) at time $\tau(-k)$. The assumption that $n(-k) = 0$ for all but finitely many k implies that the *investor* can only have finitely many open long or short positions in his stock account. However, the number of open long and/or short positions may increase from time to time. The *investor* is said to have an open long (respectively, short) position at time τ if he still owns (respectively, owes) all or part of the stock shares that were originally purchased (respectively, short sold) at a previous time τ. The only way to close a position is to sell all of what he owns and buy back all of what he owes.

Within the solvency region $\mathcal{S}_\kappa$ (to be defined later and in Subsection 7.1.4 of Chapter 7) and under the requirements of paying a fixed plus proportional transaction costs and capital gains taxes, the *investor* is allowed to consume from his *savings* account in accordance with a consumption rate process $C = \{C(t), t \geq 0\}$ and can make transactions between his *savings* and *stock* accounts according to a trading strategy $\mathcal{T} = \{(\tau(i), \zeta(i)), i = 1, 2, \ldots\}$, where $\tau(i), i = 0, 1, 2, \ldots$, denotes the sequence of transaction times and $\zeta(i)$ stands for quantities of the transaction at time $\tau(i)$.

The *investor* will follow the following set of consumption, transaction, and taxation rules (Rules (B5.1)-(B5.6)). Note that an action of the *investor* in the market is called a transaction if it involves trading of shares of the *stock* such as buying and selling.

Rule (B5.1). At the time of each transaction, the *investor* has to pay a transaction cost that consists of a fixed cost $\kappa > 0$ and a proportional transaction cost with the cost rate of $\mu \geq 0$ for both selling and buying shares of the *stock*. All the purchases and sales of any number of stock shares will be considered one transaction if they are executed at the same time instant and therefore incurs only one fixed fee, $\kappa > 0$ (in addition to a proportional transaction cost).

Rule (B5.2). Within the solvency region $\mathcal{S}_\kappa$, the *investor* is allowed to consume and to borrow money from his *savings* account for *stock* purchases. He can also sell and/or buy back some or all shares at the current price of the *stock* he bought and/or short sold at a previous time.

Rule (B5.3). The proceeds for the sales of the *stock* minus the transaction costs and capital gains taxes will be deposited in his *savings* account and the purchases of stock shares together with the associated transaction costs and capital gains taxes (if short shares of the *stock* are bought back at a profit) will be financed from his *savings* account.

Rule (B5.4). Without loss of generality it is assumed that the interest income in the *savings* account is tax-free by using the effective interest rate $\lambda > 0$, where the effective interest rate equals the interest rate paid by the bank minus the tax rate for the interest income.

Rule (B5.5). At the time of a transaction (say, $t \geq 0$), the *investor* is required to pay a capital gains tax (respectively, be paid a capital loss credit) in the amount that is proportional to the amount of profit (respectively, loss). A sale of stock shares is said to result in a profit if the current stock price $S(t)$ is higher than the base price $B(t)$ of the stock and it is a loss otherwise. The base price $B(t)$ is defined to be the price at which the stock shares were previously bought or short sold; that is, $B(t) = S(t - \tau(t))$, where $\tau(t) > 0$ is the time duration for which those shares (long or short) have been held at time t. The *investor* will also pay capital gains taxes (respectively, be paid capital loss credits) for the amount of profit (respectively, loss) by short-selling shares of the *stock* and then buying back the shares at a lower (respectively, higher) price at a later time. The tax will be paid (or the credit will be given) at the buying-back time. Throughout the end, a negative amount of tax will be interpreted as a capital loss credit. The capital gains tax and capital loss credit rates are assumed to be the same as $\beta > 0$ for simplicity. Therefore, if $|m|$ ($m > 0$ stands for buying and $m < 0$ stands for selling) shares of the stock are traded at the current price $S(t)$ and at the base $B(t) = S(t - \tau(t))$, then the amount of tax due at the transaction time is given by

$$|m|\beta(S(t) - S(t - \tau(t))).$$

Rule (B5.6). The tax and/or credit will not exceed all other gross proceeds and/or total costs of the stock shares, that is,

$$m(1 - \mu)S(t) \geq \beta m |S(t) - S(t - \tau(t))| \text{ if } m \geq 0$$

and
$$m(1 + \mu)S(t) \leq \beta m|S(t) - S(t - \tau(t))| \text{ if } m < 0,$$
where $m \in \Re$ denotes the number of shares of the stock traded, with $m \geq 0$ being the number of shares purchased and $m < 0$ being the number of shares sold.

Convention (B5.7). Throughout, we assume that $\mu + \beta < 1$.

Under the above assumption and Rules (B5.1)-(B5.6), the *investor's* objective is to seek an optimal consumption-trading strategy (C^*, T^*) in order to maximize
$$E\left[\int_0^\infty e^{-\alpha t} \frac{C^\gamma(t)}{\gamma} \, dt\right],$$
the expected utility from the total discounted consumption over the infinite time horizon, where $\alpha > 0$ represents the discount rate and $0 < \gamma < 1$ represents the *investor's* risk aversion factor.

Due to the fixed plus proportional transaction costs and the hereditary nature of the stock dynamics and inventories, the problem will be formulated as a combination of a classical control (for consumptions) and an impulse control (for the transactions) problem in infinite dimensions. A classical-impulse control problem in finite dimensions is treated in Øksendal and Sulem [ØS02] for a Black-Scholes market with fixed and proportional transaction costs without consideration of capital gains taxes. The problem treated here is the extension of existing works in the areas of the optimal consumption-investment problems in the following sense:
(i) The stock price dynamics is described by a nonlinear SHDE with unbounded but fading memory instead of an ordinary stochastic differential equation or geometric Brownian motion.

(ii) The tax basis for computing capital gains taxes or capital loss credits is based on the actual purchased price of shares of the stock.

The details of the treatment of this infinite-dimensional consumption-investment problem are given in Chapter 7. In there, the Hamilton-Jocobi-Bellman quasi-variational inequality (HJBQVI) for the value function together with its boundary conditions are derived and the verification theorem for the optimal investment-trading strategy is proved. It is also shown that the value function is a viscosity solution of the HJBQVI. Due to the complexity of the analysis involved, the uniqueness result and finite-dimensional approximations for the viscosity solution of HJBQVI are not included in the monograph.

To describe the problem and the results obtained, let
$$(X(0-), N_{0-}, S(0), S_0) = (x, \xi, \psi(0), \psi) \in \Re \times \mathbf{N} \times \Re_+ \times L^2_{\rho,+} \equiv \mathbf{S}$$
be the *investor's* initial portfolio immediately prior to $t = 0$; that is, the investor starts with $x \in \Re$ dollars in his *savings* account, the initial stock inventory,

$$\xi(\theta) = \sum_{k=0}^{\infty} n(-k)\mathbf{1}_{\{\tau(-k)\}}(\theta), \ \theta \in (-\infty, 0),$$

and the initial profile of historical stock prices $(\psi(0), \psi) \in \Re_+ \times L^2_{\rho,+}$. Within the *solvency region* $\mathbf{S}_\kappa$ (see (0.43)) the *investor* is allowed to consume from his *savings* account and can make transactions between his *savings* and *stock* accounts under Rules (B5.1)-(B5.6) and according to a consumption-trading strategy $\pi = (C, \mathcal{T})$, where the following hold:

(i) The consumption rate process $C = \{C(t), t \geq 0\}$ is a non-negative **G**-progressively measurable process such that

$$\int_0^T C(t)\, dt < \infty \ P\text{-a.s.} \ \forall T > 0.$$

(ii) $\mathcal{T} = \{(\tau(i), \zeta(i)), i = 1, 2, \ldots\}$ is a trading strategy, with $\tau(i), i = 1, 2, \ldots$, being a sequence of trading times that are **G**-stopping times such that

$$0 = \tau(0) \leq \tau(1) < \cdots < \tau(i) < \cdots$$

and

$$\lim_{i \to \infty} \tau(i) = \infty \ P\text{-a.s.},$$

and for each $i = 0, 1, \ldots,$

$$\zeta(i) = (\ldots, m(i-k), \ldots, m(i-2), m(i-1), m(i))$$

is an **N**-valued $\mathcal{G}(\tau(i))$-measurable random vector (instead of a random variable in $\Re$) that represents the trading quantities at the trading time $\tau(i)$. In the above, $\mathbf{G} = \{\mathcal{G}(t), t \geq 0\}$ is the filtration generated by the stock prices $S(\cdot) = \{S(t), t \geq 0\}$ and $m(i) > 0$ (respectively, $m(i) < 0$) is the number of stock shares newly purchased (respectively, short sold) at the current time $\tau(i)$ and at the current price of $S(\tau(i))$ and, for $k = 1, 2, \ldots, m(i-k) > 0$ (respectively, $m(i-k) < 0$) is the number of stock shares bought back (respectively, sold) at the current time $\tau(i)$ and the current price of $S(\tau(i))$ in his open short (respectively, long) position created at the previous time $\tau(i-k)$ and the base price of $S(\tau(i-k))$.

For each stock inventory ξ of the form expressed in (0.39), Rules (B5.1)-(B5.6) also dictate that the investor can purchase or short sell new shares and/or buy back (respectively, sell) all or part of what he owes (respectively, owns). Therefore, the trading quantity $\{m(-k), k = 0, 1, \ldots\}$ must satisfy the constraint set $\mathcal{R}(\xi) \subset \mathbf{N}$ defined by

$$\mathcal{R}(\xi) = \{\zeta \in \mathbf{N} \mid \zeta = \sum_{k=0}^{\infty} m(-k)\mathbf{1}_{\{\tau(-k)\}}, -\infty < m(0) < \infty, \text{ and} \qquad (0.40)$$

$$\text{either } n(-k) > 0, m(-k) \leq 0 \ \& \ n(-k) + m(-k) \geq 0$$

$$\text{or } n(-k) < 0, m(-k) \geq 0 \ \& \ n(-k) + m(-k) \leq 0 \text{ for } k \geq 1\}.$$

Define the function $H_\kappa : \mathbf{S} \to \Re$ as follows:

$$H_\kappa(x, \xi, \psi(0), \psi) = \max \Big\{ G_\kappa(x, \xi, \psi(0), \psi),$$
$$\min\{x, n(-k), k = 0, 1, 2, \ldots\} \Big\}, \qquad (0.41)$$

where $G_\kappa : \mathbf{S} \to \Re$ is the liquidating function defined by

$$G_\kappa(x, \xi, \psi(0), \psi) = x - \kappa$$
$$+ \sum_{k=0}^{\infty} \Big[\min\{(1-\mu)n(-k), (1+\mu)n(-k)\}\psi(0)$$
$$- n(-k)\beta(\psi(0) - \psi(\tau(-k))) \Big] \qquad (0.42)$$

The *solvency region* $\mathbf{S}_\kappa$ of the portfolio optimization problem is defined as

$$\mathbf{S}_\kappa = \Big\{ (x, \xi, \psi(0), \psi) \in \mathbf{S} \mid H_\kappa(x, \xi, \psi(0), \psi) \geq 0 \Big\}$$
$$= \{(x, \xi, \psi(0), \psi) \in \mathbf{S} \mid G_\kappa(x, \xi, \psi(0), \psi) \geq 0\} \cup \mathbf{S}_+, \qquad (0.43)$$

where $\mathbf{S}_+ = \Re_+ \times \mathbf{N}_+ \times \Re_+ \times M_{\rho,+}^2$ and $\mathbf{N}_+ = \{\xi \in \mathbf{N} \mid \xi(\theta) \geq 0, \forall \theta \in (-\infty, 0]\}$.

Note that within the *solvency region* $\mathbf{S}_\kappa$ there are positions that cannot be closed at all, namely those $(x, \xi, \psi(0), \psi) \in \mathbf{S}_\kappa$ such that

$$(x, \xi, \psi(0), \psi) \in \mathbf{S}_+ \text{ and } G_\kappa(x, \xi, \psi(0), \psi) < 0.$$

This is due to the insufficiency of funds to pay for the transaction costs and/or taxes and so forth. etc. Observe that the *solvency* region $\mathbf{S}_\kappa$ is an unbounded and nonconvex subset of the state space $\mathbf{S}$.

At time $t \geq 0$, the investor's portfolio in the financial market will be denoted by the quadruplet $(X(t), N_t, S(t), S_t)$, where $X(t)$ denotes the *investor's* holdings in his *savings* account, $N_t \in \mathbf{N}$ is the *inventory* of his *stock* account, and $(S(t), S_t)$ describes the profile of the unit prices of the *stock* over the past history $(-\infty, t]$ as described in (0.38).

Given the initial portfolio

$$(X(0-), N_{0-}, S(0), S_0) = (x, \xi, \psi(0), \psi) \in \mathbf{S}_\kappa$$

and applying a consumption-trading strategy $\pi = (C, \mathcal{T})$, the portfolio dynamics of $\{Z(t) = (X(t), N_t, S(t), S_t), t \geq 0\}$ can then be described as follows.

First, the *savings* account holdings $\{X(t), t \geq 0\}$ satisfies the following differential equation between the trading times:

$$dX(t) = [\lambda X(t) - C(t)] \, dt, \quad \tau(i) \leq t < \tau(i+1), \quad i = 0, 1, 2, \ldots, \qquad (0.44)$$

and the following jumped quantity at the trading time $\tau(i)$:

$$X(\tau(i)) = X(\tau(i)-) - \kappa - \sum_{k=0}^{\infty} m(i-k)\Big[(1-\mu)S(\tau(i))$$

$$-\beta(S(\tau(i)) - S(\tau(i-k)))\Big]\mathbf{1}_{\{n(i-k)>0,-n(i-k)\leq m(i-k)\leq 0\}}$$

$$-\sum_{k=0}^{\infty} m(i-k)\Big[(1+\mu)S(\tau(i))$$

$$-\beta(S(\tau(i)) - S(\tau(i-k)))\Big]\mathbf{1}_{\{n(i-k)<0,0\leq m(i-k)\leq -n(i-k)\}}. \quad (0.45)$$

As a reminder, $m(i) > 0$ (respectively, $m(i) < 0$) means buying (respectively, selling) new stock shares at $\tau(i)$ and $m(i-k) > 0$ (respectively, $m(i-k) < 0$) means buying back (respectively, selling) some or all of what owed (respectively, owned).

Second, the inventory of the *investor's* stock account at time $t \geq 0$, $N_t \in \mathbf{N}$, does not change between the trading times and can be expressed as follows:

$$N_t = N_{\tau(i)} = \sum_{k=-\infty}^{Q(t)} n(k)\mathbf{1}_{\tau(k)} \quad \text{if } \tau(i) \leq t < \tau(i+1)\, , i = 0, 1, \ldots, \quad (0.46)$$

where $Q(t) = \sup\{k \geq 0 \mid \tau(k) \leq t\}$. It has the following jumped quantity at the trading time $\tau(i)$:

$$N_{\tau(i)} = N_{\tau(i)-} \oplus \zeta(i), \quad (0.47)$$

where $N_{\tau(i)-} \oplus \zeta(i) : (-\infty, 0] \to \mathbf{N}$ is defined, for $\theta \in (-\infty, 0]$, by

$$(N_{\tau(i)-} \oplus \zeta(i))(\theta)$$

$$= \sum_{k=0}^{\infty} \hat{n}(i-k)\mathbf{1}_{\{\tau(i-k)\}}(\tau(i) + \theta)$$

$$= m(i)\mathbf{1}_{\{\tau(i)\}}(\tau(i) + \theta) + \sum_{k=1}^{\infty}\Big[n(i-k)$$

$$+m(i-k)(\mathbf{1}_{\{n(i-k)<0,0\leq m(i-k)\leq -n(i-k)\}}$$

$$+\mathbf{1}_{\{n(i-k)>0,-n(i-k)\leq m(i-k)\leq 0\}})\Big]\mathbf{1}_{\{\tau(i-k)\}}(\tau(i) + \theta). \quad (0.48)$$

Third, since the *investor* is small, the unit stock price process $\{S(t), t \geq 0\}$ will not be in anyway affected by the *investor's* action in the market and is again described as in (0.38).

If the *investor* starts with an initial portfolio

$$(X(0-), N_{0-}, S(0), S_0) = (x, \xi, \psi(0), \psi) \in \mathbf{S}_\kappa$$

the consumption-trading strategy $\pi = (C, \mathcal{T})$ is said to be *admissible* at $(x, \xi, \psi(0), \psi)$ if

$$\zeta(i) \in \mathcal{R}(N_{\tau(i)-}), \quad \forall i = 1, 2, \ldots,$$

and

$$(X(t), N_t, S(t), S_t) \in \mathbf{S}_\kappa, \quad \forall t \geq 0.$$

The class of consumption-investment strategies admissible at $(x, \xi, \psi(0), \psi) \in \mathbf{S}_\kappa$ will be denoted by $\mathcal{U}_\kappa(x, \xi, \psi(0), \psi)$.

Given the initial state $(X(0-), N_{0-}, S(0), S_0) = (x, \xi, \psi(0), \psi) \in \mathbf{S}_\kappa$, the *investor's* objective is to find an admissible consumption-trading strategy $\pi^* \in \mathcal{U}_\kappa(x, \xi, \psi(0), \psi)$ that maximizes the following expected utility from the total discounted consumption:

$$J_\kappa(x, \xi, \psi(0), \psi; \pi) = E^{x, \xi, \psi(0), \psi; \pi} \left[\int_0^\infty e^{-\alpha t} \frac{C^\gamma(t)}{\gamma} \, dt \right] \qquad (0.49)$$

among the class of admissible consumption-trading strategies $\mathcal{U}_\kappa(x, \xi, \psi(0), \psi)$, where $E^{x, \xi, \psi(0), \psi; \pi}[\cdots]$ is the expectation with respect to $P^{x, \xi, \psi(0), \psi; \pi}\{\cdots\}$, the probability measure induced by the controlled (by π) state process $\{(X(t), N_t, S(t), S_t), t \geq 0\}$ and conditioned on the initial state

$$(X(0-), N_{0-}, S(0), S_0) = (x, \xi, \psi(0), \psi).$$

In the above, $\alpha > 0$ denotes the discount factor and $0 < \gamma < 1$ indicates that the utility function $U(c) = \frac{c^\gamma}{\gamma}$, for $c > 0$, is a function of the HARA (hyperbolic absolute risk aversion) type that was considered in most of the optimal consumption-trading literature. The admissible (consumption-trading) strategy $\pi^* \in \mathcal{U}_\kappa(x, \xi, \psi(0), \psi)$ that maximizes $J_\kappa(x, \xi, \psi(0), \psi; \pi)$ is called an optimal (consumption-trading) strategy and the function $V_\kappa : \mathcal{S}_\kappa \to \Re_+$ defined by

$$V_\kappa(x, \xi, \psi(0), \psi) = \sup_{\pi \in \mathcal{U}_\kappa(x, \xi, \psi(0), \psi)} J_\kappa(x, \xi, \psi(0), \psi; \pi)$$
$$= J_\kappa(x, \xi, \psi(0), \psi; \pi^*) \qquad (0.50)$$

is called the value function of the hereditary portfolio optimization problem.

Extending a standard technique in deriving the variational HJB inequality for stochastic classical-singular and classical-impulse control problems (see Bensoussan and Lions [BL84] for stochastic impulse controls, Brekke and Øksendal [BØ98] and Øksendal and Sulem [ØS02] for stochastic classical-impulse controls of finite-dimensional diffusion processes, and Larssen [Lar02] and Larssen and Risebro [LR03] for classical and singular controls of a special class of stochastic delay equations), one can show that on the set

$$\{(x, \xi, \psi(0), \psi) \in \mathbf{S}_\kappa^\circ \mid \mathcal{M}_\kappa V_\kappa(x, \xi, \psi(0), \psi) < V_\kappa(x, \xi, \psi(0), \psi)\},$$

we have $\mathcal{A}V_\kappa = 0$, and on the set

$$\{(x, \xi, \psi(0), \psi) \in \mathbf{S}_\kappa^\circ \mid \mathcal{A}V_\kappa(x, \xi, \psi(0), \psi) < 0\},$$

we have $\mathcal{M}_\kappa V_\kappa = V_\kappa$. Therefore, we have the following HJBQVI on $\mathbf{S}_\kappa^\circ$:

$$\max\left\{\mathcal{A}V_\kappa, \mathcal{M}_\kappa V_\kappa - V_\kappa\right\} = 0 \text{ on } \mathbf{S}_\kappa^\circ, \tag{0.51}$$

where

$$\mathcal{A}\Phi = (\mathbf{A} + \lambda x \partial_x - \alpha)\Phi + \sup_{c \geq 0}\left(\frac{c^\gamma}{\gamma} - c\partial_x\Phi\right), \tag{0.52}$$

where $\mathcal{S}$ is the infinitesimal generator given by

$$\mathcal{S}\Phi(x,\xi,\psi(0),\psi) = \lim_{\epsilon\downarrow 0}\frac{\Phi(x,\xi,\psi(0),\tilde{\psi}_\epsilon) - \Phi(x,\xi,\psi(0),\psi)}{\epsilon} \tag{0.53}$$

and $\tilde{\psi} : (-\infty, \infty) \to \Re$ is the extension of $\psi : (-\infty, 0] \to \Re$ defined by

$$\tilde{\psi}(t) = \begin{cases} \psi(0) & \text{for } t \geq 0 \\ \psi(t) & \text{for } t < 0. \end{cases} \tag{0.54}$$

$\mathbf{A}$ is a certain operator that involves second order Fréchet derivatives (with respect to $\psi(0)$ and ψ) which shall be described in detail in Section 7.3 of Chapter 7, and $\mathcal{M}_\kappa\Phi$ is given as

$$\mathcal{M}_\kappa\Phi(x,\xi,\psi(0),\psi) = \sup\{\ \Phi(\hat{x},\hat{\xi},\hat{\psi}(0),\hat{\psi}) \mid \zeta \in \mathcal{R}(\xi) - \{\mathbf{0}\},$$
$$(\hat{x},\hat{\xi},\hat{\psi}(0),\hat{\psi}) \in \mathbf{S}_\kappa\}, \tag{0.55}$$

where $(\hat{x},\hat{\xi},\hat{\psi}(0),\hat{\psi})$ are defined as follows:

$$\hat{x} = x - \kappa - (m(0) + \mu|m(0)|)\psi(0)$$
$$-\sum_{k=1}^{\infty}\Big[(1+\mu)m(-k)\psi(0)$$
$$-\beta m(-k)(\psi(0) - \psi(\tau(-k)))\Big]\mathbf{1}_{\{n(-k)<0,0\leq m(-k)\leq -n(-k)\}}$$
$$-\sum_{k=1}^{\infty}\Big[(1-\mu)m(-k)\psi(0) - \beta m(-k)(\psi(0) - \psi(\tau(-k)))\Big]$$
$$\cdot\mathbf{1}_{\{n(-k)>0,-n(-k)\leq m(-k)\leq 0\}}, \tag{0.56}$$

and for all $\theta \in (-\infty, 0]$,

$$\hat{\xi}(\theta) = (\xi \oplus \zeta)(\theta)$$
$$= m(0)\mathbf{1}_{\{\tau(0)\}}(\theta)$$
$$+\sum_{k=1}^{\infty}\Big(n(-k) + m(-k)[\mathbf{1}_{\{n(-k)<0,0\leq m(-k)\leq -n(-k)\}}$$
$$+\mathbf{1}_{\{n(-k)>0,-n(-k)\leq m(-k)\leq 0\}}]\Big)\mathbf{1}_{\{\tau(-k)\}}(\theta), \tag{0.57}$$

and

$$(\hat{\psi}(0), \hat{\psi}) = (\psi(0), \psi). \tag{0.58}$$

The boundary values for the HJBQVI on $\partial \mathbf{S}_\kappa$ are given as follows:

First, the boundary of the solvency region $\partial \mathbf{S}_\kappa$ can be decomposed as follows. For each $(x, \xi, \psi(0), \psi) \in \mathbf{S}_\kappa$, let

$$I \equiv I(x, \xi, \psi(0), \psi) = \{ i \in \{0, 1, 2, \ldots\} \mid n(-i) < 0 \}.$$

For $I \subset \{0, 1, 2, \ldots\}$, let $\mathbf{S}_{\kappa, I} \subset \mathbf{S}_\kappa$ be defined as

$$\mathbf{S}_{\kappa, I} = \{(x, \xi, \psi(0), \psi) \in \mathbf{S}_\kappa \mid n(-i) < 0 \text{ for all } i \in I$$
$$\text{and } n(-i) \geq 0 \text{ for all } i \notin I\}.$$

With this interpretation,

$$\partial \mathbf{S}_\kappa = \bigcup_{I \subset \aleph} (\partial_{-,I} \mathbf{S}_\kappa \cup \partial_{+,I} \mathbf{S}_\kappa), \tag{0.59}$$

where

$$\partial_{-,I} \mathbf{S}_\kappa = \partial_{-,I,1} \mathbf{S}_\kappa \cup \partial_{-,I,2} \mathbf{S}_\kappa, \tag{0.60}$$

$$\partial_{+,I} \mathbf{S}_\kappa = \partial_{+,I,1} \mathbf{S}_\kappa \cup \partial_{+,I,2} \mathbf{S}_\kappa, \tag{0.61}$$

$$\partial_{+,I,1} \mathbf{S}_\kappa = \{(x, \xi, \psi(0), \psi) \mid G_\kappa(x, \xi, \psi(0), \psi) = 0, x \geq 0,$$
$$n(-i) < 0 \text{ for all } i \in I \ \& \ n(-i) \geq 0 \text{ for all } i \notin I\}, \tag{0.62}$$

$$\partial_{+,I,2} \mathbf{S}_\kappa = \{(x, \xi, \psi(0), \psi) \mid G_\kappa(x, \xi, \psi(0), \psi) < 0, x \geq 0,$$
$$n(-i) = 0 \text{ for all } i \in I \ \& \ n(-i) \geq 0 \text{ for all } i \notin I\}, \tag{0.63}$$

$$\partial_{-,I,1} \mathbf{S}_\kappa = \{(x, \xi, \psi(0), \psi) \mid G_\kappa(x, \xi, \psi(0), \psi) = 0, x < 0,$$
$$n(-i) < 0 \text{ for all } i \in I \ \& \ n(-i) \geq 0 \text{ for all } i \notin I\}, \tag{0.64}$$

and

$$\partial_{-,I,2} \mathbf{S}_\kappa = \{(x, \xi, \psi(0), \psi) \mid G_\kappa(x, \xi, \psi(0), \psi) < 0, x = 0,$$
$$n(-i) = 0 \text{ for all } i \in I \ \& \ n(-i) \geq 0 \text{ for all } i \notin I\}. \tag{0.65}$$

The HJBQVI (together with the boundary conditions) should be expressed as follows:

$$\text{HJBQVI}(*) = \begin{cases} \max \left\{ \mathcal{A}\Phi, \mathcal{M}_\kappa \Phi - \Phi \right\} = 0 & \text{on } \mathbf{S}_\kappa^\circ \\ \mathcal{A}\Phi = 0, & \text{on } \bigcup_{I \subset \aleph} \partial_{+,I,2} \mathbf{S}_\kappa \\ \mathcal{L}^0 \Phi = 0, & \text{on } \bigcup_{I \subset \aleph} \partial_{-,I,2} \mathbf{S}_\kappa \\ \mathcal{M}_\kappa \Phi - \Phi = 0 & \text{on } \bigcup_{I \subset \aleph} \partial_{I,1} \mathbf{S}_\kappa, \end{cases}$$

where $\mathcal{A}\Phi$, $\mathcal{L}^0\Phi$ ($\mathcal{L}^c\Phi$ with $c=0$), and $\mathcal{M}_\kappa$ are given by

$$\mathcal{A}\Phi = (\mathbf{A} + \lambda x\partial_x - \alpha)\Phi + \sup_{c\geq 0}\left(\frac{c^\gamma}{\gamma} - c\partial_x\Phi\right), \tag{0.66}$$

$$\mathcal{L}^0\Phi(x,\xi,\psi(0),\psi) \equiv (\mathbf{A} + \lambda x\partial_x - \alpha)\Phi, \tag{0.67}$$

and

$$\mathcal{M}_\kappa\Phi(x,\xi,\psi(0),\psi) \tag{0.68}$$
$$= \sup\{\Phi(\hat{x},\hat{\xi},\hat{\psi}(0),\hat{\psi}) \mid \zeta \in \mathcal{R}(\xi) - \{\mathbf{0}\}, (\hat{x},\hat{\xi},\hat{\psi}(0),\hat{\psi}) \in \mathcal{S}_\kappa\}.$$

We have the following verification theorem for the value function $V_\kappa : \mathcal{S}_\kappa \to \Re$ for our hereditary portfolio optimization problem.

Theorem (B5.1). (The Verification Theorem)
(a) Let $U_\kappa = \mathbf{S}_\kappa - \bigcup_{I\subset\aleph}\partial_{I,1}\mathbf{S}_\kappa$. Suppose that there exists a locally bounded non-negative-valued function $\Phi \in C^{1,0,2}_{lip}(\mathbf{S}_\kappa) \cap \mathcal{D}(\mathcal{S})$ such that

$$\tilde{\mathcal{A}}\Phi \leq 0 \text{ on } U_\kappa, \tag{0.69}$$

$$\Phi \geq \mathcal{M}_\kappa\Phi \text{ on } U_\kappa, \tag{0.70}$$

where

$$\tilde{\mathcal{A}}\Phi = \begin{cases} \mathcal{A}\Phi & \text{on } \mathbf{S}^\circ_\kappa \cup \bigcup_{I\subset\aleph}\partial_{+,I,2}\mathbf{S}_\kappa \\ \mathcal{L}^0\Phi & \text{on } \bigcup_{I\subset\aleph}\partial_{-,I,2}\mathbf{S}_\kappa. \end{cases}$$

Then $\Phi \geq V_\kappa$ on $\mathcal{U}_\kappa$.
(b) Define $D \equiv \{(x,\xi,\psi(0),\psi) \in U_\kappa \mid \Phi(x,\xi,\psi(0),\psi) > \mathcal{M}_\kappa\Phi(x,\xi,\psi(0),\psi)\}$. Suppose
$$\tilde{\mathcal{A}}\Phi(x,\xi,\psi(0),\psi) = 0 \text{ on } D \tag{0.71}$$
and that $\hat{\zeta}(x,\xi,\psi(0),\psi) = \hat{\zeta}_\Phi(x,\xi,\psi(0),\psi)$ exists for all $(x,\xi,\psi(0),\psi) \in \mathbf{S}_\kappa$. Define

$$c^* = \begin{cases} (\partial_x\Phi)^{\frac{1}{\gamma-1}} & \text{on } \mathbf{S}^\circ_\kappa \cup \bigcup_{I\subset\aleph}\partial_{+,I,2}\mathbf{S}_\kappa; \\ 0 & \text{on } \bigcup_{I\subset\aleph}\partial_{-,I,2}\mathbf{S}_\kappa, \end{cases} \tag{0.72}$$

and define the impulse control $\mathcal{T}^* = \{(\tau^*(i),\zeta^*(i)), i = 1,2,\ldots\}$ inductively as follows.
First, put $\tau^*(0) = 0$ and inductively

$$\tau^*(i+1) = \inf\{t > \tau^*(i) \mid (X^{(i)}(t), N^{(i)}_t, S(t), S_t)) \notin D\}, \tag{0.73}$$

$$\zeta^*(i+1) = \hat{\zeta}(X^{(i)}(\tau^*(i+1)-), N^{(i)}_{\tau^*(i+1)-}, S(\tau^*(i+1)), S_{\tau^*(i+1)}), \tag{0.74}$$

where $\{(X^{(i)}(t), N^{(i)}_t, S(t), S_t), t \geq 0\}$ is the controlled state process obtained by applying the combined control

$$\pi^*(i) = (c^*, (\tau^*(1), \tau^*(2), \ldots, \tau^*(i); \zeta^*(1), \zeta^*(2), \cdots, \zeta^*(i))), \quad i = 1, 2, \ldots.$$

Suppose $\pi^* = (C^*, T^*) \in \mathcal{U}_\kappa(x, \xi, \psi(0), \psi)$ and that

$$e^{-\delta t} \Phi(X^*(t), N_t^*, S(t), S_t) \to 0, \text{ as } t \to \infty \text{ a.s.}$$

and that the family

$$\{e^{-\alpha\tau}\Phi(X^*(\tau), N_\tau^*, S(\tau), S_\tau)) \mid \tau \text{ G-stopping times}\} \qquad (0.75)$$

is uniformly integrable. Then $\Phi(x, \xi, \psi(0), \psi) = V_\kappa(x, \xi, \psi(0), \psi)$ and π^* obtained in (0.72)-(0.74) is optimal.

Theorem (B5.2). The value function $V_\kappa : \mathbf{S}_\kappa \to \Re$ for the hereditary portfolio optimization is a viscosity solution of the HJBQVI (*).

C. Organization of the Monograph

This monograph consists of seven chapters. The content of each of these seven chapters are summarized as follows.

Chapter 1. This chapter collects and extends some of the known results on the existence, uniqueness, and Markovian properties of the strong and weak solution process for SHDEs treated in Mohammed [Moh84, Moh98] for bounded memory $0 < r < \infty$ and for SHDEs with infinite but fading memory $r = \infty$ treated in Mizel and Trutzer [MT84]. The concept of the weak solution process and its uniqueness are also introduced in this chapter. It is also shown in there that a uniqueness of the strong solution of the SHDE implies uniqueness of the weak solution. This fact will be used in proving the DDP in Chapter 3. The Markovian and strong Markovian properties for the segment processes corresponding to the SHDE with bounded and unbounded memory are reviewed in Section 1.5.

Chapter 2. This chapter reviews some of the known results and develops some of the analytic tools on a generic Banach space as well as on the specific function spaces $\mathbf{C}$ and $\mathbf{M} \equiv \Re \times L_\rho^2((-\infty, 0]; \Re)$ that are needed in the subsequent chapters. In particular, an infinitesimal generator of a semigroup of shift operators as well as Fréchet differentiation of a real-valued function defined on $\mathbf{C}$ and/or on $\mathbf{M}$ are described in detail. The main tool established in this chapter include the weak infinitesimal generators of the $\mathbf{C}$-valued Markovian solution process $\{x_s, s \in [t, T]\}$ for an equation with bounded memory and for the $\mathbf{M}$-valued Markovian solution process $\{(S(s), S_s), s \geq 0\}$ for an equation with infinite but fading memory. These two weak infinitesimal generators along with Dynkin's and Itô's formulas for the quasi-tame of the $\mathbf{C}$ and $\mathbf{M}$-valued segment processes are the main ingredients for establishing the Hamilton-Jacobi-Bellman theory for the optimal control problems treated later.

Chapter 3. This chapter formulates and provides a characterization of the value function for the optimal classical control problem as briefly described in **B1**. The existence of optimal classical control is proved in Section 3.2 via the extended class of admissible relaxed controls. The characterization is written in terms of an infinite-dimensional HJBE. The Bellman-type DDP is stated and proved in Section 3.3. Based on the DDP, the HJBE is heuristically derived in Section 3.4 for the value function under the condition that it is sufficiently smooth. However, it is known in most of optimal control problems, deterministic or stochastic, that the value functions do not meet these smoothness conditions and therefore cannot be a solution to the HBJE in the strong sense. The concept of viscosity solution is introduced in Section 3.5. In this section and Section 3.6, it is shown that the value function is the unique viscosity solution to the HJBE. The generalized verification theorem in the framework of viscosity solution is conjectured in Section 3.7. Finally, some special optimal control problems that yield a finite-dimensional HJBE along with a couple of application examples are given in Section 3.8.

Chapter 4. This chapter formulates and provides a characterization of the value function for the optimal stopping problem briefly described in **B2**. Section 4.1 gives the problem formulation. Section 4.2 addresses the existence of optimal stopping via the construction of the least harmonic majorant of the terminal reward function of an equivalent optimal stopping problem. The characterization is written in terms of an infinite-dimensional HJBVI, which is given in Section 4.3. A verification theorem that provides the method for finding the optimal stopping is given in Section 4.4. The issues of the value as the unique viscosity solution of the HJBVI are addressed in Sections 4.5 and 4.6. The organization within this chapter is parallel to that of Chapter 3.

Chapter 5. This chapter deals with computational issues of the viscosity solution of the HJBE arising from the finite horizon optimal classical control problem described in **B1** and detailed in Chapter 3. Three approximation schemes are presented: (1) semidiscretization scheme; (2) Markov chain approximations; and (3) finite difference approximation. Preliminaries and discretization notation are given in Section 5.1. The semidiscretization scheme that discretizes the temporal variable (but not the spatial variable) of the state process and then the control process is the subject of investigation of Section 5.2. The section also provides the rate of convergence of the scheme. The well-known Markov chain approximation method for the controlled diffusion processes is extended to controlled SHDEs in Section 5.3. The finite difference approximation is given in Section 5.4. The results are based on the recent works by Chang et al. [CPP07a, CPP07b, CPP07c]. The method extends the finite difference scheme obtained by Barles and Souganidis [BS91] on the controlled diffusion process in finite dimensions to the viscosity solution of the infinite-dimensional HJBE. The convergence of the schemes are proved using the Banach fixed-point theorem. A computational algorithms are also provided based on the schemes obtained.

Chapter 6. This chapter treats pricing problems of American and European options as an application of the optimal stopping described in Chapter 4. It is illustrated that the pricing function is the unique viscosity solution of a HJBVI. A computational algorithm is given for an infinite-dimensional Black-Scholes equation. Specifically, the formulation of option pricing problem in the (B, S)-market with hereditary structure is given in Section 6.1. Section 6.2 summarizes the definitions of an HJBVI. This section also contained the optimal strategy the option writer should use in order to replicate the pricing function if and when the contract holder exercises his option. Section 6.3 explores the concept of risk-neutral martingale measure under which the (B, S)-market is complete and arbitrage-free. Section 6.4 treats specifically the HJBVI characterization of the American option. As a special case, the infinite-dimensional Black-Scholes equation for the European option is also derived under the differentiability assumption of the pricing function. This is treated in Sections 6.5 and 6.6. A series solution technique for the Black-Scholes equation is obtained in Section 6.7 for computational purposes.

Chapter 7. This chapter treats an infinite time horizon hereditary portfolio optimization problem in a market that consists of one *savings* account and one *stock* account. The *savings* account compounds continuously with a fixed interest rate and the stock account keeps track of the inventories of one underlying stock whose unit price follows a nonlinear SHDE. Within the *solvency* region, the investor is allowed to consume from the *savings* account and can make transactions between the two assets subject to paying capital gains taxes as well as a fixed plus proportional transaction cost. The *investor* is to seek an optimal consumption-trading strategy in order to maximize the expected utility from the total discounted consumption. The portfolio optimization problem is formulated as an infinite-dimensional stochastic classical-impulse control problem due to the hereditary nature of the stock price dynamics and inventories. The HJBQVI together with its boundary conditions for the value function is derived under some smoothness conditions. The verification theorem for the optimal strategy is given. It is also shown that the value function is a viscosity solution of the HJBQVI. However, the uniqueness and finite-dimensional approximations of the viscosity solution are not included here.

1

Stochastic Hereditary Differential Equations

The main purpose of this chapter is to study the strong and weak solutions $x(\cdot) = \{x(s), s \in [t - r, T]\}$ for the class of stochastic hereditary differential equations (SHDEs)

$$dx(s) = f(s, x_s)\, ds + g(s, x_s)\, dW(s), \quad s \in [t, T];$$

with a bounded memory (i.e., $0 < r < \infty$). We also study the one-dimensional SHDE

$$\frac{dS(s)}{S(s)} = f(S_s)\, ds + g(S_s)\, dW(s), \quad s \geq 0,$$

with an infinite but fading memory (i.e., $r = \infty$) that describes the stock price dynamics involved in treating hereditary portfolio optimization problem in Chapter 7. It is assumed that both of these two equations are driven by a standard Brownian motion $W(\cdot) = \{W(s), s \geq 0\}$ of appropriate dimensions.

In the above two systems, we use the conventional notation

$$x_s(\theta) = x(s + \theta), \theta \in [-r, 0] \text{ for bounded memory } 0 < r < \infty,$$

$$S_s(\theta) = S(s + \theta), \theta \in (-\infty, 0] \text{ for unbounded memory } r = \infty.$$

The first equation, which is to be studied in detail later in this chapter (see (1.27)), represents the uncontrolled version of the SHDE for the stochastic control problems that are the subject of investigation in Chapters 3 and 4. The state process $S(\cdot) = \{S(s), s \in (-\infty, \infty)\}$ in the second equation (see (1.43)) represents the prices of the underlying stock that plays an important role in the hereditary portfolio optimization problem to be studied in Chapter 7.

In particular, we are interested in the existence and uniqueness of both the *strong* and *weak* solutions and the Markovian properties of their corresponding segment processes $\{x_s, s \in [t, T]\}$ for the bounded memory case and $\{(S(s), S_s), s \geq 0\}$ for the infinite memory case.

In order to make sense of a solution (*strong* or *weak*) to the above two equations, it is necessary that an appropriate initial segment be given on the initial time interval $[t-r,t]$ and $(-\infty,0]$, respectively. This requirement is well documented in the study of deterministic functional differential equations (FDEs) (see, e.g., Hale [Hal77] and Hale and Lunel [HL93]). The fundamental theory of the first equation was systematically developed in Mohammed [Moh84, Moh98] with the state space $\mathbf{C} \equiv C([-r,0]; \Re^n)$. However, the state space of (1.43) studied by Mizel and Trutzer [MT84] will be the ρ-weighted Hilbert space $\mathbf{M} \equiv \Re \times L^2_\rho((-\infty,0]; \Re)$ instead of the Banach space $\mathbf{C}$ in order to take into account of unbounded but fading memory.

This chapter is organized as follows. In Section 1.1, we either collect or prove some preliminary results in theory of stochastic processes that are needed for the development of SHDEs or stochastic control theory. These preliminary results include (i) the Gronwall inequality; (ii) regular conditional probability measure associated with a Banach space valued random variable given a sub-σ-algebra $\mathcal{G}$ of the underlying σ-algebra $\mathcal{F}$; (iii) a summary of Itô integration of an appropriate process with respect to the Brown motion $W(\cdot)$ and its relevant properties such as the Itô formula, the Girsanov transformation, and so forth. These preliminaries are established in Section 1.2. The existence and uniqueness of the solution $x(\cdot)$ of (1.27) and $S(\cdot)$ of (1.43) are developed in Section 1.3 and Section 1.4 respectively. Since the the existence and uniqueness proofs are given in Chapter II of Mohammed [Moh84] and [Moh98] (for the bounded memory case $0 \le r < \infty$) and in Mizel and Trutzer [MT84] (for unbounded but fading memory case $r = \infty$) using the standard Piccard iteration technique, we will only give an outline of the important steps involved in these two sections and refer the interested readers to [Moh84], [Moh98], and [MT87] for the details. The Markovian and strong Markovian properties of the $\mathbf{C}$-valued segment process $\{x_s, s \in [t,T]\}$ for (1.27) and the $\mathbf{M}$-valued segment process $\{(S(s), S_s), s \in [0,\infty)\}$ for (1.43) will be established in Section 1.5.

We note here that the results obtained for (1.43) carry over for a more general equation of the following form:

$$dx(s) = f(s, x(s), x_s)\, ds + g(s, x(s), x_s)\, dW(s), \quad s \ge 0,$$

where $W(\cdot) = \{W(s), s \ge 0\}$ is an m-dimensional standard Brownian motion, and f and g are appropriate functions in $\Re^n$ and $\Re^{n \times m}$, respectively, and both are defined on $[0,\infty) \times \Re^n \times L^2_\rho((-\infty,0], \Re^n)$.

The following are some of the real world problems that require a stochastic hereditary differential equation with a bounded memory to describe its quantitative models.

Example A. (Logistic Population Growth)

Consider a single population $x(s)$ at time $s \ge 0$ evolving logistically with incubation period $r > 0$. Suppose there is instantaneous migration rate at a molecular level with contributes $\gamma dW(s)$ to the growth rate per capita during the time s. Then evolution of the population is governed by the nonlinear logistic stochastic delay differential equation (SDDE):

$$dx(s) = [a - bx(s - r)]x(s) \, ds + \gamma x(s) \, dW(s), \quad s \geq 0,$$

with a given continuous initial segment

$$x(s) = \psi(s), \quad s \in [-r, 0].$$

The above SDDE falls into a special case of the (1.27):

$$dx(s) = f(x_s) \, ds + g(x_s) \, dW(s),$$

where $f(\phi) = a - b\phi(-r)$ and $g(\phi) = \gamma\phi(-r)$ for all continuous $\phi : [-r, 0] \to \Re$.

The following is an example of a physical model that can be described as a system of two-dimensional SHDEs with unbounded but fading memory. The model describes the vertical motion of a dangling spider hanging from a ceiling with a massless viscoelastic string (see Mizel and Trutzer [MT84]).

Example B. (A Dangling Spider)

A spider of mass $m > 0$ is hanging by a massless but extensible filament of length x. The forces acting on the spider are the tension T and a body force f in the x-direction. It is assumed that there exists a time independent, twice continuously differentiable potential $h(x)$ for the force f:

$$f(x) = \frac{dh(x)}{dx}.$$

For example, if the force acting on the spider is gravity force, then

$$h = m\bar{g}x, \quad f = -\bar{g}m,$$

with $\bar{g} > 0$ being the gravitational constant. It is assumed that the filament is composed of a simple viscoelastic material with constant and uniform temperature. The value of the tension $T(s)$ at time s is given by a functional $T(s) = F(x_s)$ of the history up to s, where $x_s(\theta) = x(s+\theta)$, $\theta \in (-\infty, 0]$, and $F : L_\rho^2((-\infty, 0], \Re)$ is continuous. Then by Newton's second law of motion, we have

$$m\ddot{x}(s) = f(x(s)) - F(x_s).$$

One can take $m = 1$ and

$$F(x_s) = -\int_0^\infty k(\theta)g(x_s(\theta) - x(s))d\theta.$$

For a stochastic version of the equation, we can take

$$\ddot{x}(s) = f(x(s)) - F(x_s) + \sigma(x_s)\dot{W}(s),$$

where $W(\cdot)$ is a one-dimensional standard Brownian motion, and the stochastic term $\sigma(x_s)\dot{W}(s)$ can be interpreted as irregularity in the fiber. Let $y(s) = \dot{x}(s)$, then the above second order equation can be re-written as the following system of two equation:

$$\begin{cases} dx(s) = & y(s)\,ds \\ dy(s) = (f(x(s)) - F(x_s))\,ds + \sigma(x_s)\,dW(s), & s \geq 0. \end{cases}$$

Of course, in order make sense of the above equation a initial segment $(x(s), y(s))$ has to be given for $s \in (-\infty, 0]$.

1.1 Probabilistic Preliminaries

To make this volume accessible to those readers who have been exposed to but are not very familiar with stochastic differential equations (SDEs) of any type, we include the following basic material taken from the theory of stochastic processes for the purpose of being self-contained. These concepts can be found in any books of measure-theoretic nature in this area, such as Karatzas and Shreves [KS91] and Øksendal [Øks98].

1.1.1 Gronwall Inequality

The following *Gronwall inequality* will be used in many different places in this chapter and in our subsequent developments in other chapters.

Lemma 1.1.1 *Suppose that $h \in L^1([t, T]; \Re)$ and $\alpha \in L^\infty([t, T]; \Re)$ satisfy, for some $\beta \geq 0$,*

$$0 \leq h(s) \leq \alpha(s) + \beta \int_t^s h(\lambda)\,d\lambda \quad \text{for a.e. } s \in [t, T]. \tag{1.1}$$

Then

$$h(s) \leq \alpha(s) + \beta \int_t^s \alpha(\lambda)e^{\beta(\lambda-t)}\,d\lambda \quad \text{for a.e. } s \in [t, T]. \tag{1.2}$$

If, in addition, α is nondecreasing, then

$$h(s) \leq \alpha(s)e^{\beta(s-t)} \quad \text{for a.e. } s \in [t, T]. \tag{1.3}$$

Proof. Assume that $\beta \neq 0$. Otherwise, the lemma is trivial. Define

$$z(s) := e^{-\beta(s-t)} \int_t^s h(\lambda)\,d\lambda$$

and note that z is absolutely continuous in $[t, T]$. After multiplying (1.1) by $e^{-\beta(s-t)}$ we get

$$\dot{h}(s) \leq \alpha(s)e^{-\beta(s-t)} \quad \text{for a.e. } s \in [t, T].$$

We integrate both sides and multiply by $e^{\beta(s-t)}$ to obtain

$$\int_t^s h(\lambda)\,d\lambda \le e^{\beta s}\int_t^s \alpha(\lambda)e^{-\beta\lambda}\,d\lambda \quad \text{for a.e. } s \in [t,T].$$

This inequality and (1.1) gives (1.2).   □

Let $(\Omega,\mathcal{F},P,\mathbf{F})$ be a complete filtered probability space, where $\mathbf{F} = \{\mathcal{F}(s), s \ge 0\}$ is a filtration of sub-σ-algebras of $\mathcal{F}$ that satisfies the following usual conditions:

(i) $\mathbf{F}$ is complete, that is,

$$\mathcal{N} \equiv \{A \subset \Omega \mid \exists B \in \mathcal{F} \text{ such that } A \subset B \text{ and } P(B) = 0\} \subset \mathcal{F}(0).$$

(ii) $\mathbf{F}$ is nondecreasing, that is, $\mathcal{F}(s) \subset \mathcal{F}(\tilde{s})$, $\forall 0 \le s \le \tilde{s}$.
(iii) $\mathbf{F}$ is right-continuous, i.e., $\mathcal{F}(s+) \equiv \cap_{\epsilon\downarrow 0}\mathcal{F}(s+\epsilon) = \mathcal{F}(s)$, $\forall s \ge 0$, where $\equiv$ stands for "is defined as".

For a generic Banach (or Hilbert) space $(\Xi, \|\cdot\|_\Xi)$, a sub-σ-algebra $\mathcal{G}$ of $\mathcal{F}$, let $L^2(\Omega,\Xi;\mathcal{G})$ be the collection of all Ξ-valued random variables $X : (\Omega,\mathcal{F},P) \to (\Xi,\mathcal{B}(\Xi))$ that are $\mathcal{G}$-measurable and satisfy

$$\|X\|_{L^2(\Omega,\Xi)} \equiv E\left[\|X\|_\Xi^2\right] \equiv \int_\Omega \|X(\omega)\|_\Xi\,dP(\omega) < \infty.$$

If $\mathcal{G} = \mathcal{F}$, we simply write $L^2(\Omega,\Xi)$ for $L^2(\Omega,\Xi;\mathcal{F})$ for simplicity. For $0 < T \le \infty$, let $C([0,T];L^2(\Omega,\Xi))$ be the space of all square integrable continuous Ξ-valued processes $X(\cdot) = \{X(s), s \in [0,T]\}$ such that

$$\|X(\cdot)\|_{C([0,T];L^2(\Omega,\Xi))} \equiv \sup_{s\in[0,T]} E\left[\|X(s)\|_\Xi^2\right] < \infty.$$

Let $C([0,T],L^2(\Omega,\Xi);\mathcal{G})$ be those of $X(\cdot) \in C([0,T];L^2(\Omega,\Xi))$ that are $\mathcal{G}$-measurable. A process $X(\cdot) \in C([0,T];L^2(\Omega,\Xi))$ is said to be adapted to the filtration $\mathbf{F} = \{\mathcal{F}(s), s \ge 0\}$ if $X(s)$ is $\mathcal{F}(s)$-measurable for all $s \in [0,T]$. Denote by $C([0,T],L^2(\Omega,\Xi);\mathbf{F})$ the set of such processes.

Lemma 1.1.2 $C([0,T],L^2(\Omega,\Xi);\mathbf{F})$ *is closed in* $C([0,T],L^2(\Omega,\Xi))$.

Proof. Since for each $s \in [0,T]$ the evaluation map

$$\pi(s) : C([0,T],L^2(\Omega,\Xi)) \to L^2(\Omega;\Xi)$$

defined by $\pi(s)(X(\cdot)) = X(s)$ is continuous, it is sufficient to observe that the space $C([0,T],L^2(\Omega,\Xi);\mathcal{F}(s))$ is closed in $C([0,T],L^2(\Omega,\Xi))$ for each $s \in [0,T]$. This proves the lemma.   □

1.1.2 Stopping Times

A random variable $\tau : \Omega \rightarrow [0, \infty]$ is said to be an **F**-stopping time (or simply stopping time if there is no possibility of confusion occurring) if for each $t \geq 0$,

$$\{\tau \leq t\} \in \mathcal{F}(t).$$

The following properties of stopping times can be found in Karatzas and Shreves [KS91] and Øksendal [Øks98] and, therefore, the proofs of the trivial ones will be omitted here.

Proposition 1.1.3 *The random variable* $\tau : \Omega \rightarrow [0, \infty]$ *is an* **F***-stopping time if and only if*

$$\{\tau < s\} \in \mathcal{F}(s), \quad \forall s \geq 0.$$

Proof. Since $\{\tau \leq s\} = \cap_{s+\epsilon > \lambda > s}\{\tau < \lambda\}$, any $\epsilon > 0$, we have $\{\tau \leq s\} \in \cap_{\lambda > s}\mathcal{F}(\lambda) = \mathcal{F}(s)$, so τ is a stopping time. For the converse,

$$\{\tau < s\} = \cup_{s > \epsilon > 0}\{\tau \leq s - \epsilon\} \quad \text{and} \quad \{\tau \leq s - \epsilon\} \in \mathcal{F}(s - \epsilon),$$

hence also in $\mathcal{F}(s)$.   □

Proposition 1.1.4 *Let* τ *and* τ' *be two* **F***-stopping times. Then the following are also stopping times:*
(i) $\tau \wedge \tau' \equiv \min\{\tau, \tau'\}$.
(ii) $\tau \vee \tau' \equiv \max\{\tau, \tau'\}$.
(iii) $\tau + \tau'$.
(iv) $a\tau$, *where* $a > 1$.

Proof. The proof is omitted.   □

Let $X(\cdot) = \{X(s), s \geq 0\}$ be a Ξ-valued stochastic process defined on the complete filtered probability space $(\Omega, \mathcal{F}, P, \mathbf{F})$. We define the hitting time and the exit time $\tau : \Omega \rightarrow [0, \infty]$ as follows.

Definition 1.1.5 *Let* $X(t) \notin \Lambda$ *for some* $t \in [0, T]$. *The hitting time after time* t *of the Borel set* $\Lambda \in \mathcal{B}(\Xi)$ *for* $X(\cdot)$ *is defined by*

$$\tau = \begin{cases} \inf\{s > t \mid X(s) \in \Lambda\} & \text{if } \{\cdots\} \neq \emptyset \\ \infty & \text{if } \{\cdots\} = \emptyset. \end{cases}$$

Definition 1.1.6 *Let* $X(t) \in \Lambda$ *for some* $t \in [0, T]$. *The exit time after time* t *of the Borel set* $\Lambda \in \mathcal{B}(\Xi)$ *for* $X(\cdot)$ *is defined by*

$$\tau = \begin{cases} \inf\{s > t \mid X(s) \notin \Lambda\} & \text{if } \{\cdots\} \neq \emptyset \\ \infty & \text{if } \{\cdots\} = \emptyset. \end{cases}$$

When $t = 0$, the random time τ defined in Definition 1.1.5 and Definition 1.1.6 will be called hitting time and exit time, respectively, for simplicity.

We have the following theorem.

Theorem 1.1.7 *If $X(\cdot)$ is a continuous Ξ-valued process and if Λ is an open subset of Ξ, then the hitting time of Λ for $X(\cdot)$ after time $t \geq 0$ is a stopping time.*

Proof. Let $\mathbf{Q}$ denote the set of all rational numbers. By Proposition 1.1.3, it suffices to show that $\{\tau < s\} \in \mathcal{F}(s), 0 \leq t \leq s < \infty$. However,

$$\{\tau < s\} = \bigcup_{u \in \mathbf{Q} \cap [t,s)} \{X(u) \in \Lambda\},$$

since $\Lambda \subset \Xi$ is open and $X(\cdot) = \{X(s), s \in [t,T]\}$ has right-continuous paths. Since $\{X(u) \in \Lambda\} \in \mathcal{F}(u)$, the result of the theorem follows.   □

Theorem 1.1.8 *If $\Lambda \subset \Xi$ is a closed set, then the exit time*

$$\tau = \inf\{s > t \mid X(s) \in \Lambda \text{ or } X(s-) \in \Lambda\}$$

is of Λ after time t a stopping time.

Proof. By $X(s-, \omega)$ we mean $\lim_{u \to s, u < s} X(u, \omega)$. Let

$$A_k = \left\{ \xi \in \Xi \mid d(\xi, \Lambda) < \frac{1}{k} \right\},$$

where $d(\xi, \Lambda)$ denotes the distance from the point ξ to the set Λ. Then A_k is an open set and

$$\{\tau \leq s\} = \{X(s) \in \Lambda \text{ or } X(s-) \in \Lambda\} \cup \left\{ \cap_{k \geq 1} \cup_{u \in \mathbf{Q} \cap [t,s)} \{X(u) \in A_k\} \right\}.   □$$

1.1.3 Regular Conditional Probability

We first provide the following definition of conditional expectation of a Banach space valued random variable X with respect to a sub-σ-algebra $\mathcal{G}$ of $\mathcal{F}$.

Definition 1.1.9 *Let X be a random variable defined on a probability space $(\Omega, \mathcal{F}, P)$ and taking values in a separable Banach (or Hilbert) measurable space $(\Xi, \mathcal{B}(\Xi))$. Let $\mathcal{G}$ be a sub-σ-algebra of $\mathcal{F}$. A regular conditional probability of X given the sub-σ-algebra $\mathcal{G}$ is a function $Q : \Omega \times \mathcal{B}(\Xi) \to [0,1]$ such that the following hold:*
(i) For each $\omega \in \Omega$, $Q(\omega; \cdot)$ is a probability measure on $(\Xi, \mathcal{B}(\Xi))$.
(ii) For each $A \in \mathcal{B}(\Xi)$, the mapping $\omega \mapsto Q(\omega; A)$ is $\mathcal{G}$-measurable.
(iii) For each $A \in \mathcal{B}(\Xi)$, $P[X \in A|\mathcal{G}](\omega) = Q(\omega; A)$, P-a.e. ω.

Throughout, the notation $P[X \in \cdot | \mathcal{G}]$ will be used in place of Q for the self-explanatory reason. Let $E[X|\mathcal{G}]$ be its corresponding *conditional expectation of X given $\mathcal{G}$*, that is,

$$E[X|\mathcal{G}](\omega) = \int_{\Xi} x Q(\omega; dx), \tag{1.4}$$

where the above integral is understood to be a Bochner integral (see, e.g., Diestel and Uhl [DU77]).

The following slightly different version of *regular conditional probability* will be useful later.

Proposition 1.1.10 *Let $(\Xi, \mathcal{B}(\Xi))$ and $(\Theta, \mathcal{B}(\Theta))$ be two complete, separable metric spaces and let μ be a probability measure on $(\Xi, \mathcal{B}(\Xi))$. If $X : (\Xi, \mathcal{B}(\Xi)) \to (\Theta, \mathcal{B}(\Theta))$ is a random variable (or, equivalently, X is a Borel measurable mapping) with the distribution μ^X on $(\Theta, \mathcal{B}(\Theta)))$, that is,*

$$\mu^X(B) \equiv \mu(X \in B), \quad B \in \mathcal{B}(\Theta),$$

then there exists a function $Q : \Theta \times \mathcal{B}(\Xi) \to [0, 1]$, called a regular conditional probability measure on $(\Xi, \mathcal{B}(\Xi))$ given X, such that the following hold:
(i) For each fixed $x \in \Theta$, $A \mapsto Q(x; A)$ is a probability measure $(\Xi, \mathcal{B}(\Xi))$.
(ii) For each fixed $A \in \mathcal{B}(\Xi)$, $x \mapsto Q(x, A)$ is $\mathcal{B}(\Theta)$-measurable.
(iii) For every $A \in \mathcal{B}(\Xi)$ and $B \in \mathcal{B}(\Theta)$, we have

$$P[A \cap \{X \in B\}] = \int_B Q(x, A) \mu^X(dx).$$

If $\tilde{Q} : \Theta \times \mathcal{B}(\Xi) \to [0, 1]$ is another such function, then

$$Q(\cdot, A) = \tilde{Q}(\cdot, A) \quad \mu - a.s. \text{ for each } A \in \mathcal{B}(\Xi);$$

that is, there exists a set $N \in \mathcal{B}(\Theta)$ with $\mu(N) = 0$ such that

$$Q(x, A) = \tilde{Q}(x, A) \quad \text{for all } A \in \mathcal{B}(\Xi) \text{ and } x \in \Theta - N.$$

The null set N can be chosen so that the following additional property holds:

$$Q(x, \{X \in B\}) = \mathbf{1}_B(x), \quad B \in \mathcal{B}(\Theta), x \in \Theta - N.$$

In particular,
$$Q(x, \{X = x\}) = 1, \quad \mu^X - a.e., x \in \Theta.$$

Proof. This follows from Theorem I.3.3 in Ikeda and Wanatabe [IW81] and the corollary following it. The details of the proof are omitted here. $\quad\square$

Proposition 1.1.11 *Let $(\Omega, \mathcal{F}, P; \mathbf{F})$ be a complete filtered probability space satisfying the usual conditions and let $\zeta(\cdot) = \{\zeta(s), s \geq 0\}$ be an $\mathbf{F}$-adapted Ξ-valued process. Then, for any $s \geq 0$,*

$$P[\{\omega' \mid \zeta(t, \omega') = \zeta(t, \omega)\} \mid \mathcal{F}(s)](\omega) = 1, \quad P\text{-a.s. } \omega \in \Omega, \forall t \in [0, s]. \tag{1.5}$$

If, in addition, $\zeta(\cdot)$ is continuous, that is,

$$P\left[\lim_{\epsilon \to 0} \zeta(s+\epsilon) = \zeta(s) \;\forall s \in [0,T]\right] = 1,$$

then there exists a P-null set $N \in \mathcal{F}$ (i.e., $P(N) = 0$) such that for any $s \in [0,T]$ and $\omega \notin N$,

$$\zeta(t,\omega') = \zeta(t,\omega), \;\forall t \in [0,s], \; P[\cdot \mid \mathcal{F}(s)](\omega) - a.s. \; \omega' \in \Omega. \qquad (1.6)$$

Proof. By the definition of conditional probability and expectation, for any $t \in [0, s]$,

$$
\begin{aligned}
P[\{\omega' \mid \zeta(t,\omega') = \zeta(t,\omega)\} \mid \mathcal{F}(s)](\omega) &= E[\chi_{\{\omega' \mid \zeta(t,\omega')=\zeta(t,\omega)\}} \mid \mathcal{F}(s)](\omega) \\
&= \chi_{\{\omega' \mid \zeta(t,\omega')=\zeta(t,\omega)\}} E[1 \mid \mathcal{F}(s)](\omega) \\
&= \chi_{\{\zeta(t,\omega')=\zeta(t,\omega)\}}(\omega) = 1, \; P\text{-a.s. } \omega \in \Omega,
\end{aligned}
$$

since $\chi_{\{\zeta(t,\omega')=\zeta(t,\omega)\}} \in \mathcal{F}(s)$. This proves (1.5). By continuity of the process $\zeta(\cdot)$, (1.6) follows easily. $\square$

The above result implies that under the conditional probability measure $P[\cdot \mid \mathcal{F}(s)](\omega)$, where ω is fixed, $\zeta(t)$ is almost surely a deterministic constant $\zeta(t,\omega)$ for any $t \leq s$. This fact will have important applications in dynamic programming studied in later chapters.

1.2 Brownian Motion and Itô Integrals

1.2.1 Brownian Motion

Consider a complete filtered probability space $(\Omega, \mathcal{F}, P, \mathbf{F})$ that satisfies the usual conditions. We define a standard Brownian motion as follows.

Definition 1.2.1 *A five-tuple $(\Omega, \mathcal{F}, P, \mathbf{F}, W(\cdot))$ is said to be an m-dimensional Brownian motion defined on $(\Omega, \mathcal{F}, P, \mathbf{F})$ if the m-dimensional process*

$$W(\cdot) = \{W(s) = (W_1(s), W_2(s), \ldots, W_m(s)), s \geq 0\}$$

is $\mathbf{F}$-adapted (i.e., $W(s)$ is $\mathcal{F}(s)$-measurable for each $s \geq 0$), and satisfies the following conditions:
(i) $P\{W(0) = 0\} = 1$.
(ii) For $0 \leq s < \tilde{s} < \infty$, $W(\tilde{s}) - W(s)$ is independent of $\mathcal{F}(s)$.
(iii) For $0 \leq s < \tilde{s} < \infty$, $W(\tilde{s}) - W(s)$ is an m-dimensional Gaussian random variable with mean zero and covariance matrix $(\tilde{s}-s)I^m$, where I^m is an $m \times m$ identity matrix, that is,

$$
\begin{aligned}
P\{W(\tilde{s}) - W(s) \in A\} &= P\{W(\tilde{s}-s) \in A\} \\
&= \int_A p(\tilde{s}-s, x)\, dx, \quad A \in \mathcal{B}(\Re^m),
\end{aligned}
$$

where

$$p(s,x) = \frac{1}{(2\pi s)^{(m/2)}} \exp\left[-\frac{|x|^2}{2s}\right], \qquad s > 0, x \in \Re^m.$$

For convenience, we define $p(s,x,y) = p(s,y-x)$ for $s \geq 0$ and $x, y \in \Re^m$.

From time to time we will assume for convenience that the filtration $\mathbf{F} = \{\mathcal{F}(s), s \geq 0\}$ for the Brownian motion $(\Omega, \mathcal{F}, P, \mathbf{F}; W(\cdot))$ is the P-augmented natural filtration generated by $W(\cdot)$, that is,

$$\mathcal{F}(s) = \mathcal{F}^W(s) \vee \mathcal{N}, \quad s \geq 0,$$

where

$$\mathcal{F}^W(s) = \sigma(W(\tilde{s}), 0 \leq \tilde{s} \leq s) \vee \mathcal{N} \quad \text{for each } s \geq 0,$$

$$\mathcal{N} = \{A \subset \Omega \mid \text{there exists a } B \in \mathcal{F} \text{ such that } P(B) = 0 \text{ and } A \subset B\},$$

and the symbol $\vee$ denotes the smallest σ-algebra containing $\mathcal{F}^W(s)$ and $\mathcal{N}$.

In this case, the five-tuple $(\Omega, \mathcal{F}, P, \mathbf{F}; W(\cdot))$ will be referred to as an m-dimensional Brownian stochastic basis. Note that Condition (ii) of Definition 1.2.1 indicates that all increments of $W(\cdot)$ are independent of the Brownian past. Therefore, the Brownian motion has independent, stationary, and Gaussian increments.

Although it is not a part of the definition, it can be shown (see, e.g., Karatzas and Shreves [KS91] and Øksendal [Øks98]) that an equivalent version of $\{W(s), s \geq 0\}$ can be chosen so that all samplepaths are continuous almost surely, that is,

$$P\{\lim_{\tilde{s} \to s} W(\tilde{s}) = W(s) \ \forall s \geq 0\} = 1.$$

We often drop $(\Omega, \mathcal{F}, P, \mathbf{F})$ and write $W(\cdot)$ instead of more formal description $(\Omega, \mathcal{F}, P, \mathbf{F}, W(\cdot))$ whenever there is no danger of ambiguity.

We recall the following definition of a martingale, sub-martingale, and super-martingale as follows.

Definition 1.2.2 *Given a complete filtered probability space* $(\Omega, \mathcal{F}, P, \mathbf{F})$ *that satisfies the usual conditions, an* $\mathbf{F}$-*adapted real-valued process* $\zeta(\cdot) = \{\zeta(s), s \in [0,T]\}$ *is said to be an* $\mathbf{F}$-*martingale (super-martingale, sub-martingale) if the following hold:*
(i) $E[|\zeta(s)|] < +\infty$ *for each* $s \in [0,T]$.
(ii) $E[\zeta(\lambda)|\mathcal{F}(s)] = \zeta(s)$ *(respectively,* $E[\zeta(\lambda)|\mathcal{F}(s)] \leq \zeta(s)$, *and* $E[\zeta(\lambda)|\mathcal{F}(s)] \geq \zeta(s)$), *P-a.s.*
for every $0 \leq s \leq \lambda \leq T$.

An $\mathbf{F}$-*adapted* $\Re^n$-*valued process* $\zeta(\cdot) = \{\zeta(s), s \geq 0\}$ *is said to be* $\mathbf{F}$-*martingale if each of its n components is an* $\mathbf{F}$-*martingale. The n-dimensional sub-martingale or super-martingale will be defined similarly.*

Definition 1.2.3 *An n-dimensional process* $\zeta(\cdot) = \{\zeta(s), s \geq 0\}$ *is said to be*
(i) an $\mathbf{F}$-*local martingale if there exists a nondecreasing sequence of* $\mathbf{F}$-*stopping times* $\{\tau_k\}_{k=1}^\infty$ *with*

$$P\{\lim_{k\to\infty} \tau_k = \infty\} = 1,$$

such that $\{\zeta(s \wedge \tau_k), s \geq 0\}$ is an **F**-martingale for each k;

(ii) a continuous **F**-martingale if it is an **F**-martingale and is continuous P-a.s., that is,

$$P\{\lim_{s\to t} \zeta(s) = \zeta(t), \ \forall t \geq [0,T]\} = 1;$$

(iii) a square integrable **F**-martingale if it is a martingale and

$$P\left\{ \int_0^T |\zeta(s)|^2 \, ds < \infty \right\} = 1.$$

The collection of n-dimensional martingales, continuous, local martingales, and squared-integrable martingales will be denoted by $\mathcal{M}([0,T],\Re^n)$, $\mathcal{M}^c([0,T],\Re^n)$, $\mathcal{M}^{loc}([0,T],\Re^n)$, and $\mathcal{M}^2([0,T],\Re^n)$, respectively.

The following is a characterization of a standard Brownian motion.

Proposition 1.2.4 $(\Omega, \mathcal{F}, P, \mathbf{F}, W(\cdot))$ is an m-dimensional Brownian sto-chastic basis if and only if $M_\Phi(\cdot) = \{M_\Phi(s), s \geq 0\}$ is an **F**-martingale, where

$$M_\Phi(s) = \Phi(W(s)) - \Phi(0) - \int_0^s \mathcal{L}_w \Phi(W(t)) \, dt \qquad (1.7)$$

for $\Phi \in C_0^2(\Re^m)$ (the space of real-valued twice continuously differentiable functions on $\Re^m$ with compact support) and $\mathcal{L}_w$ denotes the differential operator defined by

$$\mathcal{L}_w \Phi(x) = \frac{1}{2} \sum_{j=1}^m \partial_j^2 \Phi(x), \qquad (1.8)$$

with $\partial_j^2 \Phi(x)$ being the second-order partial derivative of $\Phi(x) = \Phi(x_1, \ldots, x_m)$ with respect to x_j, $j = 1, 2, \cdots, m$.

Proof. See [KS91] or [Øks98]. $\square$

In some occasions such as in Section 3.2, if is desirable to show that a continuous process $\xi(\cdot) = \{\xi(s), s \geq 0\}$ is a **F**-Brownian motion. In this case, we first need to show that $\xi(s) - \xi(t)$ is independent of $\mathcal{F}(t)$ for all $0 \leq t < s < \infty$ by showing that

$$E\left[H(y)(\xi(s) - \xi(t))\right] = 0,$$

where

$$y = \{\xi(t_i), 0 \leq t_1 \leq t_2 \leq \cdots \leq t_l \leq t,\}$$

for all $l = 1, 2, \ldots$, and

$$E\left[H(y)(\xi(s) - \xi(t))(\xi(s) - \xi(t))^\top\right] = (s-t)I^{(m)}.$$

This is due to the following result.

Proposition 1.2.5 *Let* $\zeta^{(i)} : (\Omega, \mathcal{F}) \to (\Re^{m_i}, \mathcal{B}(\Re^{m_i}))$ *be a sequence of random variables* $(i = 1, 2, \ldots)$ *and let* $\mathcal{G} \equiv \vee_i \sigma(\zeta^{(i)})$ *and* $X \in L^2(\Omega, \Re^m; \mathcal{F})$. *The* $E[X|\mathcal{G}] = 0$ *if and only if* $E[H(\xi^{(1)}, \xi^{(2)}, \cdots, \xi^{(i)})X] = 0$ *for any* i *and any* $H \in C_b(\Re^N)$, *where* $N = \sum_{j=1}^i jm_i$.

Proof. See Proposition 1.12 of Chapter 1 in Yong and Zhou [YZ99].   □

The following result is Lévy's famous theorem regarding modulus of continuity for the Brownian path.

Theorem 1.2.6 (Modulus of Continuity)

$$P\left\{\overline{\lim}_{\delta\downarrow 0, 0<s-t<\delta, s<1} \frac{|W(s) - W(t)|}{\sqrt{2\delta ln(\frac{1}{\delta})}} = 1\right\} = 1.$$

Proof. See Theorem 2.8 of Knight [Kni81].   □

1.2.2 Itô Integrals

For the time being, let $(\Omega, \mathcal{F}, P, \mathbf{F}, W(\cdot))$ be a one-dimensional Brownian motion. Let $L^2_{\mathbf{F}}(0, \infty; \Re)$ be the Hilbert space of all real-valued processes (also defined on $(\Omega, \mathcal{F}, P, \mathbf{F})$) $g(\cdot) = \{g(s), s \geq 0\}$ such that the following hold:
(i) $g(\cdot)$ is measurable; that is, the mapping $(s, \omega) \mapsto g(s, \omega)$ is $\mathcal{B}([0, \infty)) \times \mathcal{F}$, where $\mathcal{B}([0, \infty))$ is the Borel σ-algebra of subsets of the interval $[0, \infty)$.
(ii) $g(\cdot)$ is $\mathbf{F}$-adapted.
(iii) $g(\cdot)$ satisfies the following global integrability property:

$$E\left[\int_0^\infty |g(s)|^2 \, ds\right] < \infty.$$

We also consider the its extension to the space $L^{2,loc}_{\mathbf{F}}(0, \infty; \Re)$ of all processes $g(\cdot) = \{g(s), s \geq 0\}$ such that Conditions (i), (ii), and (iii') are satisfied for all $\forall 0 \leq a < b < \infty$.
(iii') $g(\cdot)$ satisfies the following local integrability property:

$$\int_a^b |g(s)|^2 \, ds < \infty, \quad P\text{-a.s.}, \forall 0 \leq a < b < \infty.$$

The concept of Itô integration $\int_a^b g(s) \, dW(s)$ of $g(\cdot) \in L^{2,loc}_{\mathbf{F}}(0, \infty; \Re)$ with respect to the standard Brownian motion $W(\cdot)$ is due originally to [Itô44] and is now well known within the community of stochastic analysts. Instead of repeating its detailed construction, we will only outline its construction steps and briefly review its important properties that will be used in our later developments. The readers are referred to Karatzas and Shreves [KS91] and Øksendal [Øks98] for details.

First, $\int_a^b g(s)dW(s)$ is defined as

$$\int_a^b g(s)\,dW(s) = \sum_{k\geq 0} g_k(\omega)[W(t_{k+1},\omega) - W(t_k,\omega)] \qquad (1.9)$$

for all *elementary* $g(\cdot) \in L_{\mathbf{F}}^{2,loc}(0,\infty;\Re)$ that take the following form:

$$g(s) = \sum_{k\geq 0} g_k(\omega)\chi_{[t_k,t_{k+1})}(s), \qquad (1.10)$$

where

$$t_k = t_k^{(n)} = \begin{cases} k2^{-n} & \text{if } a \leq k2^{-n} \leq b \\ a & \text{if } k2^{-n} < a \\ b & \text{if } k2^{-n} > b, \end{cases}$$

g_k are random variables that are $\mathcal{F}(t_k)$-measurable for all $k = 0,1,2,\ldots$, and $\chi_{[t_k,t_{k+1})}$ is the *indicator function* of the interval $[t_k, t_{k+1})$.

Lemma 1.2.7 (The Itô Isometry) *If* $g(\cdot) \in L_{\mathbf{F}}^{2,loc}(0,\infty;\Re)$ *is bounded and elementary then*

$$E\left[\left(\int_a^b g(s)\,dW(s)\right)^2\right] = E\left[\int_a^b g^2(s)\,ds\right]. \qquad (1.11)$$

For any $f(\cdot) \in L_{\mathbf{F}}^{2,loc}(0,\infty;\Re)$, one can use Lemma 1.2.7 to show that there exists a sequence of bounded and elementary $\phi_k(\cdot) \in L_{\mathbf{F}}^{2,loc}(0,\infty;\Re)$ such that

$$\lim_{k\to\infty} E\left[\int_a^b (f(s) - \phi_k(s))^2\,ds\right] = 0,$$

via the following three steps:

Step 1. Let $g(\cdot) \in L_{\mathbf{F}}^{2,loc}(0,\infty;\Re)$ be bounded and continuous. Then there exists *elementary* processes $\phi_k \in L_{\mathbf{F}}^{2,loc}(0,\infty;\Re)$ such that

$$\lim_{k\to\infty} E\left[\int_a^b (g(s) - \phi_k(s))^2\,ds\right] = 0.$$

Step 2. Let $h(\cdot) \in L_{\mathbf{F}}^{2,loc}(0,\infty;\Re)$ be bounded. Then there exists bounded and continuous processes $g_k(\cdot) \in L_{\mathbf{F}}^{2,loc}(0,\infty;\Re)$ such that

$$\lim_{k\to\infty} E\left[\int_a^b (h(s) - g_k(s))^2\,ds\right] = 0.$$

Step 3. Let $f(\cdot) \in L_{\mathbf{F}}^{2,loc}(0,\infty;\Re)$. Then there exists a sequence $\{h_k(\cdot)\} \subset L_{\mathbf{F}}^{2,loc}(0,\infty;\Re)$ such that $h_k(\cdot)$ is bounded for each k and

$$\lim_{k\to\infty} E\left[\int_a^b (f(s) - h_k(s))^2\, ds\right] = 0.$$

The Itô integral $\int_a^b f(s)\, dW(s)$ for any $f(\cdot) \in L_{\mathbf{F}}^{2,loc}(0,\infty;\Re)$ can then be defined by the following definition.

Definition 1.2.8 (The Itô integral) *Let* $f(\cdot) \in L_{\mathbf{F}}^{2,loc}(0,\infty;\Re)$. *Then the Itô integral,* $\int_a^b f(s)\, dW(s)$, *of* $f(\cdot)$ *with respect to the Brownian motion* $W(\cdot)$ *from* a *to* b *is defined by*

$$\int_a^b f(s)\, dW(s) = \lim_{k\to\infty} \int_a^b \phi_k(s)\, dW(s) \quad \text{(the limit is in } L^2(\Omega,\Re)), \quad (1.12)$$

where $\{\phi_k(\cdot)\} \subset L_{\mathbf{F}}^{2,loc}(0,\infty;\Re)$ *is a sequence of bounded and elementary processes such that*

$$\lim_{k\to\infty} E\left[\int_a^b (f(s) - \phi_k(s))^2\, ds\right] = 0, \quad (1.13)$$

and $\int_a^b \phi_k(s)\, dW(s)$ *is defined in (1.9) for each* k.

Now consider the m-dimensional standard Brownian motion $(\Omega, \mathcal{F}, P, \mathbf{F}, W(\cdot))$. We say that

$$f(\cdot) = (f_1(\cdot), \ldots, f_m(\cdot)) \in L_{\mathbf{F}}^{2,loc}(0,\infty;\Re^m)$$

if $f_i(\cdot) \in L_{\mathbf{F}}^{2,loc}(0,\infty;\Re)$. If $f(\cdot) \in L_{\mathbf{F}}^{2,loc}(0,\infty;\Re^m)$, then the Itô integral $\int_a^b f(s) \cdot dW(s)$ is defined in terms of one-dimensional Itô integral by

$$\int_a^b f(s) \cdot dW(s) = \sum_{j=1}^m \int_a^b f_j(s)\, dW_j(s). \quad (1.14)$$

For applications to our SHDE described in (1.27), we define the Itô integral $\int_a^b g(s)\, dW(s)$ of the matrix-valued process

$$g(\cdot) = [g_{ij}(\cdot)] \in L_{\mathbf{F}}^{2,loc}(0,\infty;\Re^{n\times m})$$

with respect to the m-dimensional Brownian motion $W(\cdot) = \{W(s) = (W_1(s), W_2(s), \ldots, W_m(s)), s \geq 0\}$ as the n-dimensional random vector

$$\int_a^b g(s)\, dW(s) = \left(\int_a^b g_1(s) \cdot dW(s), \ldots, \int_a^b g_n(s) \cdot dW(s)\right),$$

where $\int_a^b g_i(s) \cdot dW(s)$ is as defined in (1.14) for $i = 1, 2, \ldots, n$.

The linearity, isometry, and other important properties of the Itô integrals defined in Definition 1.2.8 are preserved and can be summarized as follows.

Theorem 1.2.9 *Let* $f(\cdot), g(\cdot) \in L_{\mathbf{F}}^{2,loc}(0,\infty; \Re^{n\times m})$ *and let* $0 \le a < c < b$. *Then*
(i) $\int_a^b f(s) \, dW(s) = \int_a^c f(s) \, dW(s) + \int_c^b f(s) \, dW(s)$ *P-a.s..*
(ii) $\int_a^b (\alpha f(s) + \beta g(s)) \, dW(s) = \alpha \int_a^b f(s) \, dW(s) + \beta \int_a^b g(s) \, dW(s), \, \forall \alpha, \beta \in \Re.$
(iii) $E\left[\int_a^b f(s) \, dW(s)\right] = 0.$

In addition to the trivial properties of the Itô integrals stated in Theorem 1.2.9, the following two properties will be used frequently later. The detailed proofs of these properties can be found in [KS91].

Proposition 1.2.10 *Let* $f(\cdot) \in L_{\mathbf{F}}^{2,loc}(0,\infty; \Re^m)$, *and let* $g(\cdot), \, \hat{g}(\cdot) \in L_{\mathbf{F}}^{2,loc}$ $(0,\infty; \Re^{n\times m})$. *Then, for all* $0 \le a \le b < \infty$ *and* $i,j = 1,2,\ldots,m$,

$$E\left[\int_a^b f_i(t) \, dW_i(t) \int_a^b f_j(t) \, dW_j(t)\Big|\mathcal{F}(a)\right] = \delta_{ij} \left[\int_a^b f_i(t) f_j(t) dt \Big|\mathcal{F}(a)\right],$$
(1.15)

where δ_{ij} *is defined by*

$$\delta_{ij} = \begin{cases} 0 \text{ for } i \ne j \\ 1 \text{ for } i = j \end{cases}$$

and

$$E\left[\left|\int_a^b g(t) dW(t)\right|^2 \Big|\mathcal{F}(a)\right] = E\left[\int_a^b \text{trace } [\hat{g}(t)^\top g(t)] dt \Big|\mathcal{F}(a)\right], \quad (1.16)$$

where trace A *(the trace of the* $n \times n$ *matrix* $A = [a_{ij}]$*) is defined by* trace $A = \sum_i^n a_{ii}$.

Note that $E[\cdots \mid \mathcal{F}(a)]$ is the conditional expectation of the random variable "$\cdots$" with respect to the sub-σ-algebra $\mathcal{F}(a)$ as defined in Section 1.1.1.

Theorem 1.2.11 *(Burkholder-Davis-Gundy Inequalities)*
Let $(\Omega, \mathcal{F}, P, \mathbf{F}, W(\cdot))$ *be an* m-*dimensional standard Brownian motion and let* $g(\cdot) \in L_{\mathbf{F}}^{2,loc}(0,\infty; \Re^{n\times m})$. *Then, for any* $p > 0$, *there exists a constant* $K(p) > 0$ *such that for any* $\mathbf{F}$-*stopping time* τ,

$$\frac{1}{K(p)} E\left[\left(\int_0^\tau |g(s)|^2 \, ds\right)^p\right] \le E\left[\sup_{0 \le t \le \tau} \left|\int_0^\tau g(s) \, dW(s)\right|^{2p}\right]$$

$$\le K(p) E\left[\left(\int_0^\tau |g(s)|^2 \, ds\right)^p\right]. \quad (1.17)$$

Corollary 1.2.12 *(Doob's Maximal Inequality) Let $g(\cdot) \in L_{\mathbf{F}}^{2,loc}(0, \infty; \Re^{n \times m})$. For each $T > 0$, there exists a constant K that depends on the dimension m such that*

$$E\left[\sup_{s \in [0,T]} \left|\int_0^s g(t)\, dW(t)\right|^2\right] \leq K \int_0^T E[|g(s)|^2]\, ds. \qquad (1.18)$$

The following two theorems state the martingale properties of the indefinite Itô integral process and its converse.

Theorem 1.2.13 *For each $g(\cdot) \in L_{\mathbf{F}}^{2,loc}(0, \infty; \Re^{n \times m})$, the n-dimensional indefinite Itô integral process $I(g)(\cdot)$ defined by*

$$I(g)(\cdot) = \left\{I(g)(s) = g(0) + \int_0^s g(t)\, dW(t), s \geq 0\right\}$$

is a continuous, square integrable, $\mathbf{F}$-martingale, that is,

(i) $I(g)(s)$ is $\mathcal{F}(s)$-measurable for each $s \geq 0$;
(ii) $E[|I(g)(s)|] < \infty$ for each $s \geq 0$;
(iii) $E[I(g)(s) \mid \mathcal{F}(t)] = I(g)(t)$ for all $0 < t < s$. Furthermore, $I(g)(\cdot)$ has a continuous version.

The following theorem, a converse of the previous theorem, is often referred to as a martingale representation theorem for any $\mathbf{F}$-adapted process.

Theorem 1.2.14 *(Martingale Representation Theorem) Let $(\Omega, \mathcal{F}, P, \mathbf{F}, W(\cdot))$ be an m-dimensional standard Brownian motion. Suppose $M(\cdot) = \{M(t), s \geq 0\}$ is a squared-integrable n-dimensional martingale with respect to $(\Omega, \mathcal{F}, P, \mathbf{F})$. Then there exists a unique $g(\cdot) \in L_{\mathbf{F}}^{2,loc}(0, \infty; \Re^{n \times m})$ such that*

$$M(s) = E[M(0)] + \int_0^s g(s)\, dW(s), \quad P - a.s., \ \forall s \geq 0.$$

1.2.3 Itô's Formula

Consider the n-dimensional Itô process $x(\cdot) = \{x(s), s \geq 0\}$ defined by

$$x(s) = x(0) + \int_0^s f(t)\, dt + \int_0^s g(t)\, dW(t), \quad s \geq 0, \qquad (1.19)$$

where $f(\cdot) \in L_{\mathbf{F}}^{2,loc}(0, \infty; \Re^n)$ and $g(\cdot) \in L_{\mathbf{F}}^{2,loc}(0, \infty; \Re^{m \times n})$.

Let $C^{1,2}([0, T] \times \Re^n; \Re^l)$ be the space of functions $\Phi : [0, T] \times \Re^n \to \Re^l$ ($\Phi = (\Phi_1, \Phi_2, \ldots, \Phi_l)$) that are continuously differentiable with respect to its first variable $t \in [0, T]$ and twice continuously differentiable with respect to its second variable $x = (x_1, x_2, \ldots, x_n) \in \Re^n$. The Itô formula is basically a change-of-variable formula for the process $x(\cdot) = \{x(s), s \geq 0\}$ and can be stated as follows.

Theorem 1.2.15 *Let $x(\cdot) = \{x(s), s \geq 0\}$ be the Itô process defined by (1.19). If $\Phi \in C^{1,2}([0,T] \times \Re^n; \Re^l)$, then*

$$d\Phi_k(t, x(t)) = \partial_t \Phi_k(t, x(t))\, dt + \sum_{i=1}^{n} \partial_i \Phi_k(t, x(t)) f_i(t)\, dt$$

$$+ \sum_{i=1}^{n} \sum_{j=1}^{m} \partial_i \Phi_k(t, x(t)) g_{ij}(t)\, dW_j(t)$$

$$+ \frac{1}{2} \sum_{i,j=1}^{n} \partial_{ij}^2 \Phi_k(t, x(t)) a_{ij}(t)\, dt, \tag{1.20}$$

where $a(\cdot) = [a_{ij}(\cdot)] = g^\top(\cdot) g(\cdot)$, $\partial_t \Phi_k(t,x)$ and $\partial_i \Phi_k(t,x)$ are the partial derivatives with respect to t and x_i, respectively, and $\partial_{ij}^2 \Phi_k(t,x)$ is the second partial derivative first with respect to x_j and then with respect to x_i.

Extensions of Theorem 1.2.15 to a more general (such as convex) function $\Phi : [0,T] \times \Re^n \to \Re^l$ are possible but will not be dealt with here.

1.2.4 Girsanov Transformation

Consider the m-dimensional standard Brownian motion $(\Omega, \mathcal{F}, P, \mathbf{F}; W(\cdot))$. Let $\zeta(\cdot) \in L_\mathbf{F}^{2,loc}(0,\infty; \Re^m)$ be such that

$$P\left[\int_0^T |\zeta(s)|^2\, ds < \infty \right] = 1, \quad 1 \leq i \leq m,\ 0 \leq T < \infty. \tag{1.21}$$

Then the one-dimensional Itô integral process

$$I(\zeta)(\cdot) = \left\{ I(\zeta)(s) \equiv \int_0^s \zeta(t) \cdot dW(t), s \geq 0 \right\}$$

is well defined and is a member of $\mathcal{M}^{2,loc}$. We set

$$\xi(s) = \exp\left\{ \int_0^s \zeta(t) \cdot dW(t) - \frac{1}{2} \int_0^s |\zeta(t)|^2\, dt \right\}, \quad s \geq 0. \tag{1.22}$$

From the classical Itô formula (see (1.20)), we have

$$\xi(s) = 1 + \int_0^s \xi(t)\zeta(t) \cdot dW(t), \quad s \geq 0, \tag{1.23}$$

which shows that $\xi(\cdot) = \{\xi(s), s \geq 0\}$ is a one-dimensional $\mathcal{M}^{c,loc}$ with $Z(0) = 1$.

Under certain conditions on $\zeta(\cdot)$, to be discussed later, $\xi(\cdot)$ will in fact be a martingale, and so $E[\xi(s)] = 1$, $0 \leq s < \infty$. In this case, we can define, for each $0 \leq T < \infty$, a probability measure $\tilde{P}(T)$ on the measurable space $(\Omega, \mathcal{F}(T))$ by

$$\tilde{P}(T)(A) \equiv E[\chi_A \xi(T)], \quad A \in \mathcal{F}(T). \tag{1.24}$$

The martingale property shows that the family of probability measures $\{\tilde{P}(T),$ $0 \leq T < \infty\}$ satisfies the following consistency condition:

$$\tilde{P}(T)(A) = \tilde{P}(s)(A), \quad A \in \mathcal{F}(s), \ 0 \leq s \leq T. \tag{1.25}$$

We have the following Girsanov theorem whose proof can be found in Theorem 5.1 of [KS91]:

Theorem 1.2.16 (Girsanov–Cameron–Martin) *Assume that $\xi(\cdot)$ defined by (1.23) is a martingale. For each fixed $T > 0$, define an m-dimensional process $\tilde{W}(\cdot) = \{\tilde{W}(s) = (\tilde{W}_1(s), \tilde{W}_2(s), \ldots, \tilde{W}_m(s)), 0 \leq s \leq T\}$ by*

$$\tilde{W}(s) \equiv W(s) - \int_0^s \zeta(t)\, dt, \quad 0 \leq s \leq T. \tag{1.26}$$

Then the process $(\Omega, \mathcal{F}(T), \tilde{P}(T), \{\mathcal{F}(s), 0 \leq s \leq T\}, \tilde{W}(\cdot))$ is an m-dimensional standard Brownian motion.

1.3 SHDE with Bounded Memory

Let $(\Omega, \mathcal{F}, P, \mathbf{F}, W(\cdot))$ be an m-dimensional standard Brownian motion. In this section, we study the strong and weak solutions $x(\cdot) = \{x(s), s \in [t - r, T]\}$ of the following n-dimensional stochastic hereditary differential equation (SHDE) with bounded memory $0 < r < \infty$:

$$dx(s) = f(s, x_s)\, ds + g(s, x_s)\, dW(s), \quad s \in [t, T], \tag{1.27}$$

with the initial datum $(t, x_t) = (t, \psi_t) \in [0, T] \times L^2(\Omega, \mathbf{C}; \mathcal{F}(t))$, where $f : [0, T] \times L^2(\Omega, \mathbf{C}) \to \Re^n$ and $g : [0, T] \times L^2(\Omega, \mathbf{C}) \to \Re^{n \times m}$ are continuous functions that are to be specified later.

Remark 1.3.1 *Although the relevant concepts can be easily extended or generalized to systems driven by semi-martingales such as Levy processes, we will restrain ourselves and treat only the differential systems that are driven by a certain m-dimensional standard Brownian motion $W(\cdot)$ in this monograph, for the sake of clarity in illustrating the theory and applications of optimal control of these systems.*

1.3.1 Memory Maps

We have the following lemma regarding the memory map for the SHDE (1.27) with bounded memory $0 \leq r < \infty$. We will work extensively with the Banach space $\mathbf{C} \equiv C([-r, 0]; \Re^n)$ in this case. Recall that the space $\mathbf{C}$ is a real separable Banach space equipped with the uniform topology, that is,

$$\|\phi\| = \sup_{\theta \in [-r,0]} |\phi(\theta)|, \quad \phi \in \mathbf{C}.$$

Lemma 1.3.2 *For each $0 < T < \infty$, the memory map*

$$m : [0,T] \times C([-r,T], \Re^n) \to \mathbf{C}$$

defined by

$$m(t,\phi) = \phi_t, \quad (t,\psi) \in [0,T] \times C([-r,T], \Re^n),$$

is jointly continuous, where $\phi_t : [-r,0] \to \Re^n$ is defined by $\phi_t(\theta) = \phi(t + \theta)$, $\theta \in [-r,0]$.

Proof. Let (t,ϕ), $(s,\varphi) \in [0,T] \times C([-r,T], \Re^n)$ be given. Let $\epsilon > 0$. We will find a $\delta > 0$ such that

$$\|m(t,\phi) - m(s,\varphi)\| = \|\phi_t - \varphi_s\| < \epsilon,$$

whenever $|t - s| < \delta$ and $\sup_{s,t \in [-r,T]} |\phi(t) - \varphi(s)| < \delta$. Since ϕ and φ are continuous on the compact interval $[-r,T]$, they are both uniformly continuous on $[-r,T]$ as well as the difference $\phi - \varphi$. Therefore, given the same $\epsilon > 0$, we can find $\delta_1, \delta_2 > 0$ such that for all $s,t \in [-r,T]$ with $|s - t| = |(s + \theta) - (t + \theta)| < \delta_1$, we have $|\phi(t + \theta) - \phi(s + \theta)| < \frac{\epsilon}{2}$ for all $\theta \in [-r,0]$. Similarly, with $|s - t| = |(s + \theta) - (t + \theta)| < \delta_2$, we have $|\varphi(t + \theta) - \varphi(s + \theta)| < \frac{\epsilon}{2}$ for all $\theta \in [-r,0]$. Therefore, for all $\theta \in [-r,0]$,

$$|\phi(t + \theta) - \varphi(s + \theta)| \leq |\phi(t + \theta) - \phi(s + \theta)| + |\phi(s + \theta) - \varphi(s + \theta)|$$
$$< \frac{\epsilon}{2} + \frac{\epsilon}{2} = \epsilon,$$

whenever $|t - s| < \delta \equiv (\delta_1 \wedge \delta_2)$. This completes the proof. □

We have the following corollary.

Corollary 1.3.3 *The stochastic memory map*

$$m^* : [0,T] \times L^2(\Omega, C([-r,T], \Re^n)) \to L^2(\Omega, \mathbf{C})$$

defined by $(t, x(\cdot)) \mapsto x_t$ is a continuous map.

Proof. The *stochastic memory map* m^* is related to the memory map m by the following relation:

$$m^*(t, x(\cdot)) = m(t, \cdot)x(\cdot), \quad \forall t \in [0,T] \text{ and } x(\cdot) \in L^2(\Omega, C([-r,T], \Re^n)).$$

Now, m is continuous, and $m(t, \cdot)$ is continuous and linear for each $t \in [0,T]$ Therefore, the above relation between m^* and m yields that $m^*(t, \cdot)$ is continuous and linear for all $t \in [0,T]$ and m^* is continuous. Indeed,

$$\|m^*(t, x(\cdot))\|_{L^2(\Omega, \mathbf{C})} \leq \|x_t\|_{L^2(\Omega, \mathbf{C})},$$

$$\forall x(\cdot) \in L^2(\Omega, C([-r,T], \Re^n)) \text{ and } t \in [0,T].$$ □

Lemma 1.3.4 *Suppose $x(\cdot) = \{x(s), s \in [t-r,T]\}$ is an n-dimensional (also defined on $(\Omega, \mathcal{F}, P, \mathbf{F})$) with almost all paths continuous). Assume that the restricted process $x(\cdot)|_{[t,T]} = \{x(s), s \in [t,T]\}$ is $\mathbf{F}$-adapted and $x(\theta)$ is $\mathcal{F}(t)$-measurable for all $\theta \in [t-r,t]$. Then the $\mathbf{C}$-valued process $\{x_s, s \in [t,T]\}$ defined by $x_s(\theta) = x(s+\theta)$, $\theta \in [-r,0]$, is $\mathbf{F}$-adapted, with almost all sample paths continuous, that is,*

$$P\left[\lim_{\epsilon \to 0} \|x_{s+\epsilon} - x_s\| = 0, \ \forall s \in [t,T]\right] = 1.$$

Proof. Without loss of generality, we can and will assume $t = 0$ for simplicity. Fix $s \in [0,T]$. Since the restricted process

$$x(\cdot)|_{[-r,s]} = \{x(t), t \in [-r,s]\}$$

has continuous sample paths, it induces an $\mathcal{F}$-measurable map

$$x(\cdot)|_{[-r,s]} : \Omega \to C([-r,s]; \Re^n).$$

This map is in fact $\mathcal{F}(s)$-measurable; to see this we observe that $\mathcal{B}(C([-r,s]; \Re^n))$ is generated by cylinder sets of the form

$$\cap_{i=1}^{k} \pi^{-1}(t_i)(B_i) = \{\varphi \in C([-r,s]; \Re^n) \mid \varphi(t_i) \in B_i, i = 1, 2, \ldots, k\}$$

where k is a positive integer, $B_i \in \mathcal{B}(\Re^n)$, $t_i \in [-r,s]$, and

$$\pi(t_i) : C([-r,s]; \Re^n) \to \Re^n$$

is the point evaluation map defined by $\pi(t_i)(\varphi) = \varphi(t_i)$ for each $i = 1, 2, \ldots, k$. Therefore, it is sufficient to check the $\mathcal{F}(t_i)$-measurability of $x(\cdot)|_{[-r,s]}$ on such cylinder sets. With the above notation,

$$(x(\cdot)|_{[-r,s]})^{-1}[\cap_{i=1}^{k}\pi^{-1}(t_i)(B_i)] = \cap_{i=1}^{k}\{\omega \in \Omega \mid x(t_i, \omega) \in B_i\}.$$

By hypotheses, if $t_i \in [-r,0]$, $\{\omega \in \Omega \mid x(t_i, \omega) \in B_i\} \in \mathcal{F}(0)$; if $t_i \in [0,s]$, $\{\omega \in \Omega \mid x(t_i, \omega) \in B_i\} \in \mathcal{F}(t_i)$. Hence,

$$\{\omega \in \Omega \mid x(t_i, \omega) \in B_i\} \in \mathcal{F}(t_i) \subset \mathcal{F}(s) \text{ if } 1 \le i \le k.$$

So,

$$(x(\cdot)|_{[-r,s]})^{-1}\left[\cap_{i=1}^{k}\pi^{-1}(t_i)(B_i)\right] \in \mathcal{F}(s).$$

Now, $x_s = m(s, \cdot) \circ (x(\cdot)|_{[-r,s]})$ and the deterministic memory map $m(s, \cdot) : C([-r,s]; \Re^n) \to \mathbf{C}$ is continuous, so the $\mathbf{C}$-valued random variable x_s must be $\mathcal{F}(s)$-measurable. Therefore, the $\mathbf{C}$-valued process $\{x_s, s \in [0,T]\}$ is $\mathbf{F}$-adapted. $\square$

The following corollary is a simple consequence of the above lemma.

Corollary 1.3.5 *Let* $(\Omega, \mathcal{F}, P, \mathbf{F}, W(\cdot))$ *be an m-dimensional standard Brownian motion. If* $W_0 \equiv \mathbf{0} \in \mathbf{C}$, *then, for each* $s \geq 0$,

$$\sigma(W(t), -r \leq t \leq s) = \sigma(W_t, 0 \leq t \leq s), \quad \forall s \geq 0,$$

where $W_t : [-r, 0] \to \Re^m$ *is again defined by* $W_t(\theta) = W(t + \theta)$ *for all* $t \geq 0$ *and all* $\theta \in [-r, 0]$.

The above corollary implies that the historical information of the m-dimensional standard Brownian motion $\{W(s), s \geq -r\}$ coincides with that of the $\mathbf{C}$-valued process $\{W_s, s \geq 0\}$. The same conclusion holds for any finite-dimensional continuous process $x(\cdot) = \{x(s), s \in [t - r, T]\}$ defined on $(\Omega, \mathcal{F}, P, \mathbf{F})$, where x_t is $\mathcal{F}(t)$-measurable and its restriction to the interval $[t, T]$, $x(\cdot)|_{[t,T]}$, is $\mathbf{F}$-adapted.

1.3.2 The Assumptions

In (1.27), $f : [0, T] \times L^2(\Omega; \mathbf{C}) \to L^2(\Omega; \Re^n)$ and $g : [0, T] \times L^2(\Omega; \mathbf{C}) \to L^2(\Omega; \Re^{n \times m})$ are continuous functions that satisfy the following Lipschitz continuity and linear growth condition.

Assumption 1.3.6 (Lipschitz Continuity) *There exists a constant* $K_{lip} > 0$ *such that*

$$E\left[|f(s, \varphi) - f(t, \phi)| + |g(s, \varphi) - g(t, \phi)|\right] \leq K_{lip}(|s - t| + \|\varphi - \phi\|_{L^2(\Omega; \mathbf{C})}),$$

$$\forall (s, \varphi), (t, \phi) \in [0, T] \times L^2(\Omega; \mathbf{C}).$$

Assumption 1.3.7 (Linear Growth) *There exists a constant* $K_{grow} > 0$ *such that*

$$E[|f(t, \phi)| + |g(t, \phi)|] \leq K_{grow} E[1 + \|\phi\|], \quad \forall (t, \phi) \in [0, T] \times L^2(\Omega, \mathbf{C}).$$

Remark 1.3.8 *Equation (1.27) is general enough to include the equations that contains discrete delays such as*

$$dx(s) = \hat{f}(s, x(s), x(s - r_1), x(s - r_2), \ldots, x(s - r_k)) \, ds$$
$$+ \hat{g}(s, x(s), x(s - r_1), x(s - r_2), \ldots, x(s - r_k)) \, dW(s), \quad s \in [t, T],$$

and continuous delays such as

$$dx(s) = \hat{f}(s, x(s), y_1(s), y_2(s), \ldots, y_k(s)) \, ds$$
$$+ \hat{g}(s, x(s), y_1(s), y_2(s), \ldots, y_k(s)) \, dW(s), \quad s \in [t, T], \quad (1.28)$$

where $\hat{f}$ *and* $\hat{g}$ *are* $\Re^n$- *and* $\Re^{n \times m}$-*valued functions, respectively, defined on*

$$[0, T] \times \overbrace{\Re^n \times \cdots \times \Re^n}^{(k+1)-folds},$$

$0 < r_1 < r_2 < \cdots < r_k = r$ *and* $y_i(s) = \int_{-r}^{0} h_i(\theta) x(s + \theta) \, d\theta$ *and* $h_i : [-r, 0] \to \Re^{n \times n}$, $i = 1, 2, \ldots, k$, *are some continuous functions.*

1.3.3 Strong Solution

The definition of a strong solution process of (1.27) and its pathwise uniqueness is given below.

Definition 1.3.9 *Let* $(\Omega, \mathcal{F}, P, \mathbf{F}, W(\cdot))$ *be an m-dimensional standard Brownian motion. A process* $\{x(s; t, \psi_t), s \in [t-r, T]\}$*, defined on* $(\Omega, \mathcal{F}, P, \mathbf{F})$*, is said to be a (strong) solution of (1.27) on the interval* $[t-r, T]$ *and through the initial datum* $(t, \psi_t) \in [0, T] \times L^2(\Omega, \mathbf{C}; \mathcal{F}(t))$ *if it satisfies the following conditions:*

1. $x_t(\cdot; t, \psi_t) = \psi_t$.
2. $x(s; t, \psi_t)$ *is* $\mathcal{F}(s)$*-measurable for each* $s \in [t, T]$.
3. *For* $1 \leq i \leq n$ *and* $1 \leq j \leq m$ *we have*

$$P\left[\int_t^T \left(|f_i(s, x_s)| + g_{ij}^2(s, x_s) \right) ds < \infty \right] = 1.$$

4. *The process* $\{x(s; t, \psi_t), s \in [t, T]\}$ *is continuous and satisfies the following stochastic integral equation P-a.s.:*

$$x(s) = \psi_t(0) + \int_t^s f(\lambda, x_\lambda) \, d\lambda + \int_t^s g(\lambda, x_\lambda) \, dW(\lambda), \quad s \in [t, T]. \quad (1.29)$$

In addition, the solution process $\{x(s; t, \psi_t), s \in [t-r, T]\}$ *is said to be (strongly) unique if* $\{\tilde{x}(s; t, \psi_t), s \in [t-r, T]\}$ *is also a solution of (1.27) on* $[t-r, T]$ *and through the same initial datum* (t, ψ_t)*, then*

$$P\{x(s; t, \psi_t) = \tilde{x}(s; t, \psi_t), \forall s \in [t, T]\} = 1.$$

The following two lemmas provide intermediate steps for establishing the existence and uniqueness of the strong solution for (1.27).

Lemma 1.3.10 *Assume that* $f : [0, T] \times L^2(\Omega; \mathbf{C}) \to L^2(\Omega; \Re^n)$ *and* $g : [0, T] \times L^2(\Omega; \mathbf{C}) \to L^2(\Omega; \Re^{n \times m})$ *are continuous functions that satisfy the Lipschitz continuity Assumption 1.3.6. Let* $x(\cdot) = \{x(s), s \in [t-r, T]\}$ *and* $y(\cdot) = \{y(s), s \in [t-r, T]\}$ *be two processes in* $L^2(\Omega, C([t-r, T], \Re^n))$ *such that* $x_t, y_t \in L^2(\Omega, \mathbf{C}; \mathcal{F}(t))$ *and their restrictions to the interval* $[t, T]$*,* $x(\cdot)|_{[t,T]}$ *and* $x(\cdot)|_{[t,T]}$*, are $\mathbf{F}$-adaptive. Denote*

$$\mathcal{R}(s, x(\cdot)) = \int_t^s f(\lambda, x_\lambda) \, d\lambda + \int_t^s g(\lambda, x_\lambda) \, dW(\lambda)$$
$$\equiv R(s, x(\cdot)) + Q(s, x(\cdot)). \quad (1.30)$$

Similarly,

$$\mathcal{R}(s, y(\cdot)) = \int_t^s f(\lambda, y_\lambda) \, d\lambda + \int_t^s g(\lambda, y_\lambda) \, dW(\lambda)$$
$$\equiv R(s, y(\cdot)) + Q(s, y(\cdot)).$$

Let

$$\delta(\cdot) = x(\cdot) - y(\cdot).$$

Then the following inequality holds for each $T > 0$:

$$E[\sup_{s \in [t,T]} |\mathcal{R}(s, x(\cdot)) - \mathcal{R}(s, y(\cdot))|^2] \leq AE\left[\|\delta_t\|^2\right] + BE\left[\int_t^T \|\delta_s\|^2 \, ds\right] \quad (1.31)$$

where A and B are positive constants that depend only on K_{lip} and T.

Proof. By Lemma 1.3.2 it is clear that $x_s, y_s \in \mathbf{C}$ for each $s \in [t, T]$. The $\mathbf{C}$-valued processes $\{x_s, s \in [t, T]\}$ and $\{y_s, s \in [t, T]\}$ are continuous and progressively measurable with respect to $\mathcal{B}([t, T]) \times \mathbf{F}$.

Using the Burkholder-Davis-Gundy inequality (Theorem 1.2.11) and the elementary inequalities

$$|a + b|^r \leq k_r(|a|^r + |b|^r), \ r \geq 0, \ k_r = 2^{r-1} \vee 1;$$
$$|a + b + c|^r \leq \bar{k}_r(|a|^r + |b|^r + |c|^r), \ r \geq 0, \ \bar{k}_r = 3^{r-1} \vee 1,$$

we have

$$E\left[\sup_{s \in [t,T]} |\mathcal{R}(s, x(\cdot)) - \mathcal{R}(s, y(\cdot))|^2\right]$$

$$\leq k_2 E\left[\sup_{s \in [t,T]} |R(s, x(\cdot)) - R(s, y(\cdot))|^2\right]$$

$$+ \sup_{s \in [t,T]} \left[|Q(s, x(\cdot)) - Q(s, y(\cdot))|^2\right]$$

$$\leq k_2 E\left[(T - t) \int_t^T |f(s, x_s) - f(s, y_s)|^2 ds\right]$$

$$+ KE\left[\int_t^T |g(s, x_s) - g(s, y_s)|^2 ds\right]$$

$$\leq k_2 E\left[(T - t)K_{lip}^2 \int_t^T \|x_s - y_s\|^2 ds\right]$$

$$+ k_2 K K_{lip}^2 E\left[\int_t^T \|x_s - y_s\|^2 ds\right]$$

$$\leq k_2 K_{lip}^2 ((T - t) + K) E\left[\int_t^T \|\delta_s\|^2 ds\right].$$

This proves the lemma.   □

Lemma 1.3.11 *Assume that* $f : [0,T] \times L^2(\Omega; \mathbf{C}) \to L^2(\Omega; \Re^n)$ *and* $g :$ $[0,T] \times L^2(\Omega; \mathbf{C}) \to L^2(\Omega; \Re^{n \times m})$ *are continuous functions that satisfy the linear growth Assumption 1.3.7. Let* $x(\cdot) = \{x(s), s \in [t-r, T]\}$ *be a process in* $L^2(\Omega, C([t-r, T], \Re^n))$ *such that* $x_t \in L^2(\Omega, \mathbf{C}; \mathcal{F}(t))$ *and its restriction to the interval* $[t, T]$, $x(\cdot)|_{[t,T]}$ *is* **F**-*adaptive. Denote*

$$\mathcal{R}(s, x(\cdot)) = \int_t^s f(\lambda, x_\lambda) \, d\lambda + \int_t^s g(\lambda, x_\lambda) \, dW(\lambda)$$
$$\equiv R(s, x(\cdot)) + Q(s, x(\cdot)). \tag{1.32}$$

Then, the following estimate holds for each $T > 0$:

$$E[\sup_{s \in [t,T]} |\mathcal{R}(s, x(\cdot))|] \le A + BE\left[\int_t^T |x_s|^2 \, ds\right], \tag{1.33}$$

where A *and* B *are positive constants that depend only on* K_{grow} *and* T.

Proof. Consider

$$E[\sup_{s \in [t,T]} |\mathcal{R}(s, x(\cdot))|^2] \le k_2 E\left[\sup_{s \in [t,T]} |R(s, x(\cdot))|^2 + \sup_{s \in [t,T]} |Q(s, x(\cdot))|^2\right]$$

$$\le k_2 E\left[(T-t)\int_t^T |f(s, x_s)|^2 \, ds\right]$$

$$+ KE\left[\int_t^T |g(s, x_s))|^2 \, ds\right]$$

$$\le k_2 K_{grow}^2 E\left[(T-t)\int_t^T (1 + \|x_s\|)^2 \, ds\right]$$

$$+ k_2 K_{grow}^2 KE\left[\int_t^T (1 + \|x_s\|)|^2 \, ds\right]$$

$$= k_2 K_{grow}^2 ((T-t) + K)E\left[\int_t^T (1 + \|x_s\|)|^2 \, ds\right].$$

The lemma follows with appropriate constants A and B. $\quad\square$

We have the following existence and uniqueness result for the strong solution of (1.27).

Theorem 1.3.12 *Suppose Assumptions 1.3.6 and 1.3.7 hold. Then the SHDE (1.27) has a unique strong solution* $\{x(s; t, \psi_t), s \in [t-r, T]\}$.

Sketch of Proof. The existence and uniqueness of the strong solution process, $\{x(s; t, \psi_t), s \in [t-r, T]\}$, of (1.27) can be obtained by standard Piccard iterations and applications of Lemma 1.3.10 and Lemma 1.3.11. We only outline the important steps below and referred the readers to Chapter 2 of [Moh84] for details.

Step 1. We assume that the initial process (segment) $\psi_t \in L^2(\Omega, \mathbf{C})$ is $\mathcal{F}(t)$-measurable. Note that this assumption is equivalent to saying that $\psi_t(\cdot)(\theta)$ is $\mathcal{F}(t)$-measurable for all $\theta \in [-r, 0]$, because ψ_t has almost all sample paths continuous.

Step 2. We define inductively a sequence of continuous processes $x^{(k)}(\cdot)$, $k = 1, 2, \ldots$, where

$$x^{(1)}(s, \omega) = \begin{cases} \psi_t(0) & \text{if } s \geq 0 \\ \psi_t(s) & \text{if } s \in [-r, 0], \end{cases}$$

and for $k = 1, 2, \ldots$,

$$x^{(k+1)}(s, \omega) = \begin{cases} \psi_t(0) + \int_t^s f(\lambda, x_\lambda^k) \, d\lambda + \int_t^s g(\lambda, x_\lambda^k) \, dW(\lambda) & \text{if } s \in [t, T], \\ \psi(s) & \text{if } s \in [t - r, t], \end{cases}$$

Step 3. Verify that the following three properties:

(i) $x_t^{(k)}$ is $\mathcal{F}(t)$-measurable and

$$x^{(k)}|_{[t,T]}(\cdot) \in L^2(\Omega, C([t - r, T]; \Re^n))$$

and is adapted to $\mathbf{F}$.

(ii) For each $s \in [t, T]$, $x_s^{(k)} \in L^2(\Omega, \mathbf{C})$ and is $\mathcal{F}(s)$-measurable.

(iii) For each $k = 1, 2, \ldots$, by applying Lemma 1.3.10, there exists a constant $K > 0$ such that

$$\|x^{(k+1)}(\cdot) - x^{(k)}(\cdot)\|_{L^2(\Omega, C([t-r,T], \Re^n))}$$
$$\leq (KK_{lip})^{k-1} \frac{(T - t)^{k-1}}{(k - 1)!} \|x^{(1)} - x^{(0)}\|_{L^2(\Omega, \mathbf{C})}$$

and

$$\|x_s^{(k+1)} - x_s^{(k)}\|_{L^2(\Omega, \mathbf{C})}$$
$$\leq (KK_{lip})^{k-1} \frac{(s - t)^{k-1}}{(k - 1)!} \|x^{(1)} - x^{(0)}\|_{L^2(\Omega, \mathbf{C})}.$$

We comment here that since $\psi_t \in L^2(\Omega, \mathbf{C})$ is $\mathcal{F}(t)$-measurable, then $x^{(k)}(\cdot)$ $|_{[t,T]} \in L^2(\Omega, C([t, T], \Re^n))$ is trivially $\mathbf{F}$-adapted for all $k = 1, 2, \ldots$.

Step 4. By applying Lemma 1.3.11, we prove that the sequence $\{x^{(k)}(\cdot)\}_{k=1}^\infty$ converges to some $x(\cdot) \in L_{\mathbf{F}}^2(\Omega, C([t-r, T], \Re^n))$ and that the limiting process $x(\cdot)$ satisfies (1.29).

Step 5. Prove the uniqueness of the convergent process $x(\cdot)$ by showing for any other process $\tilde{x}(\cdot)$ satisfying (1.29),

$$\|x_s - \tilde{x}_s\|^2_{L^2(\Omega,\mathbf{C})} \le KK^2_{lip} \int_t^s \|x_\lambda - \tilde{x}_\lambda\|^2_{L^2(\Omega,\mathbf{C})} \, d\lambda$$

for all $s \in [t, T]$. This implies that $x_s - \tilde{x}_s = 0$ for all $s \in [t, T]$ by applying the Gronwell inequality, Lemma 1.1.1, by taking $h(s) = \|x_s - \tilde{x}_s\|_{L^2(\Omega,\mathbf{C})}$, $\alpha = 0$, and $\beta = K_{lip}$. Therefore $x(\cdot) = \tilde{x}(\cdot)$ a.s. in $L^2(\Omega, C([t - r, T], \Re^n))$.

This proves the existence and uniqueness of the strong solution.   □

The strong solution process of (1.27) given initial datum $(t, \psi_t) \in [0, T] \times L^2(\Omega, \mathbf{C}; \mathcal{F}(t))$ will be denoted by $x(\cdot) = \{x(s; t, \psi_t), s \in [t - r, T]\}$. The corresponding $\mathbf{C}$-valued process will be denoted by $x_s = \{x_s(\cdot; t, \psi_t), s \in [t, T]\}$ (or, simply, $x_s = \{x_s(t, \psi_t), s \in [t, T]\}$), where

$$x_s(\theta; t, \psi_t) = x(s + \theta; t, \psi_t), \quad \forall \theta \in [-r, 0].$$

For each $s \in [t, T]$, we can treat $x_s(t, \cdot)$ as a map from $L^2(\Omega, \mathbf{C}; \mathcal{F}(t))$ to $L^2(\Omega, \mathbf{C}; \mathcal{F}(s))$ and prove the continuous dependence (on the initial segment process) result as follows.

Theorem 1.3.13 *Suppose Assumption 1.3.6 holds. Then the map*

$$x_s(t, \cdot) : L^2(\Omega, \mathbf{C}; \mathcal{F}(t)) \to L^2(\Omega, \mathbf{C}; \mathcal{F}(s)), \quad \forall s \in [t, T],$$

is Lipschitz continuous. Indeed, for all $s \in [t, T]$ and $\psi_t^{(1)}, \psi_t^{(2)} \in L^2(\Omega, \mathbf{C}; \mathcal{F}(t))$,

$$\|x_s(t, \psi_t^{(1)}) - x_s(t, \psi_t^{(2)})\|_{L^2(\Omega,\mathbf{C})} \tag{1.34}$$
$$\le \sqrt{2}\|\psi_t^{(1)} - \psi_t^{(2)}\|_{L^2(\Omega,\mathbf{C})} \exp(KK^2_{lip}(s - t)),$$

where K is the constant that appeared in the proof (Step 3) of Theorem 1.3.12 and $K_{lip} > 0$ is the Lipschitz constant listed in Assumption 1.3.6.

Proof. The proof is heavily dependent on the Lipschitz condition of f and g. It is straight forward but tedious and is therefore omitted here. Interested readers are referred to Theorem (3.1) on page 41 of [Moh84] for detail.   □

1.3.4 Weak Solution

The following is the definition of a weak solution for (1.27) and its corresponding uniqueness.

Definition 1.3.14 *A weak solution process of (1.27) is a six-tuple $(\Omega, \mathcal{F}, P, \mathbf{F}, W(\cdot), x(\cdot))$, where the following hold:*

 1. *$(\Omega, \mathcal{F}, P, \mathbf{F}, W(\cdot))$ is an m-dimensional standard Brownian motion.*

2. $x(\cdot) = \{x(s); s \in [t-r, T]\}$ is a continuous process that is $\mathcal{F}(s)$-measurable for each $s \in [t, T]$ and $\mathcal{F}(t)$-measurable for $s \in [t-r, t]$.

3. For $1 \leq i \leq n$ and $1 \leq j \leq m$, we have

$$P\Big[\int_t^T \Big(|f_i(s, x_s)| + g_{ij}^2(s, x_s) \Big) ds < \infty \Big] = 1,$$

4. The process $x(\cdot) = \{x(s); t-r \leq s \leq T\}$ satisfies the following stochastic integral equation P-a.s.:

$$x(s) = \psi_t(0) + \int_t^s f(\lambda, x_\lambda) \, d\lambda + \int_t^s g(\lambda, x_\lambda) \, dW(\lambda) \text{ for } s \in [t, T]. \quad (1.35)$$

Note that Definition 1.3.14 is weaker than Definition 1.3.9, because one has a choice of an m-dimensional standard Brownian motion $(\Omega, \mathcal{F}, P, \mathbf{F}, W(\cdot))$ on which the solution process $x(\cdot)$ is defined. This is in contrast to the definition of a strong solution in which the m-dimensional standard Brownian motion $(\Omega, \mathcal{F}, P, \mathbf{F}, W(\cdot))$ is given a priori. When no confusion arises, we sometimes refer to $(W(\cdot), x(\cdot))$ as the week solution instead of the formal expression $(\Omega, \mathcal{F}, P, \mathbf{F}, W(\cdot), x(\cdot))$ for notational simplicity. The probability measure

$$\mu(t, B) \equiv P[x_t \in B], \quad B \in \mathcal{B}(\mathbf{C}),$$

is called the initial distribution of the solution.

Definition 1.3.15 *The weak uniqueness of the (weak) solution is said to hold for (1.27) if, for any two weak solutions $(\Omega, \mathcal{F}, P, \mathbf{F}, W(\cdot), x(\cdot))$ and $(\tilde{\Omega}, \tilde{\mathcal{F}}, \tilde{P}, \tilde{\mathbf{F}}, \tilde{W}(\cdot), \tilde{x}(\cdot))$ with the same initial distribution*

$$P[x_t \in B] = \tilde{P}[\tilde{x}_t \in B], \quad \forall B \in \mathcal{B}(\mathbf{C}),$$

the two process $x(\cdot)|_{[t,T]} \equiv \{x(s), s \in [t, T]\}$ and $\tilde{x}(\cdot)|_{[t,T]} \equiv \{\tilde{x}(s), s \in [t, T]\}$ also have the same probability law, that is,

$$P[x(\cdot)|_{[t,T]} \in B] = \tilde{P}[\tilde{x}(\cdot)|_{[t,T]} \in B], \quad \forall B \in \mathcal{B}(C([t, T], \Re^n)).$$

Theorem 1.3.16 *If (1.27) has a strongly unique solution then it has a weakly unique solution.*

Proof. The theorem follows from the definitions of strong and weak solutions of (1.27).   □

It is well known that strong uniqueness implies weak uniqueness for SDEs (see Yamata and Watanabe [YW71] and Zronkin and Krylov [ZK81]). The above implication is also shown to be true for SHDE (1.27) by Larssen [Lar02].

Theorem 1.3.17 *Strong uniqueness implies weak uniqueness.*

Proof. Assume that $(\Omega^{(i)}, \mathcal{F}^{(i)}, P^{(i)}, \mathbf{F}^{(i)}, W^{(i)}(\cdot), x^{(i)}(\cdot))$, $i = 1, 2$, are two weak solutions of (1.27) that have the same initial distribution

$$P^{(1)}[x_t^{(1)} \in B] = P^{(2)}[x_t^{(2)} \in B], \quad B \in \mathcal{B}(\mathbf{C}). \tag{1.36}$$

Since these two solutions are defined on different probability spaces, one cannot use the assumption of strong uniqueness directly. First, we have to bring together the solutions on the same canonical space

$$(\Omega, \mathcal{F}, P)$$

while preserving their joint distributions. Once this is done, the result follows easily from the strong uniqueness. We proceed as follows. First, set $y^{(i)}(s) \equiv x^{(i)}|_{[t,T]}(s) - x^{(i)}(t)$ for $s \in [t, T]$. Then

$$x^{(i)}(s) = x_t^{(i)}(t \wedge s) + y^{(i)}(t \vee s) \text{ for } s \in [t - r.T],$$

and we may regard the ith solution as consisting of three parts $x_t^{(i)}$, $W^{(i)}(\cdot)$, and $x^{(i)}(\cdot)$. The sample paths of the triple constitute the canonical measurable space $(\Theta, \mathcal{B}(\Theta)$ defined as follows:

$$\Theta = \mathbf{C} \times C([t, T]; \mathfrak{R}^m) \times C([t, T]; \mathfrak{R}^n), \tag{1.37}$$

$$\mathcal{B}(\Theta) = \mathcal{B}(\mathbf{C}) \times \mathcal{B}(C([t, T]; \mathfrak{R}^m)) \times \mathcal{B}(C([t, T]; \mathfrak{R}^n)). \tag{1.38}$$

For $i = 1, 2$, the process $x^{(i)}(\cdot)$ defined on $(\Theta, \mathcal{B}(\Theta))$ induces a probability measure $\pi^{(i)}$ according to

$$\pi^{(i)}(A) = \mu^{(i)}[(x_t^{(i)}(\cdot), W^{(i)}(\cdot), x^{(i)}(\cdot)) \in A], \quad A \in \mathcal{B}(\Theta). \tag{1.39}$$

We denote the generic element of Θ by $\theta = (x, w, y)$. The marginal of each $\pi^{(i)}$ on the ψ-coordinate is the initial distribution μ, the marginal on the w-coordinate is the Wiener measure P^*, and the distribution of the (x, w) pair is the product measure $\mu \times P^*$. This is because $\psi_t^{(i)}$ is $\mathcal{F}^{(i)}(t)$-measurable (see Lemma II.2.1 of Mohammed [Moh84]) and $W^{(i)}(\cdot) - W^{(i)}(t)$ is independent of $\mathcal{F}^{(i)}(t)$ (see Problem 2.5.5 of Karatzas & Streves [KS91]). Also, under $\pi^{(i)}$, the initial value of the x-coordinate is zero, almost surely.

Next, we note that on $(\Theta, \mathcal{B}(\Theta), \pi^{(i)})$ there exists a regular conditional probability (see Proposition 1.1.10 for its properties) for $\mathcal{B}(\Theta)$ given $(\psi, \omega) \in \mathbf{C} \times C([0, T], \mathfrak{R}^n)$. We will be interested only in conditional probabilities of sets in $\mathcal{B}(\Theta)$ of the form $\mathbf{C} \times C([t, T]; \mathfrak{R}^m) \times F$ for $F \in \mathcal{B}(C[t, T]; \mathfrak{R}^n))$. Thus, with a slight abuse of terminology, we speak of

$$Q^{(i)}(\psi, \omega; F) : \mathbf{C} \times C([t, T]; \mathfrak{R}^m) \times \mathcal{B}(C([t, T]; \mathfrak{R}^n)) \to [0, 1], \quad i = 1, 2,$$

as the *regular conditional probability for $\mathcal{B}(C([t, T]; \mathfrak{R}^n))$ given (ψ, ω)*. According to Proposition 1.1.10, this regular conditional probability has the following properties:

(RCP1) For each fixed $\psi \in \mathbf{C}$, $\omega \in C([t, T]; \Re^m)$, the mapping $F \mapsto Q^{(i)}(\psi, \omega; F)$ is a probability measure on $(C([t, T]; \Re^n), \mathcal{B}(C([t, T]; \Re^n)))$.
(RCP2) For each fixed $F \in \mathcal{B}(C[t, T]; \Re^n)$, $(\psi, \omega) \mapsto Q^{(i)}(\psi, \omega; F)$ is $\mathcal{B}(\mathbf{C}) \otimes \mathcal{B}(C([t, T]; \Re^m))$-measurable.
(RCP3) For every $F \in \mathcal{B}(C([t, T]; \Re^n))$ and $G \in \mathcal{B}(\mathbf{C}) \otimes \mathcal{B}(C([t, T]; \Re^n)))$, we have

$$\pi^{(i)}(G \times F) = \int_G Q^{(i)}(\psi, \omega; F)\mu(d\psi)P^*(d\omega).$$

Next, we construct the probability space $(\mathbf{\Omega}, \mathcal{F}, \mathbf{P})$. Set $\mathbf{\Omega} \equiv \mathbf{\Theta} \times C([t, T]; \Re^n)$ and denote the generic elements of $\mathbf{\Omega}$ by $\boldsymbol{\omega} = (\psi, \omega, y_1, y_2)$. Let $\mathcal{F}$ be the completion of $\mathcal{B}(\mathbf{\Theta}) \otimes \mathcal{B}(C([t, T]; \Re^n))$ by the collection $\mathcal{N}$ of null sets under the probability measure

$$\mathbf{P}(d\boldsymbol{\omega}) \equiv Q^{(1)}(\psi, \omega; dy_1)Q^{(2)}(\psi, \omega; dy_2)\mu(d\psi)P^*(d\omega). \tag{1.40}$$

In the following, we endow $(\mathbf{\Omega}, \mathcal{F}, \mathbf{P})$ with a filtration $\{\mathcal{F}(t), t \in [0, T]\}$ that satisfies the usual conditions:

(i) $\mathcal{F}(0)$ is complete; that is $\mathcal{F}(0)$ contains all subsets of $A \in \mathcal{F}$ with $\mathbf{P}(A) = 0$.
(ii) $\{\mathcal{F}(s), s \geq 0\}$ is increasing; that is, $\mathcal{F}(s) \subset \mathcal{F}(\tilde{s})$ if $s < \tilde{s}$.
(iii) $\{\mathcal{F}(s), s \geq 0\}$ is right-continuous; that is,

$$\mathcal{F}(s) = \mathcal{F}(s+) \equiv \cap_{\tilde{s}>s}\mathcal{F}(\tilde{s}).$$

We first take

$$G(0) := \sigma(x(s), -r \leq s \leq 0),$$
$$G(t) := G(0) \vee \sigma(w(s), y_1(s), y_2(s), 0 \leq s \leq t),$$
$$\tilde{G}(t) := G(t) \vee \mathcal{N}.$$

Then define the filtration $\{\mathcal{F}(t), t \in [0, T]\}$ by

$$\mathcal{F}(t) := \cap_{\epsilon>0}\tilde{G}(t + \epsilon) \text{ for } t \in [0, T].$$

Now, for any $A \in \mathcal{B}(\mathbf{\Theta})$,

$$\mathbf{P}\{\boldsymbol{\omega} \in \mathbf{\Omega} \mid (\psi, \omega, y_i) \in A\}$$

$$= \int_A Q^{(i)}(\psi, \omega; dy_i1)\mu(d\psi)P^*(d\omega) \text{ (by (1.40))}$$

$$= \pi^{(i)}(A) \text{ (by RCP3 and a monotone class argument)}$$

$$= \mu^{(i)}[(x_0^{(i)}, W^{(i)}(\cdot), y^{(i)}(\cdot)) \in A] \quad i = 1, 2 \text{ (by RCP2(ii))}. \tag{1.41}$$

This means that the distribution of $(x(0 \wedge \cdot) + y_i(0 \vee \cdot), \omega)$ under P is the same as the distribution of $(X_0^{(i)}(0 \wedge \cdot) + Y^{(i)}(0 \vee \cdot), W^{(i)}(\cdot))$ under $\mu^{(i)}$. In particular, the ω-coordinate process $\{\omega(t), G(t), 0 \leq t \leq T\}$ is an m-dimensional Brownian motion on $(\Omega, \mathcal{F}, P)$, and by the following lemma (see Lemma IV.1.2 in Ikeda and Watanabe [IW81]), the same is true for $\{\omega(t), \mathcal{F}(t), 0 \leq t \leq T\}$.

To sum up, we started from two weak solutions

$$(\Omega^{(i)}, \mathcal{F}^{(i)}, P^{(i)}, W^{(i)}(\cdot), x^{(i)}(\cdot), \mathbf{F}^{(i)}), \quad i = 1, 2,$$

of (1.27) having the same initial distribution (i.e., (1.36) holds). We have constructed, on the filtered probability space $(\Omega, \mathcal{F}, P; \mathcal{F}(t), t \in [0, T]\}$, two weak solutions $(x(0 \wedge \cdot) + y^{(i)}(0 \vee \cdot), \omega(\cdot))$, $i = 1, 2$, having the same probability laws as the ones with which we started. Moreover, these weak solutions are driven by the same Brownian motion and they have the same initial path. Consequently, we may and will regard them as two different strong solutions to the SHDE

$$dx(t) = f(t, x_t)\, dt + g(t, x_t)\, dW(t)$$

on the filtered probability space $(\Omega, \mathcal{F}, P; \mathcal{F}(t), t \in [0, T]\}$ and with the same initial path $x(\cdot)$. Then strong uniqueness (see Definition 1.3.9) implies that

$$P[(x(0 \wedge \cdot) + y^{(1)}(0 \vee \cdot) = (x(0 \wedge \cdot) + y^{(2)}(0 \vee \cdot), -r \leq t \leq T] = 1,$$

or, equivalently,

$$P[\{\omega = (\psi(\cdot), \omega(\cdot), y^{(1)}(\cdot), y^{(2)}(\cdot)) \in \Omega \mid y^{(1)}(\cdot) = y^{(2)}(\cdot)\}] = 1. \qquad (1.42)$$

From (1.40) and (1.41), we see for all $A \in \mathcal{B}(\Theta)$, that

$$\begin{aligned}
\mu^{(1)}[(X_0^{(1)}(\cdot), W_0^{(1)}(\cdot), W^{(1)}(\cdot)) \in A] &= P[\{\omega \in \Omega \mid (x(\cdot), w(\cdot), y^{(1)}(\cdot)) \in A] \\
&= P[\{\omega \in \Omega \mid (x(\cdot), w(\cdot), y^{(2)}(\cdot)) \in A] \\
&= \mu^{(2)}[(X_0^{(2)}(\cdot), W_0^{(2)}(\cdot), W^{(2)}(\cdot)) \in A],
\end{aligned}$$

and this implies weak uniqueness. We thus have proved the theorem. □

1.4 SHDE with Unbounded Memory

We first note that Lemma 1.3.2 can be extended to the memory map with infinite but fading memory ($r = \infty$). In this case, we will work with the ρ-weighted Hilbert space $\mathbf{M} \equiv \Re \times L_\rho^2((-\infty, 0]; \Re)$ equipped with the inner product

$$\begin{aligned}
\langle (x, \phi), (y, \varphi) \rangle_\rho &= xy + \langle \phi, \varphi \rangle_2 \\
&= xy + \int_{-\infty}^0 \rho(\theta)\phi(\theta)\varphi(\theta)\, d\theta,
\end{aligned}$$

$$\forall (x, \phi), (y, \varphi) \in \Re \times L^2_\rho((-\infty, 0]; \Re),$$

and the Hilbertian norm $\|(x, \phi)\|_\rho = \langle (x, \phi), (x, \phi) \rangle^{1/2}_\rho$, where $\rho : (-\infty, 0] \to [0, \infty)$ is a function that satisfies the following assumptions: Let $\rho : (-\infty, 0] \to [0, \infty)$ be the *influence function with relaxation property* that satisfies the following conditions:

Assumption 1.4.1 *The function ρ satisfies the following two conditions:*

1. ρ is summable on $(-\infty, 0]$, that is,

$$0 < \int_{-\infty}^0 \rho(\theta) \, d\theta < \infty.$$

2. For every $\lambda \leq 0$, one has

$$\bar{K}(\lambda) = ess \sup_{\theta \in (-\infty, 0]} \frac{\rho(\theta + \lambda)}{\rho(\theta)} \leq \bar{K} < \infty,$$

$$\underline{K}(\lambda) = ess \sup_{\theta \in (-\infty, 0]} \frac{\rho(\theta)}{\rho(\theta + \lambda)} < \infty.$$

1.4.1 Memory Maps

Lemma 1.4.2 *For each $0 < T < \infty$, the memory map*

$$m : [0, T] \times L^2_\rho(-\infty, T; \Re) \to \mathbf{M}$$

defined by

$$m(t, \psi) = (\psi(t), \psi_t), \quad (t, \psi) \in [0, T] \times L^2_\rho(-\infty, T; \Re)$$

is jointly continuous.

Proof. For $0 \leq T \leq \infty$, we first extend the function $\rho : (-\infty, 0] \to [0, \infty)$ to a larger domain $(-\infty, T]$ by setting $\rho(t) = \rho(0)$ for all $t \in [0, T]$. Let $C_\rho((-\infty, T], \Re)$ $(0 \leq T \leq \infty)$ be the space of continuous functions $\phi : (-\infty, T] \to \Re$ such that

$$\lim_{\theta \to -\infty} \sqrt{\rho(\theta)} \phi(\theta) = 0.$$

It is clear (see Rudin [Rud71]) that the space $C_\rho((-\infty, T], \Re)$ is dense and can be continuously embedded into $L^2_\rho((-\infty, 0], \Re)$. To show that the memory map

$$m(t, \psi) = (\psi(t), \psi_t), \quad (t, \psi) \in [0, T] \times L^2_\rho(-\infty, T; \Re)$$

is jointly continuous, we let $t, s \in [0, T]$ and $\phi^{(i)} \in L^2_\rho(-r, T; \Re^n)$ for $i = 1, 2$. For any $\epsilon > 0$, we want to find $\delta > 0$ such that, if $|t - s| < \delta$ and $\|\phi^{(1)} - \phi^{(2)}\|_{\rho,2} < \delta$, then

$$|\phi^{(1)}(t) - \phi^{(2)}(s)|^2 + \|\phi_t^{(1)} - \phi_s^{(2)}\|_{\rho,2} < \epsilon.$$

Let $\varphi^{(i)} \in C_\rho((-\infty, T], \Re)$ for $i = 1, 2$ be such that

$$\|\phi^{(i)} - \varphi^{(i)}\|_{\rho,2} < \epsilon_1, \quad i = 1, 2,$$

where $\epsilon_1 > 0$ can be made arbitrarily small. Therefore, for $t \in [0, T]$,

$$
\begin{aligned}
\|\phi_t^{(i)} - \varphi_t^{(i)}\|_{\rho,2} &= \left[\int_{-\infty}^0 \rho(\theta) |\phi^{(i)}(t+\theta) - \varphi^{(i)}(t+\theta)|^2 d\theta \right]^{1/2} \\
&= \left[\int_{-\infty}^0 \frac{\rho(\theta)}{\rho(t+\theta)} \rho(t+\theta) |\phi^{(i)}(t+\theta) - \varphi^{(i)}(t+\theta)|^2 d\theta \right]^{1/2} \\
&= \sqrt{\bar{K}} \left[\int_{-\infty}^t \rho(\theta) |\phi^{(i)}(\theta) - \varphi^{(i)}(\theta)|^2 d\theta \right]^{1/2} \\
&\leq \sqrt{\bar{K}} \left[\int_{-\infty}^T \rho(s) |\phi^{(i)}(s) - \varphi^{(i)}(s)|^2 ds \right]^{1/2} \\
&= \sqrt{\bar{K}} \|\phi^{(i)} - \varphi^{(i)}\|_{\rho,2} \\
&< \epsilon_1.
\end{aligned}
$$

Taking $\delta < \epsilon_1$,

$$\|\varphi^{(2)} - \varphi^{(1)}\|_{\rho,2} \leq \|\varphi^{(2)} - \phi^{(2)}\|_{\rho,2} + \|\phi^{(2)} - \phi^{(1)}\|_{\rho,2} + \|\phi^{(1)} - \varphi^{(1)}\|_{\rho,2}$$
$$< 3\epsilon_1.$$

Since $\sqrt{\rho}\varphi^{(i)}$, $i = 1, 2$, are uniformly continuous on $(-\infty, T]$, we can choose $\delta_0 > 0$ such that, if $t, s \in (-\infty, T]$ and $|t - s| < \delta_0$, then $|\sqrt{\rho(t)}\varphi^{(i)}(t) - \sqrt{\rho(s)}\varphi^{(i)}(s)| < \epsilon_1$. Suppose $t, s \in [0, T]$ are such that $|t - s| < \delta_0$. Then

$$\|\varphi_t^{(i)} - \varphi_s^{(i)}\|_{\rho,2} = \left[\int_{-\infty}^0 \rho(\theta) |\varphi^{(i)}(t+\theta) - \varphi^{(i)}(s+\theta)|^2 d\theta \right]^{1/2}$$
$$\leq \epsilon_1.$$

Suppose $|t - s| < \delta < \delta_0 \wedge \epsilon_1$. Then the above two inequalities imply that

$$\|\varphi_t^{(2)} - \varphi_s^{(1)}\|_{\rho,2} \leq \|\varphi_t^{(2)} - \varphi_t^{(1)}\|_{\rho,2} + \|\varphi_t^{(1)} - \varphi_s^{(1)}\|_{\rho,2} < 4\epsilon_1$$
$$\leq \|\varphi^{(2)} - \varphi^{(1)}\|_{\rho,2} + \|\sqrt{\rho}\varphi_t^{(1)} - \sqrt{\rho}\varphi_s^{(1)}\|$$
$$< 3\epsilon_1 + \epsilon_1 = 4\epsilon_1.$$

Therefore,

$$\|\phi_t^{(2)} - \phi_s^{(1)}\|_{\rho,2} \leq \|\phi_t^{(2)} - \varphi_t^{(2)}\|_{\rho,2} + \|\varphi_t^{(2)} - \varphi_s^{(1)}\|_{\rho,2} + \|\varphi_s^{(1)} - \phi_s^{(1)}\|_{\rho,2} < 6\epsilon_1$$
$$< \epsilon_1 + 4\epsilon_1 + \epsilon_1 = 6\epsilon_1.$$

By taking $\epsilon_1 < \frac{\epsilon}{6}$, the result follows. The above analysis also implies that $|\phi^{(1)}(t) - \phi^{(2)}(t)|^2$ can also be made arbitrarily small. This proves the lemma.
□

The following corollary is similar to Corollary 1.3.3.

Corollary 1.4.3 *The stochastic memory map*

$$m^* : [0,T] \times L^2(\Omega, L^2_\rho((-\infty, T], \Re)) \to L^2(\Omega, L^2_\rho((-\infty, 0], \Re))$$

defined by $(t, \phi(\cdot)) \mapsto (\phi(t), \phi_t)$ *is a continuous map.*

Proof. The proof is similar to that of Corollary 1.3.3 and is therefore omitted.
□

Notation. In this section, we adopt the notation introduced at the beginning of the previous section but with $\Xi = \mathbf{M} \equiv \Re \times L^2_\rho((-\infty, 0]; \Re)$. For example, $L^2(\Omega, \mathbf{M})$ is the space of $\mathbf{M}$-valued random variables (x, Υ) defined on the probability space $(\Omega, \mathcal{F}, P)$ such that

$$\|(x, \Upsilon)\|_{L^2(\Omega, \mathbf{M})} \equiv E\left[|x|^2 + \|\Upsilon\|^2_\rho\right]$$
$$= E\left[|x|^2 + \int_{-\infty}^0 |\Upsilon(\theta)|^2 \rho(\theta)\, d\theta\right] < \infty.$$

Let $(\Omega, \mathcal{F}, P, \mathbf{F}; W(\cdot))$ be an one-dimensional standard Brownian motion. The main purpose of this section is to establish the existence and uniqueness of the strong solution process $S(\cdot) = \{S(s), s \in (-\infty, \infty)\}$ of the following very special type of one-dimensional autonomous SHDE with infinite but fading memory $(r = \infty)$:

$$\frac{dS(s)}{S(s)} = f(S_s)\, ds + g(S_s)\, dW(s), \quad s \in [0, \infty), \qquad (1.43)$$

with the initial datum $(S(0), S_0) = (\psi(0), \psi) \in L^2(\Omega, \mathbf{M})$.

In the above, the solution process $S(\cdot) = \{S(s), s \in (-\infty, \infty)\}$ will be the stock price process and (1.43) represents the price dynamics of the underlying stock to be considered in Chapter 7, where we treat a hereditary portfolio optimization problem with a fixed plus proportional transaction costs and capital-gains taxes. In (1.43), the left-hand side $dS(s)/S(s)$ represents the instantaneous return of the stock at time $s > 0$. On the right-hand side of (1.43), $f(S_s)$ and $g(S_s)$ represent respectively the growth rate and volatility of the stock at time $s \geq 0$. Note that both $f(S_s)$ and $g(S_s)$ depend explicitly on the historical prices S_s over the infinite time history $(-\infty, s]$ instead of just the price $S(s)$ at time $s \geq 0$ alone. The stock dynamics described by (1.43) is therefore referred to as a stock with an hereditary price structure. The stock growth function and the volatility function f and g are real-valued functions that are continuous on the space $L^2_\rho((-\infty, 0]; \Re)$.

Assumption 1.4.4 (Lipschitz Continuity) *There exists a constant $K_{lip} > 0$ such that*

$$|\varphi(0)f(\varphi) - \phi(0)f(\phi)| + |\varphi(0)g(\varphi) - \phi(0)g(\phi)|$$
$$\leq K_{lip}(\|(\varphi(0), \varphi) - (\phi(0), \phi)\|_M), \quad \forall (\varphi(0), \varphi), (\phi(0), \phi) \in \mathbf{M}.$$

Assumption 1.4.5 (Linear Growth) *There exists a constant $K_G > 0$ such that*

$$|\phi(0)f(\phi)| + |\phi(0)g(\phi)| \leq K_G(1 + \|(\phi(0), \phi)\|_M), \quad \forall (\phi(0), \phi) \in \mathbf{M}.$$

Strong and Weak Solutions

The definitions of strong and weak solutions of SHDE (1.43) together with their corresponding uniqueness are given as follows.

Definition 1.4.6 *Let $(\psi(0), \psi) \in \mathbf{M} \equiv \Re \times L_\rho^2((-\infty, 0]; \Re)$. An one-dimensional process $S(\cdot) = \{S(s), s \in (-\infty, \infty)\}$ is said to be a strong solution of (1.43) on the infinite interval $(-\infty, \infty)$ and through the initial datum $(\psi(0), \psi) \in \mathbf{M}$ if it satisfies the following conditions:*

1. $S(\theta) = \psi(\theta), \quad \forall \theta \in (-\infty, 0]$.
2. $\{S(s), s \in [0, \infty)\}$ *is $\mathbf{F}$-adapted on $[0, \infty)$.*
3.

$$P\left[\int_0^\infty \left(|S(t)f(S_t)| + S^2(t)g^2(S_t)\right) dt < \infty\right] = 1.$$

4. *For each $s \in [0, \infty)$, the process $\{S(s), s \in [0, \infty)\}$ satisfies the following stochastic integral equation P-a.s.:*

$$S(s) = \psi(0) + \int_0^s S(t)f(S_t)\, dt + \int_0^s S(t)g(S_t)\, dW(t). \tag{1.44}$$

Definition 1.4.7 *A strong solution, $\{S(s), s \in (-\infty, \infty)\}$, of SHDE (1.43) is said to be unique if $\{\tilde{S}(s), s \in (-\infty, \infty)\}$ is also a strong solution of (1.43) on the interval $(-\infty, \infty)$ and through the the same initial datum $(\psi(0), \psi) \in \mathbf{M}$, then*

$$P\{S(s) = \tilde{S}(s) \quad \forall s \in [0, \infty)\} = 1.$$

Definition 1.4.8 *A weak solution of (1.43) is a six-tuple*

$$(\Omega, \mathcal{F}, P, \mathbf{F}, W(\cdot), S(\cdot)),$$

where

1. $(\Omega, \mathcal{F}, P, \mathbf{F}, W(\cdot))$ *is an one-dimensional standard Brownian motion;*
2. $S(\cdot) = \{S(s); s \in [0, \infty)\}$ *is a continuous process that is $\mathcal{F}(s)$-measurable for each $s \in [0, \infty)$;*

3.
$$P\Big[\int_0^\infty \Big(|S(s)f(S_s)| + S^2(s)g^2(S_s) \Big)\, ds < \infty \Big] = 1,$$

4. *the process* $S(\cdot) = \{S(s); s \in [0,\infty)\}$ *satisfies the following stochastic integral equation* P-*a.s.*

$$x(s) = \psi(0) + \int_0^s S(t)f(S_t)\, dt + \int_0^s S(t)g(S_t)\, dW(t). \qquad (1.45)$$

First, the following lemma is needed for establishing the existence and uniqueness of the strong solution process $\{S(s), s \in (-\infty, \infty)\}$ of (1.43).

Lemma 1.4.9 *Assume that* $f, g : L^2_\rho((-\infty, 0]; \Re) \to \Re$ *satisfy the Lipschitz condition (Assumption 1.4.4). Let* $x(\cdot) = \{x(t), t \in \Re\}$ *and* $y(\cdot) = \{y(t), t \in \Re\}$ *be two* $\mathbf{F}$-*adaptive processes that are continuous for* $t \geq 0$ *and such that*

$$(x(0), x_0), (y(0), y_0) \in \mathbf{M} \ a.s.$$

Denote

$$\mathcal{R}(s, x(\cdot)) = \int_0^s x(t)f(x_t)\, dt + \int_0^s x(t)g(x_t)\, dW(t) \qquad (1.46)$$

$$\equiv R(s, x(\cdot)) + Q(s, x(\cdot)).$$

Similarly,

$$\mathcal{R}(s, y(\cdot)) = \int_0^s y(t)f(y_t)\, dt + \int_0^s y(t)g(y_t)\, dW(t)$$

$$\equiv R(s, y(\cdot)) + Q(s, y(\cdot)).$$

Let

$$\delta(\cdot) = x(\cdot) - y(\cdot).$$

Then the following inequality holds for each $T > 0$:

$$E[\sup_{s \in [0,T]} |\mathcal{R}(s, x(\cdot)) - \mathcal{R}(s, y(\cdot))|^2]$$

$$\leq AE\Big[\|(\delta(0), \delta_0)\|^2_\rho \Big] + BE\Big[\int_0^T |\delta(s)|^2\, ds \Big], \qquad (1.47)$$

where A *and* B *are positive constants that depend only on* L, T *and* $\|\rho\|_1$.

Proof. By Lemma 1.4.2 it is clear that $(x(s), x_s), (y(s), y_s) \in \mathbf{M}$ for each $s \in [0, T]$. The $\mathbf{M}$-valued processes $\{(x(s), x_s), s \geq 0\}$ and $\{(y(s), y_s), s \geq 0\}$ are continuous and progressively measurable with respect to $\mathcal{G}(0) \vee \mathcal{F}(s)$.

Using the Burkholder-Davis-Gundy inequality (Theorem 1.2.11), we have

$$E\left[\sup_{0\leq t\leq T}|\mathcal{R}(t,x(\cdot))-\mathcal{R}(t,y(\cdot))|^2\right]$$

$$\leq k_2E\left[\sup_{0\leq t\leq T}|R(t,x(\cdot))-R(t,y(\cdot))|^2+\sup_{0\leq t\leq T}|Q(t,x(\cdot))-Q(t,y(\cdot))|^2\right]$$

$$\leq k_2E\left[T\int_0^T|x(s)f(x_s)-y(s)f(y_s)|^2\,ds\right.$$

$$\left.+C\int_0^T|x(s)g(x_s)-y(s)g(y_s)|^2\,ds\right]$$

$$\leq k_2E\left[TL\int_0^T\|(x(s),x_s)-(y(s),y_s)\|_\rho^2\,ds\right]$$

$$+k_2CLE\left[\int_0^T\|(x(s),x_s)-(y(s),y_s)\|_\rho^2\,ds\right]$$

$$\leq k_2L(T+C)E\left[\int_0^T|(\delta(s),\delta_s)\|_\rho^2 ds\right]. \tag{1.48}$$

Now,

$$\|(\delta(s),\delta_s)\|_\rho^2=|\delta(s)|^2+\int_{-\infty}^0|\delta(s+\theta)|^2\rho(\theta)\,d\theta$$

and

$$\int_{-\infty}^0|\delta(s+\theta)|^2\rho(\theta)\,d\theta$$

$$=\int_{-\infty}^{-s}|\delta(s+\theta)|^2\rho(\theta)\,d\theta+\int_{-s}^0|\delta(s+\theta)|^2\rho(\theta)\,d\theta$$

$$\leq\int_{-\infty}^{-s}|\delta(s+\theta)|^2\rho(s+\theta)\frac{\rho(\theta)}{\rho(s+\theta)}\,d\theta+\int_0^s|\delta(v)|^2\rho(v-s)\,dv$$

$$\leq\bar{K}\int_{-\infty}^{-s}|\delta(s+\theta)|^2\rho(s+\theta)\,d\theta+\int_0^s|\delta(v)|^2\rho(v-s)\,dv$$

$$=\bar{K}\int_{-\infty}^0|\delta(\theta)|^2\rho(\theta)\,d\theta+\int_0^s|\delta(v)|^2\rho(v-s)\,dv,$$

where $\bar{K}$ is the constant specified in Assumption 1.4.1. Therefore,

$$\|(\delta(s),\delta_s)\|_\rho^2\leq|\delta(s)|^2+\bar{K}\int_{-\infty}^0|\delta(\theta)|^2\rho(\theta)\,d\theta$$

$$+\int_0^s|\delta(v)|^2\rho(v-s)\,dv.$$

Integrating the above two inequalities with respect to s from 0 to T and taking the appropriate power yields

$$E\left[\sup_{0\leq t\leq T}|\mathcal{R}(t,x(\cdot))-\mathcal{R}(t,y(\cdot))|^2\right]\leq k_2 L(T+C)E\left[\int_0^T\|(\delta(s),\delta_s)\|_\rho^2\,ds\right]$$

$$\leq k_2 L(T+C)E\left[\int_0^T|\delta(s)|^2\,ds\right.$$

$$+\bar{K}\int_0^T\left(\int_{-\infty}^0|\delta(\theta)|^2\rho(\theta)d\theta\right)ds$$

$$+\int_0^T\left(\int_0^s|\delta(v)|^2\rho(v-s)dv\right)ds\bigg].$$

Note that

$$\int_0^T\left(\int_0^s|\delta(v)|^2\rho(v-s)dv\right)ds\leq\left(\int_{-\infty}^0\rho(\theta)\,d\theta\right)\left(\int_0^T|\delta(s)|^2\,ds\right)$$

$$=\|\rho\|_1\left(\int_0^T|\delta(s)|^2\,ds\right).$$

It follows that

$$E\left[\sup_{0\leq t\leq T}|\mathcal{R}(t,x(\cdot))-\mathcal{R}(t,y(\cdot))|^2\right]\leq k_2 L(T+C)(1+\|\rho\|_1)E\left[\int_0^T|\delta(s)|^2\,ds\right]$$

$$\leq k_2 L(T+C)\bar{K}TE\left[\|(\delta(0),\delta_0)\|_\rho^2\right]$$

$$\leq AE\left[\|(\delta(0),\delta_0)\|_\rho^2\right]+BE\left[\int_0^T|\delta(s)|^2\,ds\right],$$

where

$$A=k_2 LT(T+C)\bar{K}\quad\text{and}\quad B=k_2 L(T+C)(1+\|\rho\|_1).$$

This proves the lemma. $\quad\square$

We will use mainly the following consequence of the previous lemma.

Corollary 1.4.10 *Under the conditions of Lemma 1.4.9, the following inequality holds for each $T>0$:*

$$E\left[\sup_{0\leq s\leq T}|\mathcal{R}(s,x(\cdot))-\mathcal{R}(s,y(\cdot))|^2\right]$$

$$\leq E[\|(\delta(0),\delta_0)\|_\rho^2+M(T)E\left[\sup_{0\leq s\leq T}|\delta(s)|^2\right],\qquad(1.49)$$

where A is the same as above, $M(T)$ depends only on T and B, and

$$M(T)=o(1)\quad\text{as }T\to 0.\qquad(1.50)$$

Proof. Inserting $|\delta(t)|\leq\sup_{0\leq t\leq T}\delta(t)$ with $0\leq t\leq T$ into (1.47), one obtains (1.49) with appropriate constants A and $M(T)$. $\quad\square$

Theorem 1.4.11 *Assume Assumptions 1.4.4 and 1.4.5 hold. Then for each* $(\psi(0), \psi) \in \mathbf{M}$, *there exists a unique nonnegative strong solution process* $\{S(t),$ $t \in (-\infty, \infty)\}$ *through the initial datum* $(\psi(0), \psi) \in \mathbf{M}$.

Sketch of a Proof. The existence and uniqueness of the strong solution process $\{S(s), s \in (-\infty, \infty)\}$ follows from the standard method of successive approximation for constructing a *strong* solution process (see Mizel and Trutzer [MT84]). A sketch of its proof is given below.

Existence: For each fixed $T > 0$, define a sequence of processes $\{S^{(k)}(s), s \in (-\infty, T]\}$, for $k = 0, 1, 2, \ldots$, as follows:

$$S^{(0)}(s, \omega) = \begin{cases} \psi(0) & \text{if } s \geq 0 \\ \psi(s) & \text{if } s \in (-\infty, 0], \end{cases}$$

and for $k = 1, 2, \ldots$,

$$S^{(k)}(s, \omega) = \begin{cases} \psi(0) + \mathcal{R}(s, S^{(k-1)(\cdot)}) & \text{if } s \in [0, T] \\ \psi(s) & \text{if } s \in (-\infty, 0], \end{cases}$$

where $\mathcal{R}(s, S^{(k-1)(\cdot)})$ is defined in (1.47) and, again, $S_s^{(k)}(\theta) = S^{(k)}(s+\theta)$, $\theta \in (-\infty, 0]$. From this definition, the conditions (see Assumptions 1.4.4 and 1.4.5 on the drift and diffusions coefficients $\phi(0)f(\phi)$ and $\phi(0)g(\phi)$), and Lemma 1.4.2, it follows that the processes $\{S^{(k)}(s), s \in [0, T]\}, k = 0, 1, \ldots$ are nonanticipative, measurable and continuous, and $(S^{(k)}(s), S_s^{(k)}) \in \mathbf{M}$, for $s \geq 0$.

By induction we will prove below that for any $T > 0$,

$$E\left[\sup_{s \in [0,T]} |S^{(k)}(s)|^p \right] < \infty.$$

First, for $k = 0$, we have $(S^{(0)}(0), S_0^{(0)}) \in \mathbf{M}$, and by the above construction for $t \in [0, T]$,

$$\|(S^{(0)}(s), S_s^{(0)})\|_{\mathbf{M}} \leq K \cdot \|(S(0), S_0)\|_{\mathbf{M}}, \quad \forall t \in [0, T]$$

for a constant $K > 0$ depending on $\bar{K}$. Hence,

$$E\left[\sup_{s \in [0,T]} |(S^{(0)}(s), S_s^{(0)})|_{\mathbf{M}}^p \right] \leq K \cdot E[\|(S(0), S_0)\|_{\mathbf{M}} < \infty.$$

For induction purposes, suppose that

$$E\left[\sup_{s \in [0,T]} |S^{(k-1)}(s)|^p \right] < \infty.$$

Appraising as in Lemma 1.4.9 and Holder's inequality, we have

$$E\Big[\sup_{t\in[0,T]}\|(S^{(k)}(t),S_t^{(k)})\|_{\mathbf{M}}^p\Big]$$

$$\leq \bar{k}_p E\Big[|\psi(0)|^p + T^{p-1}\int_0^T |S^{(k-1)}(t)f(S_t^{(k-1)})|^p\,dt$$

$$+c_p\Big(\int_0^T |S^{(k-1)}(t)g(S_t^{(k-1)})|^2\,dt\Big)^{p/2}\Big]$$

$$\leq \bar{k}_p E\Big[|\psi(0)|^p + T^{p-1}c_2\int_0^T (1+\|(S^{(k-1)}(t),(S_t^{(k-1)}))\|_{\mathbf{M}}^p\,dt$$

$$+c_2^p c_p\Big(\int_0^T (1+\|S^{(k-1)}(t),(S_t^{(k-1)})\|_{\mathbf{M}}^2\,dt\Big)^{p/2}\Big]$$

$$\leq \bar{k}_p E\Big[|\psi(0)|^p + c_2^p(T^p + c_p T^{p/2})$$

$$\times\Big(1+\sup_{t\in[0,T]}\|(S^{(k-1)}(t),(S_t^{(k-1)}))\|_{\mathbf{M}}\Big)^p\Big]$$

$$<\infty.$$

Next, using the notation form Lemma 1.4.9 and Corollary 1.4.10, we have

$$E\Big[\sup_{0\leq t\leq T}\|(\bar{k}_p E\Big[|\psi(0)|^p + T^{p-1}\int_0^T |S^{(k-1)}(t)f(S_t^{(k-1)})|^p dt$$

$$+ c_p\Big(\int_0^T |S^{(k-1)}(t)g(S_t^{(k-1)})|^2 dt\Big)^{p/2}\Big] < \infty.$$

The unique strong solution $\{S(s), s \in (-\infty, T]\}$ can be extended to the interval $(-\infty, \infty)$ by standard continuation method in existence theory of differential equation. Therefore, we only need to show that for each initial $(S(0), S_0) = (\psi(0), \psi) \in \mathbf{M}$, $S(s) \geq 0$ for each $s \geq 0$. To show this, we note that $\{s \geq 0 \mid S(s) \geq 0\} \neq \emptyset$ by sample-path continuity of the solution process and non-negativity of the initial data $(\psi(0), \psi) \in \mathbf{M}$. Now, let

$$\tau = \inf\{t \geq 0 \mid S(t) \geq 0\}.$$

If $\tau < \infty$, then $dS(t) = 0$ for all $t \geq \tau$ with $S(\tau) = 0$. This implies that $S(t) = 0$ for all $t \geq \tau$ and, hence, $S(t) \geq 0$ for all $t \geq 0$. The same conclusion holds if $\tau = \infty$.

Uniqueness

Let $\{S(s), s \in (-\infty, \infty)\}$ and $\{\tilde{S}(s), s \in (-\infty, \infty)\}$ be two (strong) solution processes of (1.43) through the initial data $(\psi(0), \psi) \in \mathbf{M}$. We need to show that

$$P\{S(s) = \tilde{S}(s), \quad \forall s \geq 0\} = 1.$$

Let $\delta(s) = S(s) - \tilde{S}(s)$, $s \geq 0$. Then $\delta(s) = 0$ $\forall s \leq 0$, and from Lemma 1.4.9, we have for all $T > 0$,

$$E\left[\sup_{s \in [0,T]} |\delta(s)|^2\right] \leq KE\left[T \int_0^T |\delta(s)|^2 \, ds\right].$$

By Grownwall inequality in Subsection 1.1.1, this shows that for all $T > 0$

$$P\{S(s) = \tilde{S}(s), \quad \forall s \in [0,T]\} = 1.$$

Therefore,

$$P\{S(s) = \tilde{S}(s), \quad \forall s \geq 0\}$$
$$\lim_{T \uparrow \infty} P\{S(s) = \tilde{S}(s) \quad \forall s \in [0,T]\} = 1. \qquad \square$$

stochastic delay equation driven by Brownian motion:

1.5 Markovian Properties

In the following, let Ξ represent either the state space $\Xi = \mathbf{C}$ for (1.27) or the state space $\Xi = \mathbf{M} \equiv \Re \times L_\rho^2((-\infty, 0]; \Re)$ for (1.43). Let $X(\cdot) = \{X(s), s \in [t, T]\}$ be either the Ξ-valued solution process of (1.27) or (1.43) with the initial state $X(t) \in \Xi$ at the initial time $t \in [0, T]$ with $T < \infty$ or $T = \infty$. In the case of (1.27), $X(s) = x_s, s \in [t, T]$ and $X(t) = \psi \in \mathbf{C}$. In the case of (1.43), the initial time is $t = 0$, the initial datum is $X(0) = (\psi(0), \psi) \in \mathbf{M}$, and $X(s) = (S(s), S_s), s \geq 0$. In either case wherever appropriate, $x(\cdot)$ is the unique strong solution process of (1.27) with $x_s(\theta) = x(s + \theta)$ for $\theta \in [-r, 0]$ and $S(\cdot) = \{S(s), s \in (-\infty, \infty)\}$ is the strong solution of (1.43) with $S_s(\theta) = S(s + \theta)$ for $\theta \in (-\infty, 0]$. The purpose of this section is to establish the Markovian and strong Markovian properties of the Ξ-valued solution process for (1.27) and (1.43).

Let $x(\cdot) = \{x(s), s \in [t - r, T]\}$ be the *strong* solution process of (1.27) on the interval $[t - r, T]$ and through the initial datum $(t, \psi) \in [0, T] \times \mathbf{C}$. With almost the same notation and hoping that there will be no ambiguity, we also use the same notation and let $S(\cdot) = \{S(s), s \in (-\infty, \infty)\}$ be the *strong* solution of (1.43) on the interval $(-\infty, \infty)$ and through the initial datum $(\psi(0), \psi) \in \mathbf{M}$. Note that the only distinction between the two is the time interval and the initial segment involved. Here, we will state and prove the Markovian and strong Markovian properties of its corresponding **C**-valued segment process $\{x_s, s \in [t, T]\}$ (where $x_s(\theta) = x(s + \theta)$ for each $\theta \in [-r, 0]$) for (1.27) and the **M**-valued segment process $\{(S(s), S_s), s \geq 0\}$ (where $S_s(\theta) = S(s + \theta)$ for each $\theta \in (-\infty, 0]$) for (1.43).

Theorem 1.5.1 *Assume that the functions* $f : [0,T] \times \mathbf{C} \to \Re^n$ *and* $g :$ $[0,T] \times \mathbf{C} \to \Re^{n \times m}$ *satisfy Assumptions 1.3.6 and 1.3.7. Then the* $\mathbf{C}$*-valued segment process* $\{x_s, s \in [t,T]\}$ *of (1.27) describes a* $\mathbf{C}$*-valued Markov process with probability transition function* $p : [0,T] \times \mathbf{C} \times [0,T] \times \mathcal{B}(\mathbf{C}) \to [0,1]$, *where* $p(t, x_t, s, B)$ *is given by*

$$p(t, x_t, s, B) \equiv P\{x_s \in B \mid x_t\}$$
$$\equiv P^{t,x_t}\{x_s \in B\}, \quad s \in [t,T], B \in \mathcal{B}(\mathbf{C}), \qquad (1.51)$$

where $P^{t,x_t}\{\cdot\}$ *is the probability law of the* $\mathbf{C}$*-valued Markov process* $\{x_s, s \in [t,T]\}$ *given the initial datum* $(t, x_t) \in [0,T] \times \mathbf{C}$ *and* $\mathcal{B}(\mathbf{C})$ *is the Borel* σ*-algebra of subsets of* $\mathbf{C}$. *Throughout, let* $E^{t,x_t}[\cdot]$ *be the expectation taken with respect to the probability law* P^{t,x_t}. *The probability transition function* $p(t, x_t, s, B)$ *defined in (1.51) has the following properties:*

(a) For any $s \geq t$, $B \in \mathcal{B}(\mathbf{C})$, *the function* $(t, x_t) \mapsto p(t, x_t, s, B)$ *is* $\mathcal{B}([0,T]) \times \mathcal{B}(\mathbf{C})$*-measurable.*
(b) For any $\tilde{s} \geq s \geq t$, $B \in \mathcal{B}(\mathbf{C})$

$$P\{x_{\tilde{s}} \in B | \mathcal{F}(s)\} = P\{x_{\tilde{s}} \in B | x_s\} = p(s, x_s, \tilde{s}, B)$$

holds a.s. on Ω *and* $\tilde{s} > s > t$.

Proof. See Theorem (1.1) on page 51 of Mohammed [Moh84]. $\square$

When the coefficient functions f and g are time-independent, *i.e.*, $f(t, \phi) \equiv f(\phi)$ and $g(t, \phi) \equiv g(\phi)$, then (1.27) reduces to the following autonomous system:

$$dx(s) = f(x_s)\,ds + g(x_s)\,dW(s). \qquad (1.52)$$

In this case, we usually assume the initial time $t = 0$. Its corresponding $\mathbf{C}$-valued process $\{x_s, s \in [0,T]\}$ is a Markov process with time-homogeneous probability transition probabilities

$$p(x_0, s, B) \equiv p(0, x_0, s, B) = p(t, x_t, t+s, B), \quad \forall s, t \geq 0, x_0 \in \mathbf{C}, B \in \mathcal{B}(\mathbf{C}).$$

For a nonautonomous equation, one can consider the modified solution process $\{(s, x_s), s \in [t,T]\}$ instead of $\{x_s, s \in [t,T]\}$ as a solution for a modified autonomous equation. We will not dwell upon this point here.

Under Assumptions 1.3.6 and 1.3.7, we can also show that the $\mathbf{C}$-valued strong solution process of (1.27) also satisfies the following strong Markov property.

Theorem 1.5.2 (Strong Markovian Property 1) *Under Assumptions 1.3.6 and 1.3.7, the* $\mathbf{C}$*-valued solution process* $\{x_s, s \in [t,T]\}$ *of (1.27) satisfies the following strong Markov property:*

$$P\{x_{\tilde{s}} \in B \mid \mathcal{F}(\tau)\} = P\{x_{\tilde{s}} \mid x_\tau\}, \quad \forall \; \mathbf{G}(t)\text{-stopping times } t \leq \tau \leq \tilde{s}. \quad (1.53)$$

The strong Markovian property also holds for the **M**-valued process $\{(S(s), S_s), s \geq 0\}$ corresponding to (1.43).

Theorem 1.5.3 (Strong Markovian Property 2) *Under Assumptions 1.4.4 and 1.4.5, the **M**-valued solution process $\{(S(s), S_s), s \in [0, T]\}$ of (1.43) satisfies the following strong Markov property: For all $\tau \in T_0^{\tilde{s}}$,*

$$P\{(S(\tilde{s}), S_{\tilde{s}}) \in B \mid \mathcal{G}(\tau)\} = P\{(S(\tilde{s}), S_{\tilde{s}}) \in B \mid (S(\tau), S_\tau)\}. \tag{1.54}$$

1.6 Conclusions and Remarks

This chapter reviews stochastic tools and probabilistic preliminaries that are needed for developments of the subsequent chapters. The existing results on stochastic hereditary differential systems (mainly taken from Mohammed [Moh84] and Mizel and Trutzer [MT84]), including existence and uniqueness of strong and weak solution as well as Markovian and strong Markovian properties of the segment processes, are presented. The proofs of the known results are given here only if it might shed light on the subject and are needed in the later developments. The theory of SHDE is still in its toddler stage requires additional research efforts for many years to come. We have not presented at all results on stability theory, one better developed area of SHDEs but not relevant to stochastic control of finite time horizon. Interested readers are referred to Mao [Mao97] for introduction and results on this topic.

2

Stochastic Calculus

The main purpose of this chapter is to study stochastic calculus for the **C**-valued Markovian solution process $\{x_s, s \in [t, T]\}$ for the SHDE (1.27) and the **M**-valued Markovian solution process $\{(S(s), S_s), s \geq 0\}$ for the SHDE (1.43). In particular, Dynkin's formulas, which play an important role in the Hamilton-Jacobi-Bellman theory of the optimal control problems, will be derived for these two solution processes.

This chapter is organized as follows. As a preparation for developments of stochastic calculus, the concepts and preliminary results of bounded linear and bilinear functionals, Fréchet derivatives, C_0-semigroups on a generic separable Banach space $(\Xi, \|\cdot\|_\Xi)$ are introduced in Section 2.1. In Sections 2.2 and 2.3 these concepts and results that are specific to the spaces $\Xi = \mathbf{C}$ and $\Xi = \mathbf{M}$ are presented. In particular, the properties of the $\mathcal{S}$ operator are established in these two sections. Dynkin's formulas in terms of a weak infinitesimal generator for the segment processes for both (1.27) and (1.43) are obtained in Section 2.4. Dynkin's formulas can be calculated when quasi-tame functions are involved. Moreover, Itô's formulas for these two systems can be obtained for quasi-tame functions. Additional conclusions for the chapter and relevant remarks are given in Section 2.5. This chapter serves as the foundation of the construction of the Hamilton-Jacobi-Bellman theory for our optimal control problems.

2.1 Preliminary Analysis on Banach Spaces

In dealing with optimal control problems that involve the **C**-valued Markovian solution process $\{x_s, s \in [t, T]\}$ of (1.27) and the **M**-valued Markov solution process $\{(S(s), S_s), s \geq 0\}$ of (1.43), we often encounter the analysis on these two infinite-dimensional spaces. Instead of introducing these analysis separately for each of these two spaces, we review or prove some of the preliminary analytic results on a generic Banach space $(\Xi, \|\cdot\|_\Xi)$ in this section. The analyses includes the concepts of Fréchet derivatives and bounded linear and bilinear functionals as well as the concept of the C_0-semigroup of bounded linear operators and their generators. The special results that are applicable to $\Xi = \mathbf{C}$ and $\Xi = \mathbf{M}$ will be discussed in Sections 2.3 and 2.4, respectively.

2.1.1 Bounded Linear and Bilinear Functionals

For the time being, let Ξ be a generic Banach space with the norm $\|\cdot\|_\Xi$, and let $\mathcal{B}(\Xi)$ be the Borel σ-algebra of subsets of Ξ; that is, $\mathcal{B}(\Xi)$ is the smallest σ-algebra of subsets of Ξ that contains all open (and hence closed) subsets of Ξ. The pair $(\Xi, \mathcal{B}(\Xi))$ denotes a Borel measurable space. If Ξ is a Hilbert space (a special case of a Banach space), then the Hilbertian inner product $\langle \cdot, \cdot \rangle_\Xi : \Xi \times \Xi \to \Re$, which gives rise to the norm $\|\cdot\|_\Xi$, will normally be given or specified.

Let $(\Xi, \|\cdot\|_\Xi)$ and $(\Theta, \|\cdot\|_\Theta)$ be two Banach spaces or Hilbert spaces. A mapping $L : \Xi \to \Theta$ is said to be a bounded linear operator from Ξ to Θ if it satisfies the following linearity and boundedness (continuity) conditions:

(i) (Linearity Condition) for $\forall a, b \in \Re$ and $\forall \mathbf{x}, \mathbf{y} \in \Xi$,

$$L(a\mathbf{x} + b\mathbf{y}) = aL\mathbf{x} + bL\mathbf{y}.$$

(ii) (Boundedness Condition) There exists a constant $K(L) > 0$, dependent on L only, such that

$$\|L\mathbf{x}\|_\Theta \leq K(L)\|\mathbf{x}\|_\Xi, \quad \forall \mathbf{x} \in \Xi.$$

Let $\mathcal{L}(\Xi, \Theta)$ denote the Banach space of bounded linear operators from Ξ into Θ equipped with the operator norm $\|\cdot\|_{\mathcal{L}(\Xi,\Theta)}$ defined by

$$\|L\|_{\mathcal{L}(\Xi,\Theta)} = \sup_{\mathbf{x} \neq \mathbf{0}} \frac{\|L\mathbf{x}\|_\Theta}{\|\mathbf{x}\|_\Xi}$$
$$= \sup_{\|\mathbf{x}\|_\Xi = 1} \|L\mathbf{x}\|_\Theta, \ L \in \mathcal{L}(\Xi, \Theta).$$

In this case,

$$\|L\mathbf{x}\|_\Theta \leq \|L\|_{\mathcal{L}(\Xi,\Theta)}\|\mathbf{x}\|_\Xi, \quad \forall L \in \mathcal{L}(\Xi, \Theta), \mathbf{x} \in \Xi.$$

If $\Xi = \Theta$, we simply write $\mathcal{L}(\Xi)$ for $\mathcal{L}(\Xi, \Xi)$ and any $L \in \mathcal{L}(\Xi)$ is called a bounded linear operator on Ξ. If $\Theta = \Re$, then $L \in \mathcal{L}(\Xi, \Re)$ will be called a *bounded linear functional* on Ξ. The space of bounded linear functionals (or the topological dual of Ξ^*) will be denoted by Ξ^* and its corresponding operator norm will be denoted by $\|\cdot\|_\Xi^*$.

The mapping $L : \Xi \times \Xi \to \Re$ is said to be a bounded bilinear functional if $L(\mathbf{x}, \cdot)$ and $L(\cdot, \mathbf{y})$ are both in Ξ^* for each $\mathbf{x}$ and $\mathbf{y}$ in Ξ. The space of bounded bilinear functionals on Ξ will be denoted by $\Xi^\dagger$. Again, $\Xi^\dagger$ is a real separable Banach space under the operator norm $\|\cdot\|_\Xi^\dagger$ defined by

$$\|L\|_\Xi^\dagger = \sup_{\mathbf{y} \neq \mathbf{0}} \frac{\|L(\cdot, \mathbf{y})\|_\Xi^*}{\|\mathbf{y}\|_\Xi} = \sup_{\mathbf{x} \neq \mathbf{0}} \frac{\|L(\mathbf{x}, \cdot)\|_\Xi^*}{\|\mathbf{x}\|_\Xi}.$$

2.1.2 Fréchet Derivatives

Let $\Psi : \Xi \to \Re$ be a Borel measurable function. The function Ψ is said to be Fréchet differentiable at $\mathbf{x} \in \Xi$ if for each $\mathbf{y} \in \Xi$,

$$\Psi(\mathbf{x} + \mathbf{y}) - \Psi(\mathbf{x}) = D\Psi(\mathbf{x})(\mathbf{y}) + o(\mathbf{y}),$$

where $D\Psi : \Xi \to \Xi^*$ and $o : \Xi \to \Re$ is a function such that

$$\frac{o(\mathbf{x})}{\|\mathbf{x}\|_\Xi} \to 0 \text{ as } \|\mathbf{x}\|_\Xi \to 0.$$

In this case, $D\Psi(\mathbf{x}) \in \Xi^*$ is called the Fréchet derivative of Ψ at $\mathbf{x} \in \Xi$. The function Ψ is said to be continuously Fréchet differentiable at $\mathbf{x} \in \Xi$ if its Fréchet derivative $D\Psi : \Xi \to \Xi^*$ is continuous under the operator norm $\|\cdot\|_\Xi^*$. The function Ψ is said to be twice Fréchet differentiable at $\mathbf{x} \in \Xi$ if its Fréchet derivative $D\Psi(\mathbf{x}) : \Xi \to \Re$ exists and there exists $D^2\Psi(\mathbf{x}) : \Xi \times \Xi \to \Re$ such that for each $\mathbf{y}, \mathbf{z} \in \Xi$,

$$D^2\Psi(\mathbf{x})(\cdot, \mathbf{z}), D^2\Psi(\mathbf{x})(\mathbf{y}, \cdot) \in \Xi^*$$

and

$$D\Psi(\mathbf{x} + \mathbf{y})(\mathbf{z}) - D\Psi(\mathbf{x})(\mathbf{z}) = D^2\Psi(\mathbf{x})(\mathbf{y}, \mathbf{z}) + o(\mathbf{y}, \mathbf{z}).$$

Here, $o : \Xi \times \Xi \to \Re$ is such that

$$\frac{o(\mathbf{y}, \mathbf{z})}{\|\mathbf{y}\|_\Xi} \to 0 \text{ as } \|\mathbf{y}\|_\Xi \to 0$$

and

$$\frac{o(\mathbf{y}, \mathbf{z})}{\|\mathbf{z}\|_\Xi} \to 0 \text{ as } \|\mathbf{z}\|_\Xi \to 0.$$

In this case, the bounded bilinear functional $D^2\Psi(\mathbf{x}) : \Xi \times \Xi \to \Re$ is the second Fréchet derivative of Ψ at $\mathbf{x} \in \Xi$.

As usual, the function $\Psi : \Xi \to \Re$ is said to be Fréchet differentiable (respectively, twice Fréchet differentiable) if Ψ is Fréchet differentiable (twice Fréchet differentiable) at every $\mathbf{x} \in \Xi$.

For computation of the Fréchet derivatives of the function $\Psi : \Xi \to \Re$, we introduce the concept of Gâteaux derivatives as follows. The function $\Psi : \Xi \to \Re$ is said to be k-times Gâteaux differentiable at $\mathbf{x} \in \Xi$ for $k = 1, 2, \ldots$ if the following derivatives exist:

$$\Psi^{(1)}(\mathbf{x})(\mathbf{y}) \equiv \frac{d}{dt}\Psi(\mathbf{x} + t\mathbf{y})|_{t=0}, \ \mathbf{y} \in \Xi$$

and for $i = 2, 3, \ldots, k$,

$$\Psi^{(i)}(\mathbf{x})(\mathbf{y}^1, \mathbf{y}^2, \ldots, \mathbf{y}^i) = \frac{d}{dt}\Psi^{(i-1)}(\mathbf{x} + t\mathbf{y}^i)(\mathbf{y}^1, \mathbf{y}^2, \ldots, \mathbf{y}^{i-1})|_{t=0}.$$

Note that every Fréchet derivative $D\Psi(\phi)$ is also a Gâteaux derivative and

$$\Psi^{(1)}(\mathbf{x})(\mathbf{y}) = D\Psi(\mathbf{x})(\mathbf{y}), \quad \mathbf{y} \in \Xi .$$

Conversely, if the Gâteaux derivative $\Psi^{(1)}(\mathbf{x})$ of Ψ exists for all $\mathbf{x}$ in a neighborhood $U(\mathbf{x})$ of $\mathbf{x} \in \Xi$ and if $\Psi^{(1)} : U(\mathbf{x}) \subset \Xi \to \Re$ is continuous, then $\Psi^{(1)}(\mathbf{x})$ is also the Fréchet derivative at $\mathbf{x}$.

Throughout, let $C^{1,2}([a,b] \times \Xi)$ be the space of functions $\Psi(t,\mathbf{x})$ that are continuously differentiable with respect to $t \in [a,b]$ and twice continuously Fréchet differentiable with respect to $\mathbf{x} \in \Xi$. Denote the derivative of Ψ with respect to t by $\partial_t \Psi$. Of course, $\partial_t \Psi$ at the end points a (respectively, b) will be understood to be the right-hand (respectively, the left-hand) derivative. The first and second Fréchet derivatives of Ψ with respect to $\mathbf{x}$ will be denoted by $D_{\mathbf{x}}\Psi(t,\mathbf{x})$ and $D^2_{\mathbf{x}}\Psi(t,\mathbf{x})$ (or simply $D\Psi(t,\mathbf{x})$ and $D^2\Psi(t,\mathbf{x})$ when there is no danger of ambiguity), respectively. $D^2\Psi$ is said to be globally Lipschitz on $[a,b] \times \Xi$ on $[a,b]$ if there exists a constant $K > 0$ that depends on Ψ only, such that

$$\|D^2\Psi(t,\mathbf{x}) - D^2\Psi(t,\mathbf{y})\|_{\Xi}^{\dagger} \leq K\|\mathbf{x} - \mathbf{y}\|_{\Xi}, \quad \forall t \in [a,b], \mathbf{x}, \mathbf{y} \in \Xi.$$

The space of $\Psi \in C^{1,2}([a,b] \times \Xi)$ with $D^2\Psi$ being globally Lipshitz on $[a,b] \times \Xi$ will be denoted by $C^{1,2}_{lip}([a,b] \times \Xi)$.

If $\Phi \in C^{1,2}([a,b] \times \Xi)$, the following version of Taylor's formula holds (see, e.g., Lang [Lan83]):

Theorem 2.1.1 *If $\Phi \in C^{1,2}([a,b] \times \Xi)$, then*

$$\Phi(t,\mathbf{y}) = \Phi(s,\mathbf{x}) + \partial_s\Phi(s,\mathbf{x})(t-s) + D\Phi(s,\mathbf{x})(\mathbf{y}-\mathbf{x})$$
$$+ \int_0^1 (1-u)D^2\Phi(s,\mathbf{x}+u(\mathbf{y}-\mathbf{x}))(\mathbf{y}-\mathbf{x},\mathbf{y}-\mathbf{x})\,du.$$

2.1.3 $\mathcal{C}_0$-Semigroups

Again, let Ξ be a generic real separable Banach or Hilbert space with the norm $\|\cdot\|_{\Xi}$ and let $\mathcal{L}(\Xi)$ be the space of bounded linear operators on Ξ.

Definition 2.1.2 *A family $\{T(t), t \geq 0\} \subset \mathcal{L}(\Xi)$ is said to be a strongly continuous semigroups of bounded linear operators on Ξ or, in short, a $\mathcal{C}_0(\Xi)$-semigroup if it satisfies the following conditions:*
(i) $T(0) = I$ (the identity operator).
(ii) $T(s) \circ T(t) = T(t) \circ T(s) = T(s+t)$ for $s, t \geq 0$.
(iii) For any $\mathbf{x} \in \Xi$, the map $t \mapsto T(t)\mathbf{x}$ is continuous for all $t \geq 0$.
(iv) there exist constants $M \geq 1$ and $\beta > 0$ such that

$$\|T(t)\|_{\mathcal{L}(\Xi)} \leq Me^{\beta t}, \quad \forall t \geq 0. \tag{2.1}$$

The $\mathcal{C}_0(\Xi)$-semigroup is said to be a contraction if there exists a constant $k \in [0,1)$ such that

$$\|T(t)\mathbf{x} - T(t)\mathbf{y}\|_\Xi \leq k\|\mathbf{x} - \mathbf{y}\|_\Xi, \quad \forall t \geq 0, \mathbf{x}, \mathbf{y} \in \Xi.$$

One can associate such a $\mathcal{C}_0$-semigroup $\{T(t), t \geq 0\} \subset \mathcal{L}(\Xi)$ with its generator $\mathbf{A}$ defined by the following relation:

$$\mathbf{A}\mathbf{x} = \lim_{t \downarrow 0} \frac{1}{t}(T(t)\mathbf{x} - \mathbf{x}), \ t \geq 0. \tag{2.2}$$

The domain $\mathcal{D}(\mathbf{A})$ of $\mathbf{A}$ is defined by the set

$$\mathcal{D}(\mathbf{A}) = \left\{ \mathbf{x} \in \Xi \mid \lim_{t \downarrow 0} \frac{1}{t}(T(t)\mathbf{x} - \mathbf{x}) \quad \text{exists} \right\}. \tag{2.3}$$

We have the following results regarding the *generator* $\mathbf{A}$ and its other relationship to the $\mathcal{C}_0$-semigroup $\{T(t), t \geq 0\}$.

Theorem 2.1.3 *Let* $\mathbf{A} : \mathcal{D}(\mathbf{A}) \subset \Xi \to \Xi$ *be the generator of a $\mathcal{C}_0$-semigroup $\{T(t), t \geq 0\}$ of bounded linear operators on a Banach space $(\Xi, \| \cdot \|_\Xi)$. Then the following hold:*
(i) For any $t > 0$ and $\mathbf{x} \in \Xi$,

$$\int_0^t T(s)\mathbf{x}\,ds \in \mathcal{D}(\mathbf{A}) \ \text{and} \ T(t)\mathbf{x} - \mathbf{x} = \mathbf{A}\int_0^t T(s)\mathbf{x}\,ds. \tag{2.4}$$

(ii) If $\mathbf{x} \in \mathcal{D}(\mathbf{A})$, then $T(t)\mathbf{x} \in \mathcal{D}(\mathbf{A})$ and $\mathbf{A}T(t)\mathbf{x} = T(t)\mathbf{A}\mathbf{x}$. Furthermore,

$$T(t)\mathbf{x} - \mathbf{x} = \int_0^t \mathbf{A}T(s)\mathbf{x}\,ds = \int_0^t T(s)\mathbf{A}\mathbf{x}\,ds.$$

Proof.
(i) It follows from the definition that

$$\lim_{h \downarrow 0} \frac{T(h) - I}{h} \int_0^t T(s)\mathbf{x}\,ds \ \text{(by linearity)}$$

$$= \lim_{h \downarrow 0} \frac{1}{h} \int_0^t (T(s+h) - T(s))\mathbf{x}\,ds \ \text{(by (ii) of Definition 2.1.2)}$$

$$= \lim_{h \downarrow 0} \frac{1}{h} \int_t^{t+h} (T(s) - I)\mathbf{x}\,ds \ \text{(by change of integration variable)}$$

$$= T(t)\mathbf{x} - \mathbf{x} \ \text{(by continuity of } T(t)\mathbf{x} \text{ with respect to } t\text{)}.$$

This shows that $\int_0^t T(s)\mathbf{x}\,ds \in \mathcal{D}(\mathbf{A})$ and $T(t)\mathbf{x} - \mathbf{x} = \mathbf{A}\int_0^t T(s)\mathbf{x}\,ds$ for any $t > 0$ and $\mathbf{x} \in \Xi$.
(ii) Let $\mathbf{x} \in \mathcal{D}(\mathbf{A})$ and $t > 0$. Then

$$\mathbf{A}T(t)\mathbf{x} = \lim_{h \downarrow 0} \frac{(T(h) - I)T(t)\mathbf{x}}{h} = \lim_{h \downarrow 0} T(t)\left(\frac{(T(h) - I)\mathbf{x}}{h}\right) = T(t)\mathbf{A}\mathbf{x}.$$

This shows that $T(t)\mathbf{x} \in \mathcal{D}(\mathbf{A})$ and the right-hand derivative

$$\left(\frac{d}{dt}\right)^+ T(t)\mathbf{x} = \mathbf{A}T(t)\mathbf{x} = T(t)\mathbf{A}\mathbf{x}.$$

On the other hand, for $t > 0$

$$\left\| \frac{T(t - h)\mathbf{x} - T(t)\mathbf{x}}{-h} - T(t)\mathbf{A}\mathbf{x} \right\|_{\Xi}$$

$$= \left\| T(t - h)\left(\frac{T(h)\mathbf{x} - \mathbf{x}}{h} - \mathbf{A}\mathbf{x}\right) + T(t - h)\mathbf{A}\mathbf{x} - T(t)\mathbf{A}\mathbf{x} \right\|_{\Xi}$$

$$\leq \|T(t - h)\|_{\mathcal{L}(\Xi)} \left\| \frac{T(h)\mathbf{x} - \mathbf{x}}{h} - \mathbf{A}\mathbf{x} \right\|_{\Xi} + \|T(t - h)\mathbf{A}\mathbf{x} - T(t)\mathbf{A}\mathbf{x}\|_{\Xi}$$

$$\to 0 \text{ as } h \downarrow 0.$$

This proves (ii).   □

Corollary 2.1.4 *If* $\mathbf{A} : \mathcal{D}(\mathbf{A}) \subset \Xi \to \Xi$ *is the generator of a* $\mathcal{C}_0$-*semigroup* $\{T(t), t \geq 0\}$ *of bounded linear operators on a Banach space* $(\Xi, \| \cdot \|_{\Xi})$, *then* $\mathbf{A}$ *is a linear operator from the dense subspace* $\mathcal{D}(\mathbf{A}) \subset \Xi$ *to* Ξ.

Proof. It is easy to see that $\mathcal{D}(\mathbf{A})$ is a subspace of Ξ and $\mathbf{A}$ is a linear operator, since $\mathbf{0} \in \mathcal{D}(\mathbf{A})$ and $\mathbf{x}, \mathbf{y} \in \mathcal{D}(\mathbf{A})$ imply that $a\mathbf{x} + b\mathbf{y} \in \mathcal{D}(\mathbf{A})$ for all $a, b \in \Re$.

For any $\mathbf{x} \in \Xi$, we show that there exists a sequence $\{\mathbf{x}_k\} \subset \mathcal{D}(\mathbf{A})$ such that $\mathbf{x}_k \to \mathbf{x}$ as $k \to \infty$. In fact, for any $t > 0$,

$$\frac{1}{t} \int_0^t T(s)\mathbf{x} \, ds \in \mathcal{D}(\mathbf{A}) \text{ and } \frac{1}{t} \int_0^t T(s)\mathbf{x} \, ds \to T(0)\mathbf{x} = \mathbf{x} \text{ as } t \downarrow 0.$$

Hence, $\mathcal{D}(\mathbf{A})$ is a dense subset of Ξ.   □

Definition 2.1.5 *The operator* $\mathbf{A} : \mathcal{D}(\mathbf{A}) \subset \Xi \to \Xi$ *defined by (2.2) is called a closed operator if for* $\{\mathbf{x}_k\} \subset \mathcal{D}(\mathbf{A})$ *such that* $\mathbf{x}_k \to \mathbf{x}$, $\mathbf{A}\mathbf{x}_k \to \mathbf{y}$ *we have* $\mathbf{x} \in \mathcal{D}(\mathbf{A})$ *and* $\mathbf{y} = \mathbf{A}\mathbf{x}$.

Corollary 2.1.6 *If* $\mathbf{A} : \mathcal{D}(\mathbf{A}) \subset \Xi \to \Xi$ *is the generator of a* $\mathcal{C}_0$-*semigroup* $\{T(t), t \geq 0\}$ *of bounded linear operators on a Banach space* $(\Xi, \| \cdot \|_{\Xi})$, *then* $\mathbf{A}$ *is a closed operator.*

Proof. By (2.4), we have

$$T(t)\mathbf{x}_k - \mathbf{x}_k = \int_0^t T(s)\mathbf{A}\mathbf{x}_k \, ds.$$

Hence,

$$\lim_{k \to \infty} T(t)\mathbf{x}_k - \mathbf{x}_k = T(t)\mathbf{x} - \mathbf{x}$$

$$= \lim_{k \to \infty} \int_0^t T(s)\mathbf{A}\mathbf{x}_k \, ds$$

$$= \int_0^t T(s)\mathbf{A}\mathbf{x} \, ds.$$

This proves that $\mathbf{x} \in \mathcal{D}(\mathbf{A})$ and $\mathbf{y} = \mathbf{A}\mathbf{x}$.   □

Definition 2.1.7 *Let* $\mathbf{A} : \mathcal{D}(\mathbf{A}) \subset \Xi \to \Xi$ *be a closed operator on* Ξ. *The resolvent set* $\rho(\mathbf{A})$ *of* $\mathbf{A}$ *is the collection of all* $\lambda \in \Re$ *such that* $\lambda I - \mathbf{A}$ *(I is the identity operator) is invertible, its range* $\mathcal{R}(\lambda I - \mathbf{A}) = \Xi$, *and* $R(\lambda) \equiv (\lambda - \mathbf{A})^{-1} \in \mathcal{L}(\Xi)$. *For each* $\lambda \in \rho(\mathbf{A})$, $R(\lambda)$ *is called the resolvent of* $\mathbf{A}$ *at* λ.

The proofs of the following two theorems can be found in Theorem 1.2.10 and Theorem 1.2.11 of Kallingpur and Xiong [KX95].

Theorem 2.1.8 *Let* $\mathbf{A} : \mathcal{D}(\mathbf{A}) \subset \Xi \to \Xi$ *be a* C_0*-semigroup* $\{T(t), t \geq 0\}$ *of bounded linear operators on a Banach space* $(\Xi, \|\cdot\|_\Xi)$. *Let* M *and* β *be given by (2.1); then* $(\beta, \infty) \subset \rho(\mathbf{A})$,

$$R(\lambda) = \int_0^\infty e^{-\lambda t} T(t) \, dt, \tag{2.5}$$

and

$$\|(R(\lambda))^k\|_{\mathcal{L}(\Xi)} \leq M(\lambda - \beta)^{-k}, \quad k = 1, 2, \ldots, \lambda > \beta. \tag{2.6}$$

Theorem 2.1.9 *Let* $\mathbf{A} : \mathcal{D}(\mathbf{A}) \subset \Xi \to \Xi$ *be a generator of a densely defined closed linear operator on a Banach space* Ξ *such that the interval* $(\beta, \infty) \subset \rho(\mathbf{A})$ *and*

$$\|(\lambda I - \mathbf{A})^{-k}\|_{\mathcal{L}(\Xi)} \leq M(\lambda - \beta)^{-k}, \quad k = 1, 2, \ldots, \lambda > \beta.$$

Then there exists a unique C_0*-semigroup* $\{T(t), t \geq 0\}$ *of bounded linear operators on* Ξ *with generator* $\mathbf{A}$ *such that (2.1) holds.*

2.1.4 Bounded and Continuous Functionals on Banach Spaces

In this subsection we will explore some properties of bounded and real-valued continuous functions on a Banach space $(\Xi, \|\cdot\|_\Xi)$ and some strong and weak convergence results for a family or sequence of such functions.

Let $C_b(\Xi)$ be the real Banach space of bounded and continuous (but not necessarily linear) functions $\Phi : \Xi \to \Re$ under the sup-norm $\|\cdot\|_{C_b(\Xi)}$ defined by

$$\|\Phi\|_{C_b(\Xi)} = \sup_{\mathbf{x} \in \Xi} |\Phi(\mathbf{x})|.$$

The topology on the space $C_b(\Xi)$ generated by the sup-norm $\|\cdot\|_{C_b(\Xi)}$ is referred to as the strong topology.

We also define a weak topology on $C_b(\Xi)$ as follows: Let $\mathcal{M}(\Xi)$ be the Banach space of all finite *regular measures* on the Borel measurable space $(\Xi, \mathcal{B}(\Xi))$ equipped with the total variation norm defined by

$$\|\mu\|_{\mathcal{M}(\Xi)} = \sup\{\mu(B) \mid B \in \mathcal{B}(\Xi)\}, \quad \mu \in \mathcal{M}(\Xi).$$

Then it can be shown that (see, e.g., Rudin [Rud71]) there is a continuous bilinear pairing $\langle \cdot, \cdot \rangle_{(C_b, \mathcal{M})} : C_b(\Xi) \times \mathcal{M}(\Xi) \to \Re$ defined by

$$\langle \Phi, \mu \rangle_{(C_b, \mathcal{M})} = \int_{\Xi} \Phi(\mathbf{x}) \, d\mu(\mathbf{x}), \quad \Phi \in C_b(\Xi), \mu \in \mathcal{M}(\Xi).$$

A family $\{\Phi(t), t \geq 0\}$ in $C_b(\Xi)$ is said to converge weakly to $\Phi \in C_b(\Xi)$ as $t \downarrow 0$ if

$$\lim_{t \downarrow 0} \langle \Phi(t), \mu \rangle_{(C_b, \mathcal{M})} = \langle \Phi, \mu \rangle_{(C_b, \mathcal{M})}, \quad \forall \mu \in \mathcal{M}(\Xi).$$

In this case, we write

$$\Phi = (w) \lim_{t \downarrow 0} \Phi(t).$$

The following result is due originally to Dynkin [Dyn65]:

Proposition 2.1.10 *For each $t > 0$, let $\Phi(t) \in C_b(\Xi)$ and let $\Phi \in C_b(\Xi)$. Then*

$$\Phi = (w) \lim_{t \downarrow 0} \Phi(t)$$

if and only if the set $\{\|\Phi(t)\|_{C_b(\Xi)}, t > 0\}$ is bounded in $\Re$ and

$$\lim_{t \downarrow 0} \Phi(t)(\mathbf{x}) = \Phi(\mathbf{x}), \quad \forall \mathbf{x} \in \Xi.$$

Proof. For each $\mathbf{x} \in \Xi$, let $\delta_{\mathbf{x}} \in \mathcal{M}(\Xi)$ be the Dirac measure concentrated at $\mathbf{x}$ and defined by

$$\delta_{\mathbf{x}}(B) = \begin{cases} 1 \text{ if } \mathbf{x} \in B \\ 0 \text{ if } \mathbf{x} \notin B \end{cases}$$

for all $B \in \mathcal{B}(\Xi)$. For each $t > 0$, define $\hat{\Phi}(t) \in \mathcal{M}^*(\Xi)$ by

$$\hat{\Phi}(t)(\mu) = \langle \Phi(t), \mu \rangle_{(C_b, \mathcal{M})}, \quad \forall \mu \in \mathcal{M}(\Xi),$$

where $\mathcal{M}^*(\Xi)$ is the topological dual of $\mathcal{M}(\Xi)$. If $\Phi = (w) \lim_{t \downarrow 0} \Phi(t)$, then

$$\Phi(\mathbf{x}) = \langle \Phi, \delta_{\mathbf{x}} \rangle_{(C_b, \mathcal{M})} = \lim_{t \downarrow 0} \langle \Phi(t), \delta_{\mathbf{x}} \rangle_{(C_b, \mathcal{M})} = \lim_{t \downarrow 0} \Phi(t)(\mathbf{x}), \quad \forall \mathbf{x} \in \Xi,$$

and the set $\{\hat{\Phi}(t)(\mu), t > 0\}$ is bounded for each $\mu \in \mathcal{M}(\Xi)$. By the uniform boundedness principle (see, e.g., Siddiqi [Sid04]) $\{\|\hat{\Phi}(t)\|_{\mathcal{M}^*(\Xi)}, t > 0\}$ is bounded. However, $\|\hat{\Phi}(t)\|_{\mathcal{M}^*(\Xi)} = \|\Phi(t)\|_{C_b(\Xi)}$ for each $t > 0$, so $\{\|\Phi(t)\|_{C_b(\Xi)}, t > 0\}$ is bounded.

Conversely, suppose $\{\|\Phi(t)\|_{C_b(\Xi)}, t > 0\}$ is bounded and

$$\lim_{t\downarrow 0} \Phi(t)(\mathbf{x}) = \Phi(\mathbf{x}), \quad \forall \mathbf{x} \in \Xi.$$

By the dominated convergence theorem,

$$\begin{aligned}
\langle \Phi, \mu \rangle_{(C_b, \mathcal{M})} &= \int_{\Xi} \Phi(\mathbf{x}) \, d\mu(\mathbf{x}) \\
&= \lim_{t\downarrow 0} \int_{\Xi} \Phi(t)(\mathbf{x}) \, d\mu(\mathbf{x}) \\
&= \lim_{t\downarrow 0} \langle \Phi(t), \mu \rangle_{(C_b, \mathcal{M})}, \quad \forall \mu \in \mathcal{M}(\Xi).
\end{aligned}$$

Thus, $\Phi = (w)\lim_{t\downarrow 0} \Phi(t)$. This proves the proposition.   □.

Lifting from the space Ξ to a higher level $C_b(\Xi)$ (i.e., with a little abuse of notation, letting Ξ above be the Banach space $C_b(\Xi)$), let us consider a C_0-semigroup of bounded linear operators $\{\Gamma(t), t \geq 0\}$ on space $C_b(\Xi)$ (i.e., $\Gamma(t) \in \mathcal{L}(C_b(\Xi))$ for each $t \geq 0$). The definition of a C_0-semigroup given in Definition 2.1.2 still applies in the context of the Banach space $C_b(\Xi)$:
(i) $\Gamma(0) = I$ (the identity operator on $C_b(\Xi)$).
(ii) $\Gamma(s) \circ \Gamma(t) = \Gamma(t) \circ \Gamma(s) = \Gamma(s + t)$ for $t, s \geq 0$.
(iii) For any $\Phi \in C_b(\Xi), \|\Gamma(t)\Phi - \Phi\|_{C_b(\Xi)} \to 0$ as $t \downarrow 0$.

Note that Condition (iii) implies that $\Gamma(t)$ converges to $\Gamma(0)$ in the operator norm $\|\cdot\|_{\mathcal{L}(C_b(\Xi))}$.

$$\begin{aligned}
\lim_{t\downarrow 0} \|\Gamma(t) - \Gamma(0)\|_{\mathcal{L}(C_b(\Xi))} &= \lim_{t\downarrow 0} \|\Gamma(t) - I\|_{\mathcal{L}(C_b(\Xi))} \\
&\equiv \lim_{t\downarrow 0} \sup_{\Phi \neq 0} \frac{\|\Gamma(t)\Phi - \Phi\|_{C_b(\Xi)}}{\|\Phi\|_{C_b(\Xi)}} \\
&= \lim_{t\downarrow 0} \sup_{\|\Phi\|_{C_b(\Xi)}=1} \|\Gamma(t)\Phi - \Phi\|_{C_b(\Xi)}.
\end{aligned}$$

Therefore, by the dominated convergence theorem, it is easy to see that

$$(w)\lim_{t\downarrow 0} \Gamma(t)(\Phi) = \Phi, \quad \forall \Phi \in C_b(\Xi),$$

or, equivalently,

$$\lim_{t\downarrow 0} \langle \Gamma(t)(\Phi) - \Phi, \mu \rangle_{(C(\Xi), \mathcal{M}(\Xi))} = 0, \quad \mu \in \mathcal{M}(\Xi).$$

The *weak infinitesimal generator* $\boldsymbol{\Gamma}$ of the C_0-semigroup of bounded linear operators $\{\Gamma(t), t \geq 0\}$ on $C_b(\Xi)$ is defined by

$$\boldsymbol{\Gamma}(\Phi) = (w)\lim_{t\downarrow 0} \frac{\Gamma(t)\Phi - \Phi}{t}.$$

Let $\mathcal{D}(\mathbf{\Gamma})$, the domain of the *weak infinitesimal generator* $\mathbf{\Gamma}$, be the set of all $\Phi \in C_b(\Xi)$ for which the above weak limit exists.

The following properties can be found in [Dyn65] (Vol. 1, Chapter I §6, pp. 36-43).

Theorem 2.1.11 *Let* $\{\Gamma(t), t \geq 0\} \subset \mathcal{L}(C_b(\Xi))$, *and let* $C_b^0(\Xi) \subset C_b(\Xi)$ *be defined by*

$$C_b^0(\Xi) = \left\{ \Phi \in C_b(\Xi) \mid \lim_{t \downarrow 0} \Gamma(t)(\Phi) = \Phi \right\}.$$

Then the following hold:

(i) $\mathcal{D}(\mathbf{\Gamma}) \subset C_b^0(\Xi)$ *and* $\mathcal{D}(\mathbf{\Gamma})$ *is weakly dense in* $C_b(\Xi)$; *that is, for each* $\Phi \in C_b(\Xi)$, *there exists a sequence* $\{\Phi_k\} \subset \mathcal{D}(\mathbf{\Gamma})$ *such that*

$$\lim_{k \downarrow \infty} \langle \Phi_k, \mu \rangle_{(C(\Xi), \mathcal{M}(\Xi))} = \langle \Phi, \mu \rangle_{(C_b(\Xi), \mathcal{M}(\Xi))}, \quad \forall \mu \in \mathcal{M}(\Xi),$$

and

$$\Gamma(t)(\mathcal{D}(\mathbf{\Gamma})) \subset \mathcal{D}(\mathbf{\Gamma}), \quad \forall t \geq 0.$$

(ii) If $\Phi \in \mathcal{D}(\mathbf{\Gamma})$ *and* $t > 0$, *then the weak derivative*

$$(w)\frac{d}{dt}\Gamma(t)(\Phi) \equiv (w) \lim_{h \to 0} \frac{1}{h}[\Gamma(t+h)(\Phi) - \Gamma(t)(\Phi)]$$

exists and

$$(w)\frac{d}{dt}\Gamma(t)(\Phi) = \Gamma(t)(\mathbf{\Gamma}(\Phi)),$$

$$\Gamma(t)(\Phi) - \Phi = \int_0^t \Gamma(u)(\mathbf{\Gamma}(\Phi))\,du, \quad \forall t > 0.$$

(iii) $\mathbf{\Gamma}$ *is weakly closed; that is,* $\{\Phi(k)\}_{k=1}^\infty \subset \mathcal{D}(\mathbf{\Gamma})$ *is weakly convergent and* $\{\mathbf{\Gamma}(\Phi(k))\}_{k=1}^\infty$ *is also weakly convergent. Then*

$$(w) \lim_{k \to \infty} \Phi(k) \in \mathcal{D}(\mathbf{\Gamma}) \text{ and } (w) \lim_{k \to \infty} \mathbf{\Gamma}(\Phi(k)) = \mathbf{\Gamma}((w) \lim_{k \to \infty} \Phi(k)).$$

(iv) For each $\lambda > 0$, *the operator* $\lambda I - \mathbf{\Gamma}$ *(note: I denotes the identity operator) is a bijection of* $\mathcal{D}(\mathbf{\Gamma})$ *onto* $C_b(\Xi)$. *Indeed, the resolvent* $R(\lambda; \mathbf{\Gamma}) \equiv (\lambda I - \mathbf{\Gamma})^{-1}$ *and satisfies the relation:*

$$(\lambda I - \mathbf{\Gamma})^{-1}(\Phi) = \int_0^\infty e^{-\lambda t} \Gamma(t)(\Phi)\,dt, \quad \forall \Phi \in C_b(\Xi).$$

The resolvent $R(\lambda; \mathbf{\Gamma}) = (\lambda I - \mathbf{\Gamma})^{-1}$ *is bounded and linear and its operator norm* $\|R(\lambda; \mathbf{\Gamma})\|_{\mathcal{L}(C_b(\Xi))} \leq \frac{1}{\lambda}$ *for all* $\lambda > 0$.

(v) (w)$\lim_{\lambda \to \infty} \lambda R(\lambda)(\Phi) = \Phi, \; \forall \Phi \in C_b(\Xi)$.

In dealing with SHDE (1.27) with bounded memory $(0 < r < \infty)$ and SHDE (1.43) with infinite but fading memory $(r = \infty)$ and their associated stochastic control problems, we will, in the following two sections, specifically apply the results obtained in the preceding subsection to the Banach space $(\Xi, \| \cdot \|_\Xi)$ that takes the form of $\Xi = \mathbf{C}$ and $\Xi = \mathbf{M}$.

2.2 The Space C

The analyses reviewed for a general generic Banach space $(\Xi, \|\cdot\|_\Xi)$ in the previous section will be specialized to the space $\mathbf{C} \equiv C([-r, 0]; \Re^n)$ in this section.

For $-\infty < a < b \leq \infty$, the interval $[a, b]$ will be interpreted as $[a, b] \cap \Re$. The intervals $(a, b]$, (a, b), and $[a, b)$ will be interpreted similarly. We use the notation $\bar{\Re}_+$ for the compactified interval $[0, \infty]$.

For $-\infty < a < b \leq \infty$, let $C([a, b]; \Re^n)$ be the space of continuous functions $\phi : [a, b] \to \Re^n$. Note that when $b < \infty$, the space $C([a, b]; \Re^n)$ is a real separable Banach space equipped with the uniform topology defined by the the sup-norm $\|\cdot\|_\infty$, where

$$\|\phi\|_\infty = \sup_{t \in [a,b]} |\phi(t)|.$$

If $b = \infty$, the space $C([0, \infty); \Re^n)$ will be equipped with the topology defined by the metric:

$$d(\omega_1, \omega_2) = \sum_{k=1}^{\infty} \frac{1}{2^k} \max_{0 \leq t \leq k} (|\omega_1(t) - \omega_2(t)| \wedge 1). \qquad (2.7)$$

Let $L^2(a, b; \Re^n)$ be the space of Lebesque square-integrable functions $\phi : [a, b] \to \Re^n$. It is clear that $L^2(a, b; \Re^n)$ is a real separable Hilbert space with the inner product $\langle \cdot, \cdot \rangle_2$ defined by

$$\langle \phi, \varphi \rangle_2 = \int_a^b \phi(t) \cdot \varphi(t) \, dt$$

and the Hilbertian norm defined by $\|\phi\|_2 = \langle \phi, \phi \rangle_2^{1/2}$.

We have the following theorem.

Theorem 2.2.1 *For $-\infty < a < b < \infty$, the space $C([a, b]; \Re^n)$ can be continuously and densely embedded into the space $L^2(a, b; \Re^n)$.*

Proof. Consider the linear injection $i : C([a, b]; \Re^n) \to L^2(a, b; \Re^n)$ defined by

$$i(\varphi) = \varphi \in L^2(a, b; \Re^n), \quad \varphi \in C([a, b]; \Re^n).$$

Then

$$\|i(\varphi)\|_2 = \left[\int_a^b |\varphi(t)|^2 \, dt \right]^{1/2}$$
$$\leq \sqrt{b - a} \|\varphi\|_\infty.$$

This shows that the injection i is continuous.

The fact that $C([a, b]; \Re^n)$ is dense in $L^2(a, b; \Re^n)$ is a well-known result that can be found in any standard real analysis text (see, e.g., [Rud71]) whose proof will be omitted here. $\square$

Readers are reminded of the notation of these two spaces as follows. For $0 < r < \infty$, let $\mathbf{C} \equiv C([-r,0]; \Re^n)$ be the space of continuous functions $\phi : [-r,0] \to \Re^n$ equipped with the uniform topology, that is,

$$\|\phi\|_{\mathbf{C}} = \sup_{\theta \in [-r,0]} |\phi(\theta)|, \quad \phi \in \mathbf{C}.$$

To avoid the frequent usage of the cumbersome subscript of the sup-norm $\|\cdot\|_{\mathbf{C}}$ throughout this monograph, we simply denote $\|\cdot\|$ as $\|\cdot\|_{\mathbf{C}}$, that is,

$$\|\phi\| = \sup_{\theta \in [-r,0]} |\phi(\theta)|, \quad \phi \in \mathbf{C}.$$

Similarly, we denote $\|\cdot\|^* = \|\cdot\|_{\mathbf{C}}^*$ and $\|\cdot\|^\dagger = \|\cdot\|_{\mathbf{C}}^\dagger$ as the operator norms in $\mathbf{C}^*$ and $\mathbf{C}^\dagger$, respectively.

Let $BV([a,b]; \Re^n)$ be the space of functions $\phi : [a,b] \to \Re^n$ that are of bounded variation on the interval $[a,b]$.

In the following, we will study $\mathbf{C}^*$ and $\mathbf{C}^\dagger$ in a little bit more detail than for the general Ξ^* and $\Xi^\dagger$.

We first note that $\mathbf{C}^*$ has the following representation (see, e.g., Dunford and Schwartz [DS58]).

Theorem 2.2.2 $\Phi \in \mathbf{C}^*$ *if and only if there exists a unique regular finite measure* $\eta : [-r,0] \to \Re$ *such that*

$$\Phi(\phi) = \int_{-r}^{0} \phi(\theta) \cdot d\eta(\theta), \quad \forall \phi \in \mathbf{C},$$

where the above integral is to be interpreted as a Lebesque-Stieltjes integral.

Define the function $\mathbf{1}_{\{0\}} : [-r,0] \to \Re$ by

$$\mathbf{1}_{\{0\}}(\theta) = \begin{cases} 0 \text{ for } \theta \in [-r,0) \\ 1 \quad \text{for } \theta = 0. \end{cases}$$

Let $\mathbf{B}$ be the vector space of all simple functions of the form $v\mathbf{1}_{\{0\}}$, where $v \in \Re^n$. Clearly, $\mathbf{C} \cap \mathbf{B} = \{\mathbf{0}\}$, where $\mathbf{0}$ is the zero function in $\mathbf{C}$. We form the direct sum $\mathbf{C} \oplus \mathbf{B}$ and equip it with the norm also denoted by $\|\cdot\|_{\mathbf{C} \oplus \mathbf{B}}$ when there is no danger of ambiguity, where

$$\|\phi + v\mathbf{1}_{\{0\}}\|_{\mathbf{C} \oplus \mathbf{B}} = \|\phi\|_{\mathbf{C}} + |v|$$
$$= \sup_{\theta \in [-r,0]} |\phi(\theta)| + |v|, \quad \phi \in \mathbf{C}, v \in \Re^n.$$

Note that $\mathbf{C} \oplus \mathbf{B}$ is the space of all functions $\xi : [-r,0] \to \Re^n$ that are continuous on $[-r,0)$ and possibly with a *jump discontinuity* at 0 by the vector amount $v \in \Re^n$. The following two lemmas are due to Mohammed [Moh84].

Lemma 2.2.3 *Let $\Gamma \in \mathbf{C}^*$; that is, $\Gamma : \mathbf{C} \to \Re$ is a bounded linear functional on $\mathbf{C}$. Then Γ has a unique (continuous) linear extension $\bar{\Gamma} : \mathbf{C} \oplus \mathbf{B} \to \Re$ satisfying the following weak continuity property:*
(W1). If $\{\xi^{(k)}\}_{k=1}^{\infty}$ is a bounded sequence in $\mathbf{C}$ such that $\xi^{(k)}(\theta) \to \xi(\theta)$ as $k \to \infty$ for each $\theta \in [-r, 0]$ for some $\xi \in \mathbf{C} \oplus \mathbf{B}$, then $\Gamma(\xi^{(k)}) \to \bar{\Gamma}(\xi)$ as $k \to \infty$. Furthermore, the extension map $\mathbf{C}^* \to (\mathbf{C} \oplus \mathbf{B})^*$, $\Gamma \mapsto \bar{\Gamma}$ is a linear isometric injective map.

Proof. We prove the lemma first for $n = 1$. Suppose $\Gamma \in C^*[-r, 0]$. By Riesz Representation Theorem 2.2.2, there is a unique regular finite measure $\eta : \mathcal{B}([-r, 0]) \to \Re$ such that

$$\Gamma(\phi) = \int_{-r}^{0} \phi(\theta) \cdot d\eta(\theta), \quad \forall \phi \in \mathbf{C}.$$

Define $\bar{\Gamma} : C[-r, 0] \oplus \mathbf{B_1} \to \Re$ by

$$\bar{\Gamma}(\phi + v\mathbf{1}_{\{0\}}) = \Gamma(\phi) + v\eta(\{0\}), \quad \phi \in \mathbf{C}, v \in \Re,$$

where $\mathbf{B_1} = \{v\mathbf{1}_{\{0\}}, v \in \Re\}$. Note that $\bar{\Gamma}$ is clearly a continuous linear extension of Γ.

Let $\{\xi^{(k)}\}_{k=1}^{\infty}$ be a bounded sequence in $C[-r, 0]$ such that $\xi^{(k)}(\theta) \to \xi(\theta)$ as $k \to \infty$ for each $\theta \in [-r, 0]$ for some $\xi \equiv \phi + v\mathbf{1}_{\{0\}} \in C[-r, 0] \oplus \mathbf{B_1}$. By the dominated convergence theorem,

$$\begin{aligned}
\lim_{k \to \infty} \Gamma(\xi^{(k)}) &= \lim_{k \to \infty} \int_{-r}^{0} \xi^{(k)}(\theta) \, d\eta(\theta) \\
&= \int_{-r}^{0} \phi(\theta) \, d\eta(\theta) + \int_{-r}^{0} v\mathbf{1}_{\{0\}}(\theta) \, d\eta(\theta) \\
&= \Gamma(\phi) + v\eta(\{0\}) \\
&= \bar{\Gamma}(\phi + v\mathbf{1}_{\{0\}}).
\end{aligned}$$

The map $\Pi : \mathbf{C}^* \to (\mathbf{C} \oplus \mathbf{B})^*$, $\Gamma \mapsto \bar{\Gamma}$, is clearly linear.

Higher dimension $n > 1$ can be reduced to the one-dimensional situation as follows:

Write $\Gamma \in \mathbf{C}^*$ in the form

$$\Gamma(\phi) = \sum_{i=1}^{n} \Gamma^{(i)}(\phi_i),$$

where

$$\phi = (\phi_1, \phi_2, \dots, \phi_n) \in \mathbf{C}, \ \phi_i \in C[-r, 0], \ i = 1, 2, \dots, n,$$

and

$$\Gamma^{(i)}(\varphi) = \Gamma(0, \dots, 0, \varphi, 0, \dots, 0) \text{ with } \varphi \in C[-r, 0] \text{ occupying the } i\text{th place.}$$

Hence, $\Gamma^{(i)} \in C^*[-r, 0]$ for $i = 1, 2, \ldots, n$.

Write $\mathbf{B}$ as the n-fold Cartesian product of $\mathbf{B}_1$:

$$\mathbf{B} = \mathbf{B}_1 \times \ldots \times \mathbf{B}_1,$$

by taking

$$v\mathbf{1}_{\{0\}} = (v_1\mathbf{1}_{\{0\}}, v_2\mathbf{1}_{\{0\}}, \ldots, v_n\mathbf{1}_{\{0\}}), \quad \forall v = (v_1, v_2, \ldots, v_n) \in \Re^n.$$

Let $\bar{\Gamma}^{(i)} \in (C[-r, 0] \oplus \mathbf{B}_1)^*$ be the extension of $\Gamma^{(i)}$ described earlier and satisfying (W1). It is easy to verify that

$$\mathbf{C} \oplus \mathbf{B} = (C[-r, 0] \oplus \mathbf{B}_1) \times \cdots \times (C[-r, 0] \oplus \mathbf{B}_1);$$

that is,

$$\phi + v\mathbf{1}_{\{0\}} = (\phi_1 + v_1\mathbf{1}_{\{0\}}, \phi_2 + v_2\mathbf{1}_{\{0\}}, \ldots, \phi_n + v_n\mathbf{1}_{\{0\}}).$$

Define $\bar{\Gamma} \in (\mathbf{C} \oplus \mathbf{B})^*$ by

$$\bar{\Gamma}(\phi + v\mathbf{1}_{\{0\}}) = \sum_{i=1}^{n} \bar{\Gamma}^{(i)}(\phi_i + v_i\mathbf{1}_{\{0\}})$$

when $\phi = (\phi_1, \phi_2, \ldots, \phi_n)$, $v = (v_1, v_2, \ldots, v_n)$. Since $\bar{\Gamma}^{(i)}$ is a continuous linear extension of $\Gamma^{(i)}$, $\bar{\Gamma}$ is a continuous linear extension of Γ. Let $\{\xi^{(k)}\}_{k=1}^{\infty}$ be a bounded sequence in $\mathbf{C}$ such that $\xi^{(k)}(\theta) \to \xi(\theta)$ as $k \to \infty$ for each $\theta \in [-r, 0]$ for some $\xi \equiv \phi + v\mathbf{1}_{\{0\}} \in \mathbf{C} \oplus \mathbf{B}$. Let $\{\xi^{(k)}\} = (\xi_1^{(k)}, \xi_2^{(k)}, \ldots, \xi_n^{(k)})$, $\xi = (\xi_1, \xi_2, \ldots, \xi_n)$, $\xi_i^{(k)} \in C[-r, 0]$, $\xi_i \in C[-r, 0] \oplus \mathbf{B}_1$, $i = 1, 2, \ldots, n$. Hence $\{\xi_i^{(k)}\}_{k=1}^{\infty}$ is bounded in $C[-r, 0]$ and

$$\xi_i^{(k)}(\theta) \to \xi_i(\theta) \text{ as } k \to \infty, \ \theta \in [-r, 0], \ i = 1, 2, \ldots, n.$$

Therefore,

$$\lim_{k \to \infty} \Gamma(\xi^{(k)}) = \lim_{k \to \infty} \sum_{i=1}^{n} \Gamma^{(i)}(\xi_i^{(k)})$$

$$= \sum_{i=1}^{n} \bar{\Gamma}^{(i)}(\xi_i) = \bar{\Gamma}(\xi).$$

Therefore, $\bar{\Gamma}$ satisfies (W1).

To prove uniqueness, let $\tilde{\Gamma} \in (\mathbf{C} \oplus \mathbf{B})^*$ be any continuous extension of Γ satisfying (W1). For any $v\mathbf{1}_{\{0\}} \in \mathbf{B}$, choose a bounded sequence $\{\xi_0^{(k)}\}_{k=1}^{\infty}$ in $C[-r, 0]$ such that $\xi_0^{(k)}(\theta) \to v\mathbf{1}_{\{0\}}$ as $k \to \infty$, for all $\theta \in [-r, 0]$. For example, take

$$\xi_0^{(k)}(\theta) = \begin{cases} (k\theta + 1)v & \text{for } -\frac{1}{k} \le \theta \le 0 \\ 0 & \text{for } -r \le \theta < -\frac{1}{k}. \end{cases}$$

Note that $\|\xi_0^{(k)}\| = |v|$ for all $k \ge 1$. Also by (W1), one has

$$\begin{aligned} \tilde{\Gamma}(\phi + v\mathbf{1}_{\{0\}}) &= \tilde{\Gamma}(\phi) + \tilde{\Gamma}(v\mathbf{1}_{\{0\}}) \\ &= \Gamma(\phi) + \lim_{k \to \infty} \Gamma(\xi_0^{(k)}) \\ &= \Gamma(\phi) + \bar{\Gamma}(v\mathbf{1}_{\{0\}}) \\ &= \bar{\Gamma}(\phi + v\mathbf{1}_{\{0\}}), \quad \forall \phi \in \mathbf{C}. \end{aligned}$$

Thus, $\tilde{\Gamma} = \bar{\Gamma}$.

Define

$$\Pi : \mathbf{C}^* \to (\mathbf{C} \oplus \mathbf{B})^*, \quad \Gamma \mapsto \bar{\Gamma}.$$

Since the extension map Π is linear in the one-dimensional case, it follows from the representation of $\bar{\Gamma}$ in terms of $\bar{\Gamma}^{(i)}$ is also linear. However, $\bar{\Gamma}$ is an extension of Γ, so $\|\bar{\Gamma}\|^* \ge \|\Gamma\|^*$. Conversely, let $\xi = \phi + v\mathbf{1}_{\{0\}} \in \mathbf{C} \oplus \mathbf{B}$ and construct $\{\xi_0^{(k)}\}_{k=1}^{\infty}$ in $C[-r,0]$ as earlier. Then

$$\bar{\Gamma}(\xi) = \lim_{k \to \infty} \Gamma(\phi + \xi_0^{(k)}).$$

However,

$$\begin{aligned} |\Gamma(\phi + \xi_0^{(k)})| &\le \|\Gamma\|^* \|\phi + \xi_0^{(k)}\| \\ &\le \|\Gamma\|^* [\|\phi\| + \|\xi_0^{(k)}\|] \\ &= \|\Gamma\|^* [\|\phi\| + |v|] \\ &= \|\Gamma\|^* \|\xi\|, \quad \forall k \ge 1. \end{aligned}$$

Hence,

$$|\bar{\Gamma}(\xi)| = \lim_{k \to \infty} |\Gamma(\phi + \xi_0^{(k)})| \le \|\Gamma\|^* \|\xi\|, \quad \forall \xi \in \mathbf{C} \oplus \mathbf{B}.$$

Thus, $\|\bar{\Gamma}\|^* \le \|\Gamma\|^*$. So $\|\bar{\Gamma}\|^* = \|\Gamma\|^*$ and Π is an isometry map. □

Lemma 2.2.4 *Let $\Gamma \in \mathbf{C}^{\dagger}$; that is, $\Gamma : \mathbf{C} \times \mathbf{C} \to \Re$ is a bounded bilinear functional on $\mathbf{C}$. Then Γ has a unique (continuous) linear extension $\bar{\Gamma} : (\mathbf{C} \oplus \mathbf{B}) \times (\mathbf{C} \oplus \mathbf{B}) \to \Re$ satisfying the following weak continuity property: (W2). If $\{\xi^{(k)}\}_{k=1}^{\infty}$ and $\{\zeta^{(k)}\}_{k=1}^{\infty}$ are bounded sequences in $\mathbf{C}$ such that $\xi^{(k)}(\theta) \to \xi(\theta)$ and $\zeta^{(k)}(\theta) \to \zeta(\theta)$ as $k \to \infty$ for every $\theta \in [-r,0]$ for some $\xi, \zeta \in \mathbf{C} \oplus \mathbf{B}$, then $\Gamma(\xi^{(k)}, \zeta^{(k)}) \to \bar{\Gamma}(\xi, \zeta)$ as $k \to \infty$.*

Proof. Here, we also deal first with the one-dimensional case. Write the continuous bilinear map $\Gamma : C[-r,0] \times C[-r,0] \to \Re$ as a continuous linear map $\Gamma : C[-r,0] \to C^*[-r,0]$. Since $C^*[-r,0]$ is weakly complete (see §IV. 13.22 of Dunford and Schwartz [DS58]). $\Gamma : C[-r,0] \to C^*[-r,0]$ is weakly compact (see Theorem (I.4.2) of [DS58]). Hence, there is a unique measure

$\eta : \mathcal{B}([-r,0]) \rightarrow C^*[-r,0]$ (of finite semivariation $\|\eta\|([-r,0]) < \infty$) such that for all $\phi \in C[-r,0]$,

$$\Gamma(\phi) = \int_{-r}^{0} \phi(\theta) \, d\eta(\theta) \text{ (see Theorem (I.4.1) of [DS58])}.$$

Using an argument similar to that used in the proof of Lemma 2.2.3, the above integral representation of Γ implies the existence of a unique continuous linear extension

$$\hat{\Gamma} : C[-r,0] \oplus \mathbf{B}_1 \rightarrow C^*[-r,0]$$

satisfying (W1). To prove this, one needs the dominated convergence for vector-valued measures (see §IV.10.10, Theorem (I.3.1)(iv) of [DS58]).

Define $\bar{\Gamma} : C[-r,0] \oplus \mathbf{B}_1 \rightarrow (C[-r,0] \oplus \mathbf{B}_1)^*$ by $\bar{\Gamma} = \Pi \circ \Gamma$, where Π is the extension isometry of Lemma 2.2.3:

$$\bar{\Gamma}(\phi + v\mathbf{1}_{\{0\}}) = \overline{\hat{\Gamma}(\phi + v\mathbf{1}_{\{0\}})}, \quad \phi \in C[-r,0], v \in \Re.$$

Clearly, $\bar{\Gamma}$ gives a continuous bilinear extension of Γ to $(C[-r,0] \oplus \mathbf{B}_1) \times (C[-r,0] \oplus \mathbf{B}_1)$. To prove that $\bar{\Gamma}$ satisfies (W2), let $\{\xi^{(k)}\}_{k=1}^{\infty}$ and $\{\zeta^{(k)}\}_{k=1}^{\infty}$ be bounded sequences in $C[-r,0]$ such that $\xi^{(k)}(\theta) \rightarrow \xi(\theta)$ and $\zeta^{(k)}(\theta) \rightarrow \zeta(\theta)$ as $k \rightarrow \infty$ for every $\theta \in [-r,0]$ for some $\xi, \zeta \in C[-r,0] \oplus \mathbf{B}_1$. By Lemma 2.2.3 for $\hat{\Gamma}$, we get $\hat{\Gamma}(\xi) = \lim_{k \rightarrow \infty} \Gamma(\xi^{(k)})$. Now, for any $k = 1, 2, \ldots$,

$$|\Gamma(\xi^{(k)})(\zeta^{(k)}) - \bar{\Gamma}(\xi)(\zeta)| \leq |\Gamma(\xi^{(k)})(\zeta^{(k)}) - \hat{\Gamma}(\xi)(\zeta^{(k)})| + |\hat{\Gamma}(\zeta)(\zeta^{(k)}) - \overline{\hat{\Gamma}(\xi)}(\zeta)|$$

$$\leq \|\Gamma(\xi^{(k)}) - \hat{\Gamma}(\xi)\|^* \|\zeta^{(k)}\| + |\hat{\Gamma}(\xi)(\zeta^{(k)}) - \overline{\hat{\Gamma}(\xi)}(\zeta)|.$$

However, by Lemma 2.2.3 for $\hat{\Gamma}(\xi)$ we have

$$\lim_{k \rightarrow \infty} |\hat{\Gamma}(\zeta)(\zeta^{(k)}) - \overline{\hat{\Gamma}(\xi)}(\zeta)| = 0.$$

Since $\{\|\zeta^{(K)}\|\}_{k=1}^{\infty}$ is bounded, it follows from the last inequality that

$$\lim_{k \rightarrow} |\Gamma(\xi^{(k)})(\zeta^{(k)}) - \bar{\Gamma}(\xi)(\zeta)| = 0.$$

When $n > 1$, we use coordinates as in Lemma 2.2.3 to reduce to the one-dimensional case. Indeed, write any continuous bilinear map $\Gamma : \mathbf{C} \times \mathbf{C} \rightarrow \Re$ as the sum of continuous bilinear maps $\mathbf{C} \times \mathbf{C} \rightarrow \Re$ in the following way:

$$\Gamma((\phi_1, \ldots, \phi_n)(\varphi_1, \ldots, \varphi_n)) = \sum_{i,j=1}^{n} \Gamma^{(ij)}(\phi_i, \varphi_j),$$

where

$$(\phi_1, \cdots, \phi_n), (\varphi_1, \cdots, \varphi_n) \in \mathbf{C}, \ \phi_i, \varphi_i \in C[-r,0], \ i = 1, 2, \cdots, n,$$

and $\Gamma^{(ij)} : C[-r,0] \times C[-r,0] \to \Re$ is the continuous bilinear map defined by

$$\Gamma^{(ij)}(\varsigma, \gamma) = \Gamma((0,\ldots,0,\varsigma,0,\ldots,0),(0,\ldots,0,\gamma,0,\ldots,0))$$

for $\varsigma, \gamma \in C[-r,0]$ occupying the ith and jth places, respectively, $1 \le i, j \le n$.

Now, extend each $\Gamma^{(ij)}$ continuously to a bilinear map $\bar{\Gamma}^{(ij)} : (C[-r,0] \times \mathbf{B}_1) \times (C[-r,0] \times \mathbf{B}_1) \to \Re$ satisfying (W2). Then define

$$\bar{\Gamma} : (\mathbf{C} \oplus \mathbf{B}) \times (\mathbf{C} \oplus \mathbf{B}) \to \Re$$

by

$$\bar{\Gamma}(\xi, \zeta) = \sum_{i,j=1}^{n} \bar{\Gamma}^{(ij)}(\xi_i, \zeta_j),$$

where

$$\xi = (\xi_1, \ldots, \xi_n), \zeta = (\zeta_1, \ldots, \zeta_n) \in \mathbf{C} \oplus \mathbf{B}, \ \xi_i, \zeta_i \in C[-r,0] \oplus \mathbf{B}_1, 1 \le i, j \le n.$$

It is then easy to see that $\bar{\Gamma}$ is a continuous bilinear extension of Γ satisfying (W2).

Finally, we prove uniqueness. Let $\tilde{\Gamma} : \mathbf{C} \oplus \mathbf{B} \to (\mathbf{C} \oplus \mathbf{B})^*$ be any continuous bilinear extension of Γ satisfying (W2). Take

$$\xi = \phi + v\mathbf{1}_{\{0\}}, \zeta = \varphi + w\mathbf{1}_{\{0\}} \in \mathbf{C} \oplus \mathbf{B}, \ \text{where } \phi, \varphi \in \mathbf{C}, v, w \in \Re^n.$$

Choose bounded sequences $\{\xi_0^{(k)}\}_{k=1}^{\infty}, \{\zeta_0^{(k)}\}_{k=1}^{\infty}$ in $C[-r,0]$ such that

$$\xi_0^{(k)}(\theta) \to v\mathbf{1}_{\{0\}}(\theta), \zeta_0^{(k)}(\theta) \to w\mathbf{1}_{\{0\}}(\theta) \text{ as } k \to \infty, \theta \in [-r,0];$$

$$\|\xi_0^{(k)}\| = |v|, \|\zeta_0^{(k)}\| = |w|, \quad \forall k \ge 1.$$

Let

$$\xi^{(k)} = \phi + \xi_0^{(k)} \text{ and } \zeta^{(k)} = \varphi + \zeta_0^{(k)}.$$

Then $\{\xi^{(k)}\}_{k=1}^{\infty}$ and $\{\zeta^{(k)}\}_{k=1}^{\infty}$ are bounded sequences in $\mathbf{C}$ such that

$$\xi^{(k)}(\theta) \to \xi(\theta) \text{ and } \zeta^{(k)}(\theta) \to \zeta(\theta) \text{ as } k \to \infty \ \forall \theta \in [-r,0].$$

Therefore, by (W2) for $\tilde{\Gamma}$ and $\bar{\Gamma}$, one gets

$$\tilde{\Gamma}(\xi)(\zeta) = \lim_{k \to \infty} \Gamma(\xi^{(k)})(\zeta^{(k)})\bar{\Gamma}(\xi)(\zeta).$$

This proves that $\tilde{\Gamma} = \bar{\Gamma}$. □

We first define for each $\phi \in \mathbf{C}$ and for all $0 < a \le \infty$ the function $\tilde{\phi} : [-r,\infty) \to \Re^n$, an extension of ϕ from the interval $[-r,0]$ to the interval $[-r,a]$, by

$$\tilde{\phi}(t) = \begin{cases} \phi(0) & \text{for } t \geq 0 \\ \phi(t) & \text{for } t \in [-r, 0). \end{cases} \tag{2.8}$$

Again using the convention, we define, for each $t \geq 0$, $\tilde{\phi}_t : [-r, 0] \to \Re^n$ by

$$\tilde{\phi}_t(\theta) = \tilde{\phi}(t + \theta), \quad \theta \in [-r, 0],$$

and

$$\tilde{T}(t)\phi = \tilde{\phi}_t, \quad t \geq 0. \tag{2.9}$$

Theorem 2.2.5 *The family of bounded linear operators $\{\tilde{T}(t), t \geq 0\}$ defined in (2.9) form a C_0-semigroup of operators on $\mathbf{C}$.*

Proof. It is clear that for each $t \geq 0$, $\tilde{T}(t) : \mathbf{C} \to \mathbf{C}$ defined in (2.9) is a linear operator. To show that it is bounded, we note that $\tilde{T}(t)\phi(\theta) = \phi(0)$ for all $\theta \in [-r, 0]$ with $t + \theta \geq 0$ and $\tilde{T}(t)\phi(\theta) = \phi(t + \theta)$ for all $\theta \in [-r, 0]$ with $t + \theta \in [-r, 0)$. Therefore, for all $t \geq 0$,

$$\begin{aligned} \|\tilde{T}(t)\phi\| &= \sup_{\theta \in [-r, 0]} |\tilde{T}(t)\phi(\theta)| \\ &\leq \sup_{\theta \in [-r, 0]} |\phi(\theta)| = \|\phi\|, \quad \forall \phi \in \mathbf{C}. \end{aligned}$$

We then show that the family $\{\tilde{T}(t), t \geq 0\}$ satisfies Conditions (i)-(iii) of Definition 2.1.2.

It is clear that $\tilde{T}(0)\phi = \phi$ for all $\phi \in \mathbf{C}$, i.e., $\tilde{T}(0) = \mathbf{I}$ (the identity operator). Now, for $s, t \geq 0$, $\phi \in \mathbf{C}$, and $\theta \in [-r, 0]$, we have

$$\begin{aligned} \tilde{T}(s) \circ \tilde{T}(t)\phi(\theta) &= \tilde{T}(s)(\tilde{\phi}_t)(\theta) \\ &= \begin{cases} (\tilde{\tilde{\phi}}_s)(0) = \phi(0) & \text{for } t + \theta \geq 0 \\ (\tilde{\tilde{\phi}}_s)(t + \theta) & \text{for } -r \leq t + \theta < 0. \end{cases} \\ &= \begin{cases} \phi(0) & \text{for } t + \theta \geq 0 \\ \tilde{\phi}_s(0) = \phi(0) & \text{for } t + s + \theta \geq 0 \\ \tilde{\phi}_s(t + \theta) = \phi(t + s + \theta) & \text{for } -r \leq t + s + \theta < 0. \end{cases} \\ &= (\tilde{\phi}_{t+s})(\theta); \end{aligned}$$

that is, $\tilde{T}(s) \circ \tilde{T}(t) = \tilde{T}(t) \circ \tilde{T}(s) = \tilde{T}(t + s)$.

Let $\phi \in \mathbf{C}$ be given. For each $a > 0$, the function $\tilde{\phi} : [-r, a] \to \Re^n$ as defined in (2.8) is uniformly continuous. Therefore, for each $\epsilon > 0$, there exists a $\delta > 0$ such that $0 < t < \delta$ implies that

$$|\tilde{\phi}(t + \theta) - \tilde{\phi}(\theta)| < \epsilon, \quad \forall \theta \in [-r, 0].$$

Therefore, $\|\tilde{T}(t)\phi - \phi\| < \epsilon$ whenever $0 < t < \delta$. This shows that

$$\lim_{t \downarrow 0} \|\tilde{T}(t)\phi - \phi\| = 0.$$

This proves that the family of bounded linear operators $\{\tilde{T}(t), t \geq 0\}$ forms a C_0-semigroup of operators on **C**.   □

The infinitesimal generator $\tilde{A}$ and its domain $\mathcal{D}(\tilde{A})$ for the semigroup $\{\tilde{T}(t), t \geq 0\}$ is given as follows.

Theorem 2.2.6 *The domain of the infinitesimal generator $\tilde{A}$, $\mathcal{D}(\tilde{A})$, of $\{\tilde{T}(t), t \geq 0\}$ is given by*

$$\mathcal{D}(\tilde{A}) = \{\phi \mid \dot{\phi} \in \mathbf{C}, \dot{\phi}(0) = 0\},$$

and $\tilde{A}\phi = \dot{\phi}$.

Proof. Let $\phi \in \mathcal{D}(\tilde{A})$ and put $\varphi = \tilde{A}(\phi)$. From the definition of $\tilde{T}(t)$ and

$$0 = \lim_{t \downarrow 0} \left\| \frac{1}{t}(\tilde{T}(t)\phi - \phi) - \varphi \right\|$$

$$= \lim_{t \downarrow 0} \sup_{\theta \in [-r, 0]} \left| \frac{1}{t}\left((\tilde{T}(t)\phi(\theta) - \phi(\theta)) - \varphi(\theta)\right)\right|,$$

it follows that ϕ is right-differentiable on $[-r, 0)$ with $\dot{\phi}(\theta+) = \varphi(\theta)$ and, moreover, that necessarily $\varphi(0) = 0$. The continuity of φ implies that actually ϕ is differentiable and $\dot{\phi} = \varphi$.

Conversely, suppose that ϕ is continuously differentiable on $[-r, 0]$ and $\dot{\phi}(0) = 0$. Define $\phi(t) = \phi(0)$ for $t \geq 0$; then

$$\left|\frac{1}{t}\left(\phi(t + \theta) - \phi(\theta)\right) - \dot{\phi}\right| = \left|\frac{1}{t}\int_0^t \left(\dot{\phi}(\theta + s) - \dot{\phi}(\theta)\right)ds\right|$$

converges, as $t \downarrow 0$, to zero uniformly for $\theta \in [-r, 0]$.   □

Define the family of shift operators $\{T(t), t \geq 0\}$ on $C_b(\mathbf{C})$ by

$$T(t)\Psi(\phi) = \Psi(\tilde{T}(t)\phi), \quad \Psi \in C_b(\mathbf{C}), \ \phi \in \mathbf{C}. \tag{2.10}$$

The following result establishes that the family of shift operators $\{T(t), t \geq 0\}$ forms a contractive C_0-semigroup on $C_b(\mathbf{C})$.

Proposition 2.2.7 *The family of shift operators $\{T(t), t \geq 0\}$ forms a contractive C_0-semigroup on $C_b(\mathbf{C})$ such that for each $\phi \in \mathbf{C}$,*

$$\lim_{t \to 0+} T(t)(\Phi)(\phi) = \Phi(\phi), \quad \forall \Phi \in C_b(\mathbf{C}).$$

Proof. Let $s, t \geq 0, \phi \in \mathbf{C}, \Phi \in C_b(\mathbf{C})$, and $\theta \in [-r, 0]$. Then from the definition of $\{T(t), t \geq 0\}$ and Theorem 2.2.5, we have

$$T(t)(T(s)(\Phi))(\phi) = T(t)\Phi(\tilde{T}(s)\phi) = \Phi(\tilde{T}(t) \circ \tilde{T}(s)\phi) = \Phi(\tilde{T}(t + s)\phi).$$

Hence,

$$T(t)(T(s)(\Psi))(\phi) = T(t+s)(\Psi)(\phi), \quad \forall s, t \geq 0, \ \phi \in \mathbf{C}.$$

Therefore, $T(t) \circ T(s) = T(t+s)$.

Since $\lim_{t \downarrow 0} \tilde{T}(\phi) = \phi$, it is clear that $\lim_{t \downarrow 0} T(t)(\Phi)(\phi) = \Phi(\phi)$ for each $\phi \in \mathbf{C}, \Phi \in C_b(\mathbf{C})$.

The fact that $T(t) : C_b(\mathbf{C}) \to C_b(\mathbf{C})$ is a contractive map for each $t \geq 0$ can be easily proved and is therefore omitted here.   □

In the following, we define the infinitesimal generator $\mathcal{S} : C_b(\mathbf{C}) \to C_b(\mathbf{C})$ of the C_0-semigroup $\{T(t), t \geq 0\}$ by

$$(\mathcal{S}\Phi)(\phi) = \lim_{t \downarrow 0} \frac{T(t)\Phi(\phi) - \Phi(\phi)}{t} \tag{2.11}$$

$$= \lim_{t \downarrow 0} \frac{\Phi(\tilde{T}(t)\phi) - \Phi(\phi)}{t}$$

$$= \lim_{t \downarrow 0} \frac{\Phi(\tilde{\phi}_t) - \Phi(\phi)}{t}, \quad \phi \in \mathbf{C}.$$

Let $\hat{\mathcal{D}}(\mathcal{S})$ be the set of all $\Phi \in C_b(\mathbf{C})$ such that the above limit exists for each $\phi \in \mathbf{C}$.

Remark 2.2.8 *The Dynkin's formula (see Theorem 2.4.1) for any functional $\Phi : \mathbf{C} \to \Re$ for the $\mathbf{C}$-valued Markovian solution process $\{x_s, s \in [0,T]\}$ for (1.27) requires that $\Phi \in C_{lip}^2(\mathbf{C})$. It is important to point out here that this condition, although smooth enough, is not sufficient to ensure that $\Phi \in \hat{\mathcal{D}}(\mathcal{S})$. To illustrate this, we consider only the case where $n = 1$ and choose $\Phi : C[-r, 0] \to \Re$ such that $\Phi(\phi) = \phi(-\bar{r})$ for any $\phi \in C[-r, 0]$, where $\bar{r} \in (0, r)$. Then*

$$D\Phi(\phi)(\varphi) = \varphi(-\bar{r}),$$

$$D^2\Phi(\phi)(\psi, \varphi) = 0,$$

and $D^2\Phi : C[-r, 0] \to C^{\dagger}[-r, 0]$ is globally Lipschitz (i.e., $\Phi \in C_{lip}^2(C[-r, 0])$). However,

$$\mathcal{S}(\Phi)(\phi) = \lim_{t \downarrow 0} \frac{1}{t} \left[\Phi(\tilde{\phi}_t) - \Phi(\phi) \right] = \lim_{t \downarrow 0} \frac{\tilde{\phi}(t - \bar{r}) - \phi(-\bar{r})}{t},$$

which exists only if ϕ has a right derivative at $-\bar{r}$.

Lemma 2.2.9 *For the Fréchet differentiable $\Phi \in C_b(\mathbf{C})$, $\Phi : D(\mathcal{A}) \to \Re$, the infinitesimal generator $\mathcal{S} : \hat{\mathcal{D}}(\mathcal{S}) \subset C_b(\mathbf{C}) \to C_b(\mathbf{C})$ for the contractive C_0-semigroup $\{T(t), t \geq 0\}$ is given by*

$$\mathcal{S}(\Phi)(\phi) = D\Phi(\phi)(\tilde{A}(\phi)).$$

Proof. The proof is straightforward and given as follows:

$$(\mathcal{S}\Phi)(\phi) = \lim_{t\downarrow 0} \frac{\Phi(\tilde{\phi}_t) - \Phi(\phi)}{t}$$

$$= \lim_{t\downarrow 0} \frac{1}{t}\left[\Phi(\phi + \tilde{\phi}_t - \phi) - \Phi(\phi)\right]$$

$$= \lim_{t\downarrow 0} \frac{1}{t}\left[D\Phi(\phi)(\tilde{\phi}_t - \phi) + o(\tilde{\phi}_t - \phi)\right]$$

$$= \lim_{t\downarrow 0} \left[D\Phi(\phi)\left(\frac{\tilde{\phi}_t - \phi}{t}\right) + \frac{o(\tilde{\phi}_t - \phi)}{\|\tilde{\phi}_t - \phi\|}\left\|\frac{\tilde{\phi}_t - \phi}{t}\right\|\right]$$

$$= D\Phi(\phi)\left(\lim_{t\downarrow 0} \frac{\tilde{\phi}_t - \phi}{t}\right) + \lim_{t\downarrow 0} \frac{o(\tilde{\phi}_t - \phi)}{\|\tilde{\phi}_t - \phi\|}\left\|\frac{\tilde{\phi}_t - \phi}{t}\right\|$$

$$= D\Phi(\phi)(\tilde{A}(\phi)). \qquad \square$$

Example. As an example of a function $\Phi : \mathbf{C} \to \Re$ where $\Psi \in \hat{\mathcal{D}}(\mathcal{S})$, let

$$\Psi(\phi) = a\phi^2(0) + b\int_{\theta=-r}^{0} \phi^2(\theta)\, d\theta,$$

where a and b are real-valued constants. Then $\Phi \in \hat{\mathcal{D}}(\mathcal{S})$, with

$$(\mathcal{S}\Psi)(\phi) = \lim_{t\downarrow 0} \frac{\Phi(\tilde{\phi}_t) - \Phi(\phi)}{t}$$

$$= b\left(\phi^2(0) - \phi^2(-r)\right).$$

More examples of $(\mathcal{S}\Phi)(\phi)$ can be found in Section 3.6 of Chapter 3, when we prove that the value function is a unique viscosity solution of the HJBE.

2.3 The Space M

Recall that $\mathbf{M} \equiv \Re \times L_\rho^2((-\infty, 0]; \Re)$ is the ρ-weighted Hilbert space equipped with the Hilbertian inner product $\langle \cdot, \cdot \rangle_\rho : \mathbf{M} \times \mathbf{M} \to \Re$ defined by

$$\langle (x, \phi), (y, \varphi) \rangle_{\mathbf{M}} = xy + \int_{-\infty}^{0} \phi(\theta)\varphi(\theta)\rho(\theta)\, d\theta.$$

As usual, the Hilbertian norm $\|\cdot\|_{\mathbf{M}} : \mathbf{M} \to [0, \infty)$ is defined by

$$\|(x, \phi)\|_{\mathbf{M}} = \sqrt{\langle (x, \phi), (x, \phi) \rangle_{\mathbf{M}}}.$$

Remark 2.3.1 *Since we are working with* $\mathbf{M}$ *in which two functions in* $L_\rho^2((-\infty, 0]; \Re)$ *are equal if there differ only on a set of Lebesque measure zero, we therefore can and will assign* $\phi(0) = x$ *for any* $x \in \Re$ *and write a typical element in* $\mathbf{M}$ *as* $(\phi(0), \phi)$ *instead of* (x, ϕ).

In the following, we consider the differential calculus outlined in Section 2.2 for the space $\varXi = \mathbf{M}$.

First, for the benefit of the readers who are not familiar with the theory of infinite-dimensional Hilbert space, we note that $\mathbf{M}^*$, the space of bounded linear functionals on $\mathbf{M}$, can be identified with $\mathbf{M}$ by the Riesz Representation Theorem restated here.

Theorem 2.3.2 $\varPhi \in \mathbf{M}^*$ *if and only if there exists a unique* $(\varphi(0), \varphi) \in \mathbf{M}$ *such that*

$$\varPhi(\phi(0), \phi) = \langle (\varphi(0), \varphi), (\phi(0), \phi) \rangle_{\mathbf{M}}$$
$$\equiv \varphi(0)\phi(0) + \int_{-\infty}^{0} \varphi(\theta)\phi(\theta)\rho(\theta)\, d\theta, \quad \forall (\phi(0), \phi) \in \mathbf{M}.$$

Second, if $\varPhi \in C^2(\mathbf{M})$, then the actions of the first-order Fréchet derivative $D\varPhi(\phi(0), \phi)$ and the second-order Fréchet $D^2\varPhi(\phi(0), \phi)$ can be expressed as

$$D\varPhi(\phi(0), \phi)(\varphi(0), \varphi) = \varphi(0)\partial_x\varPhi(\phi(0), \phi) + D_\phi\varPhi(\phi(0), \phi)\varphi$$

and

$$D^2\varPhi(\phi(0), \phi)((\varphi(0), \varphi), (\varsigma(0), \varsigma))$$
$$= \varphi(0)\partial_x^2\varPhi(\phi(0), \phi)\varsigma(0) + \varsigma(0)\partial_x D_\phi\varPhi(\phi(0), \phi)\varphi$$
$$+ \varphi(0)\partial_x D_\phi\varPhi(\phi(0), \phi)\varsigma + D_\phi^2\varPhi(\phi(0), \phi)(\varphi, \varsigma),$$

where $\partial_x\varPhi$ and $\partial_x^2\varPhi$ are respectively the first- and second-order partial derivative of $\varPhi$ with respect to its first variable $x \in \Re$, $D_\phi\varPhi$ and $D_\phi^2\varPhi$ are respectively the first- and second-order Fréchet derivatives with respect to its second variable $\phi \in L_\rho^2((-\infty, 0]; \Re))$, and $\partial_x D_\phi\varPhi$ is the second-order derivative first with respect to $\phi \in L_\rho^2((-\infty, 0]; \Re)$ in the Fréchet sense and then with respect to its first variable $x \in \Re$.

2.3.1 The Weighting Function ρ

Throughout the end of this monograph, let $\rho : (-\infty, 0] \to [0, \infty)$ be an *influence function with relaxation property* and it satisfies Assumption 1.4.1 in Chapter 1, which is repeated here for convenience.

Assumption 2.3.3 *1. ρ is summable on $(-\infty, 0]$, that is,*

$$0 < \int_{-\infty}^{0} \rho(\theta)\, d\theta < \infty.$$

2. For every $\lambda \leq 0$, one has

$$\bar{K}(\lambda) = ess \sup_{\theta \in (-\infty, 0]} \frac{\rho(\theta + \lambda)}{\rho(\theta)} \leq \bar{K} < \infty,$$

$$\underline{K}(\lambda) = ess \sup_{\theta \in (-\infty, 0]} \frac{\rho(\theta)}{\rho(\theta + \lambda)} < \infty.$$

Proposition 2.3.4 *If $\rho : (-\infty, 0] \to [0, \infty)$ satisfies Assumption 2.3.3, then the following conditions are satisfied:*
1. *ρ is essentially bounded on $(-\infty, 0]$.*
2. *ρ is essentially strictly positive on $(-\infty, 0)$.*
3. *$\theta\rho(\theta) \to 0$ as $\theta \to -\infty$.*

Proof. See Coleman and Mizel [CM68].   □

Example 2.3.5 *The following functions $\rho : (-\infty, 0] \to [0, \infty)$ satisfy Assumption 2.3.3:*
(i) Let $\rho(\theta) = e^\theta$. Then

$$0 < \int_{-\infty}^0 \rho(\theta)\, d\theta = \int_{-\infty}^0 e^\theta\, d\theta = 1$$

$$\bar{K}(\lambda) = ess \sup_{\theta \in (-\infty, 0]} \frac{e^{\theta+\lambda}}{e^\theta} = e^\lambda \le 1, \quad \forall \lambda \le 0,$$

and

$$\underline{K}(\lambda) = ess \sup_{\theta \in (-\infty, 0]} \frac{e^\theta}{e^{\theta+\lambda}} = e^{-\lambda} < \infty, \quad \forall \lambda \le 0.$$

(ii) Let $\rho(\theta) = \frac{1}{1+\theta^2}$, $-\infty < \theta \le 0$. Then

$$0 < \int_{-\infty}^0 \rho(\theta)\, d\theta = \int_{-\infty}^0 \frac{1}{1+\theta^2}\, d\theta = \arctan(0) - \arctan(-\infty) = \pi/2 < \infty,$$

$$\bar{K}(\lambda) = ess \sup_{\theta \in (-\infty, 0]} \frac{1+\theta^2}{1+(\theta+\lambda)^2} \le 1,$$

and

$$\underline{K}(\lambda) = ess \sup_{\theta \in (-\infty, 0]} \frac{1+(\theta+\lambda)^2}{1+\theta^2}.$$

Since the function

$$\gamma(\theta) = \frac{1+(\theta+\lambda)^2}{1+\theta^2}$$

achieves its maximum at $\theta = \frac{-\lambda - \sqrt{\lambda^2+4}}{2}$, we have

$$\underline{K}(\lambda) = \frac{4 + \lambda(\lambda - \sqrt{\lambda^2+4})}{4 + \lambda(\lambda + \sqrt{\lambda^2+4})} < \infty.$$

Therefore, both examples satisfy Assumption 2.3.3.

Note that the influence function $\rho : (-\infty, 0] \to [0, \infty)$ plays the role of a weighting function for the Hilbert space $L^2_\rho((-\infty, 0]; \Re)$ (and hence the space **M**).

2.3.2 The $\mathcal{S}$-Operator

In the following, we develop the family of shift operators for the space $\mathbf{M}$ that are parallel to those for the space $\mathbf{C}$ in the previous section. The details are given only if they are significantly different from those for the space $\mathbf{C}$.

For each $\phi \in L^2_\rho((-\infty, 0]; \Re)$, define $\tilde{\phi} : (-\infty, \infty) \to \Re$ by

$$\tilde{\phi}(t) = \begin{cases} \phi(0) \text{ for } t \in [0, \infty) \\ \phi(t) \ t \in (-\infty, 0). \end{cases}$$

Then for each $\theta \in (-\infty, 0]$ and $t \geq 0$,

$$\tilde{\phi}_t(\theta) = \tilde{\phi}(t + \theta)$$

$$= \begin{cases} \phi(0) & \text{for } t + \theta \geq 0 \\ \phi(t + \theta) & \text{for } t + \theta < 0. \end{cases}$$

Consider the family of shift operators $\{\tilde{T}(t), t \geq 0\}$, $\tilde{T}(t) : \mathbf{M} \to \mathbf{M}$ defined by

$$\tilde{T}(t)(\phi(0), \phi) = (\phi(0), \tilde{\phi}_t), \quad t \geq 0, (\phi(0), \phi) \in \mathbf{M}. \tag{2.12}$$

Referring to Definition 2.1.2 for a C_0-semigroup of bounded linear operators, we prove the following result for the family of shift operators $\{\tilde{T}(t), t \geq 0\}$ on the space $\mathbf{M}$.

Theorem 2.3.6 *The family $\{\tilde{T}(t), t \geq 0\}$ defined in (2.12) forms a C_0-semigroup of bounded linear operators on* $\mathbf{M}$.

Proof. It is clear that for each $t \geq 0$, the map $\tilde{T}(t) : \mathbf{M} \to \mathbf{M}$ defined in (2.12) is a linear operator. To show that it is bounded, we observe from Assumption 2.3.3 and (2.12) that

$$\|\tilde{T}(t)(\phi(0), \phi)\|^2_M = \|(\phi(0), \tilde{\phi}_t)\|^2_M$$

$$= |\phi(0)|^2 + \int_{-\infty}^0 |\tilde{\phi}_t(\theta)|^2 \rho(\theta) \, d\theta$$

$$= |\phi(0)|^2 + \int_{-\infty}^0 |\tilde{\phi}(t + \theta)|^2 \rho(\theta) \, d\theta$$

$$= |\phi(0)|^2 + \int_{-\infty}^{-t} |\phi(t + \theta)|^2 \rho(\theta) \, d\theta + \int_{-t}^0 |\phi(0)|^2 \rho(\theta) d\theta$$

$$\leq |\phi(0)|^2 + \int_{-\infty}^{-t} |\phi(t + \theta)|^2 \rho(t + \theta) \frac{\rho(\theta)}{\rho(t + \theta)} \, d\theta$$

$$+ |\phi(0)|^2 \int_{-\infty}^0 \rho(\theta) \, d\theta$$

Therefore,

$$\|\tilde{T}(t)(\phi(0), \phi)\|_M^2$$

$$\leq |\phi(0)|^2 + \bar{K} \int_{-\infty}^{-t} |\phi(t+\theta)|^2 \rho(t+\theta)\, d\theta + |\phi(0)|^2 \|\rho\|_1\, d\theta$$

$$\leq |\phi(0)|^2 + \bar{K} \int_{-\infty}^{0} |\phi(\theta)|^2 \rho(\theta)\, d\theta + |\phi(0)|^2 \|\rho\|_1$$

$$\leq K\left(|\phi(0)|^2 + \int_{-\infty}^{0} |\phi(\theta)|^2 \rho(\theta)\, d\theta\right)$$

$$= K\|(\phi(0), \phi)\|_\rho^2,$$

where

$$\|\rho\|_1 = \int_{-\infty}^{0} \rho(\theta)\, d\theta,$$

$\bar{K}$ is the positive constant in Assumption 2.3.3, and

$$K = \max\{1 + \|\rho\|_1, \bar{K}\} < \infty.$$

This shows that for each $t \geq 0$, $\tilde{T}(t) : \mathbf{M} \to \mathbf{M}$ is a bounded operator.

We then show that the family $\{\tilde{T}(t), t \geq 0\}$ satisfies Conditions (i)-(iii) of Definition 2.1.2. It is clear that $\tilde{T}(0)(\phi(0), \phi) = (\phi(0), \phi)$ for all $(\phi(0), \phi) \in \mathbf{M}$ i.e., $\tilde{T}(0) = I$ (the identity operator on $\mathbf{M}$). Now, for $s, t \geq 0$, $(\phi(0), \phi) \in \mathbf{M}$, and $\theta \in (-\infty, 0]$ we have

$$\tilde{T}(s) \circ \tilde{T}(t)(\phi(0), \phi(\theta))$$

$$= \tilde{T}(s)(\phi(0), \tilde{\phi}_t(\theta))$$

$$= \begin{cases} (\phi(0), \phi(0)) & \text{for } t + \theta \geq 0 \\ (0, \tilde{\phi}_s(t+\theta)) & \text{for } -\infty < t + \theta < 0. \end{cases}$$

$$= \begin{cases} (\phi(0), \phi(0)) & \text{for } t + \theta \geq 0 \\ (\phi(0), \phi(0)) & \text{for } t + s + \theta \geq 0 \\ (\phi(0), \tilde{\phi}_s(t+\theta)) = (\phi(0), \phi(t+s+\theta)) & \text{for } -\infty < t + s + \theta < 0. \end{cases}$$

$$= (\phi(0), \tilde{\phi}_{t+s}(\theta))$$

$$= \tilde{T}(t+s)(\phi(0), \phi(\theta));$$

that is, $\tilde{T}(s) \circ \tilde{T}(t) = \tilde{T}(t+s)$.

Let $(\phi(0), \phi) \in \mathbf{M}$ be given.

$$\lim_{t \downarrow 0} \|\tilde{T}(t)(\phi(0), \phi) - (\phi(0), \phi)\|_M^2 = \lim_{t \downarrow 0} \|(\phi(0), \tilde{\phi}_t) - (\phi(0), \phi)\|_M^2$$

$$= \lim_{t \downarrow 0} \int_{-\infty}^{0} |\tilde{\phi}(t+\theta) - \phi(\theta)|^2 \rho(\theta)\, d\theta$$

$$= \lim_{t \downarrow 0} \int_{-\infty}^{-t} |\phi(t+\theta) - \phi(\theta)|^2 \rho(\theta)\, d\theta$$

$$+ \lim_{t \downarrow 0} \int_{-t}^{0} |\phi(0) - \phi(\theta)|^2 \rho(\theta)\, d\theta$$

$$= 0.$$

Therefore,

$$\lim_{t \downarrow 0} \|\tilde{T}(t)(\phi(0), \phi) - (\phi(0), \phi)\|_M = 0.$$

This proves that the family of bounded linear operators $\{\tilde{T}(t), t \geq 0\}$ forms a $\mathcal{C}_0$-semigroup of operators on $\mathbf{M}$. $\square$

The infinitesimal generator $\tilde{A}$ for the semigroup $\{\tilde{T}(t), t \geq 0\}$ is defined by

$$\tilde{A}(\phi(0), \phi) = \lim_{t \downarrow 0} \frac{\tilde{T}(t)(\phi(0), \phi) - (\phi(0), \phi)}{t} \qquad (2.13)$$

and its domain $\mathcal{D}(\tilde{A})$ for the semigroup $\{\tilde{T}(t), t \geq 0\}$ is given as follows.

Theorem 2.3.7 $\mathcal{D}(\tilde{A})$, *the domain of* $\tilde{A}$, *is given by*

$$\mathcal{D}(\tilde{A}) = \{(\phi(0), \phi) \in \mathbf{M} \mid (\phi(0), \dot{\phi}) \in \mathbf{M}, \ \dot{\phi}(0) = 0\},$$

and

$$\tilde{A}(\phi(0), \phi) = (0, \dot{\phi}), \quad (\phi(0), \phi) \in \mathcal{D}(\tilde{A}),$$

where $\dot{\phi} : (-\infty, 0] \to \Re$ *denotes the derivative of the function* $\phi : (-\infty, 0] \to \Re$.

Proof. Let $(\phi(0), \phi) \in \mathcal{D}(\tilde{A})$ and put $(\varphi(0), \varphi) = \tilde{A}(\phi(0), \phi)$. We will prove that $\varphi(0) = 0$ and $\varphi = \dot{\phi}$. From the definition of $\tilde{A}$ and working with $\|\cdot\|_M^2$ instead of $\|\cdot\|_M$ for convenience, we have

$$0 = \lim_{t \downarrow 0} \left\| \frac{\tilde{T}(t)(\phi(0), \phi) - (\phi(0), \phi)}{t} - (\varphi(0), \varphi) \right\|_{\mathbf{M}}^2$$

$$= \lim_{t \downarrow 0} \left\| \frac{(0, \tilde{\phi}_t - \phi)}{t} - (\varphi(0), \varphi) \right\|_{\mathbf{M}}^2.$$

Therefore,

$$0 = \lim_{t \downarrow 0} \left[|\varphi(0)|^2 + \int_{-\infty}^0 \left| \frac{\tilde{\phi}_t(\theta) - \phi(\theta)}{t} - \varphi(\theta) \right|^2 \rho(\theta)\, d\theta \right]$$

$$= |\varphi(0)|^2 + \lim_{t \downarrow 0} \int_{-\infty}^0 \left| \frac{\tilde{\phi}(t + \theta) - \phi(\theta)}{t} - \varphi(\theta) \right|^2 \rho(\theta)\, d\theta.$$

Since both terms on the right-handside of the last expression are non-negative, in order to have zero as a sum we must have both terms being zero; that is $\varphi(0) = 0$ and

$$0 = \lim_{t \downarrow 0} \int_{-\infty}^0 \left| \frac{\tilde{\phi}(t + \theta) - \phi(\theta)}{t} - \varphi(\theta) \right|^2 \rho(\theta)\, d\theta.$$

These two requirements together imply that ϕ is right-differentiable on $(-\infty, 0)$ with $\dot{\phi}(\theta+) = \varphi(\theta)$ and, moreover, that necessarily $\varphi(0) = 0$. The continuity of φ implies that actually ϕ is differentiable and $\dot{\phi} = \varphi$.

Conversely, suppose that ϕ is continuously differentiable on $(-\infty, 0]$ and $\dot{\phi}(0) = 0$. Define $\phi(t) = \phi(0)$ for $t \geq 0$; then

$$\left| \frac{1}{t}(\phi(t+\theta) - \phi(\theta)) - \dot{\phi} \right| = \left| \frac{1}{t} \int_0^t \left(\dot{\phi}(\theta+s) - \dot{\phi}(\theta) \right) ds \right|$$

converges, as $t \uparrow 0$, to zero uniformly for $\theta \in [-r, 0]$. $\square$

Let $C_b(\mathbf{M})$ be the real Banach space of bounded continuous functions $\Psi : \mathbf{M} \to \Re$ under the sup-norm $\| \cdot \|_{C_b(\mathbf{M})}$ defined by

$$\|\Psi\|_{C_b(\mathbf{M})} = \sup_{(\phi(0),\phi)\in\mathbf{M}} |\Psi(\phi(0),\phi)|$$

and let $\mathcal{M}(\mathbf{M})$ be the Banach space of all finite regular measures on $(\mathbf{M}, \mathcal{B}(\mathbf{M}))$. The concepts of weak topology on $C_b(\mathbf{M})$ and weak convergence of a family $\{\Phi(t), t \geq 0\}$ in $C_b(\mathbf{M})$ to $\Phi \in C_b(\mathbf{M})$ are similarly defined as in previous section for $C_b(\mathbf{C})$.

Let $\{T(t), t \geq 0\}$ be a contractive $\mathcal{C}_0$-semigroup of bounded linear operators from $C_b(\mathbf{M})$ to $C_b(\mathbf{M})$ defined by

$$\begin{aligned} T(t)(\Psi)(\psi(0), \psi) &= \Psi(\tilde{T}(t)(\psi(0), \psi)) \\ &= \Psi(\psi(0), \tilde{\psi}_t) \quad \forall \Psi \in C_b(\mathbf{M}), \quad (\psi(0), \psi) \in \mathbf{M}. \end{aligned}$$

Because of the boundedness of $\Psi \in C_b(\mathbf{M})$, it follows that $T(t)(\Psi) : \mathbf{M} \to \Re$ is bounded for all $t \geq 0$. Also, for each $t \geq 0$, the map $\tilde{T}(t) : \mathbf{M} \to \mathbf{M}$ is uniformly continuous. Therefore, $T(t) \in C_b(\mathbf{M})$. The family of operators $\{T(t), t \in [0, T]\}$ satisfies the following property.

Proposition 2.3.8 *The family $\{T(t), t \geq 0\}$ is a weakly continuous contraction semigroup in $C_b(\mathbf{M})$.*

Proof. To show that $\{T(t), t \geq 0\}$ is weakly continuous, observe that if $(\psi(0), \psi) \in \mathbf{M}$, then

$$\begin{aligned} \lim_{t\downarrow 0} \|(\phi(0), \tilde{\phi}_t) - (\phi(0), \phi)\|_{\mathbf{M}}^2 &= \lim_{t\downarrow 0} \|\tilde{\phi}_t - \phi\|_{\rho,2}^2 \\ &\equiv \lim_{t\downarrow 0} \int_{-\infty}^0 |\tilde{\phi}(t+\theta) - \phi(\theta)|^2 \rho(\theta)\, d\theta \\ &= 0. \end{aligned}$$

Therefore, if $\Psi \in C_b(\mathbf{M})$,

$$\lim_{t\downarrow 0} T(t)(\Psi)((\phi(0), \phi)) = \lim_{t\downarrow 0} \Psi((\phi(0), \tilde{\phi}_t)) = \Psi((\phi(0), \phi)).$$

Next, it is clear from the definition that, for each $t \geq 0$, $T(t)$ is a bounded linear operator on $C_b(\mathbf{M})$, and since

$$\|T(t)\|_{\mathcal{L}(C_b(\mathbf{M}))} \equiv \sup\{\|T(t)(\Psi)\|_M \mid \Psi \in C_b(\mathbf{M}), \|\Psi\|_{C_b(\mathbf{M})} = 1\} \leq 1,$$

it is true in fact that $\|T(t)\|_{\mathcal{L}(C_(\mathbf{M}))} = 1$ for all $t > 0$.

To show the semigroup property, let $t, s \geq 0$, $(\phi(0), \phi) \in \mathbf{M}$ and $\Psi \in C_b(\mathbf{M})$ be given. Then

$$\begin{aligned}
T(t)(T(s)(\Psi))(\phi(0), \phi) &= T(t)(\Psi)(\tilde{T}(s)(\phi(0), \phi)) \\
&= \Psi(\tilde{T}(t) \circ \tilde{T}(s)(\phi(0), \phi)) \\
&= \Psi(\tilde{T}(t + s)(\phi(0), \phi)) \\
&= T(t + s)(\Psi)(\phi(0), \phi). \qquad \square
\end{aligned}$$

We denote the infinitesimal generator associated with the semigroup as

$$\mathcal{S}(\Psi)(\phi(0), \phi) = \lim_{t \downarrow 0} \frac{\Psi(\phi(0), \tilde{\phi}_t) - \Psi(\phi(0), \phi)}{t}, \quad (\phi(0), \phi) \in \mathbf{M}. \qquad (2.14)$$

Let $\hat{\mathcal{D}}(\mathcal{S})$ be the set of all $\Psi \in C_b(\mathbf{M})$ such that the above limit exists for each $(\phi(0), \phi) \in \mathbf{M}$.

2.4 Itô and Dynkin Formulas

Let $\mathcal{D}(\mathcal{S})$ be the collection of $\Phi : [0, T] \times \Xi \to \Re$ such that $\Phi(t, \cdot) \in \hat{\mathcal{D}}(\mathcal{S})$ for each $t \in [0, T]$, where Ξ is either $\Xi = \mathbf{C}$ or $\Xi = \mathbf{M}$. In this section, we investigate Itô's and Dynkin's formulas for $\Phi(s, x_s)$, where $\Phi \in C_{lip}^{1,2}([0, T] \times \mathbf{C}) \cap \mathcal{D}(\mathcal{S})$ and $\{x_s, s \in [t, T]\}$ is the $\mathbf{C}$-valued segment process for (1.27). Similar formulas will be developed for $\Phi(S(s), S_s)$, where $\Phi \in C_{lip}^2(\mathbf{M}) \cap \hat{\mathcal{D}}(\mathcal{S})$ and $\{(S(s), S_s), s \geq 0\}$ is the $\mathbf{M}$-valued segment process for (1.43).

The details of the formulas will be given in the next two subsections.

2.4.1 $\{x_s, s \in [t, T]\}$.

Let $\{x_s, s \in [t, T]\}$ be the $\mathbf{C}$-valued segment process for (1.27) with the initial datum $(t, \psi) \in [0, T] \times \mathbf{C}$. In this subsection, we will establish the infinitesimal generator that characterizes the (strong) Markovian property of $\{x_s, s \in [t, T]\}$. If $\Psi \in C_{lip}^2(\mathbf{C})$, we recall from Lemmas 2.2.3 and 2.2.4 that its first- and second-order Fréchet derivatives $D\Psi(\psi) \in \mathbf{C}^*$ and $D^2\Psi(\psi) \in \mathbf{C}^\dagger$ at $\psi \in \mathbf{C}$ can be extended to $\overline{D\Psi(\psi)} \in (\mathbf{C} \oplus \mathbf{B})^*$ and $\overline{D^2\Psi(\psi)} \in (\mathbf{C} \oplus \mathbf{B})^\dagger$. With this in mind, we state the following.

Theorem 2.4.1 *Let $\{x_s(\cdot; t, \psi), s \in [t, T]\}$ be the $\mathbf{C}$-valued segment process for (1.27) through the initial datum $(t, \psi) \in [0, T] \times \mathbf{C}$. If Φ is such that $\Phi \in C_{lip}^{1,2}([0, T] \times \mathbf{C}) \cap \mathcal{D}(\mathcal{S})$, then*

$$\lim_{\epsilon \downarrow 0} \frac{E[\Phi(t + \epsilon, x_{t+\epsilon}(\cdot; t, \psi))] - \Phi(t, \psi)}{\epsilon} = (\partial_t + \mathbf{A})\Phi(t, \psi), \qquad (2.15)$$

where

$$\mathbf{A}\Phi(t, \psi) = \mathcal{S}\Phi(t, \psi) + \overline{D\Phi(t, \psi)}(f(t, \psi)\mathbf{1}_{\{0\}})$$

$$+ \frac{1}{2} \sum_{j=1}^{m} \overline{D^2\Phi(t, \psi)}(g(t, \psi)\mathbf{e}_j \mathbf{1}_{\{0\}}, g(t, \psi)\mathbf{e}_j \mathbf{1}_{\{0\}}), \qquad (2.16)$$

$\mathbf{e}_j, j = 1, 2, \ldots, m$, *is the jth unit vector of the standard basis in* $\Re^m$, *and* $\mathcal{S}\Phi(t, \psi)$ *is given by (2.11).*

Sketch of Proof. In the following, we will only give a sketch of the proof. The detailed proof for the autonomous SHDE with bounded drift and diffusion can be found in Theorem (3.2) of Mohammed [Moh84].

Note that

$$\lim_{\epsilon \downarrow 0} \frac{E[\Phi(t + \epsilon, x_{t+\epsilon}(\cdot; t, \psi)) - \Phi(t, \psi)]}{\epsilon}$$

$$= \lim_{\epsilon \downarrow 0} \frac{E[\Phi(t + \epsilon, x_{t+\epsilon}(\cdot; t, \psi)) - \Phi(t, x_{t+\epsilon}(\cdot; t, \psi))]}{\epsilon}$$

$$+ \lim_{\epsilon \downarrow 0} \frac{E[\Phi(t, x_{t+\epsilon}(\cdot; t, \psi)) - \Phi(t, \psi)]}{\epsilon}.$$

Since $x_{t+\epsilon}(\cdot; t, \psi) \to x_t(\cdot, t, \psi) = \psi$ in $\mathbf{C}$, the first limit on the left-handside of the last expression becomes

$$\lim_{\epsilon \downarrow 0} \frac{E[\Phi(t + \epsilon, x_{t+\epsilon}(\cdot; t, \psi)) - \Phi(t, x_{t+\epsilon}(\cdot; t, \psi))]}{\epsilon} = \partial_t \Phi(t, \psi).$$

We therefore need to establish that

$$\lim_{\epsilon \downarrow 0} \frac{E[\Phi(t, x_{t+\epsilon}(\cdot; t, \psi)) - \Phi(t, \psi)]}{\epsilon} = \mathbf{A}\Phi(t, \psi).$$

By Taylor's formula (see Theorem 2.1.1) and using the simpler notation $x_{t+\epsilon} = x_{t+\epsilon}(\cdot; t, \psi)$, we have P-a.s.

$$\Phi(t, x_{t+\epsilon}) - \Phi(t, \psi) = \Phi\left(t, \tilde{\psi}_{t+\epsilon}\right) - \Phi(t, \psi)$$

$$+ D\Phi\left(t, \tilde{\psi}_{t+\epsilon}\right)\left(x_{t+\epsilon} - \tilde{\psi}_{t+\epsilon}\right) + R_2(t + \epsilon),$$

where

$$R_2(t + \epsilon) = \int_0^1 (1 - \lambda)D^2\Phi\left(\tilde{\psi}_{t+\epsilon} + \lambda(x_{t+\epsilon} - \tilde{\psi}_{t+\epsilon})\right)$$

$$\times \left(x_{t+\epsilon} - \tilde{\psi}_{t+\epsilon}, x_{t+\epsilon} - \tilde{\psi}_{t+\epsilon}\right) d\lambda.$$

Taking the expectation, we obtain

$$
\frac{1}{\epsilon} E\left[\varPhi\left(t, x_{t+\epsilon}\right) - \varPhi(t, \psi)\right] = \frac{1}{\epsilon}\left[\varPhi\left(t, \tilde{\psi}_{t+\epsilon}\right) - \varPhi(t, \psi)\right]
$$
$$
+ \frac{1}{\epsilon} E\left[D\varPhi\left(t, \tilde{\psi}_{t+\epsilon}\right)\left(x_{t+\epsilon} - \tilde{\psi}_{t+\epsilon}\right)\right]
$$
$$
+ \frac{1}{\epsilon} E[R_2(t + \epsilon)].
$$

The outline of the rest of the proof is sketched in the following steps:

Step 1. It is clear from the definition of the operator $\mathcal{S}$ (see (2.11)) that

$$
\lim_{\epsilon \downarrow 0} \frac{\varPhi\left(t, \tilde{\psi}_{t+\epsilon}\right) - \varPhi(t, \psi)}{\epsilon} = \mathcal{S}\varPhi(t, \psi).
$$

Step 2. By the continuity and local boundedness of $D\varPhi$, we have

$$
\lim_{\epsilon \downarrow 0} \frac{1}{\epsilon} E\left[D\varPhi\left(t, \tilde{\psi}_{t+\epsilon}\right)\left(x_{t+\epsilon} - \tilde{\psi}_{t+\epsilon}\right)\right] = \frac{1}{\epsilon} E\left[D\varPhi(t, \psi)\left(x_{t+\epsilon} - \tilde{\psi}_{t+\epsilon}\right)\right]
$$
$$
= \overline{D\varPhi(t, \psi)} f(t, \psi) \mathbf{1}_{\{0\}}.
$$

Step 3. From Lemma (3.5) and the proof of Theorem (3.2) of [Moh84], one can prove that

$$
\frac{1}{\epsilon} E[R_2(t+\epsilon)] = \int_0^1 (1 - \lambda) \lim_{\epsilon \downarrow 0} \frac{1}{\epsilon} E\left[D^2\varPhi(t, \psi)\left(x_{t+\epsilon} - \tilde{\psi}_{t+\epsilon}, x_{t+\epsilon} - \tilde{\psi}_{t+\epsilon}\right)\right] d\lambda
$$
$$
= \frac{1}{2} \sum_{j=1}^{m} \overline{D^2\varPhi(t, \psi)}(g(t, \psi)\mathbf{e}_j \mathbf{1}_{\{0\}}, g(t, \psi)\mathbf{e}_j \mathbf{1}_{\{0\}}).
$$

This completes the sketch of the proof.   □

We define a tame and quasi-tame function $\varPhi : \mathbf{C} \to \Re$ next.

Definition 2.4.2 *A function* $\varPhi : \mathbf{C} \to \Re$ *is said to be tame if there is an integer $k > 0$ such that*

$$
\varPhi(\psi) = h\left(\eta(-r_k), \eta(-r_{k-1}), \dots, \eta(-r_1)\right)
$$

for all $\eta \in \mathbf{C}$, where $h(\cdot) : (\Re^n)^k \to \Re$ is continuous and $0 \leq r_1 < r_2 < \cdots < r_k \leq r$.

The following theorem is due to Mohammed [Moh84].

Theorem 2.4.3 *The class of tame functions* $\varPhi : \mathbf{C} \to \Re$ *defined by Definition 2.4.2 is weakly dense in $C_b(\mathbf{C})$.*

Proof. See Theorem IV.4.1 of Mohammed [Moh84].   □

Definition 2.4.4 *A function* $\Phi : \mathbf{C} \to \Re$ *is said to be quasi-tame if there is an integer* $k > 0$ *such that*

$$\Phi(\eta) = h\left(\int_{-r}^0 \alpha_1(\eta(\theta))\beta_1(\theta)\,d\theta, ..., \int_{-r}^0 \alpha_{k-1}(\eta(\theta))\beta_{k-1}(\theta)\,d\theta, \eta(0)\right)$$

for all $\eta \in \mathbf{C}$, *where the following hold:*
(i) $h(\cdot) : (\Re^n)^k \equiv \Re^n \times \cdots \times \Re^n \to \Re$ *is twice continuously differentiable and bounded.*
(ii) $\alpha_i : \Re^n \to \Re^n$, $i = 1, 2, ..., k-1$, *is* C^∞ *and bounded.*
(iii) $\beta_i : [-r, 0] \to \Re$, $i = 1, 2, ..., k-1$, *is piecewise* C^1 *and each of its derivative* $\dot{\beta}_i$ *is absolutely integrable over* $[-r, 0]$.

We will use the function $q : \mathbf{C} \to (\Re^n)^k$ defined by

$$q(\phi) = \left(\int_{-r}^0 \alpha_1(\phi(\theta))\beta_1(\theta)\,d\theta, ..., \int_{-r}^0 \alpha_{k-1}(\phi(\theta))\beta_{k-1}(\theta)\,d\theta, \phi(0)\right)$$

so that a quasi-tame function $\Phi(\phi) = h \circ q(\phi) = h(q(\phi))$. The collection of quasi-tame functions defined $\Phi : \mathbf{C} \to \Re$ defined in Definition 2.4.4 will be denoted by $Q(\mathbf{C}, \Re)$.

Proposition 2.4.5 *Every tame function* $\Phi : \mathbf{C} \to \Re$ *can be approximated by a sequence of quasi-tame functions* $\{\Phi^{(k)}\}_{k=1}^\infty \subset Q(\mathbf{C}, \Re)$.

Proof. Without loss of generality, we will assume, in order to simplify the notation, that the tame function $\Phi : \mathbf{C} \to \Re$ takes the following form $\Phi(\phi) = h(\phi(-\bar{r}), \phi(0))$, where $h : \Re^n \times \Re^n \to \Re$ and $0 < \bar{r} \le r$, i.e., there is only one delay.

We will approximate the function $\Phi(\phi) = h(\phi(-\bar{r}), \phi(0))$ by a sequence of quasi-tame functions $\{\Phi^{(k)}\}_{k=1}^\infty \subset Q(\mathbf{C}, \Re)$ as follows. Throughout the end of this proof, we define for each $\phi \in \mathbf{C}$ and each $k = 1, 2, ...$, the function $\phi^{(k)}(-\bar{r}; \gamma)$ by

$$\phi^{(k)}(-\bar{r}; \gamma) = k \int_{-r}^0 \gamma(k(-\bar{r} - \varsigma))\phi(\varsigma)d\varsigma,$$

where $\gamma : \Re \to [0, \infty)$ is the mollifier defined by

$$\gamma(\varsigma) = \begin{cases} 0 & \text{if } \varsigma \ge 2 \\ c\exp\left\{\frac{1}{(\varsigma-1)^2-1}\right\} & \text{if } 0 < \varsigma < 2, \end{cases}$$

and $c > 0$ is the constant chosen so that $\int_0^2 \gamma(\varsigma)\,d\varsigma = 1$. It is clear that γ is C^∞ and that

$$\lim_{k\to\infty} \Phi^{(k)}(\phi) \equiv \lim_{k\to\infty} h(\phi^{(k)}(-\bar{r}; \gamma)), \phi(0))$$
$$= h(\phi(-\bar{r}), \phi(0))$$
$$= \Phi(\phi),$$

since $\lim_{k\to\infty} \phi^{(k)}(-\bar{r}; \gamma) = \phi(-\bar{r})$. This proves the proposition. $\square$

If $\Phi : [0,T] \times \mathbf{C} \to \Re$ is a quasi-tame function as defined in Definition 2.4.4, then $\mathcal{S}_q\Phi(t,\psi)$ and $\mathbf{A}_q\Phi(t,\psi)$ can be explicitly expressed as follows.

Theorem 2.4.6 *Every quasi-tame function on $\mathbf{C}$ is in $\mathcal{D}(\mathcal{S}_q)$. Indeed, if Φ : $\mathbf{C} \to \Re$ is of the form $\Phi(\psi) = h(q(\psi)), \psi \in \mathbf{C}$, where $h : [0,T] \times (\Re)^k \to \Re$ and $q : \mathbf{C} \to (\Re^n)^k$ are as given in Definition 2.4.4, then*

$$\mathcal{S}_q\Phi(\psi) = \sum_{j=1}^{k-1} \left[\nabla_j h(t, q(\psi)) \Big(\alpha_j(\psi(0))\beta_j(0) - \alpha_j(\psi(-r))\beta_j(-r) \right.$$
$$\left. - \int_{-r}^{0} \alpha_j(\psi(\theta))\dot{\beta}_j(\theta)\, d\theta \Big) \right],$$

where ∇_j is the gradient operator of $h(x_1, \ldots, x_k)$ with respect to $x_j \in \Re^n$ for $j = 1, 2, \ldots, k$.

Proof. In proving the first and second statements, we will assume that $\Phi = h \circ q$, where $h : (\Re^n)^2 \to \Re$ is C^∞-bounded and

$$q(\psi) = \left(\int_{-r}^{0} \alpha(\psi(\theta))\beta(\theta)\, d\theta, \psi(0) \right)$$

for some C^∞-bounded map $\alpha : \Re^n \to \Re^n$ and a (piecewise) C^1 function $\beta : [-r,0] \to \Re$.

Let $0 < t < r$ and consider the expression

$$\frac{1}{t}[\Phi(\tilde{\psi}_t) - \Phi(\psi)] = \frac{1}{t} \left[h(q(\tilde{\psi}_t)) - h(q(\psi)) \right]$$
$$= \int_{0}^{1} Dh\left((1-\lambda)q(\psi) + \lambda q(\tilde{\psi}_t) \right) \left[\frac{1}{t}\left(q(\tilde{\psi}_t) - q(\psi) \right) \right] d\lambda$$

using the meanvalue theorem for h, with $0 \le \lambda \le 1$. However,

$$\frac{1}{t}\left(q(\tilde{\psi}_t) - q(\psi) \right) = \left(\frac{1}{t} \int_{-r}^{-t} \alpha(\psi(t+\theta))\beta(\theta)\, d\theta + \frac{1}{t} \int_{-t}^{0} \alpha(\psi(0))\beta(\theta)\, d\theta \right.$$
$$\left. - \frac{1}{t} \int_{-t}^{0} \alpha(\psi(\theta))\beta(\theta)\, d\theta, 0 \right)$$
$$= \left(\frac{1}{t} \int_{t-r}^{0} \alpha(\psi(\theta)) \frac{1}{t}\left(\beta(\theta-t) - \beta(\theta) \right) d\theta \right.$$
$$\left. + \alpha(\psi(0)) \frac{1}{t} \int_{-t}^{0} \beta(\theta)\, d\theta - \frac{1}{t} \int_{-r}^{t-r} \alpha(\psi(\theta))\beta(\theta)\, d\theta, 0 \right).$$

Now, α is bounded and β is (piecewise) C^1, so all three terms in the above expression are bounded in t and ψ. Moreover, letting $t \downarrow 0$, we obtain

$$\frac{1}{t}\left(q(\tilde{\psi}_t) - q(\psi) \right) = \left(\alpha(\psi(0))\beta(0) - \alpha(\psi(-r))\beta(-r) - \int_{-r}^{0} \alpha(\psi(\theta))\dot{\beta}(\theta)\, d\theta, 0 \right).$$

As the quantity $(1 - \lambda)q(\psi) + \lambda q(\tilde{\psi}_t)$ is bounded in ψ and continuous in (t, λ), it follows from the above three expressions and the dominated convergence theorem that $\varPhi \in \mathcal{D}(\mathcal{S}_q)$ and

$$
\begin{aligned}
\mathcal{S}_q\varPhi(\psi) &= D\varPhi(\psi)\left[\lim_{t\downarrow 0}\frac{1}{t}\left(q(\tilde{\psi}_t) - q(\psi)\right)\right] \\
&= \nabla_1 h(q(\psi))\bigg[\alpha(\psi(0))\beta(0) - \alpha(\psi(-r))\beta(-r) \\
&\qquad - \int_{-r}^{0}\alpha(\psi(\theta))\dot{\beta}(\theta)\,d\theta\bigg].\square
\end{aligned}
$$

Theorem 2.4.7 *Every quasi-tame function on* $\mathbf{C}$ *is in* $\mathcal{D}(\mathbf{A}_q)$*. Indeed, if* $\varPhi :$ $\mathbf{C} \to \Re$ *is of the form* $\varPhi(\psi) = h(q(\psi)), \psi \in \mathbf{C}$*, where* $h : (\Re^n)^k \to \Re$ *and* $q : \mathbf{C} \to (\Re^n)^k$ *are as given in Theorem 2.4.7, then*

$$
\begin{aligned}
\mathbf{A}_q\varPhi(\psi) &= \sum_{j=1}^{k-1}\nabla_j h(q(\psi))\bigg(\alpha_j(\psi(0))\beta_j(0) - \alpha_j(\psi(-r))\beta_j(-r) \\
&\qquad - \int_{-r}^{0}\alpha_j(\psi(\theta))\dot{\beta}_j(\theta)d\theta\bigg) + \nabla_k h(t, q(\psi))(f(t, \psi)) \\
&\qquad + \frac{1}{2}\text{trace}\left[\nabla_k^2 h(t, q(\psi)) \circ (g(t, \psi) \times g(t, \psi))\right],
\end{aligned}
$$

where ∇_j *is the gradient operator of* $h(x_1, \ldots, x_k)$ *with respect to* $x_j \in \Re^n$ *for* $j = 1, 2, \ldots, k$.

Sketch of Proof. To prove $\varPhi = h \circ q \in \mathcal{D}(\mathbf{A}_q)$, we prove $\varPhi = h \circ q \in C_{lip}^2(\mathbf{C}) \cap \mathcal{D}(\mathcal{S}_q)$. First, it is not hard to see that $\varPhi \in Q(\mathbf{C}, \Re)$ is C^∞. Also, by applying the chain rule and differentiating under the integral sign, one gets

$$
\begin{aligned}
D\varPhi(\psi)(\phi) &= \nabla h(q(\psi))\bigg(\int_{-r}^{0}\nabla\alpha_1(\psi(\theta))(\phi(\theta))\beta_1(\theta)\,d\theta, \ldots, \\
&\qquad \int_{-r}^{0}\nabla\alpha_{k-1}(\psi(\theta))(\phi(\theta))\beta_{k-1}(\theta)\,d\theta, \phi(0)\bigg),
\end{aligned}
$$

$$
\begin{aligned}
D^2\varPhi(\psi)(\phi, \varphi) &= \nabla^2 h(q(\psi))(\nabla q(\psi)\phi, \nabla q(\psi)\varphi) \\
&\quad + \nabla h(q(\psi))\bigg(\int_{-r}^{0}\nabla^2\alpha_1(\psi(\theta))(\phi(\theta), \varphi(\theta))\beta_1(\theta)\,d\theta, \ldots, \\
&\qquad \int_{-r}^{0}\nabla^2\alpha_{k-1}(\psi(\theta))(\phi(\theta), \varphi(\theta))\beta_{k-1}(\theta)\,d\theta, 0\bigg)
\end{aligned}
$$

for all $\psi, \phi, \varphi \in \mathbf{C}$.

Since all derivatives of h, $\alpha_j, 1 \le j \le k - 1$, are bounded, it is easy to see, from the above formulas, that $\Phi \in C^2_{lip}(\mathbf{C}) \cap \mathcal{D}(\mathcal{S}_q)$. From the above two formulas it is easy to see that the unique weakly continuous extensions $\overline{D\Phi(\psi)}$ and $\overline{D^2\Phi(\psi)}$ of $D\Phi(\psi)$ and $D^2\Phi(\psi)$ to $\mathbf{C} \oplus \mathbf{B}$ are given by

$$\overline{D\Phi(\psi)}(v\mathbf{1}_{\{0\}}) = \nabla h(q(\psi))(0, 0, \dots, v) = \nabla_k h(q(\psi)) \cdot v,$$

$$\overline{D^2\Phi(\psi)}(v\mathbf{1}_{\{0\}}, w\mathbf{1}_{\{0\}}) = \nabla^2 h(q(\psi))((0, 0, \dots, v), (0, 0, \dots, v))$$
$$= \nabla_k^2 h(q(\psi))(v, w)$$

for all $v, w \in \Re^n$. The given formula for $\mathbf{A}_q\Phi(\psi)$ now follows directly from Theorem 2.4.1 and Theorem 2.4.6.   □

Theorem 2.4.8 *If $\Phi : \mathbf{C} \to \Re$ is a quasi-tame function as defined in Definition 2.4.4, then we have the following Itô's formula:*

$$\Phi(x_s) = \Phi(\psi_t) + \int_t^s \mathbf{A}_q\Phi(x_\lambda)\, d\lambda + \int_t^s \nabla_k h(q(x_\lambda)) \cdot g(\lambda, x_\lambda)\, dW(\lambda), \quad (2.17)$$

where $\mathbf{A}_q$ is as defined in Theorem 2.4.7.

Proof. Let $y(\cdot) = \{y(s), s \in [t, T]\}$ be defined as $y(s) = q(x_s)$. Applying Itô formula (see Theorem 1.2.15), we obtain (2.17).   □

2.4.2 $\{(S(s), S_s), s \ge 0\}$.

Again, consider the $\mathbf{M}$-valued segment process $\{(S(t), S_t), t \ge 0\}$, where $S_t(\theta) = S(t + \theta), \theta \in (-\infty, 0]$) for the nonlinear SHDE (1.43) with the initial historical price function $(S(0), S_0) = (\psi(0), \psi) \in \mathbf{M}$. We have the following result for its weak infinitesimal generator $\mathbf{A}$. Other results for equations of a similar nature can be found in Arriojas [Arr97] and Mizel and Trutzer [MT84].

Theorem 2.4.9 *If $\Phi \in C^2_{lip}(\mathbf{M}) \cap \mathcal{D}(\mathcal{S})$ and $\{(S(s), S_s), s \ge 0\}$ is the $\mathbf{M}$-valued segment process for (1.43) with an initial historical price function $(\psi(0), \psi) \in \mathbf{M}$, then*

$$\lim_{t\downarrow 0} \frac{E[\Phi(S(t), S_t) - \Phi(\psi(0), \psi)]}{t} = \mathbf{A}\Phi(\psi(0), \psi), \quad (2.18)$$

where

$$\mathbf{A}\Phi(\psi(0), \psi) = \mathcal{S}\Phi(\psi(0), \psi) + \frac{1}{2}\partial_x^2\Phi(\psi(0), \psi)\psi^2(0)g^2(\psi)$$
$$+ \partial_x\Phi(\psi(0), \psi)\psi(0)f(\psi), \quad (2.19)$$

$\mathcal{S}(\Phi)(\psi(0), \psi)$ is as defined in (2.14), and $\partial_x\Phi(\psi(0), \psi)$ and $\partial_x^2\Phi(\psi(0), \psi)$ are the first and second partial derivatives of $\Phi(\psi(0), \psi)$ with respect to its first variable $x = \psi(0)$.

From a glance at (2.19), it seems that $\mathbf{A}\Phi(\psi(0),\psi)$ requires only the existence of the first- and second-order partial derivatives $\partial_x\Phi$ and $\partial_x^2\Phi$. However, detailed derivations of the formula reveal that a stronger condition than $\Phi \in C^2_{lip}(\mathbf{M})$ is required.

We have the following Dynkin's formula (see Mizel and Trutzer [MT84], Arriojas [Arr97], and Kolmanovskii and Shaikhet [KS96] for similar derivations).

Theorem 2.4.10 *Let $\Phi \in C^2_{lip}(\mathbf{M}) \cap \hat{\mathcal{D}}(\mathcal{S})$. Then*

$$E[e^{-\alpha\tau}\Phi(S(\tau),S_\tau)] = \Phi(\psi(0),\psi) + E\left[\int_0^\tau e^{-\alpha t}(\mathbf{A}-\alpha I)\Phi(S(t),S_t)dt\right] \quad (2.20)$$

for all $P-a.s.$ finite $\mathbf{G}$-stopping time τ, where the operators $\mathbf{A}$ and $\mathcal{S}$ are as described in (2.19) and (2.14).

The function $\Phi \in C^2_{lip}(\mathbf{M}) \cap \hat{\mathcal{D}}(\mathcal{S})$ that has the following special form is referred to as a *quasi-tame function*

$$\Phi(\phi(0),\phi) = h(q(\phi(0),\phi)), \quad (2.21)$$

where $h : \Re^{k+1} \to \Re$ is a function $h(x,y_1,y_2,\ldots,y_k)$ such that its partial derivatives $\partial_x h, \partial_x^2 h, \partial_{y_i} h, i = 1,2,\ldots,k$ exist and are continuous, and $\forall(\phi(0),\phi) \in \mathbf{M}$:

$$q(\phi(0),\phi) = \left(\phi(0), \int_{-\infty}^0 \eta_1(\phi(\theta))\lambda_1(\theta)d\theta, \cdots, \int_{-\infty}^0 \eta_k(\phi(\theta))\lambda_k(\theta)d\theta\right) \quad (2.22)$$

for some positive integer k and some functions $q \in C(\mathbf{M}, \Re^{k+1})$, $\eta_i \in C^\infty(\Re)$, $\lambda_i \in C^1((-\infty,0])$ with

$$\lim_{\theta\to-\infty}\lambda_i(\theta) = \lambda_i(-\infty) = 0$$

for $i = 1,2,\ldots,k$, and $h \in C^\infty(\Re^{k+1})$ of the form $h(x,y_1,y_2,\ldots,y_k)$.

The collection of quasi-tame functions $\Phi : \mathbf{M} \to \Re$ will be denoted by $Q(\mathbf{M},\Re)$. Again the infinitesimal generators $\mathcal{S}$ and $\mathbf{A}$ that apply to $Q(\mathbf{M},\Re)$ will be denoted by $\mathcal{S}_q$ and $\mathbf{A}_q$, respectively.

We have the following Itô's formula in case $\Phi \in C_b(\mathbf{M})$ is a quasi-tame function in the sense defined above.

Theorem 2.4.11 *Let $\{(S(s),S_s),s \geq 0\}$ be the $\mathbf{M}$-valued solution process corresponding to (1.43) with an initial historical price function $(\psi(0),\psi) \in \mathbf{M}$. If $\Phi \in C(\mathbf{M})$ is a quasi-tame function, then $\Phi \in \mathcal{D}(\mathbf{A}_q)$ and*

$$e^{-\alpha\tau}\Phi(S(\tau),S_\tau) = \Phi(\psi(0),\psi) + \int_0^\tau e^{-\alpha t}(\mathbf{A}_q - \alpha I)\Phi(S(t),S_t)\,dt$$

$$+ \int_0^\tau e^{-\alpha t}\partial_x\Phi(S(t),S_t)S(t)f(S_t)\,dW(t) \quad (2.23)$$

for all finite **G**-*stopping times* τ, *where* I *is the identity operator. Moreover, if* $\Phi \in C(\mathbf{M})$ *is the form described in (2.21) and (2.22), then*

$$
\mathbf{A}_q\Phi(\psi(0), \psi) = \sum_{i=1}^{k} \partial_{y_i} h(q(\psi(0), \psi)) \left(\eta_i(\psi(0))\lambda_i(0) - \int_{-\infty}^{0} \eta_i(\psi(\theta))\dot{\lambda}_i(\theta)\, d\theta \right)
$$
$$
+ \partial_x h(q(\psi(0), \psi))\psi(0)f(\psi)
$$
$$
+ \frac{1}{2}\partial_x^2 h(q(\psi(0), \psi))\psi^2(0)g^2(\psi), \qquad (2.24)
$$

where $\partial_{y_i} h(x, y_1, \ldots, y_k)$, $\partial_x h(x, y_1, \ldots, y_k)$, *and* $\partial_x^2 h(x, y_1, \ldots, y_k)$ *are partial derivatives of* $h(x, y_1, \cdots, y_k)$ *with respect to its appropriate variables.*

Proof. Itô's formula for a quasi-tame function $\Phi : \Re \times L^2([-r, 0]) \to \Re$ for the $\Re \times L^2([-r, 0])$ solution process $\{(x(t), x_t), t \geq 0\}$ of a stochastic function differential equation with a bounded delay $r > 0$ is obtained in Arriojas [Arr97] (the same result can also be obtained from Mohammed [Moh84, Moh98] with some modifications). The same arguments can be easily extended to the stochastic hereditary differential equation with unbounded memory (1.43) considered in this monograph. To avoid further lengthening the exposition, we omit the proof here. $\square$.

In the following, we will prove that the above Itô formula also holds for any tame function $\Phi : C(-\infty, 0] \to \Re$ of the form

$$
\Phi(\phi) = h(\pi(\phi)) = h(\phi(0), \phi(-\theta_1), \ldots, \phi(-\theta_k)), \qquad (2.25)
$$

where $C(-\infty, 0]$ is the space continuous functions $\phi : (-\infty, 0] \to \Re$ equipped with uniform topology (see (2.7)), $0 < \theta_1 < \theta_2 < \cdots < \theta_k < \infty$, and $\Psi(x, y_1, \ldots, y_k)$ is such that $\Psi \in C^\infty(\Re^{k+1})$.

Theorem 2.4.12 *Let* $\{(S(s), S_s), s \geq 0\}$ *be the* **M**-*valued segment process for (1.43) with an initial historical price function* $(\psi(0), \psi) \in \mathbf{M}$. *If* $\Phi : C(-\infty, 0] \to \Re$ *is a tame function defined by (2.25), then* $\Phi \in \mathcal{D}(\mathbf{A}_q)$ *and*

$$
e^{-\alpha\tau} h(S(\tau), S(\tau - \theta_1), \ldots, S(\tau - \theta_k))
$$
$$
= h(\psi(0), \psi(-\theta_1), \ldots, \psi(-\theta_k))
$$
$$
+ \int_0^\tau e^{-\alpha t}(\mathbf{A}_q - \alpha I)h(S(t), S(t - \theta_1), \ldots, S(t - \theta_k))\, dt
$$
$$
+ \int_0^\tau e^{-\alpha t}\partial_x h(S(t), S(t - \theta_1), \ldots, S(t - \theta_k))S(t)f(S_t)\, dW(t) \quad (2.26)
$$

for all finite **G**-*stopping times* τ, *where*

$$
\mathbf{A}_q h(\psi(0), \psi(-\theta_1), \cdots, \psi(-\theta_k))
$$
$$
= \partial_x h(\psi(0), \psi(-\theta_1), \cdots, \psi(-\theta_k))\psi(0)f(\psi)
$$
$$
+ \frac{1}{2}\partial_x^2 h(\psi(0), \psi(-\theta_1), \cdots, \psi(-\theta_k))\psi^2(0)g^2(\psi), \qquad (2.27)
$$

with $\partial_x h$ and $\partial_x^2 h$ being the first- and second-order derivatives with respect to x of $h(x, y_1, \ldots, y_k)$.

Proof. Without loss of generality, we will assume in order to simplify the notation that $h : \Re \times \Re \to \Re$ with $h(x, y)$ and there is only one delay in the function $h(\phi(0), \phi(-\bar{\theta}))$ for some fixed $\bar{\theta} \in (0, \infty)$.

We will approximate the $\Phi(\phi) = h(\phi(-\bar{r}), \phi(0))$ function by a sequence of quasi-tame functions $\{\Phi^{(k)}\}_{k=1}^\infty \subset Q(\mathbf{C}, \Re)$ as follows. Throughout the end of this proof, we define for each $\phi \in \mathbf{C}$ and each $k = 1, 2, \ldots$ the function $\phi^{(k)}(-\bar{r}; \gamma)$ by

$$\phi^{(k)}(-\bar{r}; \gamma) = k \int_{-r}^{0} \gamma(k(-\bar{r} - \varsigma))\phi(\varsigma)\, d\varsigma,$$

where $\gamma : \Re \to [0, \infty)$ is the mollifier defined by

$$\gamma(\varsigma) = \begin{cases} 0 & \text{if } |\varsigma| \geq 1 \\ c \exp\left\{\frac{1}{(\varsigma)^2 - 1}\right\} & \text{if } 0 < \varsigma < 2 \end{cases}$$

and $c > 0$ is the constant chosen so that $\int_{-\infty}^{\infty} \gamma(\varsigma)\, d\varsigma = 1$. It is clear that γ is C^∞ and that

$$\begin{aligned} \lim_{k \to \infty} \Phi^{(k)}(\phi) &\equiv \lim_{k \to \infty} h(\phi^{(k)}(-\bar{r}; \gamma)), \phi(0)) \\ &= h(\phi(-\bar{r}), \phi(0)) \\ &= \Phi(\phi), \end{aligned}$$

since $\lim_{k \to \infty} \phi^{(k)}(-\bar{r}; \gamma) = \phi(-\bar{r})$. Moreover,

$$\mathcal{S}_q h(\phi(0), \phi^{(k)}(-\bar{r}; \gamma))$$

$$= \partial_y h(\phi(0), -\phi(\bar{r}))\bigg(\phi(0)k\gamma(-k\bar{r})$$

$$- 2k^3 \int_{-\infty}^{0} \phi(\theta)\gamma(k(\bar{r} + \theta))\frac{\bar{r} + \theta}{(k^2(-\bar{r} - \theta)^2 - 1)^2}\, d\theta\bigg),$$

and by the Lebesque dominating convergence theorem, we have

$$\lim_{k \to \infty} \mathcal{S}_q h(\phi(0), \phi^{(k)}(-\bar{r}; \gamma)) = 0.$$

Therefore, for any almost surely finite **G**-stopping time τ, we have from Theorem 2.4.11 and sample path convergence property of the Itô integrals (see Section 1.2 in Chapter 1) that

$$e^{-\alpha\tau} h(S(\tau), S(\tau - \bar{r}))$$

$$= \lim_{k \to \infty} e^{-\alpha\tau} h(S(\tau), S_\tau^{(k)}(-\bar{r}; \gamma))$$

Therefore,

$$
e^{-\alpha\tau}h(S(\tau), S(\tau - \bar{r}))
$$

$$
= \lim_{k\to\infty} \Big[h(\psi(0), \psi^{(k)}(-\bar{r}; \gamma))
$$

$$
+ \int_0^\tau e^{-\alpha s}(\mathbf{A}_q - \alpha I)h(S(s), S_s^{(k)}(-\bar{r}; \gamma))ds
$$

$$
+ \int_0^\tau e^{-\alpha s}\partial_x h(S(s), S_s^{(k)}(-\bar{r}; \gamma))S(s)f(S_s)dW(s)\Big].
$$

$$
= h(\psi(0), \psi(-\bar{r})) + \int_0^\tau e^{-\alpha s}\Big(\partial_x h(S(s), S(s - \bar{r}))S(s)f(S_s)
$$

$$
+ \frac{1}{2}\partial_x^2 h(S(s), S(s - \bar{r}))S^2(s)g^2(S_s)\Big) ds
$$

$$
+ \int_0^\tau e^{-\alpha s}\partial_x h(S(s), S(s - \bar{r}))S(s)f(S_s)\, dW(s). \qquad \square
$$

2.5 Martingale Problem

In this section, we consider a **C**-valued martingale problem. Much of the material presented here are mainly due to Kallianpur and Mandal [KM02].

Without loss of generality for formulating the martingale problem, we can and will consider the following autonomous SHDE throughout the end of this section.

$$
dx(s) = f(x_s)\, ds + g(x_s)\, dW(s), \quad s \geq 0, \tag{2.28}
$$

with the initial segment $\psi \in L^2(\Omega, \mathbf{C}; \mathcal{F}(0))$ at the initial time $t = 0$.

As before, we will assume that the functions $f : \mathbf{C} \to \Re^n$ and $g : \mathbf{C} \to \Re^{n\times m}$ satisfy the Lipschitz condition: There exists a constant $K_{lip} > 0$ such that

$$
|f(\phi) - f(\varphi)| + |g(\phi) - g(\varphi)|
$$
$$
\leq K_{lip}\|\phi - \varphi\|, \quad \forall \phi, \varphi \in \mathbf{C}. \tag{2.29}
$$

Recall from Proposition 2.1.10 that a sequence $\{\Phi^{(k)}\}_{k=1}^\infty \subset C_b(\Xi)$ converges weakly to $\Phi \in C_b(\Xi)$ if and only if the sequence $\{\Phi^{(k)}\}_{k=1}^\infty$ is bounded pointwise, that is,

$$
\sup_k \|\Phi^{(k)}\|_{C_b(\mathbf{C})} < \infty
$$

and

$$
\lim_{t\downarrow 0} \Phi(t)(\mathbf{x}) = \Phi(\mathbf{x}), \quad \forall \mathbf{x} \in \Xi.
$$

In this case, we write $\Phi = (bp)\lim_{k\to\infty} \Phi^{(k)}$.

Let us consider an operator $\Gamma : \mathcal{D}(\Gamma) \subset C_b(\mathbf{C}) \to \Re$. We list the conditions that might be applicable to the operator Γ:

C1. There exists $\Phi \in C(\mathbf{C})$ such that

$$|\Gamma \Psi(\phi)| \le K_\Psi \Phi(\phi), \quad \forall \Psi \in \mathcal{D}(\Gamma), \phi \in \mathbf{C},$$

where K_Ψ is a constant depending on Ψ.

C2. There exists a countable subset $\Psi^{(k)} \subset \mathcal{D}(\Gamma)$ such that

$$\{(\Psi, \Phi^{-1} \Gamma \Psi) \mid \Psi \in \mathcal{D}(\Gamma)\} \subset \overline{(\{(\Psi^{(k)}, \Phi^{-1} \Gamma \Psi^{(k)}) \mid k \ge 1\})},$$

where $\overline{\{\cdots\}}$ denotes the closure of the set $\{\cdots\}$ in the bounded pointwise (bp) convergence topology.

C3. $\mathcal{D}(\Gamma)$ is an algebra that separates points in $\mathbf{C}$ and contains the constant functions.

Let μ be a probability measure on $(\mathbf{C}, \mathcal{B}(\mathbf{C}))$. We define a solution to the martingale problem (see Mandal and Kallianpur [MK02]) as follows.

Definition 2.5.1 *A $\mathbf{C}$-valued process $X(\cdot) = \{X(s), s \in [0, T]\}$ defined on some complete filtered probability space $(\Omega, \mathcal{F}, P, \mathbf{F})$ is said to be a solution to the martingale problem for (Γ, μ) if:*
(i) $P \circ X^{-1}(0) = \mu$.

(ii) $\int_0^s E[\Phi(X(t))]dt < \infty$, for all $s \in [0, T]$.

(iii) for all $\Psi \in \mathcal{D}(\Gamma)$,

$$M^\Psi(s) \equiv \Psi(X(s)) - \Psi(X(0)) - \int_0^s (\Gamma)(X(t))dt$$

is an $\mathbf{F}$-martingale. The martingale problem for (Γ, μ) is said to be well posed in a class of $\mathbf{C}$-valued processes $\mathcal{C}$ if $X^{(i)}(\cdot)$, $i = 1, 2$, are two solutions to the martingale problem for (Γ, μ), then $X^{(1)}(\cdot)$ and $X^{(2)}(\cdot)$ have the same probability laws.

The following result states that the uniqueness of the solution of a martingale problem always implies the Markovian property.

Lemma 2.5.2 *Suppose that the operator Γ satisfies Conditions **C1** and **C2**. Furthermore, assume that the martingale problem for (Γ, δ_ψ) is well posed in the class of r.c.l.l. (right continuous with left hand limits) solutions for every $\psi \in \mathbf{C}$, where δ_ψ is the Dirac measure at ψ. Then the solution $X(\cdot)$ to the martingale problem for (Γ, μ) is a $\mathbf{F}$-Markov process. Furthermore, if $\mathbf{A}$ is the infinitesimal generator of $X(\cdot)$, then $\mathcal{D}(\Gamma) \subset \mathcal{D}(\mathbf{A})$ and Γ and $\mathbf{A}$ coincide on $\mathcal{D}(\Gamma)$.*

Proof. This is a very specialized result. We therefore omit the proof here and refer interested readers to Theorems IV.4.2 and IV.4.6 of Ethier and Kurtz [EK86] and Remark 2.1 of Horowitz and Karandikar [HK90].

We recall below the definition (see Definition 2.4.4) of a quasi-tame function $\Phi \in Q(\mathbf{C}, \Re)$ and its infinitesimal generators $\mathbf{A}_q$ as given in Theorem 2.4.7.

$$\mathbf{A_q}\Phi(\psi) = \mathcal{S}_q\Phi(\psi) + \nabla_k h(t, q(\psi))(f(t, \psi))$$
$$+ \frac{1}{2} trace \left[\nabla_k^2 h(t, q(\psi)) \circ (g(t, \psi)g^\top(t, \psi)) \right], \qquad (2.30)$$

where

$$\mathcal{S}_q\Phi(\psi) = \sum_{j=1}^{k-1} \nabla_j h(q(\psi)) \left(\alpha_j(\psi(0))\beta_j(0) - \alpha_j(\psi(-r))\beta_j(-r) \right)$$
$$- \int_{-r}^{0} \alpha_j(\psi(\theta))\dot{\beta}_j(\theta) d\theta \qquad (2.31)$$

and ∇_j is the gradient operator of $h(x_1, \ldots, x_k)$ with respect to $x_j \in \Re^n$ for $j = 1, 2, \ldots, k$.

Proposition 2.5.3 *Suppose that the operator $\mathbf{A}_q$ is as defined in (2.30). Then $\mathbf{A}_q$ satisfies conditions C1-C3.*

Proof. Suppose that $\Phi \in \mathcal{D}(\mathbf{A}_q)$ is a quasi-tame function and is given by $\Phi(\phi) = h(q(\phi))$, where $h(\cdot) : (\Re^n)^k \to \Re$ is twice continuously differentiable and bounded,

$$q(\phi) = \left(\int_{-r}^{0} \alpha_1(\phi(\theta))\beta_1(\theta) d\theta, \ldots, \int_{-r}^{0} \alpha_{k-1}(\phi(\theta))\beta_{k-1}(\theta)\, d\theta, \phi(0) \right),$$

and
(i). $\alpha_i : \Re^n \to \Re^n$, $i = 1, 2, \ldots, k-1$, is C^∞ and bounded;
(ii). $\beta_i : [-r, 0] \to \Re$, $i = 1, 2, \ldots, k-1$, is piecewise C^1 and each of its derivative $\dot{\beta}_i$ is absolutely integrable over $[-r, 0]$. Then for $\phi \in \mathbf{C}$,

$$|\mathbf{A_q}\Phi(\phi)| = \sum_{j=1}^{k-1} \left[|\nabla_j h(q(\psi))| \right.$$
$$\times \left| (\alpha_j(\psi(0))\beta_j(0) - \alpha_j(\psi(-r))\beta_j(-r)) \right.$$
$$\left. - \int_{-r}^{0} \alpha_j(\psi(\theta))\dot{\beta}_j(\theta) d\theta \right| \right]$$
$$+ |\nabla_k h(t, q(\psi))(f(t, \psi))| + \frac{1}{2}|\nabla_k^2 h(t, q(\psi))||(g(t, \psi))|^2$$
$$\leq K\Psi(\phi),$$

where K is a constant depending on h, α_j, β_j, $j = 1, 2, \ldots, k-1$ and

$$\Psi(\phi) = 1 + |f(\phi)| + |g(\phi)|^2.$$

Therefore, **C1** is satisfied by $\mathbf{A}_q$.

To see that **C2** holds, note that

$$C_b^\infty((\Re^n)^k) \text{ is separable in } |h| + |\nabla h| + |\nabla^2 h| \text{ norm,}$$
$$C_b^\infty(\Re^n) \text{ is separable in } |\alpha| \text{ norm,}$$
$$C^1([-r,0],\Re^n) \text{ is separable in } |\beta| + |\dot\beta| \text{ norm.}$$

This will imply the existence of a countable set of quasi-tame functions $\{\Phi^{(j)}\}_{j=1}^\infty \subset \mathcal{D}(\mathbf{A}_q)$ such that

$$\{(\Phi, \Psi^{-1}(\mathbf{A}_q\Phi) \mid \Phi \in \mathcal{D}(\mathbf{A}_q)\} \subset \overline{(\{\Phi^{(j)}, \Psi^{-1}(\mathbf{A}_q\Phi^{(j)}), j \geq 1\})}$$

This proves that **C2** holds.

That $\mathcal{D}(\mathbf{A}_q)$ is an algebra follows from Mohammed [Moh84]. It is also easy to check that $\mathcal{D}(\mathbf{A}_q)$ separates points in **C** and contains the constant functions which implies that $\mathbf{A}_q$ satisfies **C3**.   □

We therefore have the following theorem from Lemma 2.5.2 and the previous proposition.

Theorem 2.5.4 *Let $\{x_s, s \in [0,T]\}$ be the **C**-valued segment process for (2.28). If $\Phi : \mathbf{C} \to \Re$ is a quasi-tame function defined by Definition 2.4.4. Then*

$$M^\Phi(s) \equiv \Phi(x_s) - \Phi(\psi) - \int_0^s (\mathbf{A}_q\Phi)(x_t)dt$$

*is an **F**-martingale.*

We establish the **C**-valued segment process $\{x_s, s \in [0,T]\}$ as the unique solution to the martingale problem below.

Theorem 2.5.5 *Suppose $\psi \in L^2(\Omega, \mathbf{C})$ and the operator $\mathbf{A}_q$ is as defined in (2.30). Then the martingale problem for $(\mathbf{A}_q, \psi)$ is well posed.*

Proof. Let $\{X(s), s \in [0,T]\}$ be a **C**-valued process defined on a complete filtered probability space $(\hat\Omega, \hat{\mathcal{F}}, \hat{P}, \hat{\mathbf{F}})$, and be a progressively measurable solution to the $(\mathbf{A}_q, \psi)$-martingale problem, that is, for a quasi-tame function Φ defined by Definition 2.4.4, $\{\Phi(X(s)), s \in [0,T]\}$ is a semimartingale, given by

$$\Phi(X(s)) = \Phi(X(0)) + \int_0^s (\mathbf{A}_q)(X(t))\,dt + M^\Phi(s) \tag{2.32}$$

and $X(0) = \psi$. We shall show that $X(s) = \hat{x}_s$ for some n-dimensional continuous process $\{\hat{x}(s), s \in [-r,T]\}$ satisfying a SHDE of the form (2.28) in the weak sense. Then by the uniqueness of the solution to (2.28) we will have that the distribution of $X(s)$ is the same as that of x_s, proving the well-posedness of the martingale problem for $(\mathbf{A}_q, \psi)$.

From (2.32) it follows that

$$\langle M^{\Phi} \rangle_s = \int_0^s \Gamma(\Phi, \Phi)(X(t))\, dt, \tag{2.33}$$

where

$$\Gamma(\Phi, \Phi) = \mathbf{A}_q(\Phi^2) - 2\Phi(\mathbf{A}_q\Phi). \tag{2.34}$$

Applying Itô's formula (see Theorem 2.4.8) to the semimartingale process $\{\Phi(X(s)), s \in [0,T]\}$, we have for $\gamma \in C^1[0,T]$,

$$\gamma(s)\Phi(X(s)) = \gamma(0)\Phi(X(0)) + \int_0^s \dot{\gamma}(t)\Phi(X(t))\, dt$$

$$+ \int_0^s \gamma(t)(\mathbf{A}_q\Phi)(X(t))dt + \int_0^s \gamma(t)dM^{\Phi}(t). \tag{2.35}$$

Now, suppose $\alpha \in C_b^{\infty}(\Re^n)$, $\beta \in C^1[-r,0]$. Let $\Delta = \Delta(\alpha, \beta)$ be a bound for the integral $\int_{-r}^0 \alpha(\phi(\theta))\beta(\theta)d\theta$, $\phi \in \mathbf{C}$. Suppose $\varsigma_{\Delta} : \Re \to [0,1]$ is a C_b^{∞}-bump function such that

$$\varsigma_{\Delta}(x) = 1, \quad \text{if } |x| \leq \Delta,$$

$$0 < \varsigma_{\Delta}(x) \leq 1, \quad \text{if } \Delta < |x| < \Delta + 1,$$

$$\varsigma_{\Delta}(x) = 0, \quad \text{if } |x| \geq \Delta + 1.$$

Suppose $h(x) = \sum_{i=1}^n x_i \varsigma_{\Delta}(x_i)$ for $x = (x_1, x_2, \ldots, x_n) \in \Re^n$ so that $h \in C_b^{\infty}(\Re^n)$ and $h(x) = \sum_{i=1}^n x_i$ for $x = (x_1, x_2, \ldots, x_n) \in [-\Delta, \Delta]^n$. Let $h^* \in C_b^{\infty}(\Re^n \times \Re^n)$ by defined by $h^*(x,y) = h(x)$. Consider a quasi-tame function Φ of the form defined in Definition 2.4.4 with $k = 2$, given by

$$\Phi(\phi) = h^* \left(\int_{-r}^0 \alpha(\phi(\theta))\beta(\theta)d\theta, \phi(0) \right)$$

$$= h \left(\int_{-r}^0 \alpha(\phi(\theta))\beta(\theta)d\theta \right)$$

$$= \sum_{i=1}^n \left(\int_{-r}^0 \alpha(\phi_i(\theta))\beta(\theta)d\theta \right) \tag{2.36}$$

Then, from Theorem 2.4.7, we have

$$\mathbf{A}_q\Phi(\phi) = \sum_{i=1}^n \left[\alpha(\phi_i(0))\beta(0) - \alpha(\phi_i(-r))\beta(-r) - \int_{-r}^0 \alpha(\phi_i(\theta))\dot{\beta}(\theta)d\theta \right],$$

and similarly by chain rule,

$$(\mathbf{A}_q \Phi^2)(\phi)$$

$$= 2h \left(\sum_{i=1}^{n} \int_{-r}^{0} \alpha(\phi_i(\theta))\beta(\theta)d\theta \right) \times \sum_{j=1}^{n} \left[\partial_j h \left(\sum_{i=1}^{n} \int_{-r}^{0} \alpha(\phi_i(\theta))\beta(\theta)d\theta \right) \right.$$

$$\left. \times \left(\alpha(\phi_j(0))\beta(0) - \alpha(\phi_j(-r))\beta(-r) - \int_{-r}^{0} \alpha(\phi_j(\theta))\dot{\beta}(\theta)d\theta \right) \right]$$

$$= 2\Phi(\theta)(\mathbf{A}_q \Phi(\phi)).$$

This shows that

$$\Gamma(\Phi, \Phi)(\phi) = (\mathbf{A}_q \Phi^2)(\phi) - 2\Phi(\phi)(\mathbf{A}_q \Phi(\phi)) = 0$$

and hence, from (2.33), $\langle M^\Phi \rangle_s = 0$. Therefore, $M^\Phi(s) = 0$, a.s. for all $s \in [0, T]$. From (2.35), we then have $0 \le t \le s \le T$,

$$\gamma(s)\Phi(X(s)) = \gamma(t)\Phi(X(t)) + \int_t^s \dot{\gamma}(\lambda)\Phi(X(\lambda))\, d\lambda + \int_t^s \gamma(\lambda)(\mathbf{A}_q \Phi)(X(\lambda))\, d\lambda.$$

Since $X(s)$ is **C**-valued, we write $X(s, \cdot) : [-r, 0] \to \Re^n$ as a continuous function. Using the special forms of Φ and $\mathbf{A}_q \Phi$ given above, we have

$$\int_{-r}^{0} \alpha(X(s, \theta))\gamma(s)\beta(\theta) - \int_{-r}^{0} \alpha(X(t, \theta))\gamma(t)\beta(\theta)$$

$$= \int_t^s \int_{-r}^{0} \alpha(X(\lambda, \theta))\dot{\gamma}(\lambda)\beta(\theta)\, d\lambda$$

$$+ \int_t^s \gamma(\lambda) \left[\alpha(X(\lambda, 0))\beta(0) - \alpha(X(\lambda, -r))\beta(-r) - \int_{-r}^{0} \alpha(X(\lambda, \theta))\dot{\beta}(\theta)d\theta \right] d\lambda$$

$$= \int_t^s \int_{-r}^{0} \alpha(X(\lambda, \theta))\dot{\gamma}(\lambda)\beta(\theta)\, d\lambda$$

$$+ \int_t^s \alpha(X(\lambda, 0))\gamma(\lambda)\beta(0)d\lambda - \int_t^s \alpha(X(\lambda, -r))\gamma(\lambda)\beta(-r)\, d\lambda$$

$$+ \int_t^s \int_{-r}^{0} \alpha(X(\lambda, \theta))[\dot{\gamma}(\lambda)\beta(\theta) - \gamma(\lambda)\dot{\beta}(\theta)]d\theta\, d\lambda.$$

Letting

$$G(s, \theta) = \gamma(s)\beta(\theta), \quad \forall(s, \theta) \in [0, T] \times [-r, 0],$$

we may rewrite the above equation in the following form

$$\int_{-r}^{0} \alpha(X(s, \theta))G(s, \theta)\, d\theta - \int_{-r}^{0} \alpha(X(t, \theta))G(t, \theta)\, d\theta$$

$$= \int_t^s \alpha(X(\lambda, 0))G(\lambda, 0)\, d\lambda - \int_t^s \alpha(X(\lambda, -r))G(\lambda, -r)\, d\lambda$$

$$+ \int_t^s \int_{-r}^{0} \alpha(X(\lambda, \theta))(\partial_\lambda G - \partial_\theta G)d\theta\, d\lambda. \tag{2.37}$$

By linearity (2.37), this holds for all functions G of the form $G(s,\theta) = \sum_{i=1}^{l} \gamma_i(s)\beta_i(\theta)$, where $\gamma_i \in C^1[0,T]$, $\beta_i \in C^1[-r,0]$, $i = 1,2,\ldots,l$. Then, by standard limiting arguments, it can be shown that (2.37) holds for all $G \in C^{1,1}([0,T] \times [-r,0])$.

Define the process $\hat{x}(\cdot) = \{\hat{s}(s), s \in [-r,T]\}$ by

$$\hat{x}(s) = \begin{cases} X(s,0) & \text{for } s \in [0,T] \\ \psi(s) & \text{for } s \in [-r,0]. \end{cases}$$

To show that $X(s) = \hat{x}_s$, it suffices to show that for $s_1, s_2 \in [0,T]$, $\theta_1, \theta_2 \in [-r,0]$,

$$X(s_1,\theta_1) = X(s_2,\theta_2), \quad \text{if } s_1 + \theta_1 = s_2 + \theta_2.$$

For, if $s \geq 0$, $-r \leq \theta \leq 0$,

$$\hat{x}_s(\theta) = \hat{x}(s+\theta) = \begin{cases} X(s+\theta,0) & \text{for } 0 \leq s+\theta \leq T \\ X(0,s+\theta) & \text{for } -r \leq s+\theta \leq 0. \end{cases} = X(s,\theta).$$

First, let us consider the case when $-r < \theta_2 < \theta_1 < 0$. It suffices to show that for some $0 < \delta < \theta_2 + r$,

$$X(s,\theta) = X(s',\theta + s - s'), \quad \theta_2 \leq \theta \leq \theta_1 \quad \text{whenever } s < s' < s + \delta. \quad (2.38)$$

Because, if (2.38) holds, then letting j to be the largest integer smaller than $\lfloor 2(s_2 - s_1)/\delta \rfloor$, we have

$$X(s_1,\theta_1) = X(s_1 + \theta, \theta_1 - \delta/2) = X(s_1 + \delta, \theta_1 - \delta)$$
$$= \cdots = X(s_1 + j\delta/2, \theta_1 - j\delta/2) = X(s_2 + \theta_1 - \theta_2) = X(s_2,\theta_2).$$

Note that $\theta_2 \leq \theta_1 - i\delta/2 \leq \theta_1$ for all $i = 1, 2, \cdots, j$.

To prove (2.38), we suppose that $\epsilon > 0$ is such that $-r < \theta_2 - \epsilon < \theta_1 + \theta < 0$. Let $\delta = \min\{r + \theta_2 - \epsilon, -(\theta_1 + \epsilon)\}$ and $\beta^* \in C^1[-r,0]$ be supported on the interval $[\theta_2 - \epsilon, \theta_1 + \epsilon]$. Fix an $s \in [0,T]$ and let $s < s' < s + \delta$. Taking $G(\lambda,\theta) = \beta^*(\lambda + \theta - s)$ in (2.37), we see that for $s \leq \lambda \leq s'$,

$$G(\lambda,0) = \beta^*(\lambda - s) = 0, \text{ since } \lambda - s \geq 0 > \theta_1 + \epsilon.$$
$$G(\lambda,-r) = \beta^*(\lambda - r - s) = 0,$$
$$\text{since } \lambda - r - s \leq s' - s - r < \delta - r \leq \theta_2 - \epsilon.$$
$$(\partial_\lambda - \partial_\theta)G = 0.$$

Hence from (2.37)

$$\int_{-r}^{0} \alpha(X(s',\theta))G(s',\theta)\,d\theta = \int_{-r}^{0} \alpha(X(s,\theta))G(s,\theta)\,d\theta$$
$$\Rightarrow \int_{-r}^{0} \alpha(X(s',\theta))\beta^*(s' + \theta - s)\,d\theta = \int_{-r}^{0} \alpha(X(s,\theta))\beta^*(\theta)\,d\theta$$
$$\Rightarrow \int_{-r}^{0} \alpha(X(s',t + s - s'))\beta^*(t)\,dt = \int_{-r}^{0} \alpha(X(s,t))\beta(t)\,dt, \quad (2.39)$$

putting $t = s' + \theta - s$. Note that during the change of variable of integration in (2.39), the boundary points for t lie outside the support of β^* because $s' - r - s < \delta - r \leq \theta_2 - \epsilon$ and $s' - s \geq 0 > \theta_1 + \epsilon$.

Since (2.39) holds for all $\beta^* \in C^1[\theta_2 - \epsilon, \theta_1 + \epsilon]$, we have for any $t \in (\theta_2 - \epsilon, \theta_1 + \epsilon)$,

$$\alpha(X(s', t + s - s')) = \alpha(X(s, t)).$$

But this being true for all $\alpha \in C_b^\infty(\Re^n, \Re^n)$, we have $X(s', t+s-s') = X(s,t)$, $\forall t \in [\theta_2, \theta_1]$, which is (2.39). Hence, we have proved (2.38) when $-r < \theta_2 < \theta_1 < 0$.

If $-r \leq \theta_2 < \theta_1 < 0$, then take a sequence $\theta_2^{(j)} > -r$ which decreases to θ_2. Then, for large j so that $\theta_1 - \theta_2 + \theta_2^{(j)} < 0$, we have

$$X(s_2, \theta_2^{(j)}) = X(s_1, \theta_1 - \theta_2 + \theta_2^{(j)})$$

since $s_1 + \theta_1 - \theta_2 + \theta_2^{(j)} = s_2 + \theta_1 - \theta_2 + \theta_2^{(j)} = s_2 + \theta_2^{(j)}$.

Taking the limit as $j \to \infty$, by continuity of $X(s, \cdot) : [-r, 0] \to \Re^n$, we then have $X(s_2, \theta_2) = X(s_1, \theta_1)$.

If $-r < \theta_2 < \theta_1 \leq 0$, then, taking a sequence $\theta_1^{(j)} < 0$ increasing to θ_1, we have for large j (so that $\theta_2 - \theta_1 + \theta_1^{(j)} > -r$),

$$X(s_2, \theta_2 - \theta_1 + \theta_1^{(j)}) = X(s_1, \theta_1^{(j)}),$$

since $s_2 + \theta_2 - \theta_1 + \theta_1^{(j)} = s_1 + \theta_1 - \theta_1 + \theta_1^{(j)} = s_1 + \theta_1^{(j)}$.

Again by continuity of $X(s, \cdot)$, taking the limit as $j \to \infty$, we get $X(s_2, \theta_2) = X(s_1, \theta_1)$

Finally, if $\theta_2 = -r$ and $\theta_1 = 0$ then $X(s_2, -r) = X(s_1, \theta_1)$.

Thus, we have proved that $X(s, \cdot) = \hat{x}_s$. It is easy to check that $\{\hat{x}(s), s \in [-r, T]\}$ is a continuous process and hence, so is $\{X(s, \cdot), s \in [0, T]\}$. Now, it remains to show that the process $\hat{x}(\cdot)$ satisfies a SHDE of the form (2.28).

For $F \in C_b^\infty(\Re^n)$, taking $\Phi(\phi) = F(\phi(0))$ in (2.32), we have

$$F(X(s,0)) - F(X(0,0)) - \int_0^s \nabla F(X(t,0)) \cdot f(X(t))dt$$

$$-\frac{1}{2} \int_0^s g(X(t)) \cdot \nabla^2 F(X(t,0))g(X(t))dt$$

is a martingale. That is,

$$F(\hat{x}(s)) - F(\hat{x}(0)) - \int_0^s \nabla F(\hat{x}_t)) \cdot f(\hat{x}_t)dt$$

$$-\frac{1}{2} \int_0^s g(\hat{x}_t) \cdot \nabla^2 F(\hat{x}(t))g(\hat{x}_t)dt$$

is a martingale for all $F \in C_b^\infty(\Re^n)$. Then, using standard arguments (see, e.g., Theorems 13.55 and 14.80 of Jacod [Jac79] and Theorem 4.5.2 of Strook and Varadhan [SV79]), we conclude that $\hat{x}(\cdot)$ satisfies a SHDE of the form (2.28) for some Brownian motion $\tilde{W}(\cdot)$. This implies that the law of $\hat{x}(\cdot)$ and $X(s) = \hat{x}_s$ is uniquely determined. Thus, the martingale problem for $(\mathbf{A}_q, \psi)$ is well posed.   □

Remark 2.5.6 *In the course of proving Theorem 2.5.5, we have proved that, for any probability measure μ on $(\mathbf{C}, \mathcal{B}(\mathbf{C}))$, the martingale measurable solution for $(\mathbf{A}_q, \mu)$ is well posed in the class of progressively measurable solutions and any progressively measurable solution $(\mathbf{A}_q, \mu)$ has a continuous modification. In particular, the following two conditions (i) and (ii) hold for $\mathbf{A}_q$.*
(i) The martingale problem for $(\mathbf{A}_q, \delta_\psi)$ is well posed in the class of r.c.l.l. solutions for every $\psi \in \mathbf{C}$. (ii) For all probability measure μ on $(\mathbf{C}, \mathcal{B}(\mathbf{C}))$, any progressively measurable solution to the martingale problem for $(\mathbf{A}_q, \mu)$ admits a r.c.l.l. modification, where r.c.l.l. stands for right-continuous with left limits.

We have the following theorem.

Theorem 2.5.7 *Consider the $\mathbf{C}$-valued segment process $\{x_s, s \in [0, T]\}$ for (2.28), with f and g satisfying the Lipschitz condition (2.29). Then*
(i) $\{x_s, s \in [0, T]\}$ is a Markov process.
(ii) $Q(\mathbf{C}, \Re) \subset \mathcal{D}(\mathbf{A})$, the domain of the weak infinitesimal generator $\mathbf{A}$ of the $\mathbf{C}$-valued Markov process $\{x_s, s \in [0, T]\}$ and the restriction of $\mathbf{A}$ on $Q(\mathbf{C}, \Re)$ is the same as $\mathbf{A}_q$.
(iii) $Q(\mathbf{C}, \Re)$ is weakly dense in $C_b(\mathbf{C})$.

Proof. By Theorem 2.5.4, $\{x_s, s \in [0, T]\}$ is a well posed solution for $(\mathbf{A}_q, \psi)$. Therefore, Conclusions (i) and (ii) follow from Lemma 2.5.2 and Proposition 2.5.3. Since the class of tame functions is weakly dense in $C_b(\mathbf{C})$ and every tame function can be approximated by a sequence of quasi-tame functions, the class of quasi-tame functions $Q(\mathbf{C}, \Re)$ is weakly dense in $C_b(\mathbf{C})$. This proves (iii).   □

2.6 Conclusions and Remarks

The main results presented in this chapter are Dynkin's formulas for $\Phi(s, x_s)$ for the $\mathbf{C}$-valued segment process $\{x_s, s \in [t, T]\}$ for (1.27) and $\Phi(S(s), S_s)$ for the $\mathbf{M}$-valued segment process $\{(S(s), S_s), s \geq \infty\}$ for (1.43) provided the Φ satisfies appropriate smooth conditions. When Φ is a quasi-tame function in appropriate spaces, $\mathcal{S}\Phi$ and $\mathbf{A}\Phi$ can be explicitly computed. These concepts and results are needed for treating problems in optimal classical control (Chapter 3), optimal stopping (Chapter 4), option pricing (Chapter 6), and hereditary portfolio optimization (Chapter 7). We note here that Itô's formula for $\Phi(x(s), x_s))$, $\Phi : \times \Re^n \times L^2([-r, 0]; \Re^n)$ for the solution process $x(\cdot)$ of the equation

$$dx(s) = f(x(s), x_s)ds + g(x(s), x_s)dW(s)$$

in the L^2-setting has just recently been obtained by Yan and Mohammed [YM05] using Malliavin calculus for anticipating integrands (see, e.g., Malliavin [Mal97], Nualart and Pardoux [NP88], and Nualart [Nua95])). Itô' formula for stochastic systems with pure delays has been established by Hu et al. [HMY04] for applications of discrete-time approximations of stochastic delay equations. However, the Itô's formula for the **C**-valued segment process for (1.27) is not yet available and it remains an open problem for the community of stochastic analysts. The source of difficulty in obtaining such a formula is due to non-differentiability in the Fréchet sense of the sup-norm $\|\cdot\|$ for the Banach space **C**. It is conjectured that Itô's formula in the L^2-setting can be generalized to the **M**-valued segment process $\{(S(s), S_s), s \geq 0\}$ for (1.43).

3

Optimal Classical Control

Let $0 < T < \infty$ be the given and fixed terminal time and let $t \in [0, T]$ be a variable initial time for the optimal classical control problem treated in this chapter. The main theme of this chapter is to consider the infinite-dimensional optimal classical control problem over a finite time horizon $[t, T]$. The dynamics of the process $x(\cdot) = \{x(s), s \in [t - r, T]\}$ being controlled are governed by a stochastic hereditary differential equation (SHDE) with a bounded memory of duration $0 < r < \infty$ and are taking values in the Banach space $\mathbf{C} = C([-r, 0]; \Re^n)$. The formulation of the control problem is given in Section 3.1. The value function $V : [0, T] \times \mathbf{C} \to \Re$ of the optimal classical control problem is written as a function of the initial datum $(t, \psi) \in [0, T] \times \mathbf{C}$. The existence of optimal control is proved in Section 3.2. In there, we consider an optimizing sequence of stochastic relaxed control problems with its corresponding sequence of value functions that converges to the value function of our original optimal control problem. Since the regular optimal control is a special case of optimal relaxed control, the existence of optimal control is therefore established. The Bellman-type dynamic programming principle (DPP) originally due to Larssen [Lar02] is derived and proved in Section 3.3. Based on the DDP, an infinite-dimensional Hamilton-Jacobi-Bellman equation (HJBE) is heuristically derived in Section 3.4 for the value function under the condition that it is sufficiently smooth. This HJBE involves a first- and second-order Fréchet derivatives with respect to spatial variable $\psi \in \mathbf{C}$ as well as an $\mathcal{S}$-operator that is unique only to SHDE. However, it is known in most optimal control problems, deterministic or stochastic, that the value functions, although can be proven to be continuous, do not meet these smoothness conditions and, therefore, cannot be a solution to the HBJE in the classical sense. To overcome this difficulty, the concept of viscosity solution to the infinite-dimensional HJBE is introduced in Section 3.5. Section 3.6 concerns the comparison principle between a super-viscosity solution and a sub-viscosity solution. Based on this comparison principle, it is shown that the value function is the unique viscosity solution to the HJBE. Due to the lack of smoothness of the value function, a classical verification theorem will not be useful in characterizing the optimal control. A generalized verification theorem in the framework of a viscosity solution is stated without a proof in Section 3.7. In Section 3.8, we prove that, under some special conditions on the controlled SHDE and the value function, the HJBE can take a finite-dimensional form in which only regular partial derivatives but not Fréchet derivatives are involved. Two application examples in this special form are also illustrated in this section.

We give the following example as a motivation for studying optimal classical control problems. Two other completely worked-out examples are given in Subsection 3.8.3.

Example. (Optimal Advertising Problem) (see Gossi and Marinelli [GM04] and Gossi et al. [GMS06])

Let $y(\cdot) = \{y(s), s \in [0,T]\}$ denote the stock of advertising goodwill of the product to be launched. The process $y(\cdot)$ is described by the following one-dimensional controlled stochastic hereditary differential equation:

$$dy(s) = \left[a_0 y(s) + \int_{-r}^{0} a_1(\theta)y(s+\theta)\, d\theta + b_0 u(s) \right.$$
$$\left. + \int_{-r}^{0} b_1(\theta)u(s+\theta)\, d\theta \right] ds + \sigma dW(s), \quad s \in [0,T],$$

with the initial conditions $y_0 = \psi \in C[-r,0]$ and $u_0 = \phi \in L^2([-r,0])$ at initial time $t = 0$.

In the above $(\Omega, \mathcal{F}, P, \mathbf{F}, W(\cdot))$ denotes an one-dimensional standard Brownian motion and the control process $u(\cdot) = \{u(s), s \in [0,T]\}$ denotes the advertising expenditures as a process in $L^2([0,T], \Re_+; \mathbf{F})$, the space of square integrable non-negative processes adapted to $\mathbf{F}$. Moreover, it is assumed that the following conditions are satisfied:

(i) $a_0 \leq 0$ denotes a constant factor of image deterioration of the product in absence of advertising.

(ii) $a_1(\cdot) \in L^2([-r,0], \Re)$ is the distribution of the forgetting time.

(iii) $b_0 \geq 0$ denotes the effective constant of instantaneous advertising effect.

(iv) $b_1(\cdot) \in L^2([-r,0], \Re_+)$ is the density function of the time lag between the advertising expenditure $u(\cdot)$ and the corresponding effect on the goodwill level.

(v) $\psi(\cdot)$ and $\phi(\cdot)$ are non-negative and represent, respectively, the histories of goodwill level and the advertising expenditure before time zero.

The objective of this optimal advertising problem is to seek an advertising strategy $u(\cdot)$ that maximizes the objective functional

$$J(\psi, \phi; u(\cdot)) = E\left[\Psi(y(T)) - \int_{0}^{T} L(u(s))\, ds \right],$$

where $\Psi : [0, \infty) \to [0, \infty)$ is a concave utility function with polynomial growth at infinity, $L : [0, \infty) \to [0, \infty)$ is a convex cost function which is superlinear at infinity, that is,

$$\lim_{u \to \infty} \frac{L(u)}{u} = \infty.$$

The above objective functional accounts for the balance between an utility of terminal goodwill $\Psi(y(T))$ and overall functional of advertising expenditures $\int_0^T L(u(s)) \, ds$ over the period. Note that this model example involves the histories of both the state and control processes. The general theory for optimal control of stochastic systems with delays in both state and control processes has yet to be developed. If $b_1(\cdot) = 0$, then there is no aftereffect of previous advertising expenditures on the goodwill level. In this case, it is a special case of what to be developed in this chapter.

3.1 Problem Formulation

3.1.1 The Controlled SHDE

In the following, let $(\Omega, \mathcal{F}, P, \mathbf{F}, W(\cdot))$ be a certain m-dimensional Brownian stochastic basis.

Consider the following controlled SHDE with a bounded memory (or delay) of duration $0 < r < \infty$:

$$dx(s) = f(s, x_s, u(s)) \, ds + g(s, x_s, u(s)) \, dW(s), \quad s \in [t, T], \qquad (3.1)$$

with the given initial data $(t, x_t) = (t, \psi) \in [0, T] \times \mathbf{C}$ and defined on a certain m-dimensional Brownian stochastic basis $(\Omega, \mathcal{F}, P, \mathbf{F}, W(\cdot))$ that is yet to be determined.

In (3.1), the following is understood:
(i) The drift $f : [0, T] \times \mathbf{C} \times U \to \Re^n$ and the diffusion coefficient $g : [0, T] \times \mathbf{C} \times U \to \Re^{n \times m}$ are deterministic continuous functions.
(ii) U, the control set, is a complete metric space and is typically a subset of an Euclidean space.
(iii) $u(\cdot) = \{u(s), s \in [t, T]\}$ is a U-valued $\mathbf{F}$-progressively measurable process that satisfies the following conditions:

$$E\left[\int_t^T |u(s)|^2 \, ds\right] < \infty. \qquad (3.2)$$

Note that the control process $u(\cdot) = \{u(s), s \in [t, T]\}$ defined on

$$(\Omega, \mathcal{F}, P, \mathbf{F}, W(\cdot))$$

is said to be progressively measurable if $u(\cdot) = \{u(s), t \leq s \leq T\}$ in U is $\mathbf{F}$-adapted (i.e., $u(s)$ is $\mathcal{F}(s)$-measurable for every $s \in [t, T]$), and for each $a \in [t, T]$ and $A \in \mathcal{B}(U)$, the set $\{(s, \omega) \mid t \leq s \leq a, \omega \in \Omega, u(s, \omega) \in A\}$ belongs to the product σ-field $\mathcal{B}([t, a]) \otimes \mathcal{F}(a)$; that is, if the mapping

$$(s, \omega) \mapsto u(s, \omega) : ([t, a] \times \Omega, \mathcal{B}([t, a]) \otimes \mathcal{F}(a)) \to (U, \mathcal{B}(U))$$

is measurable, for each $a \in [t, T]$.

Let $L : [0, T] \times \mathbf{C} \times U \to \Re$ and $\Psi : \mathbf{C} \to \Re$ be two deterministic continuous functions that represent the instantaneous and terminal reward functions, respectively, for the optimal classical control problem.

Assumption 3.1.1 *The assumptions on the functions f, g, L, and Ψ are stated as follows:*
(A3.1.1) (Lipschitz Continuity) *The maps $f(t, \phi, u)$, $g(t, \phi, u)$, $L(t, \phi, u)$, and $\Psi(\phi)$ are Lipschitz on $[0, T] \times \mathbf{C} \times U$ and Hölder continuous in $t \in [0, T]$: There is a constant $K_{lip} > 0$ such that*

$$|f(t, \phi, u) - f(s, \varphi, v)| + |g(t, \phi, u) - g(s, \varphi, v)|$$
$$+ |L(t, \phi, u) - L(s, \varphi, v)| + |\Psi(\phi) - \Psi(\varphi)|$$
$$\leq K_{lip}(\sqrt{|t - s|} + \|\phi - \varphi\| + |u - v|),$$

$$\forall s, t \in [0, T], u, v \in U, \text{ and } \phi, \varphi \in \mathbf{C}.$$

(A3.1.2) (Linear Growth) *There exists a constant $K_{grow} > 0$ such that*

$$|f(t, \phi, u)| + |g(t, \phi, u)| \leq K_{grow}(1 + \|\phi\|)$$

and

$$|L(t, \phi, u)| + |\Psi(\phi)| \leq K_{grow}(1 + \|\phi\|_2)^k, \quad \forall (t, \phi) \in [0, T] \times \mathbf{C}, \ u \in U.$$

(A3.1.3) The initial function ψ belongs to the space $L^2(\Omega, \mathbf{C}; \mathcal{F}(t))$ of $\mathcal{F}(t)$-measurable elements in $L^2(\Omega, \mathbf{C})$ such that

$$\|\psi\|_{L^2(\Omega; \mathbf{C})}^2 \equiv E[\|\psi\|^2] < \infty.$$

Condition (A3.1.2) in Assumption 3.1.1 stipulates that both L and Ψ satisfy a polynomial growth in $\phi \in \mathbf{C}$ under the norm L^2-norm $\| \cdot \|_2$ instead of the sup-norm $\| \cdot \|$. This stronger requirement is needed in order to show that the uniqueness of the viscosity solution of the HJBE in Section 3.6.

The solution process of the controlled SHDE (3.1) is given next.

Definition 3.1.2 *Given an m-dimensional Brownian stochastic basis $(\Omega, \mathcal{F}, P, \mathbf{F}, W(\cdot))$ and the control process $u(\cdot) = \{u(s), s \in [t, T]\}$, a process $x(\cdot; t, \psi_t, u(\cdot)) = \{x(s; t, \psi_t, u(\cdot)), s \in [t - r, T]\}$ is said to be a (strong) solution of the controlled SHDE (3.1) on the interval $[t - r, T]$ and through the initial datum $(t, \psi) \in [0, T] \times L^2(\Omega, \mathbf{C}; \mathcal{F}(t))$ if it satisfies the following conditions:*

1. $x_t(\cdot; t, \psi_t, u(\cdot)) = \psi_t(\cdot)$, *P-a.s.*
2. $x(s; t, \psi_t, u(\cdot))$ *is $\mathcal{F}(s)$-measurable for each $s \in [t, T]$;*

3. *For $1 \leq i \leq n$ and $1 \leq j \leq m$, we have*

$$P\left[\int_t^T \Big(|f_i(s, x_s, u(s))| + g_{ij}^2(s, x_s, u(s))\Big) ds < \infty\right] = 1.$$

4. *The process $\{x(s; t, \psi_t, u(\cdot)), s \in [t, T]\}$ is continuous and satisfies the following stochastic integral equation P-a.s.:*

$$x(s) = \psi_t(0) + \int_t^s f(\lambda, x_\lambda, u(\lambda)) \, d\lambda + \int_t^s g(\lambda, x_\lambda, u(\lambda)) \, dW(\lambda).$$

In addition, the solution process $\{x(s; t, \psi_t, u(\cdot)), s \in [t - r, T]\}$ is said to be (strongly) unique if $\{y(s; t, \psi, u(\cdot)), s \in [t - r, T]\}$ is also a solution of (3.1) on $[t - r, T]$ with the control process $u(\cdot)$ and through the same initial datum (t, ψ_t); then

$$P\{x(s; t, \psi, u(\cdot)) = y(s; t, \psi, u(\cdot)), \ \forall s \in [t, T]\} = 1.$$

Theorem 3.1.3 *Let $(\Omega, \mathcal{F}, P, \mathbf{F}, W(\cdot))$ be an m-dimensional Brownian motion and let $u(\cdot) = \{u(s), s \in [t, T]\}$ be a control process. Then for each initial datum $t, \psi_t) \in [0, T] \times L^2(\Omega, \mathbf{C}; \mathcal{F}(t))$, the controlled SHDE (3.1) has a unique strong solution process*

$$x(\cdot; t, \psi_t, u(\cdot)) = \{x(s; t, \psi_t, u(\cdot)), \quad s \in [t, T]\}$$

under Assumption 3.1.1. The following holds:

1. *The map $(s, \omega) \mapsto x(s; t, \psi_t, u(\cdot))$ belongs to the space $L^2(\Omega, C([t - r, T]; \Re^n)); \mathcal{F}(s))$, and the map $(t, \omega) \mapsto x_s(\cdot; t, \psi_t, u(\cdot))$ belongs to the space $L^2(\Omega, \mathbf{C}; \mathcal{F}(s))$. Moreover, there exists constants $K_b > 0$ and $k \geq 1$ such that*

$$E[\|x_s(\cdot; t, \psi_t, u(\cdot))\|^2] \leq K_b(1 + E[\|\psi_t\|^2])^k, \quad \forall s \in [t, T] \text{ and } u(\cdot) \in \mathcal{U}[t, T]. \tag{3.3}$$

2. *The map $\psi_t \mapsto x_s(\cdot; t, \psi_t, u(\cdot))$ is Lipschitz; that is, there is a constant $K > 0$ such that for all $s \in [t, T]$ and $\psi_t^{(1)}, \psi_t^{(2)} \in L^2(\Omega, \mathbf{C}; \mathcal{F}(t))$,*

$$E[\|x_s(\cdot; t, \psi_t^{(1)}, u(\cdot)) - x_s(\cdot; t, \psi_t^{(2)}, u(\cdot))\|] \leq KE[\|\psi_t^{(1)} - \psi_t^{(2)}\|]. \tag{3.4}$$

Proof. Let the random functions $\tilde{f} : [0, T] \times \mathbf{C} \times \Omega \to \Re^n$ and $\tilde{g} : [0, T] \times \mathbf{C} \times \Omega \to \Re^{n \times m}$ be defined as follows:

$$\tilde{f}(s, \phi, \omega) = f(s, \phi, u(s, \omega))$$

and

$$\tilde{g}(s, \phi, \omega) = g(s, \phi, u(s, \omega)),$$

where $u(\cdot) = \{u(s), s \in [t, T]\}$ is the control process. If the functions f and g satisfy Assumption 3.1.1, then the random functions $\tilde{f}$ and $\tilde{g}$ defined above satisfy Assumptions 1.3.6 and 1.3.7 of Chapter 1 and, therefore, the controlled system (3.1) has a unique *strong* solution process on $[t-r,T]$ and through the initial datum $(t,\psi) \in [0,T] \times L^2(\Omega, \mathbf{C}; \mathcal{F}(t))$, which is denoted by

$$x(\cdot;t,\psi_t,u(\cdot)) = \{x(s;t,\psi_t,u(\cdot)), s \in [t,T]\}$$

(or simply $x(\cdot)$ when there is no danger of ambiguity) according to Theorem 1.3.12 of Chapter 1.   □

Remark 3.1.4 *It is clear from the appearance of (3.1) that the use of the term "classical control" (as opposed to "impulse control" in Chapter 7) becomes self-explanatory. This is due to the fact that an application of the control $u(s)$ at time $s \in [t,T]$ will only change the rate of the drift and the diffusion coefficient and, therefore, the pathwise continuity of controlled state process $x(\cdot) = \{x(s), s \in [t-r,T]\}$ will not be affected by this action.*

Definition 3.1.5 *The (classical) control process $u(\cdot)$ is an Markov (or feedback) control if there exist a Borel measurable function $\eta : [0,T] \times \mathbf{C} \to U$ such that*

$$u(s) = \eta(s,x_s),$$

where $\{x_s, s \in [t,T]\}$ is the $\mathbf{C}$-valued Markov process corresponding to the solution process $\{x(s), s \in [t-r,T]\}$ of the following feedback equation:

$$dx(s) = f(s,x_s,\eta(s,x_s))\,ds + g(s,x_s,\eta(s,x_s))\,dW(s) \qquad (3.5)$$

with the initial data $(t,x_t) = (t,\psi) \in [0,T] \times \mathbf{C}$.

3.1.2 Admissible Controls

Definition 3.1.6 *For each $t \in [0,T]$, a six-tuple $(\Omega, \mathcal{F}, P, \mathbf{F}, W(\cdot), u(\cdot))$ is said to be an admissible control if it satisfies the following conditions:*

1. *$(\Omega, \mathcal{F}, P, \mathbf{F}, W(\cdot))$ is a certain m-dimensional Brownian stochastic basis;*
2. *$u : [t,T] \times \Omega \to U$ is an $\mathbf{F}$-adapted and is right-continuous at the initial time t; that is, $\lim_{s\downarrow t} u(s) = u(t)$ (say $= u \in U$).*
3. *Under the control process $u(\cdot) = \{u(s), s \in [t,T]\}$, (3.1) admits a unique strong solution $x(\cdot;t,\psi,u(\cdot)) = \{x(s;t,\psi,u(\cdot)), s \in [t,T]\}$ on $(\Omega, \mathcal{F}, P, \mathbf{F}, W(\cdot))$ and through each initial datum $(t,\psi) \in [0,T] \times \mathbf{C}$.*
4. *The $\mathbf{C}$-valued segment process $\{x_s(t,\psi,u(\cdot)), s \in [t,T]\}$ defined by*

$$x_s(\theta;t,\psi,u(\cdot)) = x(s+\theta;t,\psi,u(\cdot)), \quad \theta \in [-r,0],$$

is a strong Markov process with respect to the Brownian stochastic basis $(\Omega, \mathcal{F}, P, \mathbf{F}, W(\cdot))$.
5. *The control process $u(\cdot)$ is such that*

$$E\left[\int_t^T |L(s,x_s(\cdot;t,\psi,u(\cdot)),u(s))|\,ds + |\Psi(x_T(\cdot;t,\psi,u(\cdot)))|\right] < \infty,$$

where $L : [0,T] \times \mathbf{C} \times U \to \Re$ and $\Psi : \mathbf{C} \to \Re$ represent the instantaneous and the terminal reward functions, respectively.

The collection of *admissible controls* $(\Omega, \mathcal{F}, P, \mathbf{F}, W(\cdot), u(\cdot))$ over the interval $[t,T]$ will be denoted by $\mathcal{U}[t,T]$.

We will write $u(\cdot) \in \mathcal{U}[t,T]$ or formally the 6-tuple $(\Omega, \mathcal{F}, P, \mathbf{F}, W(\cdot), u(\cdot)) \in \mathcal{U}[t,T]$ interchangeably, whenever there is no danger of ambiguity.

Remark 3.1.7 *Definition 3.1.6 defines a weak formulation of an admissible control in that the Brownian stochastic basis $(\Omega, \mathcal{F}, P, \mathbf{F}, W(\cdot))$ is not predetermined and in fact is a part of the ingredients that constitute an admissible control. This is contrary to the strong formulation of an admissible control in which the Brownian stochastic basis $(\Omega, \mathcal{F}, P, \mathbf{F}, W(\cdot))$ is predetermined and given.*

Remark 3.1.8 *To avoid using the yet-to-be-developed Itô formula for the the $\mathbf{C}$-valued process $\{x_s(t, \psi, u(\cdot)), s \in [t,T]\}$ in the development of the HJB theory, we make additional requirement in Condition 3 of Definition 3.1.6 that it is a strong Markov process. This requirement is not a stringent one. In fact, the class of admissible controls $\mathcal{U}[t,T]$ defined in Definition 3.1.6 includes all Markov (or feedback) control (see Definition 3.1.5), where $\eta : [0,T] \times \mathbf{C} \to U$ is Lipschitz with respect to the segment variable; that is, there exists a constant $K > 0$ such that*

$$|\eta(t,\phi) - \eta(t,\varphi)| \leq \|\phi - \varphi\|, \quad \forall (t,\phi), (t,\varphi) \in [0,T] \times \mathbf{C}.$$

Throughout, we assume that the functions $f : [0,T] \times \mathbf{C} \times U \to \Re^n$, $g : [0,T] \times \mathbf{C} \times U \to \Re^{n \times m}$, $L : [0,T] \times \mathbf{C} \times U \to \Re$, and $\Psi : \mathbf{C} \to \Re$ satisfy Assumption 3.1.1.

Given an admissible control $u(\cdot) \in \mathcal{U}[t,T]$, let

$$x(\cdot; t, \psi, u(\cdot)) = \{x(s; t, \psi, u(\cdot)), s \in [t-r,T]\}$$

be the solution of (3.1) through the initial datum $(t,\psi) \in [0,T] \times \mathbf{C}$. We again consider the corresponding $\mathbf{C}$-valued segment process $\{x_s(\cdot; t, \psi, u(\cdot)), s \in [t,T]\}$. For notational simplicity, we often write $x(s) = x(s; t, \psi, u(\cdot))$ and $x_s = x_s(\cdot; t, \psi, u(\cdot))$ for $s \in [t,T]$ whenever there is no danger of ambiguity.

3.1.3 Statement of the Problem

Given any initial datum $(t,\psi) \in [0,T] \times \mathbf{C}$ and any admissible control $u(\cdot) \in \mathcal{U}[t,T]$, we define the objective functional

$$J(t,\psi; u(\cdot)) \equiv E\Bigg[\int_t^T e^{-\alpha(s-t)} L(s, x_s(\cdot; t, \psi, u(\cdot)), u(s)) \, ds$$

$$+ e^{-\alpha(T-t)} \Psi(x_T(\cdot; t, \psi, u(\cdot))) \Bigg], \qquad (3.6)$$

where $\alpha \geq 0$ denotes a discount factor.

For each initial datum $(t, \psi) \in [0, T] \times \mathbf{C}$, the optimal control problem is to find $u(\cdot) \in \mathcal{U}[t, T]$ so as to maximize the objective functional $J(t, \psi; u(\cdot))$ defined above. In this case, the value function $V : [0, T] \times \mathbf{C} \rightarrow \Re$ is defined to be

$$V(t, \psi) = \sup_{u(\cdot) \in \mathcal{U}[t, T]} J(t, \psi; u(\cdot)). \tag{3.7}$$

The control process $u^*(\cdot) = \{u^*(s), s \in [t, T]\} \in \mathcal{U}[t, T]$ is said to be an optimal control for the optimal classical control problem if

$$V(t, \psi) = J(t, \psi; u^*(\cdot)). \tag{3.8}$$

The (strong) solution process

$$x^*(\cdot; t, \psi, u^*(\cdot)) = \{x^*(s; t, \psi, u^*(\cdot)), s \in [t - r, T]\}$$

of (3.1) corresponding to the optimal control $u^*(\cdot)$) will be called the optimal state process corresponding to $u^*(\cdot)$. The pair $(u^*(\cdot), x^*(\cdot))$ will be called the optimal control-state pair.

The characterizations of the value function $V : [0, T] \times \mathbf{C} \rightarrow \Re$ and the optimal control-state pair $(u^*(\cdot), x^*(\cdot))$ will normally constitute an solution to the control problem. The optimal classical control problem, Problem (OCCP), is now formally formulated as follows.

Problem (OCCP). For each initial datum $(t, \psi) \in [0, T] \times \mathbf{C}$:

1. Find an $u^*(\cdot) \in \mathcal{U}[t, T]$ that maximizes $J(t, \psi; u(\cdot))$ defined in (3.6) among $\mathcal{U}[t, T]$.
2. Characterize the value function $V : [0, T] \times \mathbf{C} \rightarrow \Re$ defined in (3.7).
3. Identify the optimal control-state pair $(u^*(\cdot), x^*(\cdot))$.

3.2 Existence of Optimal Classical Control

In the class $\mathcal{U}[t, T]$ of admissible controls it may happen that there does not exist an optimal control. The following artificial example of Kushner and Dupuis [KD01, p.86] shows that an optimal control does not exist even for a controlled deterministic equation without a delay.

Example. Consider the following one-dimensional controlled deterministic equation:

$$\dot{x}(s) = f(x(s), u(s)) \equiv u(s), \quad s \geq 0$$

with the control set $U = [-1, 1]$. Starting from the initial state $x(0) = x \in \Re$, the objective is to find an admissible (deterministic) control $u(\cdot) = \{u(s), s \geq 0\}$ that minimizes the following discounted cost functional over the infinite time horizon:

$$J(x; u(\cdot)) = \int_0^\infty e^{-\alpha s}[x^2(s) + (u^2(s) - 1)^2]\, ds.$$

Again, let $V : \Re \to \Re$ be the value function of the control problem defined by

$$V(x) = \inf_{u(\cdot) \in \mathcal{U}[0,\infty)} J(x; u(\cdot)).$$

Note that $V(0) = 0$ and, in general, $V(x) = x^2/\alpha$ for all $x \in \Re$. To see this, define the sequence of controls $u^{(k)}(\cdot)$ by

$$u^{(k)}(s) = (-1)^j \text{ on the half-open interval } [j/k, (j+1)/k), \ j = 0, 1, 2, \ldots.$$

It is not hard to see that $J(0; u^{(k)}(\cdot)) \to 0$ as $k \to \infty$. In a sense, when $x(0) = 0$, the optimal control $u^*(\cdot)$ wants to take values ± 1. However, it is easy to check that $u^*(\cdot)$ does not satisfy Definition 3.1.6. Therefore, an optimal control $u^*(\cdot)$ does not exist as defined.

In order to establish the existence of an optimal control for Problem (OCCP), we will enlarge the class of controls, allowing the so-called relaxed controls, so that the existence of an optimal (relaxed) control is guaranteed, and the supremum of the expected objective functional over this new class of controls coincides with the value function $V : [0, T] \times \mathbf{C} \to \Re$ of the original optimal classical control problem defined by (3.7). The idea to show the existence of an optimal relaxed control is to consider a maximizing sequence of admissible relaxed controls $\{\mu^{(k)}(\cdot, \cdot)\}_{k=1}^\infty$ on the Borel measurable space $([0, T] \times U, \mathcal{B}([0, T] \times U))$ and the corresponding sequence of objective functionals $\{\hat{J}(t, \psi; \mu^{(k)}(\cdot, \cdot))\}_{k=1}^\infty$. By the fact that $[0, T] \times U$ is compact (and hence the maximizing sequence of admissible relaxed controls $\{\mu^{(k)}(\cdot, \cdot)\}_{k=1}^\infty$ is compact in the Prohorov metric) and the fact that $\hat{J}(t, \psi; \mu(\cdot, \cdot))$ is upper semicontinuous in admissible relaxed controls $\mu(\cdot, \cdot)$, one can show that the sequence $\{\mu^{(k)}(\cdot, \cdot)\}_{k=1}^\infty$ converges weakly to an admissible relaxed control $\mu^*(\cdot, \cdot)$. This $\mu^*(\cdot, \cdot)$ can be shown to be optimal among the class of admissible relaxed controls and that its value function coincides with the value function of Problem (OCCP). We also prove that an optimal (classical) control exists if the value function $V(t, \psi)$ is finite for each initial datum $(t, \psi) \in [0, T] \times \mathbf{C}$.

We recall the concept and characterizations of weak convergence of probability measures without proofs as follows. The detail can be found in Billingsley [Bil99].

Let (Ξ, d) be a generic metric space with the Borel σ-algebra denoted by $\mathcal{B}(\Xi)$. Let $\mathcal{P}(\Xi)$ (or simply $\mathcal{P}$ whenever there is no ambiguity) be the collection of probability measures defined on $(\Xi, \mathcal{B}(\Xi))$. We will equip the space $\mathcal{P}(\Xi)$ with the Prohorov metric $\pi : \mathcal{P}(\Xi) \times \mathcal{P}(\Xi) \to [0, \infty)$ defined by

$$\pi(\mu, \nu) = \inf\{\epsilon > 0 \mid \mu(A^\epsilon) \leq \nu(A) + \epsilon \text{ for all closed } A \in \mathcal{B}(\Xi)\}, \quad (3.9)$$

where A^ϵ is the ϵ-neighborhood of $A \in \mathcal{B}(\Xi)$, that is,

$$A^\epsilon = \{\mathbf{y} \in \Xi \mid d(\mathbf{x}, \mathbf{y}) < \epsilon \text{ for some } \mathbf{x} \in \Xi\}.$$

If $\mu^{(k)}, k = 1, 2, \ldots$, is a sequence in $\mathcal{P}(\Xi)$, we say that the sequence $\mu^{(k)}, k = 1, 2, \ldots$, converges weakly to $\mu \in \mathcal{P}(\Xi)$ and is to be denoted by $\mu^{(k)} \Rightarrow \mu$ as $k \to \infty$ if

$$\lim_{k \to \infty} \int_{\Xi} \Phi(\mathbf{x}) \mu^{(k)}(d\mathbf{x}) = \int_{\Xi} \Phi(\mathbf{x}) \mu(d\mathbf{x}), \quad \forall \Phi \in C_b(\Xi), \tag{3.10}$$

where $C_b(\Xi)$ is the space of bounded and continuous functions $\Phi : \Xi \to \Re$ equipped with the sup-norm:

$$\|\Phi\|_{C_b(\Xi)} = \sup_{\mathbf{x} \in \Xi} |\Phi(\mathbf{x})|, \quad \Phi \in C_b(\Xi).$$

In the case where $\mu^{(k)}$ and $\mu \in \mathcal{P}(\Xi)$ are probability measures induced by Ξ-valued random variables $X^{(k)}$ and X, respectively, we often say that $X^{(k)}$ converges weakly to X and is to be denoted by $X^{(k)} \Rightarrow X$ as $k \to \infty$. A direct consequence of the definition of weak convergence is that $X^{(k)} \Rightarrow X$ implies that $\Phi(X^{(k)}) \Rightarrow \Phi(X)$ for any continuous function Φ from Ξ to another metric space.

We state the following results without proofs. The details of these results can be found in [Bil99].

Theorem 3.2.1 *If Ξ is complete and separable, then $\mathcal{P}(\Xi)$ is complete and separable under the Prohorov metric. Furthermore, if Ξ is compact, then $\mathcal{P}(\Xi)$ is compact.*

Let $\{\mu^{(\lambda)}, \lambda \in \Lambda\} \subset \mathcal{P}(\Xi)$, where Λ is an arbitrary index set.

Definition 3.2.2 *The collection of probability measure $\{\mu^{(\lambda)}, \lambda \in \Lambda\}$ is called tight if for each $\epsilon > 0$, there exists a compact set $K_\epsilon \subset \Xi$ such that*

$$\inf_{\lambda \in \Lambda} \mu^{(\lambda)}(K_\epsilon) \geq 1 - \epsilon. \tag{3.11}$$

If the measures $\mu^{(\lambda)}$ are the induced measures defined by some random variables $X^{(\lambda)}$, then we also refer to the collection $\{X^{(\lambda)}\}$ as tight. Condition (3.11) then reads (in the special case where all the random variables are defined on the same space)

$$\inf_{\lambda \in \Lambda} P\{X^{(\lambda)} \in K_\epsilon\} \geq 1 - \epsilon.$$

Theorem 3.2.3 (Prohorov's Theorem) *If Ξ is complete and separable, then a set $\{\mu^{(\lambda)}, \lambda \in \Lambda\} \subset \mathcal{P}(\Xi)$ has compact closure in the Prohorov metric if and only if $\{\mu^{(\lambda)}, \lambda \in \Lambda\}$ is tight.*

Remark 3.2.4 *Let Ξ_1 and Ξ_2 be two complete and separable metric spaces, and consider the space $\Xi = \Xi_1 \times \Xi_2$ with the usual product space topology. For $\{\mu^{(\lambda)}, \lambda \in \Lambda\} \subset \mathcal{P}(\Xi)$, let $\{\mu_i^{(\lambda)}, \lambda \in \Lambda\} \subset \mathcal{P}(\Xi_i)$, for $i = 1, 2$, be defined by taking $\mu_i^{(\lambda)}$ to be the marginal distribution of $\mu^{(\lambda)}$ on Ξ_i. Then $\{\mu^{(\lambda)}, \lambda \in \Lambda\}$ is tight if and only if $\{\mu_1^{(\lambda)}, \lambda \in \Lambda\}$ and $\{\mu_2^{(\lambda)}, \lambda \in \Lambda\}$ are tight.*

Theorem 3.2.5 *Let Ξ be a metric space and let $\mu^{(k)}, k = 1, 2, \ldots$, and μ be probability measures in $\mathcal{P}(\Xi)$ satisfying $\mu^{(k)} \Rightarrow \mu$. Let Φ be a real-valued measurable function on Ξ and define $\mathbf{D}(\Phi)$ to be the measurable set of points at which Φ is not continuous. Let $X^{(k)}$ and X be Ξ-valued random variables defined on a probability space $(\Omega, \mathcal{F}, P)$ that induce the measures $\mu^{(k)}$ and μ on Ξ, respectively. Then $\Phi(X^{(k)}) \Rightarrow \Phi(X)$ whenever $P\{X \in \mathbf{D}(\Phi)\} = 0$.*

Theorem 3.2.6 (Aldous Criterion) *Let $X^{(k)}(\cdot) = \{X^{(k)}(t), t \in [0, T]\}, k = 1, 2, \ldots$, be a sequence of Ξ-valued continuous processes (defined on the same filtered probability space $(\Omega, \mathcal{F}, P, \mathbf{F})$). Then the sequence $\{X^{(k)}(\cdot)\}_{k=1}^{\infty}$ converges weakly if and only if the following condition is satisfied: Given $k = 1, 2, \ldots$, any bounded $\mathbf{F}$-stopping time τ, and $\delta > 0$, we have*

$$E^{(k)}[\|X^{(k)}(\tau + \delta) - X^{(k)}(\tau)\|_{\Xi}^2 \mid \mathcal{F}^{(k)}(\tau)] \leq 2K^2\delta(\delta + 1).$$

We also recall the following *Skorokhod representation theorem* that is often used to prove convergence with probability 1. The proof can be found in Ethier and Kurtz [EK86].

Theorem 3.2.7 (Skorokhod Representation Theorem) *Let Ξ be a separable metric space and assume the probability measures $\{\mu^{(k)}\}_{k=1}^{\infty} \subset \mathcal{P}(\Xi)$ converges weakly to $\mu \in \mathcal{P}(\Xi)$. Then there exists a probability space $(\tilde{\Omega}, \tilde{\mathcal{F}}, \tilde{P})$ on which there are defined Ξ-valued random variables $\{\tilde{X}^{(k)}\}_{k=1}^{\infty}$ and $\tilde{X}$ such that for all Borel sets $B \in \mathcal{B}(\Xi)$ and all $k = 1, 2, \ldots$,*

$$\tilde{P}\{\tilde{X}^{(k)} \in B\} = \mu^{(k)}(B), \qquad \tilde{P}\{\tilde{X} \in B\} = \mu(B)$$

and such that

$$\tilde{P}\{\lim_{k \to \infty} \tilde{X}^{(k)} = \tilde{X}\} = 1.$$

3.2.1 Admissible Relaxed Controls

We first define a *deterministic relaxed control* as follows.

Definition 3.2.8 *A deterministic relaxed control is a positive measure μ on the Borel σ-algebra $\mathcal{B}([0, T] \times U)$ such that*

$$\mu([0, t] \times U) = t, \quad t \in [0, T]. \tag{3.12}$$

For each $G \in \mathcal{B}(U)$, the function $t \mapsto \mu([0, t] \times G)$ is absolutely continuous with respect to Lebesque measure on $([0, T], \mathcal{B}([0, T]))$ by virtue of (3.12). Denote by $\dot{\mu}(\cdot, G) = \frac{d}{dt}\mu([0, t] \times G)$ any Lebesque density of $\mu([0, t] \times G)$. The family of densities $\{\dot{\mu}(\cdot, G), G \in \mathcal{B}(U)\}$ is a probability measure on $\mathcal{B}(U)$ for each $t \in [0, T]$, and

$$\mu_{u(\cdot)}(B) = \int_0^T \int_U \mathbf{1}_{\{(t,u)\in B\}}\mu(dt, du) \tag{3.13}$$

$$= \int_0^T \int_U \mathbf{1}_{\{(t,u)\in B\}}\dot{\mu}(t, du)\, dt, \quad \forall B \in \mathcal{B}([0, T] \times U).$$

Denote by $\mathcal{R}$ the space of deterministic relaxed controls that is equipped with the *weak compact topology* induced by the following notion of convergence: A sequence $\{\mu^{(k)}, k = 1, 2, \ldots\}$ of relaxed controls converges (weakly) to $\mu \in \mathcal{R}$ if

$$\lim_{k \to \infty} \int_{[0,T] \times U} \gamma(t, u) \, d\mu^{(k)}(t, u) = \int_{[0,T] \times U} \gamma(t, u) \, d\mu(t, u), \qquad (3.14)$$

$$\forall \gamma \in C_c([0, T] \times U),$$

where $C_c([0, T] \times U)$ is the space of all real-valued continuous functions on $[0, T] \times U$ having compact support. Note that if U is compact then $C_c([0, T] \times U) = C([0, T] \times U)$. Under the *weak-compact* topology defined above, $\mathcal{R}$ is a (sequentially) compact space; that is, every sequence in $\mathcal{R}$ has a subsequence that converges to an element in $\mathcal{R}$ in the sense of (3.14).

Now, we introduce a suitable filtration for $\mathcal{R}$ as follows. We first identify each $\mu \in \mathcal{R}$ as a linear functional on $C([0, T] \times U)$ in the following way:

$$\mu(\varsigma) \equiv \int_0^T \int_U \varsigma(t, u) \mu(dt, du), \quad \forall \varsigma \in C([0, T] \times U).$$

For any $\varsigma \in C([0, T] \times U)$ and $t \in [t, T]$, define $\varsigma^t \in C([0, T] \times U)$ by

$$\varsigma^t(s, u) \equiv \varsigma(s \wedge t, u).$$

Since $C([0, T] \times U)$ is separable (and therefore has a countable dense subset), we may let $\{\varsigma^{(k)}\}_{k=1}^\infty$ be countable dense subset (with respect to the uniform norm). It is easy to see that $\{\varsigma^{(k),t}\}_{k=1}^\infty$ is dense in the set $\{\varsigma^t \mid \varsigma \in C([0, T] \times U)\}$. Define

$$\mathcal{B}_s(\mathcal{R}) \equiv \sigma\{\{\mu \in \mathcal{R} \mid \mu(\varsigma^t) \in B\} \mid \varsigma \in C([0, T] \times U), t \in [0, s], B \in \mathcal{B}(\Re)\}.$$

One can easily show that $\mathcal{B}_s(\mathcal{R})$ can be generated by cylinder sets of the following form:

$$\sigma\{\{\mu \in \mathcal{R} \mid \mu(\varsigma^{(k),t}) \in (a, b)\} \mid s \geq t \in \mathbf{Q}, k = 1, 2, \ldots, a, b \in \mathbf{Q}\}. \qquad (3.15)$$

Definition 3.2.9 *A relaxed control process is an $\mathcal{R}$-valued random variable μ, defined on a Brownian stochastic basis $(\Omega, \mathcal{F}, P, \mathbf{F}, W(\cdot))$, such that the mapping $\omega \mapsto \mu([0, t] \times G)(\omega)$ is $\mathcal{F}(t)$-measurable for all $t \in [0, T]$ and $G \in \mathcal{B}(U)$.*

Using the relaxed control process $\mu(\cdot, \cdot) \in \mathcal{R}$, the controlled state equation can be written as

$$dx(s) = \int_U f(s, x_s, u)\dot{\mu}(s, du) \, ds + \int_U g(s, x_s, u)\dot{\mu}(s, du) \, dW(s), \quad s \in [t, T],$$

or, equivalently, in the form of the stochastic integral equation:

$$x(s) = \psi(0) + \int_t^s \int_U f(\lambda, x_\lambda, u)\dot{\mu}(\lambda, du)\, d\lambda \tag{3.16}$$

$$+ \int_t^s \int_U g(\lambda, x_\lambda, u)\dot{\mu}(\lambda, du)\, dW(\lambda), \quad s \in [t, T],$$

with the initial datum $(t, \psi) \in [0, T] \times \mathbf{C}$. The objective functional can be written as

$$\hat{J}(t, \psi; \mu(\cdot, \cdot)) = E\Big[\int_U \int_t^T L(s, x_s(\cdot; t, \psi, \mu(\cdot, \cdot)), u)\dot{\mu}(s, du)\, ds$$

$$+ \Psi(x_T(\cdot; t, \psi, \mu(\cdot, \cdot))) \Big], \tag{3.17}$$

where $\{x(s; t, \psi, \mu(\cdot, \cdot)), s \in [t, T]\}$ is the solution process of (3.16) when the relaxed control process $\mu(\cdot, \cdot) \in \mathcal{R}$ is applied.

We now define an admissible relaxed control as follows.

Definition 3.2.10 *For each initial datum* $(t, \psi) \in [0, T] \times \mathbf{C}$, *a six-tuple* $(\Omega, \mathcal{F}, P, \mathbf{F}, W(\cdot), \mu(\cdot, \cdot))$ *is said to be an admissible relaxed control at* $(t, \psi) \in [0, T] \times \mathbf{C}$ *if it satisfies the following conditions:*

1. $(\Omega, \mathcal{F}, P, \mathbf{F}, W(\cdot))$ *is a certain m-dimensional Brownian stochastic basis.*
2. $\mu(\cdot, \cdot) \in \mathcal{R}$ *is a relaxed control defined on the Brownian stochastic basis* $(\Omega, \mathcal{F}, P, \mathbf{F}, W(\cdot))$.
3. *Under the relax control process* $\mu(\cdot, \cdot)$, *(3.16) admits a unique strong solution* $x(\cdot; t, \psi, \mu(\cdot, \cdot)) = \{x(s; t, \psi, \mu(\cdot, \cdot)), s \in [t, T]\}$ *and through each initial datum* $(t, \psi) \in [0, T] \times \mathbf{C}$.
4. *The control process* $\mu(\cdot, \cdot)$ *is such that*

$$E\Big[\int_U \int_t^T |L(s, x_s(\cdot; t, \psi, \mu(\cdot, \cdot)), u)|\dot{\mu}(du, s)\, ds$$

$$+ |\Psi(x_T(\cdot; t, \psi, \mu(\cdot, \cdot)))| \Big] < \infty,$$

The collection of admissible relaxed controls $(\Omega, \mathcal{F}, P, \mathbf{F}, W(\cdot), \mu(\cdot, \cdot))$ *over the interval* $[t, T]$ *will be denoted by* $\hat{\mathcal{U}}[t, T]$. *Again, when there is no ambiguity, we often write* $\mu(\cdot, \cdot) \in \hat{\mathcal{U}}[t, T]$ *instead of* $(\Omega, \mathcal{F}, P, \mathbf{F}, W(\cdot), \mu(\cdot, \cdot))$.

The optimal relaxed control problem can be stated as follows.
Problem (ORCP) Find an optimal relaxed control $\mu^*(\cdot, \cdot) \in \hat{\mathcal{U}}[t, T]$ that maximizes (3.17) subject to (3.16).

We again define the value function $\hat{V} : [0, T] \times \mathbf{C} \to \Re$ for the Problem (ORCP) by

$$\hat{V}(t, \psi) = \sup_{\mu(\cdot, \cdot) \in \hat{\mathcal{U}}[t, T]} \hat{J}(t, \psi; \mu(\cdot, \cdot)). \tag{3.18}$$

We have the following existence theorem for Problem (ORCP).

Theorem 3.2.11 *Let Assumption 3.1.1 hold. Given any initial datum $(t, \psi) \in [0, T] \times \mathbf{C}$, then Problem (ORCP) admits an optimal relaxed control $\mu^*(\cdot, \cdot)$, and its value function $\hat{V}$ coincides with the value function V of Problem (OCCP).*

We will postpone proof of Theorem 3.2.11 until the end of the next subsection.

3.2.2 Existence Result

For the existence of an optimal control for Problem (OCCP), we need the following Roxin condition:
(Roxin's Condition). For every $(t, \psi) \in [0, T] \times \mathbf{C}$, the set

$$(f, gg^\top, L)(t, \psi, U) \equiv \Big\{ \big(f_i(t, \psi, u), (gg^\top)_{ij}(t, \psi, u), L(t, \psi, u) \big) \mid$$
$$u \in U, i, j = 1, 2, \ldots, n \Big\}$$

is convex in $\Re^{n+nn+1}$.

The main purpose of this subsection is to prove the existence theorem.

Theorem 3.2.12 *Let Assumption 3.1.1 and the Roxin condition hold. Given any initial datum $(t, \psi) \in [0, T] \times \mathbf{C}$, then Problem (OCCP) admits an optimal classical control $u^*(\cdot) \in \mathcal{U}[t, T]$ if the value function $V(t, \psi)$ is finite.*

Proof. Without loss of generality, we can and will assume that $t = 0$ in the following for notational simplicity. The proof is similar to that of Theorem 2.5.3 in Yong and Zhou [YZ99] and will be carried out by the following steps:
Step 1. Since $V(0, \psi)$ is finite, we can find a sequence of maximizing admissible controls in $\mathcal{U}[t, T]$,

$$\{(\Omega^{(k)}, \mathcal{F}^{(k)}, P^{(k)}, \mathbf{F}^{(k)}, W^{(k)}(\cdot), u^{(k)}(\cdot))\}_{k=1}^\infty,$$

such that

$$\lim_{k \to \infty} J(0, \psi; u^{(k)}(\cdot)) = V(0, \psi). \tag{3.19}$$

Let $x^{(k)}(\cdot) = \{x(\cdot; 0, \psi, u^{(k)}(\cdot)), s \in [0, T]\}$ be the solution of (3.1) corresponding to $u^{(k)}(\cdot)$. Define

$$X^{(k)}(\cdot) \equiv (x^{(k)}(\cdot), F^{(k)}(\cdot), G^{(k)}(\cdot), L^{(k)}(\cdot), W^{(k)}(\cdot)), \tag{3.20}$$

where the processes $F^{(k)}(\cdot)$, $G^{(k)}(\cdot)$, and $L^{(k)}(\cdot)$ are defined as follows:

$$F^{(k)}(s) = \int_0^s f(t, x_t^{(k)}(\cdot; 0, \psi, u^{(k)}(\cdot)), u^{(k)}(t))\, dt,$$

$$G^{(k)}(s) = \int_0^s g(t, x_t^{(k)}(\cdot; 0, \psi, u^{(k)}(\cdot)), u^{(k)}(t))\, dW^{(k)}(t),$$

and

$$L^{(k)}(s) = \int_0^s e^{-\alpha t} L(t, x_t^{(k)}(\cdot; 0, \psi, u^{(k)}(\cdot)), u^{(k)}(t))\, dt, \quad s \in [0, T].$$

Step 2. We prove the following lemma:

Lemma 3.2.13 *Assume Assumption 3.1.1 holds. Then there exists a constant $K > 0$ such that*

$$E^{(k)}\left[\left|X^{(k)}(s_1) - X^{(k)}(s_2)\right|^4\right] \le K|s_1 - s_2|^2, \quad \forall s_1, s_2 \in [0, T], \ \forall k = 1, 2, \dots,$$

where $E^{(k)}[\cdots]$ is the expectation with respect to the probability measure $P^{(k)}$.

Proof of the Lemma. Let us fix k, $0 \le s_1 \le s_2 \le T$, and consider

$$E^{(k)}\left[|F^{(k)}(s_1) - F^{(k)}(s_2)|^4\right]$$

$$\le E^{(k)}\left[\left|\int_{s_1}^{s_2} f(t, x_t^{(k)}(\cdot; 0, \psi, u^{(k)}(\cdot)), u^{(k)}(t))\, dt\right|^4\right]$$

$$\le |s_1 - s_2|^2 E^{(k)}\left[\left(\int_{s_1}^{s_2} \left|f(t, x_t^{(k)}(\cdot; 0, \psi, u^{(k)}(\cdot)), u^{(k)}(t))\right|^2 dt\right)^2\right]$$

$$\le |s_1 - s_2|^2 \int_{s_1}^{s_2} K_{grow}^2 E^{(k)}\left[\left(1 + \|x_t^{(k)}(\cdot; t, \psi, u^{(k)}(\cdot))\|\right)^2\right] dt.$$

Since $x_t^{(k)}(\cdot; 0, \psi, u^{(k)}(\cdot))$ is continuous $P^{(k)}$-a.s. in t on the compact interval $[0, T]$, it can be shown that there exists a constant $K > 0$ such that

$$\int_{s_1}^{s_2} K_{grow}^2 E^{(k)}\left[\left(1 + \|x_t^{(k)}(\cdot; 0, \psi, u^{(k)}(\cdot))\|\right)^2\right] dt < K.$$

Therefore,

$$E^{(k)}\left[|F^{(k)}(s_1) - F^{(k)}(s_2)|^4\right] \le K|s_1 - s_2|^2, \quad \forall s_1, s_2 \in [0, T], \ \forall k = 1, 2, \dots.$$

Similar conclusion holds for $L^{(k)}$; that is,

$$E^{(k)}\left[|L^{(k)}(s_1) - L^{(k)}(s_2)|^4\right] \le K|s_1 - s_2|^2, \quad \forall s_1, s_2 \in [0, T], \ \forall k = 1, 2, \dots.$$

We consider

$$E^{(k)}\left[|G^{(k)}(s_1) - G^{(k)}(s_2)|^4\right]$$

$$\le E^{(k)}\left[\left|\int_{s_1}^{s_2} g(t, x_t^{(k)}(\cdot; 0, \psi, u^{(k)}(\cdot)), u^{(k)}(t))\, dW^{(k)}(t)\right|^4\right]$$

$$\leq K_1(s_2 - s_1) \left(\int_{s_1}^{s_2} E^{(k)} \left[\left| g(t, x_t^{(k)}(\cdot; 0, \psi, u^{(k)}(\cdot)), u^{(k)}(t)) \right|^4 \right] dt \right)^2$$

for some constant $K_1 > 0$

$$\leq |s_1 - s_2|^2 \int_{s_1}^{s_2} K_1 K_{grow}^2 \left(1 + \| x_t^{(k)}(\cdot; 0, \psi, u^{(k)}(\cdot)) \| \right)^2 dt.$$

$$\leq K |s_1 - s_2|^2 \text{ for some constant } K > 0.$$

It is clear that

$$E^{(k)} \left[|W^{(k)}(s_1) - W^{(k)}(s_2)|^4 \right] = E^{(k)} \left[|W^{(k)}(s_2 - s_1)|^4 \right] \leq K |s_1 - s_2|^2,$$

since $W^{(k)}(s_2 - s_1)$ is Gaussian with mean zero and variance $I^{(m)}(s_2 - s_1)$.

The above estimates give

$$E^{(k)} \left[\left| X^{(k)}(s_1) - X^{(k)}(s_2) \right|^4 \right] \leq K |s_1 - s_2|^2, \quad \forall s_1, s_2 \in [0, T], \ \forall k = 1, 2, \dots.$$

This completes the proof of the lemma.   □

From the above lemma, we use the following well-known results to conclude that $\{(X^{(k)}(\cdot), \mu_{u^{(k)}}(\cdot, \cdot))\}_{k=1}^{\infty}$ is tight as a sequence of $C([0, T], \Re^{3n+m+1})$, since $\mathcal{R}$ is compact.

Proposition 3.2.14 *Let $\{\zeta^{(k)}(\cdot)\}_{k=1}^{\infty}$ be a sequence of d-dimensional continuous processes over $[0, T]$ on a probability space $(\Omega, \mathcal{F}, P)$ satisfying the following conditions:*

$$\sup_{k \geq 1} E[|\zeta^{(k)}(0)|^c] < \infty$$

and

$$\sup_{k \geq 1} E[|\zeta^{(k)}(t) - \zeta^{(k)}(s)|^a] \leq K |t - s|^{1+b}, \quad \forall t, s \in [0, T],$$

for some constants $a, b, c > 0$. Then $\{\zeta^{(k)}(\cdot)\}_{k=1}^{\infty}$ is tight as $C([0, T], \Re^d)$-valued random variables. As a consequence there exists a subsequence $\{k_j\}$, d-dimensional continuous processes $\{\hat{\zeta}^{(k_j)}(\cdot)\}_{j=1}^{\infty}$ and $\hat{\zeta}(\cdot)$ defined on a probability space $(\hat{\Omega}, \hat{\mathcal{F}}, \hat{P})$ such that

$$P(\zeta^{(k_j)}(\cdot) \in A) = \hat{P}(\hat{\zeta}^{(k_j)}(\cdot) \in A), \quad \forall A \in \mathcal{B}(C([0, T], \Re^d))$$

and

$$\lim_{j \to \infty} \hat{\zeta}^{(k_j)}(\cdot) \to \hat{\zeta}(\cdot) \text{ in } C([0, T], \Re^d), \ \hat{P}\text{-a.s.}$$

Corollary 3.2.15 *Let $\zeta(\cdot)$ be a d-dimensional process over $[0, T]$ such that*

$$E[|\zeta(t) - \zeta(s)|^a] \leq K |t - s|^{1+b}, \quad \forall t, s \in [0, T],$$

for some constants $a, b > 0$. Then there exists a d-dimensional continuous process that is stochastically equivalent to $\zeta(\cdot)$.

We refer the readers to Ikeda and Watanabe [IW81, pp.17-20] for a proof of the above proposition and corollary.

Step 3. Since

$$\{(X^{(k)}(\cdot), \mu_{u^{(k)}}(\cdot, \cdot))\}_{k=1}^{\infty}$$

is tight as a sequence in $C([0, T], \Re^{3n+m+1})$, by the Skorokhod representation theorem (see Theorem 3.2.7), one can choose a subsequence (still labeled as $\{k\}$) and have

$$\{(\bar{X}^{(k)}(\cdot), \bar{\mu}^{(k)}(\cdot, \cdot))\} \equiv \{(\bar{x}^{(k)}(\cdot), \bar{F}^{(k)}(\cdot), \bar{G}^{(k)}(\cdot), \bar{L}^{(k)}(\cdot), \bar{W}^{(k)}(\cdot), \bar{\mu}^{(k)}(\cdot, \cdot))\}$$

and

$$(\bar{X}(\cdot), \bar{\mu}(\cdot, \cdot)) \equiv (\bar{x}(\cdot), \bar{F}(\cdot), \bar{G}(\cdot), \bar{L}(\cdot), \bar{W}(\cdot), \bar{\mu}(\cdot, \cdot))$$

on a suitable common probability space $(\bar{\Omega}, \bar{\mathcal{F}}, \bar{P})$ such that

$$\text{law of } (\bar{X}^{(k)}(\cdot), \bar{\mu}^{(k)}(\cdot, \cdot)) = \text{law of } (X^{(k)}(\cdot), \mu^{(k)}(\cdot, \cdot)), \quad \forall k \geq 1, \quad (3.21)$$

and $\bar{P}$-a.s.,

$$\bar{X}^{(k)}(\cdot) \to \bar{X}(\cdot) \text{ uniformly on } [0, T] \quad (3.22)$$

and

$$\bar{\mu}^{(k)}(\cdot, \cdot) \to \bar{\mu}(\cdot, \cdot) \text{ weakly on } \mathcal{R}. \quad (3.23)$$

Step 4. Construct the filtration $\bar{\mathbf{F}}^{(k)} = \{\bar{\mathcal{F}}^{(k)}(s), s \geq 0\}$ and $\bar{\mathbf{F}} = \{\bar{\mathcal{F}}, s \geq 0\}$, where

$$\bar{\mathcal{F}}^{(k)}(s) = \sigma\{(\bar{W}^{(k)}(t), \bar{x}^{(k)}(t)), t \leq s\} \vee (\bar{\mu}^{(k)})^{-1}(\mathcal{B}_s(\mathcal{R}))$$

and

$$\bar{\mathcal{F}}(s) = \sigma\{(\bar{W}(t), \bar{x}(t)), t \leq s\} \vee (\bar{\mu})^{-1}(\mathcal{B}_s(\mathcal{R})), \quad s \geq 0.$$

By the definition of $\mathcal{B}_s(\mathcal{R})$ and the fact that the σ-algebra generated by the cylinder sets of $C([0, T]; \Re^d)$ coincides with $\mathcal{B}(C([0, T]; \Re^d))$, $\bar{\mathcal{F}}^{(k)}(s)$ is the σ-algebra generated by

$$\bar{W}^{(k)}(t_1), \ldots, \bar{W}^{(k)}(t_l), \bar{x}^{(k)}(t_1), \ldots, \bar{x}^{(k)}(t_l), \bar{\mu}^{(k)}(\varsigma^{(j)}, t_1), \ldots, \bar{\mu}^{(k)}(\varsigma^{(j)}, t_l),$$

$$t_1, \ldots, t_l \leq s, \ \varsigma^{(j)} \in C([0, T], U) \text{ and } j, l = 1, 2, \ldots.$$

A similar statement can be made for $\bar{\mathcal{F}}(s)$.

We need to show that $\bar{W}^{(k)}(\cdot) = \{\bar{W}^{(k)}(s), s \geq 0\}$ is an $\bar{\mathbf{F}}^{(k)}$ Brownian motion. We first note that $W^{(k)}(\cdot)$ is a Brownian motion with respect to

$$\left\{\sigma\{(W^{(k)}(t), x^{(k)}(t)), t \leq s\} \vee (\mu_{u^{(k)}}^{-1}(\mathcal{B}_s(\mathcal{R})), s \geq 0\right\}.$$

This implies that for any $0 \leq t \leq s \leq T$ and any bounded function H on $\Re^{(m+n+b)l}$ (b is a positive integer), we have

$$E^{(k)}\left[H(y^{(k)})(W^{(k)}(s) - W^{(k)}(t))\right] = 0,$$

where

$$y^{(k)} = \{W^{(k)}(t_i), x^{(k)}(t_i), \mu^{(k)}(\varsigma^{j_a, t_i})\}$$

$$\forall \leq 0 \leq t_1 \leq t_2 \leq \cdots \leq t_l \leq t, \ a = 1, 2, \ldots, b.$$

We have

$$\bar{E}^{(k)} \left[H(\bar{y}^{(k)})(\bar{W}^{(k)}(s) - \bar{W}^{(k)}(t)) \right] = 0,$$

where

$$\bar{y}^{(k)} = \{\bar{W}^{(k)}(t_i), \bar{x}^{(k)}(t_i), \bar{\mu}^{(k)}(\varsigma^{j_a, t_i})\}$$

$$\forall 0 \leq t_1 \leq t_2 \leq \cdots \leq 0 \leq t_l \leq t, a = 1, 2, \ldots, b,$$

since

$$\text{law of } (\bar{X}^{(k)}(\cdot), \bar{\mu}^{(k)}(\cdot, \cdot)) = \text{law of } (X^{(k)}(\cdot), \mu^{(k)}(\cdot, \cdot)), \quad \forall k \geq 1.$$

We therefore have

$$E^{(k)}[(W^{(k)}(s) - W^{(k)}(t)) \mid \mathcal{F}^{(k)}(t)] = 0.$$

In order to show $\bar{W}^{(k)}(\cdot)$ is an $\bar{\mathbf{F}}^{(k)}$ Brownian motion, we need

$$E^{(k)}[(W^{(k)}(s) - W^{(k)}(t))(W^{(k)}(s) - W^{(k)}(t))^\top \mid \mathcal{F}^{(k)}(t)] = (s - t)I^{(m)}.$$

This can be shown similarly.

Step 5. Again, since

$$\text{law of } (\bar{X}^{(k)}(\cdot), \bar{\mu}^{(k)}(\cdot, \cdot)) = \text{law of } (X^{(k)}(\cdot), \mu^{(k)}(\cdot, \cdot)), \quad \forall k \geq 1,$$

the following stochastic integral equation (defined on $(\bar{\Omega}, \bar{\mathcal{F}}, \bar{\mathbf{F}}^{(k)}, \bar{P})$) holds:

$$\bar{x}^{(k)}(s) = \psi(0) + \int_0^s \tilde{f}(t, \bar{x}_t^{(k)}, \bar{\mu}^{(k)}) \ dt$$

$$+ \int_0^s \tilde{g}(t, \bar{x}_t^{(k)}, \bar{\mu}^{(k)}) \, d\bar{W}^{(k)}(t),$$

where

$$\tilde{f}(t, \bar{x}_t^{(k)}, \bar{\mu}^{(k)}) = \int_U f(t, \bar{x}_t^{(k)}, u)\dot{\bar{\mu}}^{(k)}(t, du)$$

and

$$\tilde{g}(t, \bar{x}_t^{(k)}, \bar{\mu}^{(k)}) = \int_U g(t, \bar{x}_t^{(k)}, u)\dot{\bar{\mu}}^{(k)}(t, du).$$

Note that the above integrals are well defined, since $\bar{W}^{(k)}(\cdot)$ is a $\bar{\mathbf{F}}^{(k)}$ Brownian motion. Moreover, for each $s \in [0, T]$,

$$\lim_{k \to \infty} \int_0^s \tilde{f}(t, \bar{x}_t^{(k)}, \bar{\mu}^{(k)}) \, dt = \int_0^s \tilde{f}(t, \bar{x}_t, \bar{\mu}) \, dt$$

$$\lim_{k\to\infty} \int_0^s e^{-\alpha t} \tilde{L}(t, \bar{x}_t^{(k)}, \bar{\mu}^{(k)})\,dt = \int_0^s e^{-\alpha t}\tilde{L}(t, \bar{x}_t, \bar{\mu})\,dt, \qquad \bar{P}\text{-a.s.,}$$

$$e^{-\alpha(T-t)}\Psi(\bar{x}_T^{(k)}) \to e^{-\alpha(T-t)}\Psi(\bar{x}_T), \quad \bar{P}\text{-a.s.,}$$

and

$$\lim_{k\to\infty} \int_0^s \tilde{g}(t, \bar{x}_t^{(k)}, \bar{\mu}^{(k)})\,d\bar{W}^{(k)}(t)$$

$$= \int_0^s \int_U \tilde{g}(t, \bar{x}_t, \bar{\mu})\,d\bar{W}(t), \qquad \bar{P}\text{-a.s.,}$$

where

$$\tilde{f}(t, \bar{x}_t, \bar{\mu}) = \int_U f(t, \bar{x}_t, u)\dot{\bar{\mu}}(t, du),$$

$$\tilde{L}(t, \bar{x}_t^{(k)}, \bar{\mu}^{(k)}) = \int_U L(t, \bar{x}_t^{(k)}, u)\dot{\bar{\mu}}^{(k)}(t, du),$$

$$\tilde{L}(t, \bar{x}_t, \bar{\mu}) = \int_U L(t, \bar{x}_t, u)\dot{\bar{\mu}}(t, du),$$

and

$$\tilde{g}(t, \bar{x}_t, \bar{\mu}) = \int_U g(t, \bar{x}_t, u)\dot{\bar{\mu}}(t, du).$$

We have by taking the limit $k \to \infty$

$$\bar{x}(s) = \psi(0) + \int_0^s \tilde{f}(t, \bar{x}_t, \bar{\mu})\,dt$$

$$+ \int_0^s \tilde{g}(t, \bar{x}_t, \bar{\mu})\,d\bar{W}(t), \quad \forall s \in [0, T], \ \bar{P} - a.s.$$

Moreover,

$$J(0, \psi; u^{(k)}(\cdot)) = \bar{E}\left[\int_0^T e^{-\alpha t}\tilde{L}(t, \bar{x}_t, \bar{\mu}^{(k)})\,dt + e^{-\alpha T}\Psi(\bar{x}_T^{(k)}) \right]$$

$$\to \sup_{u(\cdot)\in\mathcal{U}[0,T]} J(0, \psi; u(\cdot))$$

$$\text{as } k \to \infty.$$

Step 6. Let us consider the sequence

$$\tilde{A}^{(k)}(s) \equiv (\tilde{g}\tilde{g}^\top)(s, \bar{x}_s^{(k)}, \bar{\mu}^{(k)}), \quad s \in [0, T].$$

By the Lipschitz continuity and linear growth conditions on f and g, it is tedious but straight forward to show that

$$\sup_k \bar{E}\left[\int_0^T |\tilde{A}^{(k)}(s)|^2 ds \right] < \infty.$$

Hence the sequence $\{\tilde{A}^{(k)}\}_{k=1}^{\infty}$ is weakly relatively compact in the space $L^2([0,T] \times \bar{\Omega}, \mathcal{S}^n)$, where $\mathcal{S}^n$ is the space of symmetric $n \times n$ matrices. We can then find a subsequence (still labelled by $\{k\}$) and a function $\tilde{A} \in L^2([0,T] \times \bar{\Omega}, \mathcal{S}^n)$ such that

$$\tilde{A}^{(k)} \to \tilde{A}, \quad \text{weakly on } L^2([0,T] \times \bar{\Omega}, \mathcal{S}^n). \tag{3.24}$$

Denoting by $\tilde{A}_{ij}$ the (ij) entry of the matrix $\tilde{A}$, we claim that for almost all (s, ω),

$$\underline{\lim}_{k\to\infty} \tilde{A}_{ij}^{(k)}(s,\omega) \le \tilde{A}_{ij}(s,\omega)$$
$$\le \overline{\lim}_{k\to\infty} \tilde{A}_{ij}^{(k)}(s,\omega), \quad i,j = 1,2,\ldots,n. \tag{3.25}$$

Indeed, if (3.25) is not true and on a set $A \subset [0,T] \times \bar{\Omega}$ of positive measure,

$$\underline{\lim}_{k\to\infty} \tilde{A}_{ij}^{(k)}(s,\omega) > \tilde{A}_{ij}(s,\omega),$$

then, by Fatou's lemma, we have

$$\underline{\lim}_{k\to\infty} \int_A \tilde{A}_{ij}^{(k)}(s,\omega) ds d\bar{P}(\omega) > \int_A \tilde{A}_{ij}(s,\omega) ds d\bar{P}(\omega),$$

which is a contradiction to (3.24). The same can be said for $\overline{\lim}$, which proves (3.25). Moreover, by the Lipschitz continuity and linear growth of f and g and the fact that $\bar{X}^{(k)}(\cdot) \to \bar{X}(\cdot)$ uniformly on $[0,T]$, for almost all (s,ω), we have

$$\underline{\lim}_{k\to\infty} \tilde{A}^{(k)}(s)$$
$$= \underline{\lim}_{k\to\infty} (\tilde{g}\tilde{g}^{\top})(s, \bar{x}_s^{(k)}, \bar{\mu}^{(k)}), \quad (s,\omega) \in [0,T] \times \bar{\Omega}, \tag{3.26}$$

and

$$\overline{\lim}_{k\to\infty} \tilde{A}^{(k)}(s)$$
$$= \overline{\lim}_{k\to\infty} (\tilde{g}\tilde{g}^{\top})(s, \bar{x}_s^{(k)}, \bar{\mu}^{(k)}), \quad (s,\omega) \in [0,T] \times \bar{\Omega}. \tag{3.27}$$

Then, combining (3.25), (3.26), (3.27) and the Roxin condition, we have

$$\tilde{A}(s,\omega) \in (gg^{\top})(s, \bar{x}_s(\omega), U), \quad a.e.(s,\omega). \tag{3.28}$$

Modify $\tilde{A}$ on a null set, if necessary, so that (3.28) holds for all $(s,\omega) \in [0,T] \times \bar{\Omega}$. One can similarly prove that there are $\tilde{f}$ and $\tilde{L} \in L^2([0,T] \times \bar{\Omega}, \Re)$ such that

$$\tilde{f}^{(k)} \to \tilde{f}, \tilde{L}^{(k)} \to \tilde{L}, \quad \text{weakly on } L^2([0,T] \times \bar{\Omega}, \Re), \tag{3.29}$$

and

$$\tilde{f}(s,\omega) \in f(s, \bar{x}_s(\omega), U), \quad \tilde{L}(s,\omega) \in L(s, \bar{x}_s(\omega), U), \tag{3.30}$$

$$\forall (s, \omega) \in [0, T] \times \bar{\Omega}.$$

By (3.28), (3.30) the Roxin condition, and a measurable selection theorem (see Corollary 2.26 of Li and Yong [LY91, p102]), there is a U-valued, $\bar{\mathbf{F}}$-adapted process $\bar{u}(\cdot)$ such that

$$(\tilde{f}, \tilde{A}, \tilde{L})(s, \omega) = (f, gg^\top, L)(s, \bar{x}_s(\omega), \bar{u}(s, \omega)), \tag{3.31}$$

$$\forall (s, \omega) \in [0, T] \times \bar{\Omega}.$$

Step 7. The last step is to use Roxin condition to show that there exists an m-dimensional Brownian motion defined on the filtered space $(\hat{\Omega}, \hat{\mathcal{F}}, \hat{P}, \hat{\mathbf{F}})$ which extends the filtered probability space $(\bar{\Omega}, \bar{\mathcal{F}}, \bar{P}, \bar{\mathbf{F}})$. We, then, conclude that

$$(\hat{\Omega}, \hat{\mathcal{F}}, \hat{P}, \hat{\mathbf{F}}, \hat{W}(\cdot), \bar{u}(\cdot)) \in \mathcal{U}[0, T]$$

is an optimal control.

We next prove that the Itô's integral process $\bar{I}(\tilde{g})(\cdot) = \{\bar{I}(\tilde{g})(s), s \in [0, T]\}$ is an $\bar{\mathbf{F}}$-martingale, where

$$\bar{I}(\tilde{g})(s) = \int_0^s \tilde{g}(t, \bar{x}_t, \bar{\mu}) \, d\bar{W}(t), \quad s \in [0, T].$$

To see this, once again, let $0 \le t \le s \le T$, and define

$$\bar{y}^{(k)} \equiv \{\bar{W}^{(k)}(t_i), \bar{x}^{(k)}(t_i), \bar{\mu}^{(k)}(\varsigma^{j_a, t_i})\},$$

and

$$\bar{y} \equiv \{\bar{W}(t_i), \bar{x}(t_i), \bar{\mu}(\varsigma^{j_a, t_i})\},$$

$$0 \le t_1 \le t_2 \le \cdots \le t_l \le s, \quad a = 1, 2, \ldots, b.$$

Since $\bar{I}^{(k)}(\tilde{g})(\cdot)$ is a $\bar{\mathbf{F}}^{(k)}$-martingale, for any bounded continuous function $H : \Re^{(m+n+b)l} \to \Re$, we have

$$\begin{aligned} 0 &= \bar{E}[\Phi(\bar{y}^{(k)}(\bar{I}^{(k)}(\tilde{g})(s) - \bar{I}^{(k)}(\tilde{g})(t))] \\ &\to \bar{E}[\Phi(\bar{y}(\bar{I}(\tilde{g})(s) - \bar{I}(\tilde{g})(t))], \end{aligned} \tag{3.32}$$

since $\bar{X}^{(k)}(\cdot) \to \bar{X}(\cdot)$ uniformly on $[0, T]$ and $\bar{\mu}^{(k)} \to \bar{\mu}$ weakly on $\mathcal{R}$ and by the dominated convergence theorem. This proves that $\bar{I}(\tilde{g})(\cdot)$ is an $\bar{\mathbf{F}}$-martingale. Furthermore,

$$\langle \bar{I}^{(k)}(\tilde{g}) \rangle(s) = \int_0^s \tilde{A}^{(k)}(t) dt,$$

where $\langle \bar{I}^{(k)}(\tilde{g}) \rangle$ is the quadratic variation of $\bar{I}^{(k)}(\tilde{g})(\cdot)$. Hence,

$$(\bar{I}^{(k)}(\tilde{g}))(\bar{I}^{(k)}(\tilde{g}))^\top - \int_0^s \tilde{A}^{(k)}(t) dt$$

is an $\bar{\mathbf{F}}^{(k)}$-martingale. Recalling $\tilde{A}^{(k)}(\cdot) \to \tilde{A}(\cdot)$ weakly on $L^2([0, T] \times \bar{\Omega})$, we have for any $t, s \in [0, T]$,

$$\int_t^s \tilde{A}^{(k)}(\lambda)d\lambda \;\rightarrow\; \int_t^s (gg^\top)(\lambda, \bar{x}_\lambda, \bar{u}(\lambda))d\lambda, \quad \text{weakly on } L^2(\Omega).$$

On the other hand, by the dominated convergence theorem, we have, for real-valued function H of appropriate dimension,

$$H(\bar{y}^{(k)}) \rightarrow H(\bar{y}), \quad \text{strongly on } L^2(\Omega).$$

Thus,

$$\bar{E}\left[H(\bar{y}^{(k)})\int_t^s \tilde{A}^{(k)}(\lambda)d\lambda\right] \rightarrow \bar{E}\left[H(\bar{y})\int_t^s (gg^\top)(\lambda, \bar{x}_\lambda, \bar{u}(\lambda))d\lambda\right].$$

Therefore, using an argument similar to the above, we obtain that $\bar{M}(\cdot) = \{\bar{M}(s), s \in [0,T]\}$ is an $\bar{\mathbf{F}}$-martingale, where

$$\bar{M}(s) \equiv (\bar{I}(g))(\bar{I}(g))^\top (s) - \int_0^s (gg^\top)(t, \bar{x}_t, \bar{u}(t))dt.$$

This implies that

$$\bar{I}(g)(s) = \int_0^s (gg^\top)(t, \bar{x}_t, \bar{u}(t))dt.$$

By a martingale representation theorem (see Subsection of Chapter 1), there is an extension $(\hat{\Omega}, \hat{\mathcal{F}}, \hat{\mathbf{F}}, \hat{P})$ of $(\bar{\Omega}, \bar{\mathcal{F}}, \bar{\mathbf{F}}, \hat{P})$ on which lives an m-dimensional $\hat{\mathbf{F}}$ Brownian motion $\hat{W}(\cdot) = \{\hat{W}(s), s \geq 0\}$, such that

$$\bar{I}(g)(s) = \int_0^s g(t, \bar{x}_t, \bar{u}(t))\, d\hat{W}(t).$$

Similarly, one can show that

$$\bar{F}(s) = \int_0^s f(t, \bar{x}_t, \bar{u}(t))\, dt.$$

Putting into

$$\bar{x}(s) = \psi(0) + \bar{F}(s) + \bar{I}(g)(s), \quad \forall s \in [0,T], \ \bar{P} - a.s.,$$

with

$$\bar{E}\left[\int_0^T e^{-\alpha t}(t, \bar{x}_t, \bar{u}(t))\, dt \;+ e^{-\alpha T}\Psi(\bar{x}_T)\right] = \inf_{u(\cdot)\in\mathcal{U}[0,T]} J(0, \psi; u(\cdot)),$$

we arrive at the conclusion that

$$(\hat{\Omega}, \hat{\mathcal{F}}, \hat{\mathbf{F}}, \hat{P}, \hat{W}(\cdot), \bar{u}(\cdot)) \in \mathcal{U}[0,T]$$

is an optimal control. This prove the theorem.   □

Proof of Theorem 3.2.11.

The idea in proving the existence of an optimal relaxed control $\mu^*(\cdot, \cdot)$ is to (1) observe that the space of relaxed control $\mathcal{R}$ is sequentially compact, since $[0, T] \times U$ is compact (see Theorem 3.2.1), and hence every sequence in $\mathcal{R}$ has a convergent subsequence under the weak compact topology defined by (3.14); (2) check that $\hat{J}(t, \psi; \cdot) : \mathcal{R} \to \Re$ is a (sequentially) upper semi-continuous function defined on the sequentially compact space $\mathcal{R}$; 3) provoke a classical theorem (see, e.g., Rudin [Rud71]) that states any (sequentially) upper semicontinuous real-valued function defined on a (sequentially) compact space attends a maximum in the space, and hence $\hat{J}(t, \psi; \cdot)$ attains its maximum at some point $\mu^*(\cdot, \cdot) \in \mathcal{R}$ (see, e.g., Yong and Zhou [YZ99, p.65]); and (4) show that the value function for the Problem (ORCP) coincides with that of the original Optimal Classical Control Problem (OCCP).

First, the following proposition is analogous to Theorem 10.1.1 of Kushner and Dupuis [KD01, pp.271-275] for our setting. The detail of the proof is very similar to that of Theorem 3.2.12 and, therefore, only a sketch is provided here.

Proposition 3.2.16 *Let Assumption 3.1.1 hold. Let*

$$\{(\Omega^{(k)}, \mathcal{F}^{(k)}, P^{(k)}, \mathbf{F}^{(k)}, W^{(k)}(\cdot), \mu^{(k)}(\cdot, \cdot))\}_{k=1}^{\infty}$$

be any sequence of admissible relaxed controls in $\hat{\mathcal{U}}[t, T]$. For each $k = 1, 2, \ldots$, let $\{x^{(k)}(s; t, \psi, \mu^{(k)}(\cdot, \cdot)), s \in [t - r, T]\}$ be the corresponding strong solution of (3.16) through the initial datum $(t, \psi^{(k)}) \in [0, T] \times \mathbf{C}$. Assume that the sequence of initial functions $\{\psi^{(k)}, k = 1, 2, \cdots\}$ converges to $\psi \in \mathbf{C}$. The sequence

$$\{(x^{(k)}(\cdot), W^{(k)}(\cdot), \mu^{(k)}(\cdot, \cdot)), k = 1, 2, \ldots\}$$

is tight. Denote by $(x(\cdot), W(\cdot), \mu(\cdot))$ the limit point of the sequence

$$\{(x^{(k)}(\cdot), W^{(k)}(\cdot), \mu^{(k)}(\cdot, \cdot)), k = 1, 2, \ldots\}$$

Define the filtration $\{\mathcal{H}(s), s \in [t, T]\}$ by

$$\mathcal{H}(s) = \sigma((x(\lambda), W(\lambda), \mu(\lambda, G)), t \leq \lambda \leq s, G \in \mathcal{B}(U)).$$

Then $W(\cdot)$ is is an $(\mathcal{H}(t, s), s \in [t, T])$-adapted Brownian motion, the six-tuple

$$\{(\Omega, \mathcal{F}, P, \mathbf{F}, W(\cdot), \mu(\cdot, \cdot))\}$$

is an admissible relaxed control and the process

$$x(\cdot) = \{x(s; t, \psi, \mu(\cdot, \cdot)), s \in [t - r, T]\}$$

is the strong solution process to (3.16) defined on

$$\{(\Omega, \mathcal{F}, P, \mathbf{F}, W(\cdot), \mu(\cdot, \cdot))\}$$

and with the initial datum $(t, \psi) \in [0, T] \times \mathbf{C}$.

A Sketch of Proof. Without loss of generality, we can and will assume that

$$\{(\Omega^{(k)}, \mathcal{F}^{(k)}, P^{(k)}, \mathbf{F}^{(k)}, W^{(k)}(\cdot), \mu^{(k)}(\cdot, \cdot)), k = 1, 2, \ldots\}$$

is the maximizing sequence of *admissible relaxed controls* for Problem (ORCP). We claim that the sequence of triplets

$$\{(W^{(k)}(\cdot), \mu^{(k)}(\cdot, \cdot), x^{(k)}(\cdot)), k = 1, 2, \ldots\} \tag{3.33}$$

is tight and therefore has a subsequence that is also to be denoted by (3.33), which converges weakly to some triplet $(W(\cdot), \mu(\cdot, \cdot), x(\cdot))$, where $W(\cdot)$ is a standard Brownian motion, $\mu(\cdot, \cdot)$ is the optimal relaxed control process, and $x(\cdot)$ is the optimal state process (corresponding to the optimal relaxed control process) that satisfies (3.16). Componentwise tightness implies tightness of the products (cf. Remark 3.2.4 or [Bil99,p.65]). We therefore prove the following componentwise results.

First, we observe that the sequence $\{W^{(k)}(\cdot)\}_{k=1}^{\infty}$ is tight. This is because they all have the same (Wiener) probability measure. Note that $W^{(k)}(\cdot)$ is continuous for each $k = 1, 2, \ldots$, so is its limit $W(\cdot)$. To show that $W(\cdot)$ is an m-dimensional standard Brownian motion, we will use the martingale characterization theorem in Section 1.2.1 of Chapter 1 by showing that if $\Phi \in C_0^2(\Re^m)$ (the space of real-valued twice continuously differentiable functions on $\Re^m$ and with compact support), then $M_\Phi(\cdot)$ is a $\mathbf{F}$-martingale, where

$$M_\Phi(s) \equiv \Phi(W(s)) - \Phi(0) - \int_0^s \mathcal{L}_w \Phi(W(t)) \, dt, \quad s \ge 0,$$

and $\mathcal{L}_w$ is the differential operator defined by (1.8). To prove this, we have by the fact that $W^{(k)}(\cdot)$ is $\mathbf{F}^{(k)}$-Brownian motion,

$$E\left[H(x^{(k)}(t_i), W^{(k)}(t_i), \mu^{(k)}(t_i), i \le p) \right. \tag{3.34}$$

$$\left. \times \left(\Phi(W^{(k)}(t + \lambda)) - \Phi(W^{(k)}(t)) - \int_t^{t+\lambda} \mathcal{L}_w \Phi(W^{(k)}(s)) ds \right) \right] = 0.$$

By the probability 1 convergence which is implied by the Skorokhod representation theorem,

$$E\left[\left| \int_t^{t+\lambda} \mathcal{L}_w \Phi(W^{(k)}(s)) \, ds - \int_t^{t+\lambda} \mathcal{L}_w \Phi(W(s)) \, ds \right| \right] \to 0.$$

Using this result and taking limits in (3.34) yields

$$E\left[H(x(t_i), W(t_i), \mu(t_i), i \le p) \right. \tag{3.35}$$

$$\left. \times \left(\Phi(W(t + \lambda)) - \Phi(W(t)) - \int_t^{t+\lambda} \mathcal{L}_w \Phi(W(s)) \, ds \right) \right] = 0.$$

The set of random variables $H(x(t_i), W(t_i), \mu(t_i), i \leq p)$, as $H(\cdot)$, p, and t_i vary over all possibilities, induces the σ-algebra $\mathcal{F}(t)$. Thus, (3.35) implies that

$$\left(\Phi(W(s)) - \Phi(0) - \int_0^s \mathcal{L}_w \Phi(W(t)) \, dt \right)$$

is an **F**-martingale for all Φ of chosen class. Thus $W(\cdot)$ is a standard **F**-Brownian motion.

Second, the sequence $\{(\mu^{(k)}(\cdot, \cdot), k = 1, 2, \ldots\}$ is tight, because the space $\mathcal{R}$ is (sequentially) weak compact. Furthermore, its weak limit $\mu(\cdot) \in \mathcal{R}$ and $\mu([0, t]; U) = t$ for all $t \in [0, T]$.

Third, the tightness of the sequence of processes $\{x^{(k)}(\cdot), k = 1, 2, \ldots\}$ follows from the Aldous criterion (cf. Theorem 3.2.6 or [Bil99, pp. 176-179]: Given $k = 1, 2, \ldots$, any bounded **F**-stopping time τ, and $\delta > 0$, we have

$$E^{(k)}[|x^{(k)}(\tau + \delta) - x^{(k)}(\tau)|^2 \mid \mathcal{F}^{(k)}(\tau)] \leq 2K^2 \delta(\delta + 1)$$

as a consequence of Assumption 3.1.1 and Itô's isometry. To show that its limit process $x(\cdot) = \{x(s), s \in [t, T]\}$ satisfies (3.16), we note that the weak limit $(x(\cdot), W(\cdot), \mu(\cdot, \cdot))$ is continuous on the time interval $[t, T]$. This is because it has been shown in the proof of Theorem 3.2.12 that both the pathwise convergence of the Lebesque integral

$$\lim_{k \to \infty} \int_t^s \int_U f(\lambda, x_\lambda^{(k)}, u) \dot{\mu}^{(k)}(\lambda, du) \, d\lambda = \int_t^s \int_U f(\lambda, x_\lambda, u) \dot{\mu}(\lambda, du) \, d\lambda, \quad P\text{-a.s.},$$

and of the stochastic integral

$$\lim_{k \to \infty} \int_t^s \int_U g(\lambda, x_\lambda^{(k)}, u) dW^{(k)}(\lambda) \dot{\mu}^{(k)}(\lambda, du) d\lambda$$

$$= \int_t^s \int_U g(\lambda, x_\lambda, u) \dot{\mu}(\lambda, du) dW(\lambda), \quad P\text{-a.s.}$$

We assume that the probability spaces are chosen as required by the Skorokhod representation (Theorem 3.2.7), so that we can suppose that the convergence of

$$\{(W^{(k)}(\cdot), \mu^{(k)}(\cdot, \cdot), x^{(k)}(\cdot))\}_{k=1}^\infty$$

to its limit is with probability 1 in the topology of the path spaces of the processes. Thus,

$$\int_t^s \int_U f(\lambda, x_\lambda^{(k)}, u) \dot{\mu}^{(k)}(\lambda, du) \, d\lambda \to \int_t^s \int_U f(\lambda, x_\lambda, u) \dot{\mu}(\lambda, du) d\lambda$$

and

$$\int_t^s \int_U g(\lambda, x_\lambda^{(k)}, u) \dot{\mu}^{(k)}(\lambda, du) \, dW^{(k)}(\lambda) \to \int_t^s \int_U g(\lambda, x_\lambda, u) \dot{\mu}(\lambda, du) \, dW(\lambda)$$

as $k \to \infty$ uniformly on $[t, T]$ with probability 1. The sequence $\{\mu^{(k)}(\cdot, \cdot)\}_{k=1}^{\infty}$ converges weakly. In particular, for any $\Phi \in C_b([0, T] \times U)$,

$$\int_t^T \int_U \Phi(\lambda, u) \mu^{(k)}(d\lambda, du) \to \int_t^T \int_U \Phi(\lambda, u) \mu(d\lambda, du).$$

Now, the Skorokhod representation theorem 3.2.7 and weak convergence imply that

$$\int_t^s \int_U f(\lambda, x_\lambda^{(k)}, u) \dot{\mu}^{(k)}(\lambda, du) \, d\lambda \to \int_t^s \int_U f(\lambda, x_\lambda, u) \dot{\mu}(\lambda, du) \, d\lambda$$

and

$$\int_t^s \int_U g(\lambda, x_\lambda^{(k)}, u) \dot{\mu}^{(k)}(\lambda, du) \, dW^{(k)}(\lambda) \to \int_t^s \int_U g(\lambda, x_\lambda, u) \dot{\mu}(\lambda, du) \, dW(\lambda)$$

as $k \to \infty$ uniformly on $[t, T]$ with probability 1. Since $\psi^{(k)} \in \mathbf{C}$ converges to $\psi \in \mathbf{C}$, we therefore prove that

$$x(s) = \psi(0) + \int_t^s \int_U f(\lambda, x_\lambda, u) \dot{\mu}(\lambda, du) \, d\lambda$$

$$+ \int_t^s \int_U g(\lambda, x_\lambda, u) \dot{\mu}(\lambda, du) \, dW(\lambda). \quad s \in [t, T], \qquad (3.36)$$

We next claim that the weak limit $(x(\cdot), W(\cdot), \mu(\cdot, \cdot))$ is continuous on the time interval $[t, T]$. First, $x(\cdot)$ is a continuous process; this is because both the pathwise Lebesque integral

$$\lim_{k \to \infty} \int_t^s \int_U f(\lambda, x_\lambda^{(k)}, u) \dot{\mu}^{(k)}(\lambda, du) \, d\lambda = \int_t^s \int_U f(\lambda, x_\lambda, u) \dot{\mu}(\lambda, du) \, d\lambda, \quad P\text{-a.s.},$$

and the stochastic integral

$$\lim_{k \to \infty} \int_t^s \int_U g(\lambda, x_\lambda^{(k)}, u) \dot{\mu}^{(k)}(\lambda, du) \, dW^{(k)}(\lambda)$$

$$= \int_t^s \int_U g(\lambda, x_\lambda, u) \dot{\mu}(\lambda, du) \, dW(\lambda), \quad P\text{-a.s.}$$

Similarly,

$$\int_t^T \int_U e^{\alpha(s-t)} L(s, x_s, u) \dot{\mu}^{(k)}(s, du) \, ds \to \int_t^T \int_U e^{\alpha(s-t)} L(s, x_s, u) \dot{\mu}(s, du) \, ds,$$

and

$$e^{-\alpha(T-t)} \Psi(x_T^{(k)}) \to e^{-\alpha(T-t)} \Psi(x_T)$$

as $k \to \infty$ with probability 1. $\qquad \square$

We have therefore proved the following two propositions.

Proposition 3.2.17 *Let Assumption 3.1.1 hold. Suppose the sequence of initial segment functions $\{\psi^{(k)}\}_{k=1}^{\infty} \subset \mathbf{C}$ converges to $\psi \in \mathbf{C}$. Then*

$$\lim_{k\to\infty} \hat{V}(t, \psi(k)) = \hat{V}(t, \psi).$$

Proposition 3.2.18 *Let Assumption 3.1.1 hold. Let $\mu(\cdot, \cdot) \in \hat{U}[t, T]$ be the relaxed control representation of $u(\cdot) \in \mathcal{U}[t, T]$ via (3.13). Then*

$$V(t, \psi) := \sup_{u(\cdot)\in\mathcal{U}[0,T]} J(t, \psi; u(\cdot)) = \hat{V}(t, \psi) := \sup_{\mu(\cdot,\cdot)\in\mathcal{U}[0,T]} \hat{J}(t, \psi; \mu(\cdot, \cdot)).$$

Proof of Theorem 3.2.11 The theorem follows immediately from Propositions 3.2.17 and 3.2.18.   □

3.3 Dynamic Programming Principle

3.3.1 Some Probabilistic Results

To establish and prove the dynamics programming principle (DDP), we need some probabilistic results as follows.

First, we recall that if O is a nonempty set and if $\mathcal{O}$ is a collection of subsets of O, the collection $\mathcal{O}$ is called a π-system if it is closed under the finite intersection; that is, $A, B \in \mathcal{O}$ imply that $A \cap B \in \mathcal{O}$. It is a λ-system if the following three conditions are satisfied: (i) $O \in \mathcal{O}$; (ii) $A, B \in \mathcal{O}$ and $A \subset B$ imply that $B - A \in \mathcal{O}$; and (iii) $A_i \in \mathcal{O}$, $A_i \uparrow A, i = 1, 2, \ldots$, implies that $A \in \mathcal{O}$.

The following lemmas will be used later.

Lemma 3.3.1 *Let $\mathcal{O}$ and $\tilde{\mathcal{O}}$ be two collections of subsets of O with $\mathcal{O} \subset \tilde{\mathcal{O}}$. Suppose $\mathcal{O}$ is a π-system and $\tilde{\mathcal{O}}$ is a λ-system. Then $\sigma(\mathcal{O}) \subset \tilde{\mathcal{O}}$, where $\sigma(\mathcal{O})$ is the smallest σ-algebra containing $\mathcal{O}$.*

Proof. This is the well-known *monotone class theorem*, the proof of which can be found in Lemma 1.1.2 of [YZ99].

Lemma 3.3.2 *Let $\mathcal{O}$ be a π-system on O. Let $\mathcal{H}$ be a linear space of functions from O to $\Re$ such that*

$$1 \in \mathcal{H}; \quad I_A \in \mathcal{H}, \quad \forall A \in \mathcal{O},$$

and

$$\varphi^{(i)} \in \mathcal{H} \text{ with } 0 \leq \varphi^{(i)} \uparrow \varphi, \varphi \text{ is finite} \Rightarrow \varphi \in \mathcal{H}.$$

Then $\mathcal{H}$ contains all $\sigma(\mathcal{O})$-measurable functions from O to $\Re$, where 1 is the constant function of value 1 and I_A is the indicator function of A.

Proof. Let

$$\tilde{\mathcal{O}} = \{A \subset O \mid I_A \in \mathcal{H}\}.$$

Then $\tilde{\mathcal{O}}$ is a λ-system containing $\mathcal{O}$. From Lemma 3.3.1 it can be shown that $\sigma(\mathcal{O}) \subset \tilde{\mathcal{O}}$.

Now, for any $\sigma(\mathcal{O})$-measurable function $\varphi : O \to \Re$, we set for $i = 1, 2, \ldots$

$$\varphi^{(i)} = \sum_{j \geq 0} j 2^{-i} I_{\{j 2^{-i} \leq \varphi^+(\omega) < (j+1) 2^{-i}\}},$$

where $\varphi^+ \equiv \max\{\varphi, 0\}$ denotes the positive part of φ. Clearly, $\varphi^{(i)} \in \mathcal{H}$ and $0 \leq \varphi^{(i)} \uparrow \varphi^+$. Hence, by our assumption, $\varphi^+ \in \mathcal{H}$. Similarly, $\varphi^- \in \mathcal{H}$, where $\varphi^+ \equiv \max\{-\varphi, 0\}$ denotes the negative part of φ. Thus, $\varphi \in \mathcal{H}$. This proves the lemma. $\quad\square$

First, let us introduce some notation. Define

$$C_t([0,T]; \Re^m) := \{\eta(\cdot \wedge t) \mid \eta \in C([0,T]; \Re^m)\},$$
$$\mathcal{B}_t(C([0,T]; \Re^m)) := \sigma(\mathcal{B}(C_t([0,T]; \Re^m))),$$
$$\mathcal{B}_{t+}(C([0,T]; \Re^m)) := \cap_{s>t} \mathcal{B}_s(C([0,T]; \Re^m))), \ t \in [0,T],$$

where $\sigma(\mathcal{B}(C_t([0,T]; \Re^m)))$ denotes the smallest σ-algebra in $C([0,T]; \Re^m)$ that contains $\mathcal{B}(C_t([0,T]; \Re^m))$. Clearly, both of the following are filtered measurable spaces:

$$(C([0,T]; \Re^m), \mathcal{B}(C([0,T]; \Re^m)), \{\mathcal{B}_t(C([0,T]; \Re^m))\}_{t \geq 0})$$

and

$$(C([0,T]; \Re^m), \mathcal{B}(C([0,T]; \Re^m)), \{\mathcal{B}_{t+}(C([0,T]; \Re^m))\}_{t \geq 0}).$$

However,

$$\mathcal{B}_{t+}(C([0,T]; \Re^m)) \neq \mathcal{B}_t(C([0,T]; \Re^m)), \quad \forall t \in [0,T].$$

A set $B \subset C([0,T]; \Re^m)$ is called a *Borel cylinder* if there exists a partition $\pi = \{0 \leq t_1 < t_2 < \cdots < t_j \leq T\}$ of $[0,T]$ and $A \in \mathcal{B}((\Re^m)^j)$ such that

$$B = \pi^{-1}(A) \equiv \{\xi \in C([0,T]; \Re^m) \mid (\xi(t_1), \xi(t_2), \ldots, \xi(t_j)) \in A\}. \quad (3.37)$$

For $s \in [0,T]$, let $\mathbf{C}(s)$ be the set of all Borel cylinder in $C_s([0,T]; \Re^m)$ of the form (3.37) with partition $\pi \subset [0,s]$.

Lemma 3.3.3 *The σ-algebra $\sigma(\mathbf{C}(T))$ generated by $\mathbf{C}(T)$ coincides with the Borel σ-algebra $\mathcal{B}(C[0,T]; \Re^m))$ of $C([0,T]; \Re^m)$.*

Proof. Let the partition $\pi = \{0 \leq t_1 < t_2 < \cdots < t_j \leq T\}$ of $[0,T]$ be given. We define a point-mass projection map $\Pi : C([0,T]; \Re^m) \to (\Re^m)^j$ associated with the partition π as follows:

$$\Pi(\xi) = (\xi(t_1), \xi(t_2), \ldots, \xi(t_j)), \quad \forall \xi \in C([0,T]; \Re^m).$$

Clearly, Π is continuous. Consequently, for any $A \in \mathcal{B}((\Re^m)^j)$, it follows that $\Pi^{-1}(A) \in \mathcal{B}(C([0,T]; \Re^m))$. This implies that

$$\sigma(\mathbf{C}(T)) \subset \mathcal{B}((C([0,T]; \Re^m)). \tag{3.38}$$

Next, for any $\hat{\xi} \in C([0,T]; \Re^m)$ and $\epsilon > 0$, the closed ϵ-ball $B(\hat{\xi}; \epsilon)$ in $C([0,T]; \Re^n)$ can be written as

$$B(\hat{\xi}; \epsilon) \equiv \{\xi \in C([0,T]; \Re^m) \mid \sup_{t \in [0,T]} |\xi(t) - \hat{\xi}(t)| \leq \epsilon\} \tag{3.39}$$

$$= \bigcap_{t \in \mathbf{Q} \cap [0,T]} \{\xi \in C([0,T]; \Re^m) \mid |\xi(t) - \hat{\xi}(t)| \leq \epsilon\} \in \sigma(\mathbf{C}(T)),$$

since $\{\xi \in C([0,T]; \Re^m) \mid |\xi(t) - \hat{\xi}(t)| \leq \epsilon\}$ is a Borel cylinder and $\mathbf{Q}$ is the set of all rational numbers (which is countable). Because the set of all sets in the form of the left-hand side of (3.39) is a basis of the closed (and therefore open) sets in $C([0,T]; \Re^m)$, we have

$$\mathcal{B}(C([0,T]; \Re^m)) \subset \sigma(\mathbf{C}(T)). \tag{3.40}$$

Combining (3.38) and (3.40), we obtain the conclusion of the lemma. $\qquad \square$

Lemma 3.3.4 *Let $(\Omega, \mathcal{F}, P)$ be a probability space and let $\zeta : [0,T] \times \Omega \to \Re^m$ be a continuous process. Then there exists an $\Omega_0 \in \mathcal{F}$ with $P(\Omega_0) = 1$ such that $\zeta : \Omega_0 \to C([0,T]; \Re^m)$, and for any $s \in [0,T]$,*

$$\Omega_0 \bigcap \mathcal{F}^\zeta(s) = \Omega_0 \bigcap \zeta^{-1}(\mathcal{B}_s(C([0,T]; \Re^m)), \tag{3.41}$$

where $\mathbf{F}^\zeta = \{\mathcal{F}^\zeta(s), s \in [0,T]\}$ is the filtration of sub-σ-algebras generated by the process $\zeta(\cdot)$; that is, for all $s \in [0,T]$,

$$\mathcal{F}^\zeta(s) = \sigma\{\zeta(t), 0 \leq t \leq s\}.$$

Proof. Let $t \in [0,s]$ and $A \in \mathcal{B}(\Re^m)$ be fixed. Then

$$B(t) \equiv \{\xi \in C([0,T]; \Re^m) \mid \xi(t) \in A\} \in \mathbf{C}(s)$$

and

$$\omega \in \zeta^{-1}(B(t)) \iff \zeta(\cdot, \omega) \in B(t) \iff \zeta(t, \omega) \in A$$
$$\iff \omega \in \zeta^{-1}(t, \cdot)(A).$$

Thus,

$$\zeta^{-1}(t, \cdot)(A) = \zeta^{-1}(B(t)).$$

Then by Lemma 3.3.3, we obtain (3.41). $\qquad \square$

Lemma 3.3.5 *Let $(\Omega, \mathcal{F})$ and $(\tilde{\Omega}, \tilde{\mathcal{F}})$ be two measurable spaces and let (Ξ, d) be a Polish (complete and separable) metric space. Let $\zeta : \Omega \to \tilde{\Omega}$ and $\varphi : \Omega \to \Xi$ be two random variables. Then φ is $\sigma(\zeta)$-measurable; that is,*

$$\varphi^{-1}(\mathcal{B}(\Xi)) \subset \zeta^{-1}(\tilde{\mathcal{F}}) \tag{3.42}$$

if and only if there exists a measurable map $\eta : \tilde{\Omega} \to \Xi$ such that

$$\varphi(\omega) = \eta(\zeta(\omega)), \quad \forall \omega \in \Omega. \tag{3.43}$$

Proof. We only need to prove the necessity. First, we assume that $\Xi = \Re$. For this case, set

$$\mathcal{H} \equiv \{\eta(\zeta(\cdot)) \mid \eta : \tilde{\Omega} \to \Xi \text{ is measurable}\}.$$

Then $\mathcal{H}$ is a linear space and $\mathbf{1} \in \mathcal{H}$. We note here that $\zeta^{-1}(\tilde{\mathcal{F}})$ forms a π-system; that is, it is closed under finite intersections. Also, if $A \in \sigma(\zeta) \equiv \zeta^{-1}(\tilde{\mathcal{F}})$, then for some $B \in \tilde{\mathcal{F}}$, $I_A(\cdot) = I_B(\zeta(\cdot)) \in \mathcal{H}$. Now, suppose $\eta^{(i)} : \tilde{\Omega} \to \Xi$ is measurable for $i = 1, 2, \ldots$ and $\eta^{(i)}(\zeta(\cdot)) \in \mathcal{H}$ such that $0 \le \eta^{(i)}(\zeta(\cdot)) \uparrow \xi(\cdot)$, which is finite. Set

$$A = \{\tilde{\omega} \in \tilde{\Omega} \mid \sup_i \eta^{(i)}(\tilde{\omega}) < \infty\}.$$

Then $A \in \tilde{\mathcal{F}}$ and $\zeta(\Omega) \subset A$. Define

$$\eta(\tilde{\omega}) = \begin{cases} \sup_i \eta^{(i)}(\tilde{\omega}) & \text{if } \tilde{\omega} \in A \\ 0 & \text{if } \tilde{\omega} \in \tilde{\Omega} - A. \end{cases}$$

Clearly, $\eta : \tilde{\Omega} \to \Xi$ is measurable and $\xi(\cdot) = \eta(\zeta(\cdot))$. Thus, $\xi(\cdot) \in \mathcal{H}$. By Lemma 3.3.2, $\mathcal{H}$ contains all $\sigma(\zeta)$-measurable random variables, in particular, $\varphi \in \mathcal{H}$, which lead to (3.43). This proves our conclusion for the case $U = \Re$.

Now, let (Ξ, d) be an uncountable Polish space. Then it is known that (Ξ, d) is Borel isomorphic to the Borel measurable space of real numbers $(\Re, \mathcal{B}(\Re))$; that is, there exists a bijection $h : \Xi \to \Re$ such that $h(\mathcal{B}(\Xi)) = \mathcal{B}(\Re)$. Consider the map $\tilde{\varphi} = h \circ \varphi : \Omega \to \Re$, which satisfies

$$\tilde{\varphi}^{-1}(\mathcal{B}(\Re)) = \varphi^{-1} \circ h^{-1}(\mathcal{B}(\Re)) = \varphi^{-1}(\mathcal{B}(\Xi)) \subset \zeta^{-1}(\tilde{\mathcal{F}}).$$

Thus, there exists an $\tilde{\eta} : \tilde{\Omega} \to \Re$ such that

$$\tilde{\varphi}(\omega) = \tilde{\eta}(\zeta(\omega)), \quad \forall \omega \in \Omega.$$

By taking $\eta = h^{-1} \circ \tilde{\eta}$, we obtain the desired result.

Finally, if (Ξ, d) is countable or finite, we can prove the result by replacing $\Re$ in the above by the set of natural numbers $\mathbf{N}$ or $\{1, 2, \ldots, n\}$. $\quad\square$

Later we will take Ξ to be the control set U. Let $\mathcal{A}_T^m(U)$ be the set of all $\mathcal{B}_{t+}(C([0,T];\Re^m))$-progressively measurable processes

$$\eta : [0,T] \times C([0,T];\Re^m) \to U,$$

where U is the control set, which is assumed to be a Polish (complete and separable) metric space.

Proposition 3.3.6 *Let $(\Omega, \mathcal{F}, P)$ be a complete probability space and let U be a Polish space. Let $\zeta : [0,T] \times \Omega \to \Re^m$ be a continuous process and $\mathcal{F}^\zeta(s) = \sigma(\zeta(t); 0 \le t \le s)$. Then $\varphi : [0,T] \times \Omega \to U$ is $\{\mathcal{F}^\zeta(s)\}$-adapted if and only if there exists an $\eta \in \mathcal{A}_T^m(U)$ such that*

$$\varphi(t,\omega) = \eta(t, \zeta(\cdot \wedge t, \omega)), \ t \in [0,T], \ P - a.s. \ \omega \in \Omega.$$

Proof. We prove only the "only if" part. The "if" part is clear.
For any $s \in [0,T]$, we consider a mapping

$$\theta^s(t,\omega) \equiv (t \wedge s, \zeta(\cdot \wedge s, \omega)) : [0,T] \times \Omega \to [0,s] \times C_t([0,T];\Re^m).$$

By Lemma 3.3.4, we have $\mathcal{B}([0,s]) \times \mathcal{F}^\zeta(s) = \sigma(\theta^s)$. On the other hand, $(t,\omega) \mapsto \varphi(t \wedge s, \omega)$ is $(\mathcal{B}([0,s]) \times \mathcal{F}^\zeta(s))/\mathcal{B}(U)$-measurable. Thus, by Lemma 3.3.5, there exists a measurable map

$$\eta^s : ([0,T] \times C_s([0,T];\Re^m), \mathcal{B}([0,s]) \times \mathcal{B}_s(C([0,T];\Re^m))) \to U$$

such that

$$\varphi(t \wedge s, \omega) = \eta^s(t \wedge s, \zeta(\cdot \wedge s, \omega)), \quad \forall \omega \in \Omega, t \in [0,T]. \tag{3.44}$$

Now, for any $i \ge 1$, let $0 = t_0^i < t_1^i < \cdots$ be a partition of $[0,T]$ (with the mesh $\max_{j \ge 1} |t_j^i - t_{j-1}^i| \to 0$ as $i \to \infty$) and define

$$\eta^{(i)}(t,\xi) = \eta^0(0, \xi(\cdot \wedge 0)) I_{\{0\}}(t) \tag{3.45}$$
$$+ \sum_{j \ge 1} \eta^{t_j^i}(t, \xi(\cdot \wedge t_j^i)) I_{(t_{j-1}^i, t_j^i]}(t), \quad \forall (t,\xi) \in [0,T] \times C([0,T];\Re^m).$$

For any $t \in [0,T]$, there exists j such that $t_{j-1}^i < t \le t_j^i$. Then

$$\eta^{(i)}(t, \zeta(\cdot \wedge t_j^i, \omega)) = \eta^{t_j^i}(t, \zeta(\cdot \wedge t_j^i, \omega)) = \varphi(t, \omega). \tag{3.46}$$

Now, in the case $U = \Re$, $\mathbf{N}$, $\{1, 2, \ldots, n\}$, we may define

$$\eta(t,\xi) = \overline{\lim}_{i \to \infty} \eta^{(i)}(t,\xi) \tag{3.47}$$

to get the desired result. In the case where U is a general Polish space, we need to amend the proof in the same fasion as in that of Lemma 3.3.5. $\square$

3.3.2 Continuity of the Value Function

For each $t \in [0, T]$, the following lemma shows that the value function $V(t, \cdot) : \mathbf{C} \to \Re$ is Lipschitz.

Lemma 3.3.7 *Assume Assumptions (A3.1.1)-(A3.1.3) hold. The value function V satisfies the following properties: There is a constant $K_V \geq 0$ not greater than $3K_{lip}(T + 1)e^{3T(T+4m)K_{lip}^2}$ such that for all $t \in [0, T]$ and $\phi, \tilde{\phi} \in \mathbf{C}$, we have*

$$|V(t, \phi) - V(t, \varphi)| \leq K_V \|\phi - \varphi\|.$$

Proof. Let $t \in [0, T]$ and $\phi, \varphi \in \mathbf{C}$. We have, by definition,

$$|V(t, \phi) - V(t, \varphi)| \leq \sup_{u(\cdot) \in \mathcal{U}[t,T]} |J(t, \phi; u(\cdot)) - J(t, \varphi; u(\cdot))|.$$

Let $x(\cdot)$ and $y(\cdot)$ be the solution processes of (3.1) under the control process $u(\cdot)$ but with two different initial data: (t, ϕ) and $(t, \varphi) \in [0, T] \times \mathbf{C}$, respectively. Then by (A3.1.3) of Assumption 3.1.1, we have for all $u(\cdot) \in \mathcal{U}[t, T]$,

$$|J(t, \phi; u(\cdot)) - J(t, \varphi; u(\cdot))|$$

$$\leq E\left[\int_t^T |L(s, x_s, u(s)) - L(s, y_s, u(s))| \, ds + |\Psi(x_T) - \Psi(y_T)|\right]$$

$$\leq K_{kip}(1 + T - t)E\left[\sup_{s \in [-r,T]} |x(s) - y(s)|^2\right].$$

Now,

$$E\left[\sup_{s \in [-r,T]} |x(s) - y(s)|^2\right] \leq 2E\left[\sup_{s \in [0,T]} |x(s) - y(s)|^2\right] + 2\|\phi - \varphi\|^2,$$

while Hölder's inequality, Doob's maximal inequality, Itô's isometry, Fubini's theorem and (A3.1.3) of Assumption 3.1.1 together yield

$$E\left[\sup_{s \in [0,T]} |x(s) - y(s)|^2\right]$$

$$\leq 3|\phi(0) - \varphi(0)|^2 + 3TE\left[\int_t^T |f(s, x_s, u(s)) - f(s, y_s, u(s))|^2 \, ds\right]$$

$$+ 3m \sum_{i=1}^n \sum_{j=1}^m E\left[\int_t^T |(g_{ij}(s, x_s, u(s)) - g_{ij}(s, y_s, u(s))) \, dW_j(s)|^2\right]$$

$$\leq 3|\phi(0) - \varphi(0)|^2 + 3TK_{lip}^2 \int_t^T E[\|x_s - y_s\|^2] \, ds$$

$$+ 12mE\left[\int_t^T \sum_{i=1}^n \sum_{j=1}^m |(g_{ij}(s, x_s, u(s)) - g_{ij}(s, y_s, u(s)))|^2\, ds\right]$$

$$\leq 3|\phi(0) - \varphi(0)|^2 + 3(T + 4m)K_{lip}^2 \int_t^T E\left[\sup_{\lambda \in [t-r,s]} \|x(\lambda) - y(\lambda)\|^2\right] ds.$$

Since $|\phi(0) - \varphi(0)| \leq \|\phi - \varphi\|$, Gronwall's lemma gives

$$E\left[\sup_{s \in [t-r,T]} \|x(s) - y(s)\|^2\right] \leq 8\|\phi - \varphi\|^2 e^{6T(T+4m)K_{lip}^2}.$$

Combining the above estimates, we obtain the assertion.   □

3.3.3 The DDP

The advantage of the weak formulation of the control problem will be apparent in the following lemmas and its use in the proof of the DPP. Let $t \in [0, T]$ and $u(\cdot) \in \mathcal{U}[t, T]$. Then under Assumption 3.1.1, for any $\mathbf{F}(t)$-stopping time $\tau \in [t, T)$ and $\mathcal{F}(t, \tau)$-measurable random variable $\xi : \Omega \to \mathbf{C}$, we can solve the following controlled SHDE:

$$dx(s) = f(s, x_s, u(s))\, ds + g(s, x_s, u(s))\, dW(s), \quad s \in [\tau, T], \tag{3.48}$$

with the initial function $x_\tau = \xi$ at the stopping time τ.

Lemma 3.3.8 *Let $t \in [0, T]$ and $(\Omega, \mathcal{F}, P, \mathbf{F}, W(\cdot), u(\cdot)) \in \mathcal{U}[t, T]$. Then for any $\mathbf{F}$-stopping time, $\tau \in [t, T)$, and any $\mathcal{F}(\tau)$- measurable random variable $\xi : \Omega \to \mathbf{C}$,*

$$J(\tau, \xi(\omega); u(\cdot)) = E\bigg[\int_\tau^T e^{\alpha(s-\tau)} L(s, x_s(\cdot; \tau, \xi, u(\cdot)), u(s))ds$$

$$+ e^{-\alpha(T-\tau)}\Psi(x_T(\cdot; \tau, \xi, u(\cdot))|\mathcal{F}(\tau)\bigg](\omega) \quad P\text{-a.s. } \omega.$$

Proof. Since $u(\cdot)$ is $\mathbf{F}$-adapted, where $\mathbf{F}$ is the P-augmented natural filtration generated by $W(\cdot)$, by Propostion 3.3.6 there is a function $\eta \in \mathcal{A}_T^m(U)$ such that

$$u(s, \omega) = \eta(s, W(\cdot \wedge s, \omega)), \quad P - a.s. \ \omega \in \Omega, \ \forall s \in [t, T].$$

Then (3.48) can be written as

$$dx(s) = f(s, x_s, \eta(s, W(\cdot \wedge s)))\, ds$$
$$+ g(s, x_s, \eta(s, W(\cdot \wedge s)))\, dW(s), \quad s \in [\tau, T], \tag{3.49}$$

with $x_\tau = \xi$. Due to Assumption 3.1.1, Theorem 1.3.12, and Theorem 1.3.16, this equation has a strongly unique strong solution and, therefore, weak uniqueness holds. In addition, we may write, for $s \geq \tau$,

$$u(s, \omega) = \eta(s, W(\cdot \wedge s, \omega)) = \eta(s, \tilde{W}(\cdot \wedge s, \omega) + W(\tau, \omega))$$

where $\tilde{W}(s) = W(s) - W(\tau)$. Since τ is random, $\tilde{W}(\cdot)$ is not a Brownian motion under the probability measure P. However, we may, under the weak formulation of the control problem, change the probability measure P as follows. Note first that

$$
\begin{aligned}
P\{\omega' \mid \tau(\omega') &= \tau(\omega) | \mathcal{F}(\tau)\}(\omega) \\
&= E[1_{\{\omega':\tau(\omega')=\tau(\omega)\}} | \mathcal{F}(\tau)](\omega) \\
&= 1_{\{\omega':\tau(\omega')=\tau(\omega)\}}(\omega) \\
&= 1, \ P-a.s. \ \omega \in \Omega.
\end{aligned}
$$

This means that there is an $\Omega_0 \in \mathcal{F}$ with $P(\Omega_0) = 1$, so that for any fixed $\omega_0 \in \Omega_0$, τ becomes a deterministic time $\tau(\omega_0)$; that is, $\tau = \tau(\omega_0)$ almost surely in the new probability space $(\Omega, \mathcal{F}, P(\cdot|\mathcal{F}(\tau)))$, where $P(\cdot|\mathcal{F}(\tau)))$ denotes the probability measure P restricted to the σ-sub-algebra $\mathcal{F}(\tau)$. A similar argument shows that $W(\tau)$ almost surely equals a constant $W(\tau(\omega_0), \omega)$ and also that ξ almost surely equals a constant $\xi(\omega_0)$ when we work in the probability space $(\Omega, \mathcal{F}, P(\cdot|\mathcal{F}(\tau))(\omega_0))$. So, under the measure $P(\cdot|\mathcal{F}(\tau))(\omega_0)$, for $s \geq \tau(\omega_0)$, the process $\tilde{W}(\cdot)$ will be a standard Brownian motion

$$\tilde{W}(s) = W(s) - W(\tau(\omega_0)),$$

and for any $s \geq \tau(\omega_0)$,

$$u(s, \omega) = \eta(s, \tilde{W}(\cdot \wedge s, \omega) + W(\tau(\omega_0), \omega)).$$

It follows then that $u(\cdot)$ is adapted to the filtration $\mathbf{F}(\tau(\omega_0))$ generated by the standard Brownian motion $\tilde{W}(s)$ for $s \geq \tau(\omega_0)$. Hence, by the definition of admissible controls,

$$(\Omega, \mathcal{F}, P(\cdot|\mathcal{F}(\tau))(\omega_0), \tilde{W}(\cdot), u|_{[\tau(\omega_0), T]}) \in \mathcal{U}[\tau(\omega_0), T].$$

Note that for $A \in \mathcal{B}(\mathbf{C})$,

$$P[\xi \in A | \mathcal{F}(\tau)](\omega_0) = E[1_{\{\xi \in A\}} | \mathcal{F}(\tau)](\omega_0) = E[1_{\{\xi \in A\}}] = P\{\xi \in A\}.$$

This means that the two weak solutions

$$(\Omega, \mathcal{F}, P, \mathbf{F}, x(\cdot), W(\cdot)) \text{ and } (\Omega, \mathcal{F}, P(\cdot|\mathcal{F}(\tau)(\omega_0)), \tilde{\mathbf{F}}, x(\cdot), \tilde{W}(\cdot))$$

of (3.1) have the same initial distribution. Then, by the weak uniqueness,

$$
\begin{aligned}
J(\tau, \xi(\omega); u(\cdot)) &= E^{\tau, \xi(\omega), u(\cdot)} \left[\int_\tau^T e^{-\alpha(s-\tau)} L(s, x_s, u(s)) \, ds + e^{-\alpha(T-\tau)} \Psi(x_T) \right] \\
&= E\left[\int_\tau^T e^{-\alpha(s-\tau)} L(s, x_s(\tau, \xi, u(\cdot)), u(s)) \, ds \right. \\
&\qquad \left. + e^{-\alpha(T-\tau)} \Psi(x_T(\tau, \xi, u(\cdot))) \Big| \mathcal{F}(\tau) \right](\omega), \ P-a.s. \ \omega. \qquad \square
\end{aligned}
$$

The following dynamic programming principle (DDP) is due to Larssen [Lar02].

Theorem 3.3.9 (Dynamic Programming Principle) *Let Assumption 3.1.1 hold. Then for any initial datum* $(t, \psi) \in [0, T] \times \mathbf{C}$ *and* $\mathbf{F}$-*stopping time* $\tau \in [t, T]$,

$$
V(t, \psi) = \sup_{u(\cdot) \in \mathcal{U}[t,T]} E\Big[\int_t^\tau e^{-\alpha(s-t)} L(s, x_s(\cdot; t, \psi, u(\cdot)), u(s))\, ds
$$
$$
+ e^{-\alpha(\tau-t)} V(\tau, x_\tau(\cdot; t, \psi, u(\cdot)))\Big]. \tag{3.50}
$$

Proof. Denote the right-hand side of (3.50) by $\bar{V}(t, \psi)$. Given any $\epsilon > 0$, there exists an $(\Omega, \mathcal{F}, P, \mathbf{F}, W(\cdot), u(\cdot)) \in \mathcal{U}[t, T]$ such that

$$
V(t, \psi) - \epsilon \leq J(t, \psi; u(\cdot)).
$$

Equivalently,

$$
V(t, \psi) - \epsilon \leq J(t, \psi; u(\cdot))
$$
$$
= E\Big[\int_t^T e^{-\alpha(s-t)} L(s, x_s(\cdot; t, \psi, u(\cdot)), u(s))\, ds
$$
$$
+ e^{-\alpha(T-t)} \Psi(x_T(\cdot; t, \psi, u(\cdot)))\Big]
$$
$$
= E\Big[\int_t^\tau e^{-\alpha(s-t)} L(s, x_s(\cdot; t, \psi, u(\cdot)), u(s))\, ds
$$
$$
+ \int_\tau^T e^{-\alpha(s-t)} L(s, x_s(\cdot; t, \psi, u(\cdot)), u(s))\, ds
$$
$$
+ e^{-\alpha(T-t)} \Psi(x_T(\cdot; t, \psi, u(\cdot)))\Big].
$$

Therefore,

$$
V(t, \psi) - \epsilon \leq E\Big[\int_t^\tau e^{-\alpha(s-t)} L(s, x_s(\cdot; t, \psi, u(\cdot)), u(s))\, ds
$$
$$
+ E\Big[\int_\tau^T e^{-\alpha(s-\tau)} L(s, x_s(\cdot; t, \psi, u(\cdot)), u(s))\, ds
$$
$$
+ e^{-\alpha(T-t)} \Psi(x_T(\cdot; t, \psi, u(\cdot))) \Big| \mathcal{F}(\tau)\Big]\Big]
$$
$$
= E\Big[\int_t^\tau e^{-\alpha(s-t)} L(s, x_s(\cdot; t, \psi, u(\cdot)), u(s))\, ds
$$
$$
+ e^{-\alpha(\tau-t)} J(\tau, x_\tau(\cdot; t, \psi, u(\cdot)); u(\cdot))\Big] \quad \text{(by Lemma 3.3.11)}
$$
$$
\leq E\Big[\int_t^\tau e^{-\alpha(s-t)} L(s, x_s(\cdot; t, \psi, u(\cdot)), u(s))\, ds
$$
$$
+ e^{-\alpha(\tau-t)} V(\tau, x_\tau(\cdot; t, \psi, u(\cdot)))\Big],
$$

so by taking the supremum over $u(\cdot) \in \mathcal{U}[t, T]$, we have

$$V(t, \psi) - \epsilon \leq \bar{V}(t, \psi), \quad \forall \epsilon > 0.$$

This shows that

$$V(t, \psi) \leq \bar{V}(t, \psi), \quad \forall (t, \psi) \in [0, T] \times \mathbf{C}. \tag{3.51}$$

Conversely, we want to show that

$$V(t, \psi) \geq \bar{V}(t, \psi), \quad \forall (t, \psi) \in [0, T] \times \mathbf{C}.$$

Let $\epsilon > 0$; by Lemma 3.3.7 and its proof, there is a $\tilde{\delta} = \tilde{\delta}(\epsilon)$ such that whenever

$$\|\psi - \hat{\psi}\| < \tilde{\delta},$$

$$|J(\tau, \psi; u(\cdot)) - J(\tau, \hat{\psi}; u(\cdot))| + |V(\tau, \psi) - V(\tau, \hat{\psi})| \leq \epsilon, \ \forall u(\cdot) \in \mathcal{U}[\tau, T]. \tag{3.52}$$

Now, let $\{D_j\}_{j \geq 1}$ be a *Borel partition* of $\mathbf{C}$. This means that

$$D_j \in \mathcal{B}(\mathbf{C}) \text{ for each } j, \ \cup_{j \geq 1} D_j = \mathbf{C}, \text{ and } D_i \cap D_j = \text{ if } i \neq j.$$

We also assume that the D_j are chosen so that $\|\phi - \varphi\| < \tilde{\delta}$ whenever ϕ and φ are both in D_j. Choose $\psi^{(j)} \in D_j$. For each j, there exists an $(\Omega^{(j)}, \mathcal{F}^{(j)}, P^{(j)}, W^{(j)}(\cdot), u^{(j)}(\cdot)) \in \mathcal{U}[\tau, T]$ such that

$$J(\tau, \psi^{(j)}; u^{(j)}(\cdot)) \leq V(\tau, \psi^{(j)}) + \epsilon. \tag{3.53}$$

For any $\psi \in D_j$, (3.52) implies in particular that

$$J(\tau, \psi; u^{(j)}(\cdot)) \geq J(\tau, \psi^{(j)}; u^{(j)}(\cdot)) - \epsilon \text{ and } V(\tau, \psi^{(j)}) \geq V(\tau, \psi) - \epsilon. \tag{3.54}$$

Combining the above two inequalities, we see that

$$J(\tau, \psi; u^{(j)}(\cdot)) \geq J(\tau, \psi^{(j)}; u^{(j)}(\cdot)) - \epsilon \geq V(\tau; \psi^{(j)}) - 2\epsilon \geq V(\tau, \psi) - 3\epsilon. \tag{3.55}$$

By the definition of the five-tuple

$$(\Omega^{(j)}, \mathcal{F}^{(j)}, P^{(j)}, \mathbf{F}^{(j)}, W^{(j)}(\cdot), u^{(j)}(\cdot)) \in \mathcal{U}[\tau, T],$$

there is a function $\varphi_j \in \mathcal{A}_T(U)$ such that

$$u^{(j)}(s, \omega) = \varphi^{(j)}(s, W^{(j)}(\cdot \wedge s, \omega)), \quad P^{(j)} - a.s. \ \omega \in \Omega^{(j)}, \ \forall s \in [\tau, T].$$

Now, let $(\Omega, \mathcal{F}, P, \mathbf{F}, W(\cdot), u(\cdot)) \in \mathcal{U}[t, T]$ be arbitrary. Define the new control $\tilde{u}(\cdot) = \{\tilde{u}(s), s \in [t, T]\}$, where

$$\tilde{u}(s, \omega) = u(s, \omega), \text{if } s \in [t, \tau),$$

and

$$\tilde{u}(s,\omega) = \varphi^{(j)}(s, W^{(j)}(\cdot \wedge s, \omega) - W(\tau, \omega)) \text{ if } s \in [\tau, T] \text{ and } x_s(t, \psi, u(\cdot)) \in D_j.$$

Then $(\Omega, \mathcal{F}, P, \mathbf{F}, W(\cdot), \tilde{u}(\cdot)) \in \mathcal{U}[t, T]$. Thus,

$$V(t, \psi) \geq J(t, \psi; \tilde{u}(\cdot))$$

$$= E\left[\int_t^T e^{-\alpha(s-t)} L(s, x_s(\cdot; t, \psi, \tilde{u}(\cdot)), \tilde{u}(s))\, ds \right.$$

$$\left. + e^{-\alpha(T-t)} \Psi(x_T(\cdot; t, \psi, \tilde{u}(\cdot))) \right]$$

$$\geq E\left[\int_t^\tau e^{-\alpha(s-t)} L(s, x_s(\cdot; t, \psi, u(\cdot)), u(s))\, ds \right.$$

$$+ E\left[\int_\tau^T e^{-\alpha(s-\tau)} L(s, x_s(\cdot; \tau, x_s(t, \psi, u(\cdot)), \tilde{u}(s))\, ds \right.$$

$$\left. \left. + e^{-\alpha(T-t)} \Psi(x_T(\cdot; \tau, x_\tau(t, \psi, u(\cdot)), \tilde{u}(\cdot))) \Big| \mathcal{F}(\tau) \right] \right].$$

Therefore,

$$V(t, \psi) \geq E\left[\int_t^\tau e^{-\alpha(s-t)} L(s, x_s(\cdot; t, \psi, u(\cdot)), u(s))\, ds \right.$$

$$\left. + J(\tau, x_\tau(\cdot; t, \psi, u(\cdot)); \tilde{u}(\cdot)) \right] \quad \text{(by Lemma reflem:3.3.8)}$$

$$\geq E\left[\int_t^\tau L(s, x_s(\cdot; t, \psi, u(\cdot)), u(s))\, ds \right.$$

$$\left. + V(\tau, x_\tau(\cdot; t, \psi, u(\cdot))) \right] - 3\epsilon \quad \text{(by (3.55))}.$$

Since this holds for arbitrary $(\Omega, \mathcal{F}, P, \mathbf{F}(t), W(\cdot), \tilde{u}(\cdot)) \in \mathcal{U}[t, T]$, by taking the supremum over $\mathcal{U}[t, T]$ we obtain

$$V(t, \psi) \geq \bar{V}(t, \psi) - 3\epsilon, \quad \forall \epsilon > 0. \tag{3.56}$$

Letting $\epsilon \downarrow 0$, we conclude that

$$V(t, \psi) \geq \bar{V}(t, \psi), \quad \forall (t, \psi) \in [0, T] \times \mathbf{C}.$$

This proves the DDP.   □

Assumption 3.3.10 *There exists a constant $K > 0$ such that*

$$|f(t,\phi,u)| + |g(t,\phi,u)| + |L(t,\phi,u)| + |\Psi(\phi)| \leq K, \quad \forall(t,\phi,u) \in [0,T] \times \mathbf{C} \times U.$$

In addition to its continuity in the initial function $\psi \in \mathbf{C}$ as proved in Lemma 3.3.7, the value function V has some regularity in the time variable as well. By using Theorem 3.3.9, it can be shown below that the value function V is Hölder continuous in time with respect to a parameter γ for any $\gamma \leq \frac{1}{2}$, provided that the initial segment is at least γ-Hölder continuous. Notice that the coefficients f, g, and L need not be Hölder continuous in time. Except for the role of the initial segment, the statement and proof of the following lemma are analogous to the nondelay case (see, e.g., Krylov [Kry80, p.167]). See also Proposition 2 of Fischer and Nappo [FN06] for delay case.

Lemma 3.3.11 *Assume Assumptions 3.1.1 and 3.8.1 hold. Let the initial function $\psi \in \mathbf{C}$. If ψ is γ-Hölder continuous with $\gamma \leq K(H)$, then the function $V(\cdot, \psi) : [0,T] \to \Re$ is Hölder continuous; that is, there is a constant $K(V) > 0$ depending only on $K(H)$, K (the Lipschitz constant in Assumption 3.1.1), T, and the dimensions such that for all $t, \tilde{t} \in [0,T]$*

$$|V(t,\psi) - V(\tilde{t},\psi)| \leq K(V)\left(|t - \tilde{t}|^\gamma \vee \sqrt{|t - \tilde{t}|}\right).$$

Proof. Let the initial function $\psi \in \mathbf{C}$ be γ-Hölder continuous with $\gamma \leq K(H)$. Without loss of generality, we assume that $s = t + h$ for some $h > 0$. We may also assume that $h \leq \frac{1}{2}$, because we can choose $K(V) \geq 4K(T+1)$ so that the asserted inequality holds for $|t - s| > \frac{1}{2}$. By the DDP (Theorem 3.3.9), we see that

$$|V(t,\psi) - V(s,\psi)|$$

$$= |V(t,\psi) - V(t+h,\psi)|$$

$$= \left| \sup_{u(\cdot) \in \mathcal{U}[t,T]} E\left[\int_t^{t+h} e^{-\alpha(s-t)} L(s, x_s(\cdot; t, \psi, u(\cdot)), u(s))\, ds \right.\right.$$

$$\left.\left. + e^{-\alpha h} V(t+h, x_{t+h}(t, \psi, u(\cdot)))\right]\right.$$

$$\left. - V(t+h, \psi)\right|$$

$$\leq \sup_{u(\cdot) \in \mathcal{U}[t,T]} E\left[\int_t^{t+h} e^{-\alpha(s-t)} L(s, x_s(\cdot; t, \psi, u(\cdot)), u(s))\, ds\right]$$

$$+ \sup_{u(\cdot) \in \mathcal{U}[t,T]} E\left[|e^{-\alpha h} V(t+h, x_{t+h}(\cdot; t, \psi, u(\cdot))) - V(t+h, \psi)|\right]$$

$$\leq Kh + \sup_{u(\cdot) \in \mathcal{U}[t,T]} L(V) E[\|x_{t+h}(\cdot; t, \psi, u(\cdot)) - \psi\|],$$

where K is the largest Lipschitz constant from Assumption 3.1.1 and $L(V)$ is the Lipschitz constant for V in the segment variable according to Lemma 3.3.7. Notice that $\psi = x_t(t, \psi, u(\cdot))$ for all $u(\cdot) \in \mathcal{U}[t, T]$. By the linear growth condition (Assumption 3.1.1) of f and g, the Hölder inequality, Doob-Davis-Gundy's maximal inequality (Theorem 1.2.11), and Itó's isometry, we have for arbitrary $u(\cdot)$,

$$
\begin{aligned}
E[\|x_{t+h}(\cdot; t, \psi, u(\cdot)) - \psi\|] \leq{} & \sup_{\theta \in [-r, -h]} |\psi(\theta + h) - \psi(\theta)| + \sup_{\theta \in [-r, 0]} |\psi(0) - \psi(\theta)| \\
& + E\left[\int_t^{t+h} |f(s, x_s(\cdot; t, \psi, u(\cdot)), u(s))|\, ds\right] \\
& + E\left[\left|\int_t^{t+h} g(s, x_s(\cdot; t, \psi, u(\cdot)), u(s))\, dW(s)\right|^2\right]^{\frac{1}{2}} \\
\leq{} & 2K(H)h^\gamma + Kh + 4Km\sqrt{h}.
\end{aligned}
$$

Putting everything together, we have the assertion of the lemma.   □

3.4 The Infinite-Dimensional HJB Equation

We will use the dynamic programming principle (Theorem 3.3.9) to derive the HJBE. Recall that $\{x_s(\cdot; t, \psi, u(\cdot)), s \in [t, T]$ is a $\mathbf{C}$-valued (strong) Markov process whenever $u(\cdot) \in \mathcal{U}[t, T]$. Therefore, using Theorem 2.4.1, we have the following.

Theorem 3.4.1 *Suppose that $\Phi \in C_{lip}^{1,2}([0, T] \times \mathbf{C}) \cap \mathcal{D}(\mathcal{S})$. Let $u(\cdot) \in \mathcal{U}[t, T]$ with $\lim_{s \downarrow t} u(s) = u \in U$ and $\{x_s(\cdot; t, \psi, u(\cdot)), s \in [t, T]\}$ be the $\mathbf{C}$-valued Markov process of (3.1) with the initial datum $(t, \psi) \in [0, T] \times \mathbf{C}$. Then*

$$
\lim_{\epsilon \downarrow 0} \frac{E[\Phi(t + \epsilon, x_{t+\epsilon}(t, \psi, u(\cdot)))] - \Phi(t, \psi)}{\epsilon} = \partial_t \Phi(t, \psi) + \mathbf{A}^u \Phi(t, \psi), \quad (3.57)
$$

where

$$
\mathbf{A}^u \Phi(t, \psi) = \mathcal{S}\Phi(t, \psi) + \overline{D\Phi(t, \psi)}(f(t, \psi, u)\mathbf{1}_{\{0\}}) \tag{3.58}
$$

$$
+ \frac{1}{2} \sum_{j=1}^m \overline{D^2\Phi(t, \psi)}(g(t, \psi, u)(\mathbf{e}_j)\mathbf{1}_{\{0\}}, g(t, \psi, u)(\mathbf{e}_j)\mathbf{1}_{\{0\}}),
$$

where $\mathbf{e}_j, j = 1, 2, \ldots, m$, is the jth unit vector of the standard basis in $\Re^m$.

Heuristic Derivation of the HJB Equation

Let $u \in U$. We define the Fréchet differential operator $\mathcal{A}^u$ as follows:

$$\mathcal{A}^u \Phi(t, \psi) \equiv \mathcal{S}(\Phi)(t, \psi) + \overline{D\Phi(t, \psi)}(f(t, \psi, u)\mathbf{1}_{\{0\}}) \tag{3.59}$$

$$+ \frac{1}{2} \sum_{j=1}^{m} \overline{D^2\Phi(t, \psi)}(g(t, \psi, u)\mathbf{e}_j\mathbf{1}_{\{0\}}, g(t, \psi, u)\mathbf{e}_j\mathbf{1}_{\{0\}}),$$

for any $\Phi \in C_{lip}^{1,2}([0, T] \times \mathbf{C}) \cap \mathcal{D}(\mathcal{S})$, where $\mathbf{e}_j$ is the jth vector of the standard basis in $\Re^m$.

To recall the meaning of the terms involved in this differential operator, we remind the readers of the following definitions given earlier in this volume.

First, $\mathcal{S}(\Phi)(t, \psi)$ is defined (see (2.11) of Chapter 2) as

$$\mathcal{S}(\Phi)(t, \psi) = \lim_{\epsilon \downarrow 0} \frac{\Phi(t, \tilde{\psi}_{t+\epsilon}) - \Phi(t, \psi)}{\epsilon} \tag{3.60}$$

and $\tilde{\psi} : [-r, T] \to \Re^n$ is the extension of $\psi \in \mathbf{C}$ from $[-r, 0]$ to $[-r, T]$ and is defined by

$$\tilde{\psi}(t) = \begin{cases} \psi(0) & \text{for } t \geq 0 \\ \psi(t) & \text{for } t \in [-r, 0). \end{cases}$$

Second, $D\Phi(t, \psi) \in \mathbf{C}^*$ and $D^2\Phi(t, \psi) \in \mathbf{C}^\dagger$ are the first- and second-order Fréchet derivatives of Φ with respect to its second argument $\psi \in \mathbf{C}$. In addition, $\overline{D\Phi(t, \psi)} \in (\mathbf{C} \oplus \mathbf{B})^*$ is the extension of of $D\Phi(t, \psi)$ from $\mathbf{C}^*$ to $(\mathbf{C} \oplus \mathbf{B})^*$ (see Lemma 2.2.3 of Chapter 2) and $\overline{D^2V(t, \psi)} \in (\mathbf{C} \oplus \mathbf{B})^\dagger$ is the extension of $D^2V(t, \psi)$ from $\mathbf{C}^\dagger$ to $(\mathbf{C} \oplus \mathbf{B})^\dagger$ (see Lemma 2.2.4 of Chapter 2).

Finally, the function $\mathbf{1}_{\{0\}} : [-r, 0] \to \Re$ is defined by

$$\mathbf{1}_{\{0\}}(\theta) = \begin{cases} 0 & \text{for } \theta \in [-r, 0) \\ 1 & \text{for } \theta = 0. \end{cases}$$

Without loss of generality, we can and will assume that for every $u \in U$, the domain of the generator $\mathbf{A}^u$ is large enough to contain $C_{lip}^{1,2}([0, T] \times \mathbf{C}) \cap \mathcal{D}(\mathcal{S})$.

From the DDP (Theorem 3.3.9), if we take a constant control $u(\cdot) \equiv u \in \mathcal{U}[t, T]$, then for $\forall \delta \geq 0$,

$$V(t, \psi) \geq E\Big[\int_t^{t+\delta} e^{-\alpha(s-t)} L(s, x_s(\cdot; t, \psi, v), v) \, ds$$

$$+ e^{-\alpha\delta} V(t + \delta, x_{t+\delta}(t, \psi, v)) \Big].$$

From this principle, we have

$$0 \geq \lim_{\delta \downarrow 0} \frac{1}{\delta} E\left[\int_t^{t+\delta} e^{-\alpha(s-t)} L(s, x_s(\cdot; t, \psi, u), u)\, ds \right.$$
$$\left. + e^{-\alpha\delta} V(t+\delta, x_{t+\delta}(\cdot; t, \psi, u)) - V(t, \psi) \right]$$

$$= \lim_{\delta \downarrow 0} \frac{1}{\delta} E\left[\int_t^{t+\delta} e^{-\alpha(s-t)} L(s, x_s(\cdot; t, \psi, u), u)\, ds \right.$$
$$\left. + \lim_{\delta \downarrow 0} \frac{1}{\delta} E[e^{-\alpha\delta} V(t+\delta, x_{t+\delta}(\cdot; t, \psi, u)) - e^{-\alpha\delta} V(t + x_{t+\delta}(\cdot; t, \psi, u))] \right.$$

$$+ \lim_{\delta \downarrow 0} \frac{1}{\delta} E[e^{-\alpha\delta} V(t, x_{t+\delta}(t, \psi, u)) - e^{-\alpha\delta} V(t, \psi))]$$

$$+ \lim_{\delta \downarrow 0} \frac{1}{\delta} [(e^{-\alpha\delta} - 1) V(t, \psi)]$$

$$= -\alpha V(t, \psi) + \partial_t V(t, \psi) + \mathbf{A}^u V(t, \psi) + L(t, \psi, u) \qquad (3.61)$$

for all $(t, \psi) \in [0, T] \times \mathbf{C}$, provided that $V \in C_{lip}^{1,2}([0, T] \times \mathbf{C}) \cap \mathcal{D}(\mathcal{S})$.

Moreover, if $u^*(\cdot) \in \mathcal{U}[t, T]$ is the optimal control policy that satisfies $\lim_{s \downarrow t} u^*(s) = v^*$, we should have, $\forall \delta \geq 0$, that

$$V(t, \psi) = E\left[\int_t^{t+\delta} e^{-\alpha(s-t)} L(s, x_s^*(\cdot; t, \psi, u^*(\cdot)), u^*(s))\, ds \right.$$
$$\left. + e^{-\alpha\delta} V(t+\delta, x_{t+\delta}^*(\cdot; t, \psi, u^*(\cdot))) \right], \qquad (3.62)$$

where $x_s^*(t, \psi, u^*(\cdot))$ is the $\mathbf{C}$-valued solution process corresponding to the initial datum (t, ψ) and the optimal control $u^*(\cdot) \in \mathcal{U}[t, T]$. Similarly, under the strong assumption on $u^*(\cdot)$ (including the right-continuity at the initial time t), we can get

$$0 = -\alpha V(t, \psi) + \partial_t V(t, \psi) + \mathcal{A}^{v^*} V(t, \psi) + L(t, \psi, v^*). \qquad (3.63)$$

Inequalities (3.61) and (3.62) are equivalent to the HJBE

$$0 = -\alpha V(t, \psi) + \partial_t V(t, \psi) + \max_{u \in U} [\mathcal{A}^u V(t, \psi) + L(t, \psi, u)].$$

We therefore have the following result.

Theorem 3.4.2 *Let $V : [0, T] \times \mathbf{C} \to \Re$ be the value function defined by (3.7). Suppose $V \in C_{lip}^{1,2}([0, T] \times \mathbf{C}) \cap \mathcal{D}(\mathcal{S})$. Then the value function V satisfies the following HJBE:*

$$\alpha V(t, \psi) - \partial_t V(t, \psi) - \max_{u \in U} [\mathbf{A}^u V(t, \psi) + L(t, \psi, u)] = 0 \qquad (3.64)$$

on $[0, T] \times \mathbf{C}$, and $V(T, \psi) = \Psi(\psi)$, $\forall \psi \in \mathbf{C}$, where

$$\mathcal{A}^u V(t, \psi) \equiv \mathcal{S}V(t, \psi) + \overline{DV(t, \psi)}(f(t, \psi, u)\mathbf{1}_{\{0\}})$$

$$+ \frac{1}{2}\sum_{j=1}^{m} \overline{D^2 V(t, \psi)}(g(t, \psi, u)\mathbf{e}_j \mathbf{1}_{\{0\}}, g(t, \psi, u)\mathbf{e}_j \mathbf{1}_{\{0\}}).$$

Note that it is not known that the value function V satisfies the necessary smoothness condition $V \in C_{lip}^{1,2}([0, T] \times \mathbf{C}) \cap \mathcal{D}(\mathcal{S})$. In fact, the following simple example shows that the value function does not possess the smoothness condition for it to be a classical solution of the HJBE (3.64) even for a very simple deterministic control problem.

The following is an example taken from Example 2.3 of [FS93, p.57].

Example. Consider a one-dimensional simple deterministic control problem described by $\dot{x}(s) = u(s)$, $s \in [0, 1)$, with the running cost function $L \equiv 0$ and the terminal cost function $\Psi(x) = x \in \Re$ and a control set $U = [-a, a]$ with some constant $a > 0$. Since the boundary data are increasing and the running cost is zero, the optimal control is $u^*(s) \equiv -a$. Hence, the value function is

$$V(t, x) = \begin{cases} -1 & \text{if } x + at \le a - 1 \\ x + at - a & \text{if } x + at \ge a - 1 \end{cases} \tag{3.65}$$

for $(t, x) \in [0, 1] \times \Re$. Note that the value function is differentiable except on the set $\{(t, x) \mid x + at = a - 1\}$ and it is a generalized solution of

$$-\partial_t V(t, x) + a\left|\partial x V(t, x)\right| = 0, \tag{3.66}$$

with the corresponding terminal boundary condition given by

$$V(1, x) = \Psi(x) = x, \quad x \in [-1, 1]. \tag{3.67}$$

The above example shows that in general we need to seek a weaker condition for the HJBE (3.64) such as a viscosity solution instead of a solution for HJBE (3.64) in the classical sense. In fact, it will be shown that the value function is a unique viscosity solution of the HJBE (3.64). These results will be given in Sections 3.5 and 3.6.

3.5 Viscosity Solution

In this section, we shall show that the value function $V : [0, T] \times \mathbf{C} \to \Re$ defined by Equation (3.7) is actually a viscosity solution of the HJBE (3.64).

Definitions of Viscosity Solution

First, let us define the viscosity solution of (3.64) as follows.

Definition 3.5.1 *An upper semicontinuous (respectively, lower semicontinuous) function* $w : (0,T] \times \mathbf{C} \rightarrow \Re$ *is said to be a viscosity subsolution (respectively, supersolution) of the HJBE (3.64) if*

$$w(T,\phi) \le (\ge)\Psi(\phi), \quad \forall \phi \in \mathbf{C},$$

and if, for every $\Gamma \in C_{lip}^{1,2}([0,T] \times \mathbf{C}) \cap \mathcal{D}(\mathcal{S})$ *and for every* $(t,\psi) \in [0,T] \times \mathbf{C}$ *satisfying* $\Gamma \ge (\le)w$ *on* $[0,T] \times \mathbf{C}$ *and* $\Gamma(t,\psi) = w(t,\psi)$, *we have*

$$\alpha\Gamma(t,\psi) - \partial_t\Gamma(t,\psi) - \max_{v\in U} [\mathbf{A}^v\Gamma(t,\psi) + L(t,\psi,v)] \le (\ge) \ 0. \quad (3.68)$$

We say that w is a viscosity solution of the HJBE (3.64) if it is both a viscosity supersolution and a viscosity subsolution of the HJBE (3.64).

Definition 3.5.2 *Let a function* $\Phi : [0,T] \times \mathbf{C} \rightarrow \Re$ *be given. We say that* $(p,q,Q) \in \Re \times \mathbf{C}^* \times \mathbf{C}^\dagger$ *belongs to* $D_{t+,\phi}^{1,2,+}\Phi(t,\phi)$, *the second-order superdifferential of* Φ *at* $(t,\phi) \in (0,T) \times \mathbf{C}$, *if, for all* $\psi \in \mathbf{C}$, $s \ge t$,

$$\Phi(s,\psi) \le \Phi(t,\phi) + p(s-t) + q(\psi - \phi) \quad (3.69)$$
$$+ \frac{1}{2}Q(\psi - \phi, \psi - \phi) + o(s - t + \|\psi - \phi\|^2).$$

The second-order one-sided parabolic subdifferential of Φ *at* $(t,\phi) \in (0,T) \times \mathbf{C}$, $D_{t+,\phi}^{1,2,-}\Phi(t,\phi)$, *is defined as those* $(p,q,Q) \in \Re \times \mathbf{C}^* \times \mathbf{C}^\dagger$ *such that the above inequality is reversed; that is,*

$$\Phi(s,\psi) \ge \Phi(t,\phi) + p(s-t) + q(\psi - \phi) \quad (3.70)$$
$$+ \frac{1}{2}Q(\psi - \phi, \psi - \phi) + o(s - t + \|\psi - \phi\|^2). \quad (3.71)$$

Remark 3.5.3 *The second-order one-sided parabolic subdifferential and second-order one-sided parabolic superdifferential have the following relationship:*

$$D_{t+,\phi}^{1,2,-}\Phi(t,\phi) = -D_{t+,\phi}^{1,2,+}(-\Phi)(t,\phi).$$

Lemma 3.5.4 *Let* $w : (0,T] \times \mathbf{C} \rightarrow \Re$ *and let* $(t,\psi) \in (0,T) \times \mathbf{C}$. *Then* $(p,q,Q) \in D_{t+,\phi}^{1,2,+}w(t,\psi)$ *if and only if there exists a function* $\Phi \in C^{1,2}((0,T) \times \mathbf{C})$ *such that* $w - \Phi$ *attains a strict global maximum at* (t,ψ) *relative to the set of* (s,ϕ) *such that* $s \ge t$ *and*

$$(\Phi(t,\psi), \partial_t\Phi(t,\psi), D\Phi(t,\psi), D^2\Phi(t,\psi)) = (w(t,\psi),p,q,Q). \quad (3.72)$$

Moreover, if w has polynomial growth (i.e., there exist positive constants K_p and $k \ge 1$ such that

$$|w(s,\phi)| \le K_p(1 + \|\phi\|_2)^k, \quad \forall (s,\phi) \in (0,T) \times \mathbf{C}), \quad (3.73)$$

then Φ can be chosen so that Φ $\partial_t\Phi$, $D\Phi$, and $D^2\Phi$ satisfy (3.73) (with possibly different constants K_p).

Lemma 3.5.5 *Let v be an upper-semicontinuous function on $(0,T) \times \mathbf{C}$ and let $(\bar{t}, \bar{\psi}) \in (0,T) \times \mathbf{C}$. Then $(p,q,Q) \in \bar{D}^{1,2}_{t+,\psi} v(\bar{t}, \bar{\psi})$ if and only if there exists a function $\Phi \in C^{1,2}((0,T) \times \mathbf{C})$ such that $v - \Phi \in C((0,T) \times \mathbf{C})$ attains a strict global maximum at $(\bar{t}, \bar{\psi})$ relative to the set of $(t,\psi) \in [\bar{t}, T) \times \mathbf{C}$ and*

$$(\Phi(\bar{t}, \bar{\psi}), \partial_t \Phi(\bar{t}, \bar{\psi}), D\Phi(\bar{t}, \bar{\psi}), D^2\Phi(\bar{t}, \bar{\psi})) = (v(\bar{t}, \bar{\psi}), p, q, Q). \tag{3.74}$$

Moreover, if v has polynomial growth (i.e., if there exists a constant $k \geq 1$ such that

$$|v(t,\psi)| \leq C(1 + \|\psi\|^k) \; \forall(t,\psi) \in [0,T] \times \mathbf{C}), \tag{3.75}$$

then Φ can be chosen so that Φ, $\partial_t\Phi$, $D\Phi$, and $D^2\Phi$ satisfy (3.69) under the appropriate norms and with possibly different constants C.

Proof. The proof of this lemma is an extension of Lemma 5.4 in Yong and Zhou [YZ99, Chap.4] from an Euclidean space to the infinite-dimensional space $\mathbf{C}$.

Suppose $(p,q,Q) \in \bar{D}^{1,2}_{t+,\psi} v(\bar{t}, \bar{\psi})$. Define the function $\gamma : (0,T) \times \mathbf{C} \to \Re$ as

$$\gamma(t,\psi) = \frac{1}{t - \bar{t} + \|\psi - \bar{\psi}\|^2} [v(t,\psi) - v(\bar{t}, \bar{\psi})$$
$$- p(t - \bar{t}) - q(\psi - \bar{\psi}) - \frac{1}{2}Q(\phi - \psi, \phi - \psi)] \vee 0$$
$$\text{for } (t,\psi) \neq (\bar{t}, \bar{\psi}),$$

and $\gamma(t,\psi) = 0$ otherwise.

We also define the function $\epsilon : \Re \to \Re$ by

$$\epsilon(r) = \sup\{\gamma(t,\psi) \mid (t,\psi) \in (\bar{t}, T] \times \mathbf{C}, \; t - \bar{t} + \|\psi - \bar{\psi}\|^2 \leq r\} \text{ if } r > 0$$

and $\epsilon(r) = 0$ if $r \leq 0$. Then it follows from the definition of $\bar{D}^{1,2}_{t+,\psi} v(t,\psi)$ that

$$v(t,\psi) - [v(\bar{t}, \bar{\psi}) + p(t - \bar{t}) + q(\psi - \bar{\psi}) + \frac{1}{2}Q(\phi - \psi, \phi - \psi)]$$
$$\leq (t - \bar{t} + \|\psi - \bar{\psi}\|^2)\epsilon(t - \bar{t} + \|\psi - \bar{\psi}\|^2) \; \forall(t,\psi) \in (\bar{t}, T] \times \mathbf{C}.$$

Define the function $\alpha : \Re_+ \to \Re$ by

$$\alpha(\rho) = \frac{2}{\rho} \int_0^{2\rho} \int_0^r \epsilon(\theta) \, d\theta \, dr, \rho > 0. \tag{3.76}$$

Then it is easy to see that its first-order derivative

$$\dot{\alpha}(\rho) = -\frac{2}{\rho^2} \int_0^{2\rho} \int_0^r \epsilon(\theta) \, d\theta \, dr + \frac{4}{\rho} \int_0^{2\rho} \epsilon(\theta) \, d\theta$$

and its second order-derivative

$$\ddot{\alpha}(\rho) = \frac{4}{\rho^3} \int_0^{2\rho} \int_0^r \epsilon(\theta) \, d\theta \, dr - \frac{8}{\rho^2} \int_0^{2\rho} \epsilon(\theta) \, d\theta + \frac{8}{\rho}\epsilon(2\rho).$$

Consequently,

$$|\alpha(\rho)| \le 4\rho\epsilon(2\rho), \quad |\dot{\alpha}(\rho)| \le 12\rho\epsilon(2\rho), \quad \text{and} \quad |\ddot{\alpha}(\rho)| \le \frac{32\rho\epsilon(2\rho)}{\rho}.$$

Now, we define the function $\beta : [0, T] \times \mathbf{C} \to \Re$ by

$$\beta(t, \psi) = \begin{cases} \alpha(\rho(t, \psi)) + \rho^2(t, \psi) & \text{if } (t, \psi) \ne (\bar{t}, \bar{\psi}) \\ 0 & \text{otherwise,} \end{cases}$$

where $\rho(t, \psi) = t - \bar{t} + \|\psi - \bar{\psi}\|_2$.

Finally, we define the function $\Phi : [\bar{t}, T] \times \mathbf{C} \to \Re$ by

$$\Phi(t, \psi) = v(\bar{t}, \bar{\psi}) + p(t - \bar{t}) + q(\psi - \bar{\psi})$$
$$+ \frac{1}{2}Q(\phi - \psi, \phi - \psi) + \beta(t, \psi), \quad \forall (t, \psi) \in [0, T] \times \mathbf{C}. \quad (3.77)$$

We claim that $\Phi \in C_{lip}^{1,2}([0, T] \times \mathbf{C})$ and it satisfies the following three conditions:

(i) $v(\bar{t}, \bar{\psi}) = \Phi(\bar{t}, \bar{\psi})$.

(ii) $v(t, \psi) < \Phi(t, \psi)$ for all $(t, \psi) \ne (\bar{t}, \bar{\psi})$.

(iii) $(\Phi(\bar{t}, \bar{\psi}), \partial_t \Phi(\bar{t}, \bar{\psi}), D\Phi(\bar{t}, \bar{\psi}), D^2\Phi(\bar{t}, \bar{\psi})) = (v(\bar{t}, \bar{\psi}), p, q, Q)$.

Note that (i) is trivial by the definition of Φ. The proofs for (ii) and (iii) are very similar to those of Lemma 2.7 and Lemma 2.8 of Yong and Zhou [YZ99] and are omitted here. $\square$

Proposition 3.5.6 *A function $w \in C([0, T] \times \mathbf{C})$ is a viscosity solution of the HJBE (3.64) if*

$$-p - \sup_{u \in U} G(t, \phi, u, q, Q) \le 0, \quad \forall (p, q, Q) \in D_{t+,\phi}^{1,2,+} w(t, \phi), \forall (t, \phi) \in [0, T) \times \mathbf{C},$$

$$-p - \sup_{u \in U} G(t, \phi, u, q, Q) \ge 0, \quad \forall (p, q, Q) \in D_{t+,\phi}^{1,2,-} w(t, \phi), \forall (t, \phi) \in [0, T) \times \mathbf{C},$$

and

$$w(T, \phi) = \Psi(\phi), \quad \forall \phi \in \mathbf{C},$$

where the function G is defined as

$$G(t, \phi, u, q, Q) = \mathcal{S}(\Phi)(t, \psi) + \bar{q}(f(t, \phi, v)\mathbf{1}_{\{0\}})$$
$$+ \frac{1}{2}\sum_{j=1}^m \overline{Q}(g(t, \psi, v)(\mathbf{e}_j)\mathbf{1}_{\{0\}}, g(t, \psi, v)(\mathbf{e}_j)\mathbf{1}_{\{0\}}). \quad (3.78)$$

Proof. The proposition follows immediately from Lemma 3.5.4 and Lemma 3.5.5. $\square$

For our value function V defined by (3.7), we now show that it has the following property.

Lemma 3.5.7 *Let* $V : [0, T] \times \mathbf{C} \to \Re$ *be the value function defined in (3.7).*
Then there exists a constant $k > 0$ *and a positive integer* p *such that for every*
$(t, \psi) \in [0, T] \times \mathbf{C}$,

$$|V(t, \psi)| \leq K(1 + \|\psi\|_2)^k. \tag{3.79}$$

Proof. It is clear that V has at most a polynomial growth, since L and Φ
have at most a polynomial growth. This proves the lemma. $\square$

Theorem 3.5.8 *The value function* $V : [0, T] \times \mathbf{C} \to \Re$ *defined in (3.7) is a*
viscosity solution of the HJBE:

$$\alpha V(t, \psi) - \partial_t V(t, \psi) - \max_{v \in U} [\mathbf{A}^v V(t, \psi) + L(t, \psi, v)] = 0 \tag{3.80}$$

on $[0, T] \times \mathbf{C}$, *and* $V(T, \psi) = \Psi(\psi), \forall \psi \in \mathbf{C}$, *where*

$$\mathbf{A}^v V(t, \psi) = \mathcal{S}V(t, \psi) + \overline{DV(t, \psi)}(f(t, \psi, v)\mathbf{1}_{\{0\}})$$

$$+ \frac{1}{2} \sum_{j=1}^m \overline{D^2 V(t, \psi)}(g(t, \psi, v)(\mathbf{e}_j)\mathbf{1}_{\{0\}}, g(t, \psi, v)(\mathbf{e}_j)\mathbf{1}_{\{0\}}), \tag{3.81}$$

where $\mathbf{e}_j, j = 1, 2, \ldots, m$, *is the* j*th unit vector of the standard basis in* $\Re^m$.

Proof. Let $\Gamma \in C_{lip}^{1,2}([0, T] \times \mathbf{C}) \cap \mathcal{D}(\mathcal{S})$. For $(t, \psi) \in [0, T] \times \mathbf{C}$ such that
$\Gamma \leq V$ on $[0, T] \times \mathbf{C}$ and $\Gamma(t, \psi) = V(t, \psi)$, we want to prove the viscosity
supersolution inequality, that is,

$$\alpha \Gamma(t, \psi) - \partial_t \Gamma(t, \psi) - \max_{v \in U} [\mathbf{A}^v \Gamma(t, \psi) + L(t, \psi, v)] \geq 0. \tag{3.82}$$

Let $u(\cdot) \in \mathcal{U}[t, T]$. Since $\Gamma \in C_{lip}^{1,2}([0, T] \times \mathbf{C}) \cap \mathcal{D}(\mathcal{S})$ (by virtue of Theorem 3.4.1) for $t \leq s \leq T$, we have

$$E\left[e^{-\alpha(s-t)}\Gamma(s, x_s(\cdot; t, \psi, u(\cdot)))\right] - \Gamma(t, \psi)$$

$$= E\left[\int_t^s e^{-\alpha(\xi-t)}\left(\partial_\xi \Gamma(\xi, x_\xi(\cdot; t, \psi, u(\cdot)))\right.\right.$$

$$\left.\left. + \mathbf{A}^{u(\xi)}\Gamma(\xi, x_\xi(\cdot; t, \psi, u(\cdot))) - \alpha\Gamma(\xi, x_\xi(\cdot; t, \psi, u(\cdot)))\right)d\xi\right]. \tag{3.83}$$

On the other hand, for any $s \in [t, T]$, the DDP (Theorem 3.3.9) gives,

$$V(t, \psi) = \max_{u(\cdot) \in \mathcal{U}[t,T]} E\left\{\int_t^s e^{-\alpha(\xi-t)}L(\xi, x_\xi(\cdot; t, \psi, u(\cdot)), u(\xi))d\xi\right.$$

$$\left. + e^{-\alpha(s-t)}V(s, x_s(\cdot; t, \psi, u(\cdot)))\right\}. \tag{3.84}$$

Therefore, we have

$$V(t, \psi) \geq E\left[\int_t^s e^{-\alpha(\xi-t)} L(\xi, x_\xi(\cdot; t, \psi, u(\cdot)), u(\xi))\, d\xi\right]$$
$$+ E\left[e^{-\alpha(s-t)} V(s, x_s(\cdot; t, \psi, u(\cdot)))\right]. \tag{3.85}$$

By virtue of (3.177) and using $\Gamma \leq V, \Gamma(t, \psi) = V(t, \psi)$, we can get

$$0 \geq E\left[\int_t^s e^{-\alpha(\xi-t)} L(\xi, x_\xi(\cdot; t, \psi, u(\cdot)), u(\xi))d\xi\right]$$
$$+ E\left[e^{-\alpha(s-t)} V(s, x_s(\cdot; t, \psi, u(\cdot)))\right] - V(t, \psi)$$
$$\geq E\left[\int_t^s e^{-\alpha(\xi-t)} L(\xi, x_\xi(\cdot; t, \psi, u(\cdot)), u(\xi))\, d\xi\right]$$

$$+ E\left[e^{-\alpha(s-t)} \Gamma(s, x_s(\cdot; t, \psi, u(\cdot)))\right] - \Gamma(t, \psi)$$
$$\geq E\int_t^s e^{-\alpha(\xi-t)}\left[-\alpha\Gamma(\xi, x_\xi(\cdot; t, \psi, u(\cdot))) + \partial_\xi\Gamma(\xi, x_\xi(\cdot; t, \psi, u(\cdot)))\right.$$
$$\left.+ \mathbf{A}^u\Gamma(\xi, x_\xi(\cdot; t, \psi, u(\cdot))) + L(\xi, x_\xi(\cdot; t, \psi, u(\cdot)), u(\xi))\right]d\xi.$$

Dividing both sides by $(s - t)$, we have

$$0 \leq E\left[\frac{1}{s-t}\int_t^s e^{-\alpha(\xi-t)}\left(\alpha\Gamma(\xi, x_\xi(\cdot; t, \psi, u(\cdot)))\right.\right.$$
$$- \partial_\xi\Gamma(\xi, x_\xi(\cdot; t, \psi, u(\cdot))) - \mathbf{A}^{u(\xi)}\Gamma(\xi, x_\xi(\cdot; t, \psi, u(\cdot)))$$
$$\left.\left.- L(\xi, x_\xi(\cdot; t, \psi, u(\cdot)), u(\xi))\right)d\xi\right]. \tag{3.86}$$

Now, let $s \downarrow t$ in (3.86) and $\lim_{s\downarrow t} u(s) = v$, and we obtain

$$\alpha\Gamma(t, \psi) - \partial_t\Gamma(t, \psi) - [\mathbf{A}^v\Gamma(t, \psi) + L(t, \psi, v)] \geq 0. \tag{3.87}$$

Since $v \in U$ is arbitrary, we prove that V is a viscosity supersolution.

Next, we want to prove that V is a viscosity subsolution. Let $\Gamma \in C^{1,2}_{lip}([0, T] \times \mathbf{C}) \cap \mathcal{D}(\mathcal{S})$. For $(t, \psi) \in [0, T] \times \mathbf{C}$ satisfying $\Gamma \geq V$ on $[0, T] \times \mathbf{C}$ and $\Gamma(t, \psi) = V(t, \psi)$, we want to prove that

$$\alpha\Gamma(t, \psi) - \partial_t\Gamma(t, \psi) - \max_{v\in U}[\mathbf{A}^v\Gamma(t, \psi) + L(t, \psi, v)] \leq 0. \tag{3.88}$$

We assume the contrary and try to obtain a contradiction. Let suppose that there exit $(t, \psi) \in [0, T] \times \mathbf{C}$, $\Gamma \in C^{1,2}_{lip}([0, T] \times \mathbf{C}) \cap \mathcal{D}(\mathcal{S})$, with $\Gamma \geq V$ on $[0, T] \times \mathbf{C}$ and $\Gamma(t, \psi) = V(t, \psi)$, and $\delta > 0$ such that for all control $u(\cdot) \in \mathcal{U}[t, T]$ with $\lim_{s\downarrow t} u(s) = v$,

$$\alpha \Gamma(\tau, \phi) - \partial_t \Gamma(\tau, \phi) - \mathbf{A}^v \Gamma(\tau, \phi) - L(\tau, \phi, v) \geq \delta \qquad (3.89)$$

for all $(\tau, \phi) \in N(t, \psi)$, where $N(t, \psi)$ is a neighborhood of (t, ψ). Let $u(\cdot) \in \mathcal{U}[t, T]$ with $\lim_{s \downarrow t} u(s) = v$, and t_1 such that for $t \leq s \leq t_1$, the solution $x(s; t, \psi, u(\cdot)) \in N(t, \psi)$. Therefore, for any $s \in [t, t_1]$, we have almost surely

$$\alpha \Gamma(s, x_s(t, \psi, u(\cdot))) - \partial_t \Gamma(s, x_s(\cdot; t, \psi, u(\cdot)))$$
$$- \mathbf{A}^v \Gamma(s, x_s(\cdot; t, \psi, u(\cdot))) - L(s, x_s(\cdot; t, \psi, u(\cdot)), u(s)) \geq \delta. \quad (3.90)$$

On the other hand, since $\Gamma \geq V$, using the definition of J and V, we can get

$$J(t, \psi; u(\cdot)) \leq E \left[\int_t^{t_1} e^{-\alpha(s-t)} L(s, x_s, u(s)) ds + e^{-\alpha(t_1-t)} V(t_1, x_{t_1}) \right]$$

$$\leq E \left[\int_t^{t_1} e^{-\alpha(s-t)} L(s, x_s, u(s)) ds \right.$$
$$\left. + e^{-\alpha(t_1-t)} \Gamma(t_1, x_{t_1}(t, \psi, u(\cdot))) \right].$$

Using (3.90), we have

$$J(t, \psi; u(\cdot)) \leq E \left[\int_t^{t_1} e^{-\alpha(s-t)} \left(-\delta + \alpha \Gamma(s, x_s(\cdot; t, \psi, u(\cdot))) \right. \right.$$
$$\left. - \partial_t \Gamma(s, x_s(\cdot; t, \psi, u(\cdot))) - \mathbf{A}^{u(s)} \Gamma(s, x_s(\cdot; t, \psi, u(\cdot))) \right) ds$$
$$\left. + e^{-\alpha(t_1-t)} \Gamma(t_1, x_{t_1}(t, \psi, u(\cdot))) \right]. \qquad (3.91)$$

In addition, similar to (3.86), we can get

$$E \left[e^{-\alpha(t_1-t)} \Gamma(t_1, x_{t_1}(\cdot; t, \psi, u(\cdot))) \right] - \Gamma(t, \psi)$$
$$= E \left[\int_t^{t_1} e^{-\alpha(s-t)} \left(\partial_s \Gamma(s, x_s(\cdot; t, \psi, u(\cdot))) + \mathbf{A}^{u(s)} \Gamma(s, x_s(\cdot; t, \psi, u(\cdot))) \right. \right.$$
$$\left. \left. - \alpha \Gamma(s, x_s(\cdot; t, \psi, u(\cdot))) \right) ds \right]. \qquad (3.92)$$

Therefore, we can get

$$J(t, \psi; u(\cdot)) \leq - \int_t^{t_1} e^{-\alpha(s-t)} \delta \, ds + \Gamma(t, \psi)$$
$$= -\frac{\delta}{\alpha} (1 - e^{-\alpha(t_1-t)}) + V(t, \psi)$$

Taking the supremum over all admissible controls $u(\cdot) \in \mathcal{U}[t, T]$, we have

$$V(t, \psi) \leq -\frac{\delta}{\alpha} (1 - e^{-\alpha(t_1-t)}) + V(t, \psi).$$

This contradicts the fact that $\delta > 0$. Therefore, $V(t, \psi)$ is a viscosity subsolution. This completes the proof of the theorem. $\qquad \square$

3.6 Uniqueness

In this section, we will show that the value function $V : [0,T] \times \mathbf{C} \to \Re$ of Problem (OCCP) is the unique viscosity solution of the HJBE (3.64). We first need the following comparison principle.

Theorem 3.6.1 (Comparison Principle) *Assume that $V_1(t, \psi)$ and $V_2(t, \psi)$ are both continuous with respect to the argument $(t, \psi) \in [0,T] \times \mathbf{C}$ and are respectively the viscosity subsolution and supersolution of (3.64) with at most a polynomial growth (Lemma 3.5.7). In other words, there exists a real number $\Lambda > 0$ and a positive integer $k \geq 1$ such that*

$$|V_i(t, \psi)| \leq \Lambda(1 + \|\psi\|_2)^k, \quad \forall (t, \psi) \in [0,T] \times \mathbf{C}, \quad i = 1, 2.$$

Then

$$V_1(t, \psi) \leq V_2(t, \psi), \quad \forall (t, \psi) \in [0,T] \times \mathbf{C}. \tag{3.93}$$

Before we proceed to the proof of Theorem 3.6.1, we will use the following result proven in Ekeland and Lebourg [EL76] and also in a general form in Stegall [Ste78] and Bourgain [Bou79]. The reader is also referred to Crandall et al. [CIL92] and Lions [Lio82, Lio89] for an application example of this result in a setting similar but significantly different in details from what are to be presented below. A similar proof of uniqueness of viscosity solution is also done in Chang et al. [CPP07a].

Lemma 3.6.2 *Let Φ be a bounded and upper-semicontinuous real-valued function on a closed ball B of a Banach space Ξ that has the Radon-Nikodym property. Then for any $\epsilon > 0$, there exists an element $u^* \in \Xi^*$ with norm at most ϵ, where Ξ^* is the topological dual of Ξ, such that $\Phi + u^*$ attains its maximum on B.*

Note that every separable Hilbert space $(\Xi, \|\cdot\|_\Xi)$ satisfies the Radon-Nikodym property (see, e.g., [EL76]). In order to apply Lemma 3.6.2, we will therefore restrict ourself to a subspace Ξ of the product space $\mathbf{C} \times \mathbf{C}$ ($\mathbf{C} = C([-r, 0]; \Re^n)$), which is a separable Hilbert space and dense in $\mathbf{C} \times \mathbf{C}$. A good candidates is the product space $W^{1,2}((-r, 0); \Re^n) \times W^{1,2}((-r, 0); \Re^n)$, where $W^{1,2}((-r, 0); \Re^n)$ is the Sobolev space defined by

$$W^{1,2}((-r, 0); \Re^n) = \{\phi \in \mathbf{C} \mid \phi \text{ is absolutely continuous on } (-r, 0)$$
$$\text{and } \|\phi\|_{1,2} < \infty\},$$

$$\|\phi\|_{1,2}^2 \equiv \|\phi\|_2^2 + \|\dot\phi\|_2^2,$$

and $\dot\phi$ is the derivative of ϕ in the distributional sense. Note that it can be shown that the Hilbertian norm $\|\cdot\|_{1,2}$ is weaker than the sup-norm $\|\cdot\|$; that is, there exists a constant $K > 0$ such that

$$\|\phi\|_{1,2} \leq K\|\phi\|, \quad \forall \phi \in W^{1,2}((-r, 0); \Re^n).$$

From the Sobolev embedding theorems (see, e.g., Adams [Ada75]), it is known that $W^{1,2}((-r,0); \Re^n) \subset \mathbf{C}$ and that $W^{1,2}((-r,0); \Re^n)$ is dense in $\mathbf{C}$. For more about Sobolev spaces and corresponding results, one can refer to [Ada75].

Before we proceed to the proof of the Comparison Principle, first let us establish some results that will be needed in the proof.

Let V_1 and V_2 be respectively a viscosity subsolution and supersolution of (3.64). For any $0 < \delta, \gamma < 1$, and for all $\psi, \phi \in \mathbf{C}$ and $t, s \in [0, T]$, define

$$\Theta_{\delta\gamma}(t, s, \psi, \phi) \equiv \frac{1}{\delta} \left[\|\psi - \phi\|_2^2 + \|\psi^0 - \phi^0\|_2^2 + |t - s|^2 \right] \tag{3.94}$$
$$+ \gamma \left[\exp(1 + \|\psi\|_2^2 + \|\psi^0\|_2^2) + \exp(1 + \|\phi\|_2^2 + \|\phi^0\|_2^2) \right],$$

and

$$\Phi_{\delta\gamma}(t, s, \psi, \phi) \equiv V_1(t, \psi) - V_2(s, \phi) - \Theta_{\delta\gamma}(t, s, \psi, \phi), \tag{3.95}$$

where $\psi^0, \phi^0 \in \mathbf{C}$ with $\psi^0(\theta) = \frac{\theta}{-r}\psi(-r - \theta)$ and $\phi^0(\theta) = \frac{\theta}{-r}\phi(-r - \theta)$ for $\theta \in [-r, 0]$.

Moreover, using the polynomial growth condition for V_1 and V_2, we have

$$\lim_{\|\psi\|_2, \|\phi\|_2 \to \infty} \Phi_{\delta\gamma}(t, s, \psi, \phi) = -\infty. \tag{3.96}$$

The function $\Phi_{\delta\gamma}$ is a real-valued function that is bounded above and continuous on $[0, T] \times [0, T] \times W^{1,2}((-r,0); \Re^n) \times W^{1,2}((-r,0); \Re^n)$ (since the Hilbertian norm $\| \cdot \|_{1,2}$ is weaker than the sup-norm $\| \cdot \|$). Therefore, from Lemma 3.6.2 (which is applicable by virtue of (3.96)), for any $1 > \epsilon > 0$ there exits a continuous linear functional T_ϵ in the topological dual of $W^{1,2}((-r,0); \Re^n) \times W^{1,2}((-r,0); \Re^n)$, with norm at most ϵ, such that the function $\Phi_{\delta\gamma} + T_\epsilon$ attains it maximum in $[0, T] \times [0, T] \times W^{1,2}((-r,0); \Re^n) \times W^{1,2}((-r,0); \Re^n)$ (see Lemma 3.6.2). Let us denote by

$$(t_{\delta\gamma\epsilon}, s_{\delta\gamma\epsilon}, \psi_{\delta\gamma\epsilon}, \phi_{\delta\gamma\epsilon})$$

the global maximum of $\Phi_{\delta\gamma} + T_\epsilon$ on $[0, T] \times [0, T] \times W^{1,2}((-r,0); \Re^n) \times W^{1,2}((-r,0); \Re^n)$. Without loss of generality, we assume that for any given δ, γ, and ϵ, there exists a constant $M_{\delta\gamma\epsilon}$ such that the maximum value $\Phi_{\delta\gamma} + T_\epsilon + M_{\delta\gamma\epsilon}$ is zero. In other words, we have

$$\Phi_{\delta\gamma}(t_{\delta\gamma\epsilon}, s_{\delta\gamma\epsilon}, \psi_{\delta\gamma\epsilon}, \phi_{\delta\gamma\epsilon}) + T_\epsilon(\psi_{\delta\gamma\epsilon}, \phi_{\delta\gamma\epsilon}) + M_{\delta\gamma\epsilon} = 0. \tag{3.97}$$

We have the following lemmas.

Lemma 3.6.3 $(t_{\delta\gamma\epsilon}, s_{\delta\gamma\epsilon}, \psi_{\delta\gamma\epsilon}, \phi_{\delta\gamma\epsilon})$ *is the global maximum of $\Phi_{\delta\gamma} + T_\epsilon$ in $[0, T] \times [0, T] \times \mathbf{C} \times \mathbf{C}$.*

Proof. Let $(t, s, \psi, \phi) \in [0, T] \times [0, T] \times \mathbf{C} \times \mathbf{C}$. By virtue of the density of $W^{1,2}((-r,0); \Re^n)$ in $\mathbf{C}$, we can find a sequence $(t_k, s_k, \psi_k, \phi_k)$ in $[0, T] \times [0, T] \times W^{1,2}((-r,0); \Re^n) \times W^{1,2}((-r,0); \Re^n)$ such that

$$(t_k, s_k, \psi_k, \phi_k) \to (t, s, \psi, \phi) \text{ as } k \to \infty.$$

It is known that

$$\Phi_{\delta\gamma}(t_k, s_k, \psi_k, \phi_k) + T_\epsilon(\psi_k, \phi_k) \leq \Phi_{\delta\gamma}(t_{\delta\gamma\epsilon}, s_{\delta\gamma\epsilon}, \psi_{\delta\gamma\epsilon}, \phi_{\delta\gamma\epsilon}) + T_\epsilon(\psi_{\delta\gamma\epsilon}, \phi_{\delta\gamma\epsilon}).$$

Taking the limit as k goes to ∞ in the last inequality, we obtain

$$\Phi_{\delta\gamma}(t, s, \psi, \phi) + T_\epsilon(\psi, \phi) \leq \Phi_{\delta\gamma}(t_{\delta\gamma\epsilon}, s_{\delta\gamma\epsilon}, \psi_{\delta\gamma\epsilon}, \phi_{\delta\gamma\epsilon}) + T_\epsilon(\psi_{\delta\gamma\epsilon}, \phi_{\delta\gamma\epsilon}).$$

This shows that $(t_{\delta\gamma\epsilon}, s_{\delta\gamma\epsilon}, \psi_{\delta\gamma\epsilon}, \phi_{\delta\gamma\epsilon})$ is the global maximum over $[0, T] \times [0, T] \times \mathbf{C} \times \mathbf{C}$. $\square$

Lemma 3.6.4 *For each fixed $\gamma > 0$, we can find a constant $\Lambda_\gamma > 0$ such that*

$$\|\psi_{\delta\gamma\epsilon}\|_2 + \|\psi_{\delta\gamma\epsilon}^0\|_2 + \|\phi_{\delta\gamma\epsilon}\|_2 + \|\phi_{\delta\gamma\epsilon}^0\|_2 \leq \Lambda_\gamma \tag{3.98}$$

and

$$\lim_{\epsilon\downarrow0,\delta\downarrow0} \left(\|\psi_{\delta\gamma\epsilon} - \phi_{\delta\gamma\epsilon}\|_2^2 + \|\psi_{\delta\gamma\epsilon}^0 - \phi_{\delta\gamma\epsilon}^0\|_2^2 + |t_{\delta\gamma\epsilon} - s_{\delta\gamma\epsilon}|^2 \right) = 0, \tag{3.99}$$

Proof. Noting that $(t_{\delta\gamma\epsilon}, s_{\delta\gamma\epsilon}, \psi_{\delta\gamma\epsilon}, \phi_{\delta\gamma\epsilon})$ is the global maximum of $\Phi_{\delta\gamma} + T_\epsilon$, we can obtain

$$\Phi_{\delta\gamma}(t_{\delta\gamma\epsilon}, t_{\delta\gamma\epsilon}, \psi_{\delta\gamma\epsilon}, \psi_{\delta\gamma\epsilon}) + T_\epsilon(\psi_{\delta\gamma\epsilon}, \psi_{\delta\gamma\epsilon})$$
$$+ \Phi_{\delta\gamma}(s_{\delta\gamma\epsilon}, s_{\delta\gamma\epsilon}, \phi_{\delta\gamma\epsilon}, \phi_{\delta\gamma\epsilon}) + T_\epsilon(\phi_{\delta\gamma\epsilon}, \phi_{\delta\gamma\epsilon})$$
$$\leq 2\Phi_{\delta\gamma}(t_{\delta\gamma\epsilon}, s_{\delta\gamma\epsilon}, \psi_{\delta\gamma\epsilon}, \phi_{\delta\gamma\epsilon}) + 2T_\epsilon(\psi_{\delta\gamma\epsilon}, \phi_{\delta\gamma\epsilon}).$$

It implies that

$$V_1(t_{\delta\gamma\epsilon}, \psi_{\delta\gamma\epsilon}) - V_2(t_{\delta\gamma\epsilon}, \psi_{\delta\gamma\epsilon}) - 2\gamma(\exp(1 + \|\psi_{\delta\gamma\epsilon}\|_2^2 + \|\psi_{\delta\gamma\epsilon}^0\|_2^2))$$
$$+ T_\epsilon(\psi_{\delta\gamma\epsilon}, \psi_{\delta\gamma\epsilon}) + V_1(s_{\delta\gamma\epsilon}, \phi_{\delta\gamma\epsilon}) - V_2(s_{\delta\gamma\epsilon}, \phi_{\delta\gamma\epsilon})$$
$$- 2\gamma(\exp(1 + \|\phi_{\delta\gamma\epsilon}\|_2^2 + \|\phi_{\delta\gamma\epsilon}^0\|_2^2)) + T_\epsilon(\phi_{\delta\gamma\epsilon}, \phi_{\delta\gamma\epsilon})$$
$$\leq 2V_1(t_{\delta\gamma\epsilon}, \psi_{\delta\gamma\epsilon}) - 2V_2(s_{\delta\gamma\epsilon}, \phi_{\delta\gamma\epsilon})$$
$$- \frac{2}{\delta}\left[\|\psi_{\delta\gamma\epsilon} - \phi_{\delta\gamma\epsilon}\|_2^2 + \|\psi_{\delta\gamma\epsilon}^0 - \phi_{\delta\gamma\epsilon}^0\|_2^2 + |t_{\delta\gamma\epsilon} - s_{\delta\gamma\epsilon}|^2 \right]$$
$$- 2\gamma\left(\exp(1 + \|\psi_{\delta\gamma\epsilon}\|_2^2 + \|\psi_{\delta\gamma\epsilon}^0\|_2^2) + \exp(1 + \|\phi_{\delta\gamma\epsilon}\|_2^2 + \|\phi_{\delta\gamma\epsilon}^0\|_2^2) \right)$$
$$+ 2T_\epsilon(\psi_{\delta\gamma\epsilon}, \phi_{\delta\gamma\epsilon}). \tag{3.100}$$

From the above inequality, it is easy to obtain

$$\frac{2}{\delta}\left[\|\psi_{\delta\gamma\epsilon} - \phi_{\delta\gamma\epsilon}\|_2^2 + \|\psi_{\delta\gamma\epsilon}^0 - \phi_{\delta\gamma\epsilon}^0\|_2^2 + |t_{\delta\gamma\epsilon} - s_{\delta\gamma\epsilon}|^2 \right]$$
$$\leq [V_1(t_{\delta\gamma\epsilon}, \psi_{\delta\gamma\epsilon}) - V_1(s_{\delta\gamma\epsilon}, \phi_{\delta\gamma\epsilon})] + [V_2(t_{\delta\gamma\epsilon}, \psi_{\delta\gamma\epsilon}) - V_2(s_{\delta\gamma\epsilon}, \phi_{\delta\gamma\epsilon})]$$
$$+ 2T_\epsilon(\psi_{\delta\gamma\epsilon}, \phi_{\delta\gamma\epsilon}) - [T_\epsilon(\psi_{\delta\gamma\epsilon}, \psi_{\delta\gamma\epsilon}) + T_\epsilon(\phi_{\delta\gamma\epsilon}, \phi_{\delta\gamma\epsilon})]. \tag{3.101}$$

From the polynomial growth condition of V_1 and V_2, and the fact that the norm of T_ϵ is $\epsilon \in (0, 1)$, we can find a constant $\Lambda > 0$ and a positive integer $k \geq 1$ such that

$$\frac{2}{\delta}\left[\|\psi_{\delta\gamma\epsilon} - \psi_{\delta\gamma\epsilon}\|_2^2 + \|\psi_{\delta\gamma\epsilon}^0 - \phi_{\delta\gamma\epsilon}^0\|_2^2 + |t_{\delta\gamma\epsilon} - s_{\delta\gamma\epsilon}|^2\right] \leq \Lambda(1 + \|\psi_{\delta\gamma\epsilon}\|_2 + \|\phi_{\delta\gamma\epsilon}\|_2)^k.$$
(3.102)

So,

$$\|\psi_{\delta\gamma\epsilon} - \phi_{\delta\gamma\epsilon}\|_2^2 + \|\psi_{\delta\gamma\epsilon}^0 - \phi_{\delta\gamma\epsilon}^0\|_2^2 + |t_{\delta\gamma\epsilon} - s_{\delta\gamma\epsilon}|^2 \leq \delta\Lambda(1 + \|\psi_{\delta\gamma\epsilon}\|_2 + \|\phi_{\delta\gamma\epsilon}\|_2)^k.$$
(3.103)

On the other hand, because $(t_{\delta\gamma\epsilon}, s_{\delta\gamma\epsilon}, \psi_{\delta\gamma\epsilon}, \phi_{\delta\gamma\epsilon})$ is the global maximum of $\Phi_{\delta\gamma} + T_\epsilon$, we obtain

$$\Phi_{\delta\gamma}(t_{\delta\gamma\epsilon}, s_{\delta\gamma\epsilon}, 0, 0) + T_\epsilon(0, 0) \leq \Phi_{\delta\gamma}(t_{\delta\gamma\epsilon}, s_{\delta\gamma\epsilon}, \psi_{\delta\gamma\epsilon}, \phi_{\delta\gamma\epsilon}) + T_\epsilon(\psi_{\delta\gamma\epsilon}, \phi_{\delta\gamma\epsilon})$$
(3.104)

In addition, by the definition of $\Phi_{\delta\gamma}$ and the polynomial growth condition of V_1, V_2, we can get a $\Lambda > 0$ and a positive integer $k \geq 1$ such that

$$|\Phi_{\delta\gamma}(t_{\delta\gamma\epsilon}, s_{\delta\gamma\epsilon}, 0, 0) + T_\epsilon(0, 0)| \leq \Lambda(1 + \|\psi_{\delta\gamma\epsilon}\|_2 + \|\phi_{\delta\gamma\epsilon}\|_2)^k$$

and

$$V_1(t_{\delta\gamma\epsilon}, \psi_{\delta\gamma\epsilon}) - V_2(s_{\delta\gamma\epsilon}, \phi_{\delta\gamma\epsilon}) \leq \Lambda(1 + \|\psi_{\delta\gamma\epsilon}\|_2 + \|\phi_{\delta\gamma\epsilon}\|_2)^k.$$

Therefore, by virtue of (3.104) and the definition of $\Phi_{\delta\gamma}$, we have

$$\gamma\left[\exp(1 + \|\psi_{\delta\gamma\epsilon}\|_2^2 + \|\psi_{\delta\gamma\epsilon}^0\|_2^2) + \exp(1 + \|\phi_{\delta\gamma\epsilon}\|_2^2 + \|\phi_{\delta\gamma\epsilon}^0\|_2^2)\right]$$
$$\leq V_1(t_{\delta\gamma\epsilon}, \psi_{\delta\gamma\epsilon}) - V_2(s_{\delta\gamma\epsilon}, \phi_{\delta\gamma\epsilon})$$
$$- \frac{1}{\delta}\left[\|\psi_{\delta\gamma\epsilon} - \phi_{\delta\gamma\epsilon}\|_2^2 + \|\psi_{\delta\gamma\epsilon}^0 - \phi_{\delta\gamma\epsilon}^0\|_2^2 + |t_{\delta\gamma\epsilon} - s_{\delta\gamma\epsilon}|^2\right]$$
$$- \Phi_{\delta\gamma}(t_{\delta\gamma\epsilon}, s_{\delta\gamma\epsilon}, 0, 0) + T_\epsilon(\psi_{\delta\gamma\epsilon}, \phi_{\delta\gamma\epsilon}) - T_\epsilon(0, 0)$$
$$\leq 3\Lambda(1 + \|\psi_{\delta\gamma\epsilon}\|_2 + \|\phi_{\delta\gamma\epsilon}\|_2)^k.$$
(3.105)

It follows that

$$\frac{\gamma\left[\exp(1 + \|\psi_{\delta\gamma\epsilon}\|_2^2 + \|\psi_{\delta\gamma\epsilon}^0\|_2^2) + \exp(1 + \|\phi_{\delta\gamma\epsilon}\|_2^2 + \|\phi_{\delta\gamma\epsilon}^0\|_2^2)\right]}{(1 + \|\psi_{\delta\gamma\epsilon}\|_2 + \|\phi_{\delta\gamma\epsilon}\|_2)^k} \leq 3\Lambda.$$

Consequently, there exists $\Lambda_\gamma < \infty$ such that

$$\|\psi_{\delta\gamma\epsilon}\|_2 + \|\psi_{\delta\gamma\epsilon}^0\|_2 + \|\phi_{\delta\gamma\epsilon}\|_2 + \|\phi_{\delta\gamma\epsilon}^0\|_2 \leq \Lambda_\gamma.$$
(3.106)

In order to obtain (3.99), we set δ to zero in (3.103) using the above inequality. $\square$

Now, let us introduce a functional $F : \mathbf{C} \to \Re$ defined by

$$F(\psi) \equiv \|\psi\|_2^2 \tag{3.107}$$

and the linear map $H : \mathbf{C} \to \mathbf{C}$ defined by

$$H(\psi)(\theta) \equiv \frac{\theta}{-r}\psi(-r - \theta) = \psi^0(\theta), \qquad \theta \in [-r, 0]. \tag{3.108}$$

Note that $H(\psi)(0) = \psi^0(0) = 0$ and $H(\psi)(-r) = \psi^0(-r) = -\psi(0)$. It is not hard to show that the map F is Fréchet differentiable and its derivative is given by

$$DF(\varphi)h = 2\int_{-r}^{0} \varphi(\theta) \cdot h(\theta)\, d\theta \equiv 2\langle u, h \rangle_2,$$

where $\langle \cdot, \cdot \rangle_2$ and $\| \cdot \|_2$ are the inner product and the Hilbertian norm in the Hilbert space $L^2([-r, 0]; \Re^n)$. This comes from the fact that

$$\|\psi + h\|_2^2 - \|\psi\|_2^2 = 2\langle \psi, h \rangle_2 + \|h\|_2^2,$$

and we can always find a constant $\varLambda > 0$ such that

$$\frac{\left|\|\psi + h\|_2^2 - \|\psi\|_2^2 - 2\langle \psi, h \rangle_2\right|}{\|h\|} = \frac{\|h\|_2^2}{\|h\|} \le \frac{\varLambda\|h\|^2}{\|h\|} = \varLambda\|h\|. \tag{3.109}$$

Moreover, we have

$$2\langle \psi + h, \rangle_2 - 2\langle \psi, \cdot \rangle_2 = 2\langle h, \cdot \rangle_2.$$

We deduce that F is twice differentiable and $D^2 F(u)(h, k) = 2\langle h, k \rangle_2$.

In addition, the map H is linear, thus twice Fréchet differentiable. Therefore, $DH(\psi)(h) = H(h)$ and $D^2 H(\psi)(h, k) = 0$, for all $\psi, h, k \in \mathbf{C}$.

From the definition of $\Theta_{\delta\gamma}$ and the definition of F, we obtain

$$\Theta_{\delta\gamma}(t, s, \psi, \phi) = \frac{1}{\delta}\left[F(\psi - \phi) + F(\psi^0 - \phi^0) + |t - s|^2\right]$$
$$+ \gamma[e^{1+F(\psi)+F(H(\psi))} + e^{1+F(\phi)+F(H(\phi))}].$$

The following chain rule, quoted in Theorem 5.2.5 in Siddiqi [Sid04], is needed to get the Fréchet derivatives of $\Theta_{\delta\gamma}$:

Theorem 3.6.5 (Chain Rule) *Let $\mathcal{X}, \mathcal{Y}$, and $\mathcal{Z}$ be real Banach spaces. If $S : \mathcal{X} \to \mathcal{Y}$ and $T : \mathcal{Y} \to \mathcal{Z}$ are Fréchet differentiable at $\mathbf{x}$ and $\mathbf{y} = S(\mathbf{x}) \in \mathcal{Y}$, respectively, then $U = T \circ S$ is Fréchet differentiable at $\mathbf{x}$ and its Fréchet derivative is given by*

$$D_{\mathbf{x}}U(\mathbf{x}) = D_{\mathbf{y}}T(S(\mathbf{x}))D_{\mathbf{x}}S(\mathbf{x}).$$

Given the above chain rule, we can say that $\Theta_{\delta\gamma}$ is Fréchet differentiable. Actually, for $h, k \in \mathbf{C}$, we can get

$$D_\psi \Theta_{\delta\gamma}(t,s,\psi,\phi)(h)$$
$$= \frac{2}{\delta}\Big[\langle \psi - \phi, h\rangle_2 + \langle H(\psi - \phi), H(h)\rangle_2\Big]$$
$$+ 2\gamma e^{1+F(\psi)+F(H(\psi))}[\langle \psi, h\rangle_2 + \langle H(\psi), H(h)\rangle_2]. \qquad (3.110)$$

Similarly,

$$D_\phi \Theta_{\delta\gamma}(t,s,\psi,\phi)(k)$$
$$= \frac{2}{\delta}\Big[\langle \phi - \psi, k\rangle_2 + \langle H(\phi - \psi), H(k)\rangle_2\Big]$$
$$+ 2\gamma e^{1+F(\phi)+F(H(\phi))}[\langle \phi, k\rangle_2 + \langle H(\phi), H(k)\rangle_2]. \qquad (3.111)$$

Furthermore,

$$D_\psi^2 \Theta_{\delta\gamma}(t,s,\psi,\phi)(h,k)$$

$$= \frac{2}{\delta}\Big[\langle h, k\rangle_2 + \langle H(h), H(k)\rangle_2\Big]$$

$$+ 2\gamma e^{1+F(\psi)+F(H(\psi))}\Big[2(\langle \psi, k\rangle_2 + \langle H(\psi), H(k)\rangle_2)(\langle \psi, h\rangle_2 + \langle H(\psi), H(h)\rangle_2)$$

$$+ \langle k, h\rangle_2 + \langle H(k), H(h)\rangle_2\Big]. \qquad (3.112)$$

Similarly,

$$D_\phi^2 \Theta_{\delta\gamma}(t,s,\psi,\phi)(h,k) \qquad (3.113)$$

$$= \frac{2}{\delta}\Big[\langle h, k\rangle_2 + \langle H(h), H(k)\rangle_2\Big]$$

$$+ 2\gamma e^{1+F(\phi)+F(H(\phi))}\Big[2(\langle \phi, k\rangle_2 + \langle H(\phi), H(k)\rangle_2)(\langle \phi, h\rangle_2 + \langle H(\phi), H(h)\rangle_2)$$

$$+ \langle k, h\rangle_2 + \langle H(k), H(h)\rangle_2\Big]. \qquad (3.114)$$

By the Hahn-Banach theorem (see, e.g., [Sid04]), we can extent the continuous linear functional T_ϵ to the space $\mathbf{C} \times \mathbf{C}$ and its norm is preserved. Thus, the first-order Fréchet derivatives of T_ϵ is just T_ϵ, that is,

$$D_\psi T_\epsilon(\psi,\phi)h = T_\epsilon(h,\phi),$$

$$D_\phi T_\epsilon(\psi,\phi)k = T_\epsilon(\psi,k) \quad \forall \psi,\phi,h,k \in \mathbf{C}.$$

For the second derivative, we have

$$D_\psi^2 T_\epsilon(\psi,\phi)(h,k) = 0, \qquad (3.115)$$

$$D_\phi^2 T_\epsilon(\psi,\phi)(h,k) = 0, \quad \forall \psi,\phi,h,k \in \mathbf{C}.$$

Observe that we can extend $D_\psi \Theta_{\delta\gamma}(t,s,\psi,\phi)$ and $D_\psi^2 \Theta_{\delta\gamma}(t,s,\psi,\phi)$, the first- and second-order Fréchet derivatives of $\Theta_{\delta\gamma}$ with respect to ψ, to the space $\mathbf{C} \oplus \mathbf{B}$ (see Lemma 2.2.3 and Lemma 2.2.4 in Chapter 2) by setting

$$\overline{D_\psi \Theta_{\delta\gamma}(t,s,\psi,\phi)}(h+v\mathbf{1}_{\{0\}}) \tag{3.116}$$

$$= \frac{2}{\delta}\Big[\langle\psi-\phi, h+v\mathbf{1}_{\{0\}}\rangle_2 + \langle H(\psi-\phi), H(h+v\mathbf{1}_{\{0\}})_2\Big]$$

$$+ 2\gamma e^{1+F(\psi)+F(H(\psi))}[\langle\psi, h+v\mathbf{1}_{\{0\}}\rangle_2 + \langle H(\psi), H(h+v\mathbf{1}_{\{0\}}))\rangle_2].$$

and

$$\overline{D_\psi^2 \Theta_{\delta\gamma}(t,s,\psi,\phi)}(h+v\mathbf{1}_{\{0\}}, k+w\mathbf{1}_{\{0\}})$$

$$= \frac{2}{\delta}\Big[\langle h+v\mathbf{1}_{\{0\}}, k+w\mathbf{1}_{\{0\}}\rangle_2 + \langle H(h+v\mathbf{1}_{\{0\}}), H(k+w\mathbf{1}_{\{0\}})_2\Big]$$

$$+2\gamma e^{1+F(\psi)+F(H(\psi))}\Big[2(\langle\psi, k+w\mathbf{1}_{\{0\}}\rangle_2 + \langle H(\psi), H(k+w\mathbf{1}_{\{0\}}))\rangle_2)$$

$$\times(\langle\psi, h+v\mathbf{1}_{\{0\}}\rangle_2 + \langle H(\psi), H(h+v\mathbf{1}_{\{0\}}))\rangle_2)$$

$$+\langle k+w\mathbf{1}_{\{0\}}, h+v\mathbf{1}_{\{0\}}\rangle_2 + \langle H(k+w\mathbf{1}_{\{0\}}), H(h+v\mathbf{1}_{\{0\}})_2)\Big], \tag{3.117}$$

for $v, w \in \Re^n$ and $h, k \in \mathbf{C}$.

Moreover, it is easy to see that these extensions are continuous in that there exists a constant $\Lambda > 0$ such that

$$|\langle\psi-\phi, h+v\mathbf{1}_{\{0\}}\rangle_2| \le \|\psi-\phi\|_2 \cdot \|h+v\mathbf{1}_{\{0\}}\|_2$$
$$\le \Lambda\|\psi-\phi\|_2(\|h\|+|v|), \tag{3.118}$$

$$|\langle\psi, h+v\mathbf{1}_{\{0\}}\rangle_2| \le \|\psi\|_2 \cdot \|h+v\mathbf{1}_{\{0\}}\|_2$$
$$\le \Lambda\|\psi\|_2(\|h\|+|v|), \tag{3.119}$$

$$|\langle\psi, k+w\mathbf{1}_{\{0\}}\rangle_2| \le \|\psi\|_2 \cdot \|k+w\mathbf{1}_{\{0\}}\|_2$$
$$\le \Lambda\|\psi\|_2(\|k\|+|w|), \tag{3.120}$$

and

$$|\langle k+w\mathbf{1}_{\{0\}}, h+v\mathbf{1}_{\{0\}}\rangle_2| \le \|k+w\mathbf{1}_{\{0\}}\|_2\|h+v\mathbf{1}_{\{0\}}\|_2$$
$$\le \Lambda(\|k\|+|w|)(\|h\|+|v|). \tag{3.121}$$

Similarly, we can extend the first- and second-order Fréchet derivatives of $\Theta_{\delta\gamma}$ with respect to ϕ to the space $\mathbf{C} \oplus \mathbf{B}$ and obtain similar expressions for $\overline{D_\phi \Theta_{\delta\gamma}(t,s,\psi,\phi)}(k+w\mathbf{1}_{\{0\}})$ and $\overline{D_\phi^2 \Theta_{\delta\gamma}(t,s,\psi,\phi)}(h+v\mathbf{1}_{\{0\}}, k+w\mathbf{1}_{\{0\}})$.

The same is also true for the bounded linear functional T_ϵ whose extension is still written as T_ϵ.

In addition, it is easy to verify that for any $\phi \in \mathbf{C}$ and $v, w \in \Re^n$, we have

$$\langle\phi, v\mathbf{1}_{\{0\}}\rangle_2 = \int_{-r}^0 \phi(\theta) \cdot v\mathbf{1}_{\{0\}}(\theta)d\theta = 0, \tag{3.122}$$

$$\langle w\mathbf{1}_{\{0\}}, v\mathbf{1}_{\{0\}}\rangle_2 = \int_{-r}^0 w\mathbf{1}_{\{0\}}(\theta) \cdot v\mathbf{1}_{\{0\}}(\theta)d\theta = 0, \tag{3.123}$$

$$H(v\mathbf{1}_{\{0\}}) = v\mathbf{1}_{\{-r\}}, \tag{3.124}$$

$$\langle H(\psi), H(v\mathbf{1}_{\{0\}})\rangle_2 = 0, \quad \langle H(w\mathbf{1}_{\{0\}}), H(v\mathbf{1}_{\{0\}})\rangle_2 = 0. \tag{3.125}$$

These observations will be used later.

Next, we need several lemmas about the operator S.

Lemma 3.6.6 *Given $\phi \in \mathbf{C}$, we have*

$$S(F)(\phi) = |\phi(0)|^2 - |\phi(-r)|^2, \tag{3.126}$$

$$S(F)(\phi^0) = -|\phi(0)|^2, \tag{3.127}$$

where F is the functional defined in (3.107) and S is the operator defined in (3.60).

Proof. Recall that

$$S(F)(\phi) = \lim_{t\downarrow 0}\frac{1}{t}\left[F(\tilde{\phi}_t) - F(\phi)\right] \tag{3.128}$$

for all $\phi \in \mathbf{C}$, where $\tilde{\phi} : [-r, T] \to \Re^n$ is an extension of ϕ defined by

$$\tilde{\phi}(t) = \begin{cases} \phi(t) & \text{if } t \in [-r, 0) \\ \phi(0) & \text{if } t \geq 0, \end{cases} \tag{3.129}$$

and, again, $\tilde{\phi}_t \in \mathbf{C}$ is defined by

$$\tilde{\phi}_t(\theta) = \tilde{\phi}(t + \theta), \quad \theta \in [-r, 0].$$

Therefore, we have

$$\begin{aligned}
S(F)(\phi) &= \lim_{t\to 0+}\frac{1}{t}\left[\|\tilde{\phi}_t\|_2^2 - \|\phi\|_2^2\right] \\
&= \lim_{t\to 0+}\frac{1}{t}\left[\int_{-r}^{0}|\tilde{\phi}_t(\theta)|^2\,d\theta - \int_{-r}^{0}|\phi(\theta)|^2\,d\theta\right] \\
&= \lim_{t\to 0+}\frac{1}{t}\left[\int_{-r}^{0}|\tilde{\phi}(\theta + t)|^2\,d\theta - \int_{-r}^{0}|\phi(\theta)|^2\,d\theta\right] \\
&= \lim_{t\to 0+}\frac{1}{t}\left[\int_{-r+t}^{t}|\tilde{\phi}(\theta)|^2\,d\theta - \int_{-r}^{0}|\phi(\theta)|^2\,d\theta\right] \\
&= \lim_{t\to 0+}\frac{1}{t}\left[\int_{-r+t}^{0}|\tilde{\phi}(\theta)|^2\,d\theta + \int_{0}^{t}|\tilde{\phi}(\theta)|^2\,d\theta - \int_{-r}^{0}|\phi(\theta)|^2\,d\theta\right] \\
&= \lim_{t\to 0+}\frac{1}{t}\left[\int_{-r+t}^{0}|\phi(\theta)|^2\,d\theta + \int_{0}^{t}|\phi(0)|^2\,d\theta - \int_{-r}^{0}|\phi(\theta)|^2\,d\theta\right] \\
&= \lim_{t\to 0+}\frac{1}{t}\left[\int_{-r+t}^{0}|\phi(\theta)|^2\,d\theta - \int_{-r}^{0}|\phi(\theta)|^2\,d\theta\right] \\
&\quad + \lim_{t\to 0+}\frac{1}{t}\int_{0}^{t}|\phi(0)|^2\,d\theta
\end{aligned}$$

$$= \lim_{t \to 0+} \frac{1}{t} \int_0^t |\phi(0)|^2 \, d\theta - \lim_{t \to 0+} \frac{1}{t} \left[\int_{-r}^{-r+t} |\phi(\theta)|^2 \, d\theta \right]$$

$$= |\phi(0)|^2 - |\phi(-r)|^2. \tag{3.130}$$

Similarly, we have

$$\mathcal{S}(F)(\phi^0) = |\phi^0(0)|^2 - |\phi^0(-r)|^2 = -|\phi(0)|^2. \qquad \square$$

Let $\mathcal{S}_\psi$ and $\mathcal{S}_\phi$ denote the operator $\mathcal{S}$ applied to ψ and ϕ, respectively. We have the following lemma.

Lemma 3.6.7 *Given* $\phi, \psi \in \mathbf{C}$,

$$\mathcal{S}_\psi(F)(\phi - \psi) + \mathcal{S}_\phi(F)(\phi - \psi) = |\psi(0) - \phi(0)|^2 - |\psi(-r) - \phi(-r)|^2 \tag{3.131}$$

and

$$\mathcal{S}_\psi(F)(\phi^0 - \psi^0) + \mathcal{S}_\phi(F)(\phi^0 - \psi^0) = -|\psi(0) - \phi(0)|^2, \tag{3.132}$$

where F *is the functional defined in (3.107) and* $\mathcal{S}$ *is the operator defined in (3.60).*

Proof. To proof the lemma, we need the following result, which can be easily proved by definition provided that $\psi \in \mathcal{D}(\tilde{S})$:

$$\mathcal{S}(F)(\psi) = DF(\psi)\tilde{S}(\psi), \tag{3.133}$$

where $DF(\psi)$ is the Fréchet derivative of $F(\psi)$ and $\tilde{S} : \mathcal{D}(\tilde{S}) \subset \mathbf{C} \to \mathbf{C}$ is defined by

$$\tilde{S}(\psi) = \lim_{t \downarrow 0} \frac{\tilde{\psi}_t - \psi}{t}.$$

We first assume that $\psi \in \mathcal{D}(\tilde{S})$, the domain of the operator $\tilde{S}$, consists of those $\psi \in \mathbf{C}$ for which the above limit exists. It can be shown that

$$\mathcal{D}(\tilde{S}) = \{\psi \in \mathbf{C} \mid \psi \text{ is absolutely continuous and } \dot{\psi}(0+) = 0\}.$$

In this case, we have

$$\mathcal{S}(F)(\psi) = DF(\psi)\tilde{S}(\psi) = 2(\psi|\tilde{S}(\psi)).$$

On the other hand, by virtue of Lemma 3.6.6, we have

$$\mathcal{S}(F)(\psi) = |\psi(0)|^2 - |\psi(-r)|^2.$$

Therefore, we have

$$(\psi|\tilde{S}(\psi)) = \frac{1}{2} \left[|\psi(0)|^2 - |\psi(-r)|^2 \right].$$

Since $\tilde{S}$ is a linear operator, we have

$$(\psi - \phi|\tilde{S}(\psi) - \tilde{S}(\phi)) = (\psi - \phi|\tilde{S}(\psi - \phi))$$

$$= \frac{1}{2} \left[|\psi(0) - \phi(0)|^2 - |\psi(-r) - \phi(-r)|^2 \right]. \tag{3.134}$$

Given the above results, now we can get

$$
\begin{aligned}
&\mathcal{S}_\psi(F)(\psi - \phi) + \mathcal{S}_\phi(F)(\psi - \phi) \\
&= \lim_{t\downarrow 0} \frac{1}{t}\left[\|\tilde{\psi}_t - \phi\|_2^2 - \|\psi - \phi\|_2^2 + \|\psi - \tilde{\phi}_t\|_2^2 - \|\psi - \phi\|_2^2\right] \\
&= \lim_{t\downarrow 0} \frac{1}{t}\left[\|\tilde{\psi}_t\|_2^2 - \|\psi\|_2^2 + \|\tilde{\phi}_t\|_2^2 - \|\phi\|_2^2\right. \\
&\qquad\qquad \left. -2\left[(\tilde{\psi}_t|\phi) - (\psi|\phi) + (\psi|\tilde{\phi}_t) - (\psi|\phi)\right]\right] \\
&= \mathcal{S}(F)(\psi) + \mathcal{S}(F)(\phi) - 2[(\tilde{\mathcal{S}}(\psi)|\phi) + (\psi|\tilde{\mathcal{S}}(\phi))] \\
&= 2(\psi|\tilde{\mathcal{S}}(\psi)) + 2(\phi|\tilde{\mathcal{S}}(\phi)) - 2[(\tilde{\mathcal{S}}(\psi)|\phi) + (\psi|\tilde{\mathcal{S}}(\phi))] \\
&= 2(\psi - \phi|\tilde{\mathcal{S}}(\psi - \phi)) \\
&= [|\psi(0) - \phi(0)|^2 - |\psi(-r) - \phi(-r)|^2],
\end{aligned}
$$

provided that $\psi, \phi \in \mathcal{D}(\tilde{\mathcal{S}})$.

For any $\psi, \phi \in \mathbf{C}$, one can construct sequences $\{\psi_k\}_{k=1}^\infty$ and $\{\phi_k\}_{k=1}^\infty$ in $\mathcal{D}(\tilde{\mathcal{S}})$ such that

$$
\lim_{k\to\infty} \|\psi_k - \psi\| = 0 \quad \text{and} \quad \lim_{k\to\infty} \|\phi_k - \phi\| = 0.
$$

Consequently by the linearity of the $\mathcal{S}$ operator and continuity of $F : \mathbf{C} \to \Re$, we have

$$
\begin{aligned}
\mathcal{S}_\psi(F)(\psi - \phi) + \mathcal{S}_\phi(F)(\psi - \phi) &= \lim_{k\to\infty}\left(\mathcal{S}_\psi(F)(\psi_k - \phi_k) + \mathcal{S}_\phi(F)(\psi_k - \phi_k)\right) \\
&= \lim_{k\to\infty}[|\psi_k(0) - \phi_k(0)|^2 - |\psi_k(-r) - \phi_k(-r)|^2] \\
&= [|\psi(0) - \phi(0)|^2 - |\psi(-r) - \phi(-r)|^2].
\end{aligned}
$$

By the same argument, we have

$$
\begin{aligned}
\mathcal{S}_\psi(F)(\psi^0 - \phi^0) + \mathcal{S}_\phi(F)(\psi^0 - \phi^0) &= |\psi^0(0) - \phi^0(0)|^2 - |\psi^0(-r) - \phi^0(-r)|^2 \\
&= -|\psi(0) - \phi(0)|^2. \qquad \square
\end{aligned}
$$

Lemma 3.6.8 *Given $\phi \in \mathbf{C}$, we define a new operator G as follows*

$$
G(\phi) = e^{1+F(\phi)+F(\phi^0)}. \tag{3.135}
$$

We have

$$
\mathcal{S}(G)(\phi) = (-|\phi(-r)|^2)e^{1+F(\phi)+F(\phi^0)}, \tag{3.136}
$$

where F is the functional defined in (4.88) and $\mathcal{S}$ is the operator defined in (3.60).

Proof. Recall that

$$
\mathcal{S}(G)(\phi) = \lim_{t\downarrow 0} \frac{1}{t}\left[G(\tilde{\phi}_t) - G(\phi)\right] \tag{3.137}
$$

for all $\phi \in \mathbf{C}$, where $\tilde{\phi} : [-r, T] \to \mathbb{R}^n$ is an extension of ϕ defined by

$$\tilde{\phi}(t) = \begin{cases} \phi(t) & \text{if } t \in [-r, 0) \\ \phi(0) & \text{if } t \geq 0, \end{cases} \tag{3.138}$$

and, again, $\tilde{\phi}_t \in \mathbf{C}$ is defined by

$$\tilde{\phi}_t(\theta) = \tilde{\phi}(t + \theta), \quad \theta \in [-r, 0].$$

We have

$$\mathcal{S}(G)(\phi)$$
$$= \lim_{t \to 0+} \frac{1}{t} \big[e^{1 + \int_{-r}^0 |\tilde{\phi}_t(\theta)|^2 d\theta + \int_{-r}^0 |\tilde{\phi}_t^0(\theta)|^2 d\theta}$$
$$\quad - e^{1 + \int_{-r}^0 |\phi(\theta)|^2 d\theta + \int_{-r}^0 |\phi^0(\theta)|^2 d\theta} \big]$$

$$= \lim_{t \to 0+} \frac{1}{t} \big[e^{1 + \int_{-r}^0 |\tilde{\phi}(\theta+t)|^2 d\theta + \int_{-r}^0 |\tilde{\phi}^0(\theta+t)|^2 d\theta}$$
$$\quad - e^{1 + \int_{-r}^0 |\phi(\theta)|^2 d\theta + \int_{-r}^0 |\phi^0(\theta)|^2 d\theta} \big]$$

$$= \lim_{t \to 0+} \frac{1}{t} \big[e^{1 + \int_{-r+t}^t |\tilde{\phi}(\theta)|^2 d\theta + \int_{-r+t}^t |\tilde{\phi}^0(\theta)|^2 d\theta}$$
$$\quad - e^{1 + \int_{-r}^0 |\phi(\theta)|^2 d\theta + \int_{-r}^0 |\phi^0(\theta)|^2 d\theta} \big]$$

$$= \lim_{t \to 0+} \frac{1}{t} \big[e^{1 + \int_{-r+t}^0 |\phi(\theta)|^2 d\theta + \int_0^t |\phi(0)|^2 d\theta + \int_{-r+t}^0 |\phi^0(\theta)|^2 d\theta + \int_0^t |\phi^0(0)|^2 d\theta}$$
$$\quad - e^{1 + \int_{-r}^0 |\phi(\theta)|^2 d\theta + \int_{-r}^0 |\phi^0(\theta)|^2 d\theta} \big]$$

$$= \lim_{t \to 0+} \frac{1}{t} \big[e^{1 + \int_{-r+t}^0 |\phi(\theta)|^2 d\theta + t|\phi(0)|^2 + \int_{-r+t}^0 |\phi^0(\theta)|^2 d\theta + t|\phi^0(0)|^2}$$
$$\quad - e^{1 + \int_{-r}^0 |\phi(\theta)|^2 d\theta + \int_{-r}^0 |\phi^0(\theta)|^2 d\theta} \big]. \tag{3.139}$$

Using the L'Hospital rule on the last equality, we obtain

$$\mathcal{S}(G)(\phi)$$
$$= \lim_{t \to 0+} e^{1 + \int_{-r+t}^0 |\phi(\theta)|^2 d\theta + t|\phi(0)|^2 + \int_{-r+t}^0 |\phi^0(\theta)|^2 d\theta + t|\phi^0(0)|^2} \Big(|\phi(0)|^2$$

$$-|\phi(-r+t)|^2 + |\phi^0(0)|^2 - |\phi^0(-r+t)|^2 \Big)$$
$$= (|\phi(0)|^2 - |\phi(-r)|^2 - |\phi^0(-r)|^2) e^{1 + \int_{-r}^0 |\phi(\theta)|^2 d\theta + \int_{-r}^0 |\phi^0(\theta)|^2 d\theta}$$
$$= (|\phi(0)|^2 - |\phi(-r)|^2 - |\phi(0)|^2) e^{1 + \int_{-r}^0 |\phi(\theta)|^2 d\theta + \int_{-r}^0 |\phi^0(\theta)|^2 d\theta}$$
$$= -|\phi(-r)|^2 e^{1 + F(\phi) + F(\phi^0)}. \qquad \square \tag{3.140}$$

Lemma 3.6.9 *For any $\psi, \phi \in \mathbf{C}$, we have*

$$\lim_{\epsilon \downarrow 0} |\mathcal{S}_\psi(T_\epsilon)(\psi, \phi)| = 0 \quad \text{and} \quad \lim_{\epsilon \downarrow 0} |\mathcal{S}_\phi(T_\epsilon)(\psi, \phi)| = 0. \tag{3.141}$$

Proof. We will only prove the first equality in the Lemma, since the second one can be proved similarly.

We first assume that $\psi \in \mathcal{D}(\tilde{S})$, where the operator $\tilde{S} : \mathcal{D}(\tilde{S}) \subset \mathbf{C} \to \mathbf{C}$ and $\mathcal{D}(\tilde{S})$ are defined in the proof of Lemma 3.6.8. In this case,

$$
\begin{aligned}
\lim_{\epsilon \downarrow 0} |\mathcal{S}_\psi(T_\epsilon)(\psi, \phi)| &= \lim_{\epsilon \downarrow 0} \left| \lim_{t \downarrow 0} \frac{T_\epsilon(\tilde{\psi}_t, \phi) - T_\epsilon(\psi, \phi)}{t} \right| \\
&= \lim_{\epsilon \downarrow 0} \left| (T_\epsilon) \lim_{t \downarrow 0} \left(\frac{\tilde{\psi}_t - \psi}{t} \times, \phi \right) \right| \\
&\leq \lim_{\epsilon \downarrow 0} \|T_\epsilon\| \left(\left\| \lim_{t \downarrow 0} \frac{\tilde{\psi}_t - \psi}{t} \right\| + \|\phi\| \right) \\
&\leq \lim_{\epsilon \downarrow 0} \epsilon \left(\|\tilde{S}\psi\| + \|\phi\| \right) = 0, \qquad (3.142)
\end{aligned}
$$

because T_ϵ is a bounded linear functional on $\mathbf{C} \times \mathbf{C}$ with norm equal to ϵ.

For any $\psi, \phi \in \mathbf{C}$, one can construct a sequence of

$$\psi_k \in \mathcal{D}(\tilde{S}), \quad k = 1, 2, \cdots,$$

such that

$$\lim_{k \to \infty} \|\psi_k - \psi\| = 0.$$

We have

$$\lim_{\epsilon \downarrow 0} |\mathcal{S}_\psi(T_\epsilon)(\psi_k, \phi)| = 0, \quad \forall k = 1, 2, \dots.$$

Consequently,

$$\lim_{\epsilon \downarrow 0} |\mathcal{S}_\psi(T_\epsilon)(\psi, \phi)| = 0$$

by the limit process.   □

Given all of the above results, now we are ready to prove Theorem 3.6.1.

Proof of Theorem 3.6.1. Define

$$\Gamma_1(t, \psi) \equiv V_2(s_{\delta\gamma\epsilon}, \phi_{\delta\gamma\epsilon}) + \Theta_{\delta\gamma}(t, s_{\delta\gamma\epsilon}, \psi, \phi_{\delta\gamma\epsilon}) - T_\epsilon(\psi, \phi_{\delta\gamma\epsilon}) - M_{\delta\gamma\epsilon} \quad (3.143)$$

and

$$\Gamma_2(s, \phi) \equiv V_1(t_{\delta\gamma\epsilon}, \psi_{\delta\gamma\epsilon}) - \Theta_{\delta\gamma}(t_{\delta\gamma\epsilon}, s, \psi_{\delta\gamma\epsilon}, \phi) + T_\epsilon(\psi_{\delta\gamma\epsilon}, \phi) + M_{\delta\gamma\epsilon} \quad (3.144)$$

for all $s, t \in [0, T]$ and $\psi, \phi \in \mathbf{C}$. Recall that

$$\Phi_{\delta\gamma}(t, s, \psi, \phi) = V_1(t, \psi) - V_2(s, \phi) - \Theta_{\delta\gamma}(t, s, \psi, \phi)$$

and that $\Phi_{\delta\gamma} + T_\epsilon + M_{\delta\gamma\epsilon}$ reaches its maximum value zero at $(t_{\delta\gamma\epsilon}, s_{\delta\gamma\epsilon}, \psi_{\delta\gamma\epsilon}, \phi_{\delta\gamma\epsilon})$ in $[0, T] \times [0, T] \times \mathbf{C} \times \mathbf{C}$.

By the definition of Γ_1 and Γ_2, it is easy to verify that for all ϕ and ψ, we have

$$\Gamma_1(t,\psi) \geq V_1(t,\psi), \quad \Gamma_2(s,\phi) \leq V_2(s,\phi), \quad \forall t, s \in [0,T] \text{ and } \phi, \psi \in \mathbf{C},$$

and

$$V_1(t_{\delta\gamma\epsilon}, \psi_{\delta\gamma\epsilon}) = \Gamma_1(t_{\delta\gamma\epsilon}, \psi_{\delta\gamma\epsilon}) \text{ and } V_2(s_{\delta\gamma\epsilon}, \phi_{\delta\gamma\epsilon}) = \Gamma_2(s_{\delta\gamma\epsilon}, \phi_{\delta\gamma\epsilon}).$$

Using the definitions of the viscosity subsolution of V_1 and Γ_1, we have

$$\alpha V_1(t_{\delta\gamma\epsilon}, \psi_{\delta\gamma\epsilon}) - \partial_t \Gamma_1(t_{\delta\gamma\epsilon}, \psi_{\delta\gamma\epsilon})$$
$$- \sup_{v \in U} \left[\mathbf{A}^v(\Gamma_1)(t_{\delta\gamma\epsilon}, \psi_{\delta\gamma\epsilon}) - L(t_{\delta\gamma\epsilon}, \psi_{\delta\gamma\epsilon}, v) \right] \Big\} \leq 0. \quad (3.145)$$

By the definitions of the operator $\mathbf{A}^v$ and Γ_1 and the fact that the second-order Fréchet derivatives of $T_\epsilon = 0$, we have, by combining (3.110), (3.111), (3.112), (3.113), (3.116), (3.117), (3.122), (3.123), (3.124), and (3.125),

$$\mathbf{A}^v(\Gamma_1)(t_{\delta\gamma\epsilon}, \psi_{\delta\gamma\epsilon})$$
$$= \mathcal{S}(\Gamma_1)(t_{\delta\gamma\epsilon}, \psi_{\delta\gamma\epsilon}) + \overline{D_\psi \Theta_{\delta\gamma}(\cdots)}(f(t_{\delta\gamma\epsilon}, \psi_{\delta\gamma\epsilon}, v)\mathbf{1}_{\{0\}})$$
$$\frac{1}{2} \sum_{j=1}^{m} \overline{D_\psi^2 \Theta_{\delta\gamma}(\cdots)} \Big(g(t_{\delta\gamma\epsilon}, \psi_{\delta\gamma\epsilon}, v)(\mathbf{e}_j)\mathbf{1}_{\{0\}}, g(t_{\delta\gamma\epsilon}, \psi_{\delta\gamma\epsilon}, v)(\mathbf{e}_j)\mathbf{1}_{\{0\}} \Big)$$
$$- \overline{D_\psi T_\epsilon(\psi_{\delta\gamma\epsilon}, \phi_{\delta\gamma\epsilon})}(f(t_{\delta\gamma\epsilon}, \psi_{\delta\gamma\epsilon}, v)\mathbf{1}_{\{0\}})$$
$$= \mathcal{S}(\Gamma_1)(t_{\delta\gamma\epsilon}, \psi_{\delta\gamma\epsilon}) - \overline{T_\epsilon(f(t_{\delta\gamma\epsilon}, \psi_{\delta\gamma\epsilon}, v)\mathbf{1}_{\{0\}}, \phi_{\delta\gamma\epsilon})}.$$

Note that $\Theta_{\delta\gamma}(\cdots)$ is an abbreviation for $\Theta_{\delta\gamma}(t_{\delta\gamma\epsilon}, s_{\delta\gamma\epsilon}, \psi_{\delta\gamma\epsilon}, \phi_{\delta\gamma\epsilon})$ in the above equation and the following.

Inequality (3.145) and the above equation together yield that

$$\alpha V_1(t_{\delta\gamma\epsilon}, \psi_{\delta\gamma\epsilon}) - \mathcal{S}(\Gamma_1)(t_{\delta\gamma\epsilon}, \psi_{\delta\gamma\epsilon}) - \partial_t \Gamma_1(t_{\delta\gamma\epsilon}, \psi_{\delta\gamma\epsilon}) \qquad (3.146)$$
$$- \sup_{v \in U} \left[- \overline{T_\epsilon(f(t_{\delta\gamma\epsilon}, \psi_{\delta\gamma\epsilon}, v)\mathbf{1}_{\{0\}}, \phi_{\delta\gamma\epsilon})} + L(t_{\delta\gamma\epsilon}, \psi_{\delta\gamma\epsilon}, v) \right] \leq 0.$$

Similarly, using the definitions of the viscosity supersolution of V_2 and Γ_2 and by the virtue of the same techniques similar to (3.146), we have

$$\alpha V_2(s_{\delta\gamma\epsilon}, \phi_{\delta\gamma\epsilon}) - \mathcal{S}(\Gamma_2)(s_{\delta\gamma\epsilon}, \phi_{\delta\gamma\epsilon}) - \partial_s \Gamma_2(s_{\delta\gamma\epsilon}, \phi_{\delta\gamma\epsilon})$$
$$- \sup_{v \in U} \left[\overline{T_\epsilon(\psi_{\delta\gamma\epsilon}, f(s_{\delta\gamma\epsilon}, \phi_{\delta\gamma\epsilon}, v)\mathbf{1}_{\{0\}})} + L(s_{\delta\gamma\epsilon}, \phi_{\delta\gamma\epsilon}, v) \right] \geq 0. \quad (3.147)$$

Inequality (3.146) is equivalent to

$$\alpha V_1(t_{\delta\gamma\epsilon}, \psi_{\delta\gamma\epsilon}) - \mathcal{S}(\Gamma_1)(t_{\delta\gamma\epsilon}, \psi_{\delta\gamma\epsilon}) - 2(t_{\delta\gamma\epsilon} - s_{\delta\gamma\epsilon})$$
$$- \sup_{v \in U} \left[- \overline{T_\epsilon(f(t_{\delta\gamma\epsilon}, \psi_{\delta\gamma\epsilon}, v)\mathbf{1}_{\{0\}}, \phi_{\delta\gamma\epsilon})} + L(t_{\delta\gamma\epsilon}, \psi_{\delta\gamma\epsilon}, v) \right] \leq 0. \quad (3.148)$$

Similarly, Inequality (3.147) is equivalent to

$$\alpha V_2(s_{\delta\gamma\epsilon}, \phi_{\delta\gamma\epsilon}) - \mathcal{S}(\Gamma_2)(s_{\delta\gamma\epsilon}, \phi_{\delta\gamma\epsilon}) - 2(s_{\delta\gamma\epsilon} - t_{\delta\gamma\epsilon})$$

$$- \sup_{v \in U} \left[\overline{T_\epsilon(\psi_{\delta\gamma\epsilon}, f(s_{\delta\gamma\epsilon}, \phi_{\delta\gamma\epsilon}, v)\mathbf{1}_{\{0\}})} + L(s_{\delta\gamma\epsilon}, \phi_{\delta\gamma\epsilon}, v) \right] \geq 0. \quad (3.149)$$

By virtue of (3.148) and (3.149), we obtain

$$\alpha(V_1(t_{\delta\gamma\epsilon}, \psi_{\delta\gamma\epsilon}) - V_2(s_{\delta\gamma\epsilon}, \phi_{\delta\gamma\epsilon}))$$

$$\leq \mathcal{S}(\Gamma_1)(t_{\delta\gamma\epsilon}, \psi_{\delta\gamma\epsilon}) - \mathcal{S}(\Gamma_2)(s_{\delta\gamma\epsilon}, \phi_{\delta\gamma\epsilon}) + 4(t_{\delta\gamma\epsilon} - s_{\delta\gamma\epsilon})$$

$$+ \sup_{v \in U} \left[L(t_{\delta\gamma\epsilon}, \psi_{\delta\gamma\epsilon}, v) - \overline{T_\epsilon(f(t_{\delta\gamma\epsilon}, \psi_{\delta\gamma\epsilon}, v)\mathbf{1}_{\{0\}}, \phi_{\delta\gamma\epsilon})} \right]$$

$$- \sup_{v \in U} \left[L(s_{\delta\gamma\epsilon}, \phi_{\delta\gamma\epsilon}, v) + \overline{T_\epsilon(\psi_{\delta\gamma\epsilon}, f(s_{\delta\gamma\epsilon}, \phi_{\delta\gamma\epsilon}, v)\mathbf{1}_{\{0\}})} \right]. \quad (3.150)$$

From definition (3.60) of $\mathcal{S}$, it is clear that $\mathcal{S}$ is linear and takes the value zero on constants. Recall that

$$\Gamma_1(t, \psi) = V_2(s_{\delta\gamma\epsilon}, \phi_{\delta\gamma\epsilon}) + \Theta_{\delta\gamma}(t, s_{\delta\gamma\epsilon}, \psi, \phi_{\delta\gamma\epsilon})$$

$$- T_\epsilon(\psi, \phi_{\delta\gamma\epsilon}) - M_{\delta\gamma\epsilon} \quad (3.151)$$

and

$$\Gamma_2(s, \phi) = V_1(t_{\delta\gamma\epsilon}, \psi_{\delta\gamma\epsilon}) - \Theta_{\delta\gamma}(t_{\delta\gamma\epsilon}, s, \psi_{\delta\gamma\epsilon}, \phi) \quad (3.152)$$

$$+ T_\epsilon(\psi_{\delta\gamma\epsilon}, \phi) + M_{\delta\gamma\epsilon}.$$

Thus, we have

$$\mathcal{S}(\Gamma_1)(t_{\delta\gamma\epsilon}, \psi_{\delta\gamma\epsilon}) = \mathcal{S}_\psi(\Theta_{\delta\gamma})(t_{\delta\gamma\epsilon}, s_{\delta\gamma\epsilon}, \psi_{\delta\gamma\epsilon}, \phi_{\delta\gamma\epsilon})$$

$$\mathcal{S}_\psi(T_\epsilon)(\psi_{\delta\gamma\epsilon}, \phi_{\delta\gamma\epsilon}) \quad (3.153)$$

and

$$\mathcal{S}(\Gamma_2)(s_{\delta\gamma\epsilon}, \phi_{\delta\gamma\epsilon}) = -\mathcal{S}_\phi(\Theta_{\delta\gamma})(t_{\delta\gamma\epsilon}, s_{\delta\gamma\epsilon}, \psi_{\delta\gamma\epsilon}, \phi_{\delta\gamma\epsilon})$$

$$+ \mathcal{S}_\phi(T_\epsilon)(\psi_{\delta\gamma\epsilon}, \phi_{\delta\gamma\epsilon}). \quad (3.154)$$

Therefore,

$$\mathcal{S}(\Gamma_1)(t_{\delta\gamma\epsilon}, \psi_{\delta\gamma\epsilon}) - \mathcal{S}(\Gamma_2)(s_{\delta\gamma\epsilon}, \phi_{\delta\gamma\epsilon})$$

$$= \mathcal{S}_\psi(\Theta_{\delta\gamma})(t_{\delta\gamma\epsilon}, s_{\delta\gamma\epsilon}, \psi_{\delta\gamma\epsilon}, \phi_{\delta\gamma\epsilon}) + \mathcal{S}_\phi(\Theta_{\delta\gamma})(t_{\delta\gamma\epsilon}, s_{\delta\gamma\epsilon}, \psi_{\delta\gamma\epsilon}, \phi_{\delta\gamma\epsilon})$$

$$- [\mathcal{S}_\psi(T_\epsilon)(\psi_{\delta\gamma\epsilon}, \phi_{\delta\gamma\epsilon}) + \mathcal{S}_\phi(T_\epsilon)(\psi_{\delta\gamma\epsilon}, \phi_{\delta\gamma\epsilon})]. \quad (3.155)$$

Recall that

$$\Theta_{\delta\gamma}(t, s, \psi, \phi) = \frac{1}{\delta} \left[F(\psi - \phi) + F(\psi^0 - \phi^0) + |t - s|^2 \right] + \gamma(G(\psi) + G(\phi)).$$

Therefore, we have

$$
\begin{aligned}
&\bigl(\mathcal{S}_\psi(\Theta_{\delta\gamma}) + \mathcal{S}_\phi(\Theta_{\delta\gamma})\bigr)(t_{\delta\gamma\epsilon}, s_{\delta\gamma\epsilon}, \psi_{\delta\gamma\epsilon}, \phi_{\delta\gamma\epsilon})\\
&\equiv \mathcal{S}_\psi(\Theta_{\delta\gamma})(t_{\delta\gamma\epsilon}, s_{\delta\gamma\epsilon}, \psi_{\delta\gamma\epsilon}, \phi_{\delta\gamma\epsilon})\\
&\quad + \mathcal{S}_\phi(\Theta_{\delta\gamma})(t_{\delta\gamma\epsilon}, s_{\delta\gamma\epsilon}, \psi_{\delta\gamma\epsilon}, \phi_{\delta\gamma\epsilon})\\
&= \frac{1}{\delta}[\mathcal{S}_\psi(F)(\psi_{\delta\gamma\epsilon} - \phi_{\delta\gamma\epsilon}) + \mathcal{S}_\phi(F)(\psi_{\delta\gamma\epsilon} - \phi_{\delta\gamma\epsilon})\\
&\quad + \mathcal{S}_\psi(F)(\psi^0_{\delta\gamma\epsilon} - \phi^0_{\delta\gamma\epsilon}) + \mathcal{S}_\phi(F)(\psi^0_{\delta\gamma\epsilon} - \phi^0_{\delta\gamma\epsilon})]\\
&\quad + \gamma[\mathcal{S}_\psi(G)(\psi_{\delta\gamma\epsilon}) + \mathcal{S}_\phi(G)(\phi_{\delta\gamma\epsilon})].
\end{aligned} \tag{3.156}
$$

Using Lemma 3.6.7 and Lemma 3.6.8, we deduce that

$$
\begin{aligned}
&\bigl(\mathcal{S}_\psi(\Theta_{\delta\gamma}) + \mathcal{S}_\phi(\Theta_{\delta\gamma})\bigr)(t_{\delta\gamma\epsilon}, s_{\delta\gamma\epsilon}, \psi_{\delta\gamma\epsilon}, \phi_{\delta\gamma\epsilon})\\
&= \frac{1}{\delta}\Bigl[-|\psi_{\delta\gamma\epsilon}(-r) - \phi_{\delta\gamma\epsilon}(-r)|^2\Bigr]\\
&\quad - \gamma\Bigl(|\psi_{\delta\gamma\epsilon}(-r)|^2 e^{1 + F(\psi_{\delta\gamma\epsilon}) + F(\psi^0_{\delta\gamma\epsilon})}\\
&\qquad + |\phi_{\delta\gamma\epsilon}(-r)|^2 e^{1 + F(\phi_{\delta\gamma\epsilon}) + F(\phi^0_{\delta\gamma\epsilon})}\Bigr)\\
&\le 0.
\end{aligned} \tag{3.157}
$$

Thus, by virtue of (3.155) and Lemma 3.6.9, we have

$$
\limsup_{\delta\downarrow0, \epsilon\downarrow0} \Bigl[\mathcal{S}(\Gamma_1)(t_{\delta\gamma\epsilon}, \psi_{\delta\gamma\epsilon}) - \mathcal{S}(\Gamma_2)(s_{\delta\gamma\epsilon}, \phi_{\delta\gamma\epsilon})\Bigr] \le 0. \tag{3.158}
$$

Moreover, we know that the norm of T_ϵ is less than ϵ; thus, for any $\gamma > 0$, using (3.155) and taking the lim sup on both sides of (3.150) as δ and ϵ go to zero, we obtain

$$
\begin{aligned}
&\limsup_{\epsilon\downarrow0, \delta\downarrow0} \alpha(V_1(t_{\delta\gamma\epsilon}, \psi_{\delta\gamma\epsilon}) - V_2(s_{\delta\gamma\epsilon}, \phi_{\delta\gamma\epsilon}))\\
&\le \limsup_{\epsilon\downarrow0, \delta\downarrow0} \Bigl\{ \mathcal{S}(\Gamma_1)(t_{\delta\gamma\epsilon}, \psi_{\delta\gamma\epsilon}) - \mathcal{S}(\Gamma_2)(s_{\delta\gamma\epsilon}, \phi_{\delta\gamma\epsilon})\\
&\qquad + \sup_{v\in U} [L(t_{\delta\gamma\epsilon}, \psi_{\delta\gamma\epsilon}, v) - \overline{T_\epsilon(f(t, \psi_{\delta\gamma\epsilon}, v)\mathbf{1}_{\{0\}}, \phi_{\delta\gamma\epsilon})}]\\
&\qquad - \sup_{v\in U} [L(s_{\delta\gamma\epsilon}, \phi_{\delta\gamma\epsilon}, v) + \overline{T_\epsilon(\psi_{\delta\gamma\epsilon}, f(t, \phi_{\delta\gamma\epsilon}, v)\mathbf{1}_{\{0\}})}]\Bigr\}\\
&\le \limsup_{\epsilon\downarrow0, \delta\downarrow0} \Bigl\{ \sup_{v\in U} \Bigl| [L(t_{\delta\gamma\epsilon}, \psi_{\delta\gamma\epsilon}, v) - L(s_{\delta\gamma\epsilon}, \phi_{\delta\gamma\epsilon}, v)]\Bigr|\Bigr\}.
\end{aligned} \tag{3.159}
$$

Using the Lipschitz continuity of L and Lemma 3.6.2, we see that

$$
\begin{aligned}
&\limsup_{\delta\downarrow0, \epsilon\downarrow0} \sup_{v\in U} |L(t_{\delta\gamma\epsilon}, \psi_{\delta\gamma\epsilon}, v) - L(s_{\delta\gamma\epsilon}, \phi_{\delta\gamma\epsilon}, v)|\\
&\le \limsup_{\delta\downarrow0, \epsilon\downarrow0} C\Bigl(|t_{\delta\gamma\epsilon} - s_{\delta\gamma\epsilon}| + \|\psi_{\delta\gamma\epsilon} - \phi_{\delta\gamma\epsilon}\|_2\Bigr) = 0;
\end{aligned} \tag{3.160}
$$

moreover, by virtue of (3.160), we get

$$\limsup_{\epsilon\downarrow 0,\delta\downarrow 0} \alpha(V_1(t_{\delta\gamma\epsilon},\psi_{\delta\gamma\epsilon}) - V_2(s_{\delta\gamma\epsilon},\phi_{\delta\gamma\epsilon})) \le 0. \tag{3.161}$$

Since $(t_{\delta\gamma\epsilon}, s_{\delta\gamma\epsilon}, \psi_{\delta\gamma\epsilon}, \phi_{\delta\gamma\epsilon})$ is maximum of $\Phi_{\delta\gamma} + T_\epsilon$ in $[0,T] \times [0,T] \times \mathbf{C} \times \mathbf{C}$, then, for all $(t,\psi) \in [0,T] \times \mathbf{C}$, we have

$$\Phi_{\delta\gamma}(t,t,\psi,\psi) + T_\epsilon(\psi,\psi) \le \Phi_{\delta\gamma}(t_{\delta\gamma\epsilon}, s_{\delta\gamma\epsilon}, \psi_{\delta\gamma\epsilon}, \phi_{\delta\gamma\epsilon}) + T_\epsilon(\psi_{\delta\gamma\epsilon}, \phi_{\delta\gamma\epsilon}). \tag{3.162}$$

Then we get

$$V_1(t,\psi) - V_2(t,\psi) \tag{3.163}$$
$$\le V_1(t_{\delta\gamma\epsilon}, \psi_{\delta\gamma\epsilon}) - V_2(s_{\delta\gamma\epsilon}, \phi_{\delta\gamma\epsilon})$$
$$- \frac{1}{\delta}\left[\|\psi_{\delta\gamma\epsilon} - \phi_{\delta\gamma\epsilon}\|_2^2 + \|\psi^0_{\delta\gamma\epsilon} - \phi^0_{\delta\gamma\epsilon}\|_2^2 + |t_{\delta\gamma\epsilon} - s_{\delta\gamma\epsilon}|^2 \right]$$
$$+ 2\gamma \exp(1 + \|\psi\|_2^2 + \|\psi^0\|_2^2)$$
$$- \gamma(\exp(1 + \|\psi_{\delta\gamma\epsilon}\|_2^2 + \|\psi^0_{\delta\gamma\epsilon}\|_2^2) + \exp(1 + \|\phi_{\delta\gamma\epsilon}\|_2^2 + \|\phi^0_{\delta\gamma\epsilon}\|_2^2))$$
$$+ T_\epsilon(\psi_{\delta\gamma\epsilon}, \phi_{\delta\gamma\epsilon}) - T_\epsilon(\psi,\psi)$$
$$\le V_1(t_{\delta\gamma\epsilon}, \psi_{\delta\gamma\epsilon}) - V_2(s_{\delta\gamma\epsilon}, \phi_{\delta\gamma\epsilon})$$
$$+ 2\gamma \exp(1 + \|\psi\|_2^2 + \|\psi^0\|_2^2) + T_\epsilon(\psi_{\delta\gamma\epsilon}, \phi_{\delta\gamma\epsilon}) - T_\epsilon(\psi,\psi), \tag{3.164}$$

where the last inequality comes from the fact that $\delta > 0$ and $\gamma > 0$. By virtue of (3.161), when we take the $\limsup$ on (3.164) as δ, ϵ and γ go to zero, we can obtain

$$V_1(t,\psi) - V_2(t,\psi) \le \limsup_{\gamma\downarrow 0,\epsilon\downarrow 0,\delta\downarrow 0}\left(V_1(t_{\delta\gamma\epsilon}, \psi_{\delta\gamma\epsilon}) - V_2(s_{\delta\gamma\epsilon}, \phi_{\delta\gamma\epsilon}) \right.$$
$$\left. + 2\gamma \exp(1 + \|\psi\|_2^2 + \|\psi^0\|_2^2) + T_\epsilon(\psi_{\delta\gamma\epsilon}, \phi_{\delta\gamma\epsilon}) - T_\epsilon(\psi,\psi) \right)$$
$$\le 0. \tag{3.165}$$

Therefore, we have

$$V_1(t,\psi) \le V_2(t,\psi), \quad \forall (t,\psi) \in [0,T] \times \mathbf{C}. \tag{3.166}$$

This completes the proof of Theorem 3.6.1.   □

The uniqueness of the viscosity solution of (3.64) follows directly from this theorem because any viscosity solution is both the viscosity subsolution and supersolution.

Theorem 3.6.10 *The value function $V : [0,T] \times \mathbf{C} \to \Re$ of Problem (OCCP) defined by (3.7) is the unique viscosity solution of the HJBE (3.64).*

Proof. Suppose $V_1, V_2 : [0,T] \times \mathbf{C} \to \Re$ are two viscosity solutions of the HJBE (3.64). Then they are both the viscosity subsolution and supersolution. By Theorem 3.6.1, we have

$$V_2(t, \psi) \le V_1(t, \psi) \le V_2(t, \psi), \quad \forall(t, \psi) \in [0, T] \times \mathbf{C}.$$

This shows that

$$V_1(t, \psi) = V_2(t, \psi), \quad \forall(t, \psi) \in [0, T] \times \mathbf{C}.$$

Therefore, the value function $V : [0, T] \times \mathbf{C} \to \Re$ of Problem (OCCP) is the unique viscosity solution of the HJBE (3.64).   □

3.7 Verification Theorems

In this section, conjecture on a version of the verification theorem in the framework of viscosity solutions is presented. The value function $V : [0, T] \times \mathbf{C} \to \Re$ for Problem (OCCP) has been shown to be the unique viscosity solution of the HJBE (3.64) as shown in Sections 3.5 and 3.6. The remaining question for completely solving Problem (OCCP) is the computation of the optimal state-control pair $(x^*(\cdot), u^*(\cdot))$.

The classical verification theorem reads as follows.

Theorem 3.7.1 *Let $\Phi \in C_{lip}^{1,2}([0, T] \times \mathbf{C}) \cap \mathcal{D}(\mathcal{S})$ be the (classical) solution of the HJBE (3.64). Then the following hold:*
(i) $\Phi(t, \psi) \ge J(t, \psi; u(\cdot))$ for any $(t, \psi) \in [0, T] \times \mathbf{C}$ and any $u(\cdot) \in \mathcal{U}[t, T]$.
(ii) Suppose that a given admissible pair $(x^*(\cdot), u^*(\cdot))$ for the optimal classical control problem $(OCP)(t, \psi)$ satisfies

$$0 = \partial_t \Phi(s, x_s^*) + \mathcal{A}^{u^*(s)} \Phi(s, x_s^*) + L(s, x_s^*, u^*(s)) \ P - a.s., \quad a.e. \ s \in [t, T],$$

then $(x^*(\cdot), u^*(\cdot))$ is an optimal pair for $(OCP)(t, \psi)$.

Define the Hamiltonian function $\mathcal{H} : [0, T] \times \mathbf{C} \times \mathbf{C}^* \times \mathbf{C}^\dagger \times U \to \Re$ as follows:

$$\mathcal{H}(t, \phi, q, Q, u) = \frac{1}{2} \sum_{j=1}^{m} \bar{Q}(g(t, \phi, u)\mathbf{1}_{\{0\}}\mathbf{e}_j, g(t, \phi, u)\mathbf{1}_{\{0\}}\mathbf{e}_j)$$

$$+ \bar{q}(f(t, \phi, u)\mathbf{1}_{\{0\}}) + L(t, \phi, u), \qquad (3.167)$$

where $\bar{q} \in (\mathbf{C} \oplus \mathbf{B})$ is the continuous extension of q from $\mathbf{C}^*$ to $(\mathbf{C} \oplus \mathbf{B})^*$ and $\bar{Q} \in (\mathbf{C} \oplus \mathbf{B})^\dagger$ is the continuous extension of Q from $\mathbf{C}^\dagger$ to $(\mathbf{C} \oplus \mathbf{B})^\dagger$. (see Lemma 2.2.3 and Lemma 2.2.4 for details.)

We make the following conjecture on verification theorem in the viscosity framework.

Conjecture. (The Generalized Verification Theorem). Let $\bar{V} \in C((0, T] \times \mathbf{C}, \Re)$ be the viscosity supersolution of the HJBE (3.64) satisfying the following polynomial growth condition

$$|\bar{V}(t, \psi)| \le C(1 + \|\psi\|_2^k) \text{ for some } k \ge 1, \ (t, \psi) \in (0, T) \times \mathbf{C}. \qquad (3.168)$$

and such that $\bar{V}(T, \psi) = \Psi(\psi)$. Then we have the following.

(i) $\bar{V}(t, \psi) \geq J(t, \psi; u(\cdot))$ for any $(t, \psi) \in (0, T] \times \mathbf{C}$ and $u(\cdot) \in \mathcal{U}[t, T]$.

(ii) Fix any $(t, \psi) \in (0, T) \times \mathbf{C}$. Let $(x^*(\cdot), u^*(\cdot))$ be an admissible pair for Problem (OCCP). Suppose that there exists

$$(p^*(\cdot), q^*(\cdot), Q^*(\cdot)) \in L^2_{\mathbf{F}(t)}(t, T; \Re) \times L^2_{\mathbf{F}(t)}(t, T; \mathbf{C}^*) \times L^2_{\mathbf{F}(t)}(t, T; \mathbf{C}^\dagger)$$

such that, for a.e. $s \in [t, T]$,

$$(p^*(s), q^*(s), Q^*(s)) \in \underline{D}^{1,2}_{s+,\psi} \bar{V}(s, x^*_s)), \quad P - a.s., \tag{3.169}$$

and

$$E\left[\int_t^T [p^*(s) + \mathcal{H}(s, x^*(s), q^*(s), Q^*(s); u^*(s))]ds\right] \geq 0. \tag{3.170}$$

Then $(x^*(\cdot), u^*(\cdot))$ is an optimal pair for Problem (OCCP).

3.8 Finite-Dimensional HJB Equation

It is clear that the HJBE as described in (3.64) is infinite dimensional in the sense that it is a generalized differential equation that involve a first- and second-order Fréchet derivatives of a real-valued function defined on the Banach space $\mathbf{C}$ as well as the infinitesimal generator $\mathcal{S}$. The explicit solution of this equation is not well understood in general. In this section, we investigate some special cases of the infinite-dimensional HJBE (3.64) in which only the regular partial derivatives are involved and of which explicit solutions can be found. Much of the material presented in this section can be found in Larssen and Risebro [LR03]. However, they can be shown to be a special case of (3.1) and the general HJBE (3.64) treated in the previous sections.

3.8.1 Special Form of HJB Equation

In (3.1), we consider the one-dimensional case and assume that $m = 1$ and $n = 1$. Let the controlled drift and diffusion $f, g : [0, T] \times \mathbf{C} \times U \to \Re$ be defined as follows:

$$f(t, \phi, u) = b\left(t, \phi(0), \int_{-r}^0 e^{\lambda\theta}\phi(\theta)\, d\theta, \phi(-r), u\right) \tag{3.171}$$

and

$$g(t, \phi, u) = \sigma\left(t, \phi(0), \int_{-r}^0 e^{\lambda\theta}\phi(\theta)\, d\theta, \phi(-r), u\right) \tag{3.172}$$

for all $(t, \phi, u) \in [0, T] \times \mathbf{C} \times U$, where b and σ are some real-valued functions defined on $[0, T] \times \Re \times \Re \times \Re \times U$ that satisfy the following two conditions.

Assumption 3.8.1 *There exist constants $K_1 > 0$, and $K_2 > 0$ such that for all $(t, x, y, z), (t, \bar{x}, \bar{y}, \bar{z}, u) \in [0, T] \times \Re \times \Re \times \Re$ and $u \in U$,*

$$|b(t, x, y, z, u)| + |\sigma(t, x, y, z, u)| \leq K_1(1 + |x| + |y| + |z|)^p$$

and

$$|b(t, x, y, z, u) - b(t, \bar{x}, \bar{y}, \bar{z}, u)| + |\sigma(t, x, y, z, u) - \sigma(t, \bar{x}, \bar{y}, \bar{z}, u)|$$
$$\leq K_2(|x - \bar{x}| + |y - \bar{y}| + |z - \bar{z}|).$$

It is easy to see that if Assumption 3.8.1 holds for b and σ, then Assumption 3.1.1 holds for f and g that are related through (3.171) and (3.172).

Consider the following one-dimensional control stochastic delay equation:

$$dx(s) = b(s, x(s), y(s), z(s), u(s)) \, ds$$
$$+ \sigma(s, x(s), y(s), z(s), u(s)) \, dW(s), \quad s \in (t, T], \qquad (3.173)$$

with the initial data $(t, \psi) \in [0, T] \times C[-r, 0]$, where

$$y(s) = \int_{-r}^{0} e^{\lambda \theta} x(s + \theta) \, d\theta \quad (\lambda > 0 \text{ is a given constant})$$

represents a weighted (by the factor $e^{\lambda \cdot}$) sliding average of $x(\cdot)$ over the time interval $[s - r, s]$, and $z(s) = x(s - r)$ represents the discrete delay of the state process $x(\cdot)$.

The objective of the control problem is to maximize among $\mathcal{U}[t, T]$ the following expected performance index:

$$J(t, \psi; u(\cdot)) = E\Big[\int_t^T l(s, x(s), y(s), u(s)) \, ds + h(x(T), y(T))\Big], \qquad (3.174)$$

where $l : [0, T] \times \Re \times \Re \times U \to \Re$ and $h : \Re \times \Re \to \Re$ are the instantaneous reward and the terminal reward functions, respectively, that satisfy the following assumptions:

Assumption 3.8.2 *There exist constants $K, \bar{K} > 0$ and $k \geq 1$ such that*

$$|l(t, x, y, u)| + |h(x, y)| \leq K(1 + |x| + |y|)^k$$

and

$$|l(t, x, y, u) - l(t, \bar{x}, \bar{y}, u)| + |h(x, y) - h(\bar{x}, \bar{y})| \leq \bar{K}(|x - \bar{x}| + |y - \bar{y}|),$$

for all $(t, x, y, u), (t, \bar{x}, \bar{y}, u) \in [0, T] \times \Re \times \Re \times U$.

We again define the value function $V : [0, T] \times C[-r, 0] \to \Re$ for the optimal control problem (3.173) and (3.174) is defined by

$$V(t, \psi) = \sup_{u(\cdot) \in \mathcal{U}[t, T]} J(t, \psi; u(\cdot)). \qquad (3.175)$$

As described in Problem (OCCP), the value function V may depend on the initial datum $(t, \psi) \in [0, T] \times C[-r, 0]$ in a very general and complicated way. In this section, we will show that for a certain class of systems of the form (3.173), the value function depends on the initial function only through the functional of $x \equiv \psi(0)$, $y \equiv \int_{-r}^{0} e^{\lambda\theta} \psi(\theta) \, d\theta$. Let us therefore assume with a little abuse of notation that the value function V takes the following form:

$$V(t, \psi) = \Phi\left(t, \psi(0), \int_{-r}^{0} e^{\lambda\theta} \psi(\theta) \, d\theta\right) = \Phi(t, x, y), \tag{3.176}$$

where $\Phi : [0, T] \times \Re \times \Re \to \Re$. Then the DPP (Theorem 3.3.9) takes the form

$$\Phi(t, x, y) = \sup_{u(\cdot) \in \mathcal{U}[t,T]} \left[\int_{t}^{\tau} e^{-\alpha(\tau-t)} l(s, x(s), y(s), u(s)) \, ds \right.$$
$$\left. + \Phi(\tau, x(\tau), y(\tau)) \right] \tag{3.177}$$

for all **F**-stopping times $\tau \in \mathcal{T}_t^T$ and initial datum $(t, \psi(0), \int_{-r}^{0} e^{\lambda\theta}\psi(\theta) \, d\theta) \equiv (t, x, y) \in [0, T] \times \Re^2$.

Lemma 3.8.3 (The Itô Formula) *If $\Phi \in C^{1,2,1}([0, T] \times \Re \times \Re)$, then we have the following Itô formula:*

$$d\Phi(s, x(s), y(s)) = \mathcal{L}^u \Phi(s, x(s), y(s))$$
$$+ \partial_x \Phi(s, x(s), y(s)) \sigma(s, x(s), y(s)) \, dW(s), \tag{3.178}$$

where $\mathcal{L}^u$ is the differential operator defined by

$$\mathcal{L}^u \Phi(t, x, y) = b(t, x, y, z) \partial_x \Phi(t, x, y)$$
$$+ \frac{1}{2} \sigma^2(t, x, y, z) \partial_x^2 \Phi(t, x, y)$$
$$+ (x - e^{-\lambda r} z - \lambda z) \partial_y \Phi(t, x, y). \tag{3.179}$$

Proof. Note that if $\phi \in C[t-r, T]$, then $\phi_s \in C[-r, 0]$ for each $s \in [t, T]$. Since

$$y(x_s) = \int_{-r}^{0} e^{\lambda\theta} x(s+\theta) \, d\theta \quad (\lambda \text{ constant}),$$

$$\partial_s y(x_s) = \partial_s \left(\int_{-r}^{0} e^{\lambda\theta} x(s+\theta) \, d\theta \right)$$
$$= \partial_s \left(\int_{s-r}^{s} e^{\lambda(t-s)} x(t) \, dt \right)$$
$$= x(s) - e^{-\lambda r} x(s-r) - \lambda \int_{s-r}^{s} e^{\lambda(t-s)} x(t) \, dt$$

The result follows from the classical Itô formula (see Theorem 1.2.15). □

We have the following special form of the HJBE (3.64).

Theorem 3.8.4 *If we assume that (3.176) holds and that* $\Phi \in C^{1,2,1}([0,T] \times \Re \times \Re)$, *then* Φ *solves the following HJBE:*

$$\alpha\Phi(t,x,y) - \partial_t\Phi(t,x,y) - \max_{u \in U} \left[\mathcal{L}^u\Phi(t,x,y) + l(t,x,y,u)\right] = 0, \qquad (3.180)$$

$$\forall(t,x,y) \in [0,T] \times \Re \times \Re,$$

with the terminal condition $\Phi(T,x,y) = h(x,y)$, *where* $\mathcal{L}^u$ *is the differential operator defined by (3.179).*

Note that (3.180) is also equivalent to the following:

$$\alpha\Phi(t,x,y) - \partial_t\Phi(t,x,y) + \min_{u \in U} \left[-\mathcal{L}^u\Phi(t,x,y) - l(t,x,y,u)\right] = 0, \qquad (3.181)$$

$$\forall(t,x,y) \in [0,T] \times \Re \times \Re.$$

Proof. It is clear that $\Phi : [0,T] \times \Re \times \Re$ defined above is a quasi-tame function. From Itô's formula (see Lemma 3.8.3 in Chapter 2), we have

$$dV(s,x(s),y(s)) = \mathcal{L}^u V(s,x(s),y(s))ds \qquad (3.182)$$
$$+\sigma(s,x(s),y(s),z(s))\partial_x V(s,x(s),y(s))dW(s),$$

where the differential operator $\mathcal{L}^u$ is as defined in (3.179). We use the DDP (Theorem 3.3.9) and proceed exactly as in Subsection 3.4.1 to obtain (3.180). □

3.8.2 Finite Dimensionality of HJB Equation

Theorem 3.8.4 indicates that under the assumption that the value function takes the form $\Phi \in C^{1,2,1}([0,T] \times \Re \times \Re)$, we have a finite-dimensional HJBE (3.180) in the sense that it only involves regular partial derivatives such as $\partial_x\Phi$, $\partial_y\Phi$, and $\partial_x^2\Phi$ instead of the Fréchet derivatives and the $\mathcal{S}$-operator as required in (3.64). The question that remains to be answered is under what conditions we can have the finite-dimensional HJBE (3.180). This question will be answered in this subsection.

We consider the following one-dimensional controlled SHDE:

$$dx(s) = [\mu(x(s),y(s))$$
$$+ \beta(x(s),y(s))z(s) - g(s,x(s),y(s),u(s))]\,ds$$
$$+\sigma(x(s),y(s))\,dW(s), \quad s \in (t,T], \qquad (3.183)$$

with the initial datum $(t,\psi) \in [0,T] \times C([-r,0];\Re)$, where, again, $y(s) = \int_{-r}^{0} e^{\lambda\theta}x(s+\theta)\,d\theta$ and $z(s) = x(s-r)$ are described in the previous subsection and $\mu, \beta, \sigma : \Re \times \Re \to \Re$ and $g : [0,T] \times \Re \times \Re \times U \to \Re$ are the given deterministic functions. Assume that the discount rate $\alpha = 0$ for simplicity. It will be shown in this subsection that the HJBE has a solution depending only on (t, x, y) provided that an auxiliary system of four first-order partial differential equations (PDEs) involving μ, β, g, σ, l, and h has a solution. When this is the case, the HJBE (3.180) reduces to an "effective" equation in only one spatial variable in addition to time.

For this model, the HJBE (3.180) takes the form

$$-\partial_t \Phi + \min_{u \in U} F - (x - e^{-\lambda r} z - \lambda y)\partial_y \Phi = 0, \quad \forall z \in \Re, \tag{3.184}$$

where

$$\begin{aligned} F &= F(t, x, y, z, u, \partial_x \Phi, \partial_y \Phi, \partial_x^2 \Phi) \\ &= -\mathcal{L}^u \Phi(t, x, y) - l(t, x, y, u). \end{aligned} \tag{3.185}$$

Assume that $F^* = \inf_{u \in U} F$. Then

$$-\partial_t \Phi + F^* - (x - e^{-\lambda r} z - \lambda y)\partial_y \Phi = 0, \quad \forall z \in \Re. \tag{3.186}$$

Since this holds for all z, we must have

$$\partial_z(F^* - (x - e^{-\lambda r} z - \lambda y)\partial_y \Phi) = 0.$$

Now, $\partial_{u^*} F^* = 0$ since $u^* \in U$ is a minimizer of the function F. With $\partial_z(x - e^{-\lambda r} z - \lambda y) = -e^{-\lambda r}$, this leads to $\partial_z F^* + e^{-\lambda r} \partial_y \Phi = 0$ or

$$\partial_y \Phi = -e^{\lambda r} \partial_z F^*,$$

which we insert into (3.186) to obtain

$$-\partial_t \Phi + F^* + (x - e^{-\lambda r} z - \lambda y)e^{\lambda r} \partial_z F^* = 0. \tag{3.187}$$

Here, $F^* + (x - e^{-\lambda r} z - \lambda y)e^{\lambda r} \partial_z F^*$ should not depend on z. In the following, let H and G denote generic functions that may depend on $t, x, y, u^*, \partial_x \Phi, \partial_y \Phi$, and $\partial_x^2 \Phi$ but not on z. (H and G may change from line to line in a calculation.) Then the following are equivalent:

$$F^* + e^{\lambda r} \xi \partial_z F^* = H,$$

$$e^{-\lambda r} F^* + \xi \partial_z F^* = H,$$

$$\partial_z \left(\frac{F^*}{\xi} \right) = \frac{\xi \partial_z F^* + e^{-\lambda r} F^*}{\xi^2} = \frac{H}{\xi^2},$$

where $\xi \equiv x - e^{-\lambda r} z - \lambda y$. Integrating this yields

$$\frac{F^*}{\xi} = H \int \frac{dz}{\xi^2} = -H e^{\lambda r} \int \frac{d\xi}{\xi^2} = \frac{-H e^{\lambda r}}{\xi} + G,$$

so that $F^* = H + G\xi$, which implies that F^* is linear in z; that is,

$$F^* = H + Gz,$$

where H and G are functions that do not depend on z.

Motivated by the above reasoning, we investigate more closely a modified version of (3.173) and consider

$$dx(s) = [\bar{\mu}(x(s), y(s), z(s)) - g(s, x(s), y(s), u(s))]ds$$
$$+ \bar{\sigma}(x(s), y(s), z(s))dW(s), \quad s \in (t, T], \tag{3.188}$$

with the initial datum $(t, \psi) \in [0, T] \times C[-r, 0]$. Recall the performance functional (3.174),

$$J(t, \psi; u(\cdot)) = E\left[\int_t^T l(s, x(s), y(s), u(s))\, ds + h(x(T), y(T))\right], \tag{3.189}$$

and the value function $\Phi : [0, T] \times C[-r, 0] \rightarrow \Re$ defined by (3.176),

$$\Phi(t, \psi) = \Phi\left(t, \psi(0), \int_{-r}^0 e^{\lambda\theta}\psi(\theta)\, d\theta\right) = \Phi(t, x, y). \tag{3.190}$$

It is known that if $\Phi = \Phi(t, x, y)$, then Φ satisfies the HJBE

$$-\partial_t\Phi - \bar{\mu}\partial_x\Phi - \frac{1}{2}\bar{\sigma}^2\partial_x^2\Phi - (x - e^{-\lambda r}z - \lambda y)\partial_y\Phi + F(\partial_x\Phi, x, y, t) = 0, \tag{3.191}$$

with the terminal condition

$$\Phi(T, x, y) = h(x, y), \tag{3.192}$$

where

$$F(t, x, y, p) = \inf\{(g(t, x, y, u)p - l(t, x, y, u)\}. \tag{3.193}$$

We wish to obtain conditions on $\bar{\mu}, \bar{\sigma}$, and F that ensure that (3.191) has a solution independent of z. Differentiating (3.191) with respect to z, we obtain

$$\partial_y\Phi - e^{\lambda r}\partial_z\bar{\mu}\partial_x\Phi = e^{\lambda r}\partial_z\bar{\gamma}\partial_x^2\Phi, \tag{3.194}$$

where $\bar{\gamma} = \bar{\sigma}^2/2$. Inserting this into (3.191), this equation now takes the form

$$-\partial_t\Phi - [\bar{\mu} - (z - e^{\lambda r}(x - \lambda y))\partial_z\bar{\mu}]\partial_x\Phi$$
$$-[\bar{\gamma} - (z - e^{\lambda r}(x - \lambda y))\partial_z\bar{\gamma}]\partial_x^2\Phi + F(\partial_x\Phi, x, y, t) = 0,$$

If Φ is to be independent of z, then the coefficients of $\partial_x\Phi$ and $\partial_x^2\Phi$ must be independent of z. By arguments analogous to the previous, we see that

$$\bar{\mu}(x, y, z) = \mu(x, y) + \beta(x, y)z$$

and

$$\bar{\gamma}(x, y, z) = \gamma(x, y) + \zeta(x, y)z$$

for some functions μ, β, γ, and ζ depending on x and y only. Now, since $\bar{\gamma} \geq 0$ for all (x, y, z), we must have $\zeta = 0$, and, consequently, $\partial_z \bar{\gamma} = 0$. Also note that $\partial_z \bar{\mu} = \beta$ and that (3.208) takes the form

$$\partial_y \Phi - e^{\lambda r} \beta(x, y) \partial_x \Phi = 0. \tag{3.195}$$

Using this in (3.191) we see that this equation now reads

$$- \partial_t \Phi - [\mu(x, y) + e^{\lambda r}(x - \lambda y)\beta(x, y)]\partial_x \Phi$$
$$- \frac{1}{2}\sigma^2(x, y)\partial_x^2 \Phi + F(\partial_x \Phi, x, y, t) = 0. \tag{3.196}$$

Now, we introduce new variables $\tilde{x}$ and $\tilde{y}$, such that

$$\frac{\partial}{\partial \tilde{y}} = \frac{\partial}{\partial y} - e^{\lambda r} \beta(x, y) \frac{\partial}{\partial x} \quad \text{and} \quad \frac{\partial}{\partial \tilde{x}} = \frac{\partial}{\partial x}. \tag{3.197}$$

Then (3.195) states that $\partial_{\tilde{y}} \Phi = 0$. In order to be compatible with $\partial_{\tilde{y}} \Phi = 0$, the coefficients of (3.196) and the function h must also be constant in $\tilde{y}$, or

$$\partial_y \hat{\mu} - e^{\lambda r} \beta \partial_x \hat{\mu} = 0, \tag{3.198}$$

$$\partial_y \sigma - e^{\lambda r} \beta \partial_x \sigma = 0, \tag{3.199}$$

$$e^{\lambda r} p \partial_p F \partial_x \beta + \partial_y F - e^{\lambda r} \beta \partial_x F = 0, \tag{3.200}$$

$$\partial_y h - e^{\lambda r} \beta \partial_x h = 0, \tag{3.201}$$

where

$$\hat{\mu}(x, y) = \mu(x, y) + e^{\lambda r}(x - \lambda y)\beta(x, y).$$

To see why $\partial_{\tilde{y}} F = 0$ is equivalent to (3.200), note that

$$\partial_{\tilde{y}} F = \partial_p F \partial_y (\partial_x \Phi) + \partial_y F - e^{\lambda r} \beta(\partial_p F \partial_x \Phi_x + \partial_x F)$$
$$= \partial_p F(\partial_{yx}^2 \Phi - e^{\lambda r} \beta \partial_x^2 \Phi) + \partial_y F - e^{\lambda r} \beta \partial_x F$$
$$= \partial_p F[\partial_x(\partial_y \Phi - e^{\lambda r} \beta \partial_x \Phi) + e^{\lambda r} \partial_x \beta \partial_x \Phi] + \partial_y F - e^{\lambda r} \beta \partial_x F$$
$$= e^{\lambda r} p \partial_p F \partial_x \beta + \partial_y F - e^{\lambda r} \beta \partial_x F \quad \text{by (3.195)}. \tag{3.202}$$

Conversely, if $\bar{\mu} = \mu(x, y) + \beta(x, y)z$ and $\bar{\sigma} = \sigma(x, y)$, and (3.198)-(3.200) hold, then we can find a solution of (3.195) that is independent of z.

We collect this in the following theorem.

Theorem 3.8.5 *The HJBE (3.195) with terminal condition $\Phi(T, x, y) = h(x, y)$ has a viscosity solution $\Phi = \Phi(t, x, y)$ if and only if $\bar{\mu} = \mu(x, y) + \beta(x, y)z$ and $\bar{\sigma} = \sigma(x, y)$, and (3.198)-(3.201) hold. In this case, in the coordinates given by (3.197), the HJBE (3.195) reads*

$$-\partial_t \Phi - \hat{\mu}(\tilde{x})\partial_{\tilde{x}}^2 \Phi + F(\partial_{\tilde{x}} \Phi, \tilde{x}, t) - \frac{1}{2}\sigma^2(\tilde{x})\partial_{\tilde{x}}^2 \Phi = 0 \tag{3.203}$$

with the terminal condition

$$\Phi(T, \tilde{x}) = h(\tilde{x}). \tag{3.204}$$

3.8.3 Examples

In this subsection we present two examples that satisfy the requirements (3.198)-(3.201), and also indicate why it is difficult to find more general examples that can be completely solvable.

Example 1 (Harvesting with Exponential Growth) Assume that the size $x(\cdot)$ of a population obeys the linear SHDE

$$dx(s) = (ax(s) + by(s) + cz(s) - u(s))\,ds$$
$$+(\sigma_1 x(s) + \sigma_2 y(s))dW(s), \quad s \in [t,T], \qquad (3.205)$$

with the initial datum $(t,\psi) \in [0,T] \times C[-r,0]$. We assume that $x(s) > 0$. The population is harvested at a rate $u(s) \geq 0$, and we are given the performance functional

$$J(t,\psi;u(\cdot)) = E^{t,\psi,u(\cdot)}\Big[\int_t^T \{l_1(x(s),y(s)) + l_2(u(s))\}\,ds$$
$$+h(x(T),y(T))\Big], \qquad (3.206)$$

where T is the stopping time defined by

$$T = \Big\{T_1, \inf_{s>t} \{x(s;t,\psi,u(\cdot)) = 0\}\Big\} \qquad (3.207)$$

and $T_1 > t$ is some finite deterministic time. If the value function Φ takes the form $\Phi(t,x,y)$, then Φ satisfies the HJBE

$$-\partial_s\Phi - (x - e^{-\lambda r}z - \lambda y)\partial_y\Phi - \frac{1}{2}(\sigma_1 x + \sigma_2)^2\partial_x^2\Phi$$
$$+ \inf_u \{-(ax + by + cz - u)\partial_x\Phi - l_1(x,y) - l_2(u)\} = 0 \qquad (3.208)$$

Using Theorem 3.8.5, from (3.198) and (3.199) we find that the parameters must satisfy the relations

$$\sigma_2 = \sigma_1 ce^{\lambda r}, \quad b - \lambda ce^{\lambda r} = ce^{\lambda r}(a + ce^{\lambda r}). \qquad (3.209)$$

The function F defined in (3.193) now has the form

$$F(p,x,y) = \inf_{u\in U}\{pu - l_2(u)\} - l_1(x,y) = pu^* - l_2(u^*) - l_1(x,y),$$

where u^* is the minimizer in U. Then from (3.200) we see that we must have

$$\partial_y l_1 - ce^{\lambda r}\partial_x l_1 = 0, \qquad (3.210)$$

or $l_1 = l_1(x + ce^{\lambda r}y)$. Introducing the variable $\tilde{x} = x + ce^{\lambda r}y$ and the constant $\kappa = a + ce^{\lambda r}$, we find that the "effective" equations (3.203) and (3.204) in this case will be

$$-\partial_s \Phi - (\kappa \tilde{x} - u^*)\partial_{\tilde{x}} \Phi - \frac{1}{2}\sigma_1^2 \tilde{x}^2 \partial_{\tilde{x}}^2 \Phi - l_1(\tilde{x}) - l_2(u^*) = 0, \qquad (3.211)$$

with the terminal condition

$$\Phi(T, \tilde{x}) = h(\tilde{x}), \qquad (3.212)$$

assuming h satisfies (3.201). This corresponds to the control problem without delay with system dynamics

$$d\tilde{x}(s) = (\kappa \tilde{x}(s) - u)\, ds + \sigma_1 \tilde{x}(s)\, dW(s), \quad s \in (t, T],$$

and $\tilde{x}(t) = \tilde{x} \geq 0$.

To close the discussion of this example, let us be specific and choose

$$l_1(x, y) = -c_0|x + ce^{\lambda r}y - m|, \quad l_2(u) = c_1 u - c_2 u^2, \quad h = 0, \qquad (3.213)$$

where c_0, c_1, c_2, and m are positive constants. Then (3.210) and (3.201) hold and

$$F(p, x, y) = \inf_{u \in U}\{c_2 u^2 - (c_1 - p)u\} + c_0|x + ce^{\lambda r}y - m|.$$

We solve for u and find that the optimal harvesting rate is given by

$$u^* = \max\left\{\frac{c_1 - \partial_x \Phi}{2c_2}, 0\right\}. \qquad (3.214)$$

Insert (3.213) and (3.214) into the HJBEs (3.211) and (3.212). The resulting equation is a second-order PDE that may be solved numerically and the optimal control can be found provided that the solution of the HJBE really is the value function.

Example 2 (Resource Allocation). Let $x(\cdot) = \{x(s), s \in [t - r, T]\}$ denote a population developing according to (3.205). One can think of $x(\cdot)$ as a wild population that can be caught and bred in captivity and then harvested. The population in captivity, $\hat{x}(\cdot)$, develops according to

$$d\hat{x}(s) = (\gamma \hat{x}(s) + u(s) - v(s))\, ds, \quad s \in (t, T], \qquad (3.215)$$

with $\hat{x}(t) = \hat{x} \geq 0$, where v denotes the harvesting rate. The state and control processes for the control problem are $(x(\cdot), \hat{x}(\cdot))$ and $(u(\cdot), v(\cdot))$, respectively. For this case, we consider the gain functional

$$J(t, \psi, \hat{x}; u(\cdot), v(\cdot)) = E^{t,\psi,\hat{x};u(\cdot),v(\cdot)}\left[\int_t^T (l(v(s)) - c_1 \hat{x}(s) - c_2 u^2(s))\, ds\right.$$
$$\left. + h(x(T), y(T), \hat{x}(T))\right], \qquad (3.216)$$

where T is again given by (3.207), $l(v)$ denotes the utility from consumption or sales of the animals, $c_1 \hat{x}$ models the cost of keeping the population, and $c_2 u^2$ models the cost of catch and transfer. Setting

$$V(t, \psi, \hat{x}) = \sup_{u \geq 0, v} J(t, \psi, \hat{x}; u(\cdot), v(\cdot)),$$

we find that if V takes the special form $V = \Phi(t, x, y, \hat{x})$, then

$$- \partial_t \Phi - (ax + by + cz)\partial_x \Phi - \gamma \hat{x} \partial_{\hat{x}} \Phi - \frac{1}{2}(\sigma_1 x + \sigma_2 y + \sigma_3 z)^2 \partial_x^2 \Phi$$
$$- (x - e^{-\lambda r} z - \lambda y)\partial_y \Phi + c_1 \hat{x} + F(\partial_x \Phi, \partial_{\hat{x}} \Phi) = 0 \qquad (3.217)$$

and $\Phi(T, x, y, \hat{x}) = h(x, y, \hat{x})$, where

$$F(p, q) = \inf_{u \geq 0, v \leq v_{max}} (c_2 u^2 - u(q - p) + vq - l(v)).$$

Since v is independent of z, we must demand that the parameters satisfy (3.209), and we introduce $\tilde{x}$ as before to find that $\Phi = \Phi(t, \tilde{x}, \hat{x})$ satisfies

$$-\partial_t \Phi - \kappa \tilde{x} \partial_{\tilde{x}} \Phi - \gamma \hat{x} \partial_{\hat{x}} \Phi - \frac{1}{2}\sigma_1^2 \tilde{x}^2 \partial_{\tilde{x}}^2 \Phi + c_1 \hat{x} + F(\partial_{\tilde{x}} \Phi, \partial_{\hat{x}} \Phi) = 0, \quad \text{for } t < T$$
$$(3.218)$$

and $\Phi(T, \tilde{x}, \hat{x}) = h(\tilde{x}, \hat{x})$. Again, the above PDE with the terminal condition can be solved numerically.

3.9 Conclusions and Remarks

This chapter develops the infinite-dimensional HJBE for the value function of the discounted optimal classical control problem over finite time horizon. The HJBE involves extensions of first- and second-order Fréchet derivatives as well as the shift operator, which are unique in controlled SHDEs. This distinguishes them from all other infinite-dimensional stochastic control problems such as the ones arising from stochastic partial differential equations. The main theme of this chapter is to show under very reasonable assumptions that the value function is the unique viscosity solution of the HJBE. Existence of optimal control as well as special cases that lead to a finite dimensional HJBE are demonstrated. There is no attempt to treat the ergodic controls and/or the combined classical-singular control problem. However, a combined classical-impulse control arising from a hereditary portfolio optimization problem is treated in Chapter 7 in detail.

4

Optimal Stopping

Optimal stopping problems over a finite or an infinite time horizon for Itô's diffusion processes described by stochastic differential equations (SDEs) arise in many areas of science, engineering, and finance (see, e.g., Fleming and Soner [FS93], Øksendal [Øks00], Shiryaev [Shi78], Karazas and Shreve [KS91], and references contained therein). The objective of the problem is to find a stopping time τ with respect to the filtration generated by the solution process of the SDE that maximizes or minimizes a certain expected reward or cost functional. The value function of these problems are normally expressed as a viscosity or a generalized solution of Hamilton-Jacobi-Bellman (HJB) variational inequality that involves a second-order parabolic or elliptic partial differential equation in a finite-dimensional Euclidean space.

In an attempt to achieve better accuracy and to account for the delayed effect of the state variables in the modeling of real- world problems, optimal stopping of the following stochastic delay differential equation:

$$
dx(s) = b\left(s, x(s), x(s-r), \int_{-r}^{0} e^{\lambda\theta} x(s+\theta)\, d\theta\right) ds
$$
$$
+ \sigma\left(s, x(s), x(s-r), \int_{-r}^{0} e^{\lambda\theta} x(s+\theta)\, d\theta\right) dW(s) \quad s \in [t, T].
$$

have been the subject of study in his unpublished dissertation Elsanousi [Els00].

In the above equation, b and σ are $\Re^n$, and $\Re^{n \times m}$-valued functions, respectively, defined on

$$
[0, T] \times \Re^n \times \Re^n \times \Re^n
$$

and $\lambda > 0$ is a given constant.

This chapter substantially extends the results obtained for finite- dimensional diffusion processes and stochastic delay differential equations described above and investigates an optimal stopping problem over a finite time horizon $[t, T]$ for a general system of stochastic hereditary differential equations

(SHDEs) with a bounded memory described in (4.1), where $T > 0$ and $t \in [0, T]$ respectively denote the terminal time and an initial time of the optimal stopping problem. The consideration of such a system enable us to model many real- world problems that have aftereffects.

This chapter develops its existence theory via a construction of the least superharmonic majorant of the terminal reward functional. It also derives an infinite-dimensional HJB variational inequality (HJBVI) for the value function via a dynamic programming principle (see, e.g., Theorem 3.3.9 in Chapter 3). In the content of an infinite-dimensional HJBVI, it is shown that the value function for the optimal stopping problem is the unique viscosity solution of the HJBVI. The proof of uniqueness is very similar to those in Section 3.6 in Chapter 3 and involves embedding the function space $\mathbf{C} = C([-r, 0]; \Re^n)$ into the Hilbert space $L^2([-r, 0]; \Re^n)$ and extending the concept of viscosity solution for the controlled Itô diffusion process (see, e.g., Fleming and Soner [FS93]) to an infinite-dimensional setting. As an application of the results obtained, a pricing problem is considered in Chapter 6 for American options in a financial market with one riskless bank account that grows according to a deterministic linear functional differential equation and one stock whose price dynamics follows a nonlinear stochastic functional differential equation. In there, it is shown that the option pricing can be formulated into an optimal stopping problem considered in this chapter and therefore all results obtained are applicable under very realistic assumptions.

This chapter is organized as follows. The basic assumptions and preliminary results that are needed for formulating the optimal stopping problem as well as the problem statement are contained in Section 4.1. In Section 4.2, the existence and uniqueness of the optimal stopping is developed via the concept of excessive and superharmonic functions and a construction of the least superharmonic majorant of the terminal reward functional. In Section 4.3, the HJBVI for the value function is heuristically derived using the Bellman type dynamic programming principle. In Section 4.4, the verification theorem is proved. Although continuous, the value function is not known to be smooth enough to be a classical solution of the HJBVI in general cases. It is shown in Section 4.5, however, that the value function is a viscosity solution of the HJBVI. The proof for the comparison principle as well as the necessary lemmas are given in Section 4.6. The uniqueness result for the value function being the unique viscosity solution of the HJBVI follows immediately from the comparison principle.

4.1 The Optimal Stopping Problem

Let $L^2(\Omega, \mathbf{C})$ be the space of $\mathbf{C}$-valued random variables $\Xi : \Omega \to \mathbf{C}$ such that

$$\|\Xi\|_{L^2} \equiv E[\|\Xi\|^2] \equiv \left(\int_{\Omega} \|\Xi(\omega)\|^2 \, dP(\omega) \right)^{\frac{1}{2}} < \infty.$$

Let $L^2(\Omega, \mathbf{C}; \mathcal{F}(t))$ be the space of those $\Xi \in L^2(\Omega, \mathbf{C})$ that are $\mathcal{F}(t)$-measurable.

Consider the uncontrolled SHDE described by

$$dx(s) = f(s, x_s) \, ds + g(s, x_s) \, dW(s), \quad s \in [t, T]. \tag{4.1}$$

When there is no delay (i.e., $r = 0$), it is clear that SHDE (4.1) reduces to the following Itô diffusion process described by

$$dx(s) = f(s, x(s)) \, ds + g(s, x(s)) \, dW(s), \quad s \in [t, T].$$

It is clear that (4.1) also includes the following stochastic delay differential equation (SDDE):

$$dx(s) = b\left(s, x(s), x(s - r), \int_{-r}^{0} e^{\lambda \theta} x(s + \theta) d\theta\right) ds$$
$$+ \sigma\left(s, x(s), x(s - r), \int_{-r}^{0} e^{\lambda \theta} x(s + \theta) d\theta\right) dW(s), \quad s \in [t, T]$$

and many other equations that cannot be modeled in this form as a special case.

We repeat the definition of the strong solution of (4.1) as follows.

Definition 4.1.1 *A process $\{x(s; t, \psi_t), s \in [t - r, T]\}$ is said to be a (strong) solution of (4.1) on the interval $[t-r, T]$ and through the initial datum $(t, \psi_t) \in [0, T] \times L^2(\Omega, \mathbf{C}; \mathcal{F}(t))$ if it satisfies the following conditions:*

1. *$x_t(\cdot; t, \psi_t) = \psi_t$.*
2. *$x(s; t, \psi_t)$ is $\mathcal{F}(s)$-measurable for each $s \in [t, T]$.*
3. *The process $\{x(s; t, \psi_t), s \in [t, T]\}$ is continuous and satisfies the following stochastic integral equation P-a.s.:*

$$x(s) = \psi_t(0) + \int_t^s f(\lambda, x_\lambda) \, d\lambda + \int_t^s g(\lambda, x_\lambda) \, dW(\lambda), \quad s \in [t, T]. \tag{4.2}$$

In addition, the (strong) solution process $\{x(s; t, \psi_t), s \in [t - r, T]\}$ of (4.1) is said to be (strongly) unique if $\{\tilde{x}(s; t, \psi_t), s \in [t - r, T]\}$ is also a (strong) solution of (4.1) on $[t - r, T]$ and through the same initial datum (t, ψ_t), then

$$P\{x(s; t, \psi_t) = \tilde{x}(s; t, \psi_t), \forall s \in [t, T]\} = 1.$$

Throughout, we assume that the functions $f : [0, T] \times \mathbf{C} \to \Re^n$ and $g : [0, T] \times \mathbf{C} \to \Re^{n \times m}$ are continuous functions and satisfy the following conditions to ensure the existence and uniqueness of a (strong) solution $x(\cdot) = \{x(s; t, \psi_t), s \in [t - r, T]\}$ for each $(t, \psi_t) \in [0, T] \times L^2(\Omega, C; F(t))$. (See Theorem 1.3.12 in Chapter 1).

Assumption 4.1.2   *1. (Lipschitz Continuity) There exists a constant $K_{lip} >$ 0 such that*

$$|f(t,\varphi) - f(s,\phi)| + |g(t,\varphi) - g(s,\phi)|$$
$$\leq K_{lip}(\sqrt{|t-s|} + \|\varphi - \phi\|), \quad \forall (t,\varphi), (s,\phi) \in [0,T] \times \mathbf{C}.$$

2. (Linear Growth) There exists a constant $K_{grow} > 0$ such that

$$|f(t,\phi)| + |g(t,\phi)| \leq K_{grow}(1 + \|\phi\|), \quad \forall (t,\phi) \in [0,T] \times \mathbf{C}.$$

Let $\{x(s;t,\psi_t), s \in [t,T]\}$ be the solution of (4.1) through the initial datum $(t,\psi_t) \in [0,T] \times L^2(\Omega, \mathbf{C}; \mathcal{F}(t))$.

We consider the corresponding $\mathbf{C}$-valued process $\{x_s(t,\psi_t), s \in [t,T]\}$ defined by

$$x_s(\theta; t,\psi_t) \equiv x(s+\theta; t,\psi_t), \quad \theta \in [-r,0].$$

For each $t \in [0,T]$, let $\mathbf{G}(t) = \{\mathcal{G}(t,s), s \in [t,T]\}$ be the filtration defined by

$$\mathcal{G}(t,s) = \sigma(x(\lambda; t,\psi_t), t \leq \lambda \leq s).$$

Note that it can be shown that for each $s \in [t,T]$,

$$\mathcal{G}(t,s) = \sigma(x_\lambda(t,\psi_t), t \leq \lambda \leq s)$$

due to the sample path's continuity of the process $x(\cdot) = \{x(s;t,\psi_t), s \in [t,T]\}$.

One can then establish the following Markov property (see Theorem 1.5.2 in Chapter 1).

Theorem 4.1.3 *Let Assumption 4.1.2 hold. Then the corresponding $\mathbf{C}$-valued solution process of (4.1) describes a $\mathbf{C}$-valued Markov process in the following sense: For any $(t,\psi_t) \in [0,T] \times L^2(\Omega, \mathbf{C})$, the Markovian property*

$$P\{x_s(t,\psi_t) \in B | \mathcal{G}(t,\tilde{t})\} = P\{x_s(t,\psi_t) \in B | x_{\tilde{t}}(t,\psi_t)\} \equiv p(\tilde{t}, x_{\tilde{t}}(t,\psi_t), s, B)$$

holds a.s. for $t \leq \tilde{t} \leq s$ and $B \in \mathcal{B}(\mathbf{C})$, where $\mathcal{B}(\mathbf{C})$ is the Borel σ-algebra of subsets of $\mathbf{C}$.

In the above, the function $p : [t,T] \times \mathbf{C} \times [0,T] \times \mathcal{B}(\mathbf{C}) \to [0,1]$ denotes the transition probabilities of the $\mathbf{C}$-valued Markov process $\{x_s(t,\psi_t), s \in [t,T]\}$.

A random function $\tau : \Omega \to [0,\infty]$ is said to be a $\mathbf{G}(t)$-stopping time if

$$\{\tau \leq s\} \in \mathcal{G}(t,s), \quad \forall s \geq t.$$

Let $\mathcal{T}_t^T(\mathbf{G})$ (or simply $\mathcal{T}_t^T$) be the collection of all $\mathbf{G}(t)$-stopping times $\tau \in \mathcal{T}$ such that $t \leq \tau \leq T$ a.s.. For each $\tau \in \mathcal{T}_t^T$, let the sub-σ-algebra $\mathcal{G}(t,\tau)$ of $\mathcal{F}$ be defined by

$$\mathcal{G}(t,\tau) = \{A \in \mathcal{F} \mid A \cap \{t \leq \tau \leq s\} \in \mathcal{G}(t,s) \ \forall s \in [t,T]\}.$$

The collection $\mathcal{G}(t,\tau)$ can be interpreted as the collection of events that are measurable up to the stopping time $\tau \in \mathcal{T}_t^T$. With a little bit more effort, one can also show that the corresponding $\mathbf{C}$-valued solution process $\{x_s(t,\psi_t), s \in [t,T]\}$ of (4.1) is also a strong Markov process in $\mathbf{C}$; that is,

$$P\{x_s(t,\psi_t) \in B | \mathcal{G}(t,\tau)\} = P\{x_s(t,\psi_t) \in B | x_\tau(t,\psi_t)\} \equiv p(\tau, x_\tau(t,\psi_t), s, B)$$

holds a.s. for all $\tau \in \mathcal{T}_t^T$, all deterministic $s \in [\tau, T]$, and $B \in \mathcal{B}(\mathbf{C})$.

If the drift f and the diffusion coefficient g are time independent (i.e., $f(t,\phi) \equiv f(\phi)$ and $g(t,\phi) \equiv g(\phi)$), then (4.1) reduces to the following autonomous system:

$$dx(t) = f(x_t)\,dt + g(x_t)\,dW(t). \tag{4.3}$$

In this case, we usually assume the initial datum $(t,\psi_t) = (0,\psi)$ and denote the solution process of (4.3) through $(0,\psi)$ and on the interval $[-r,T]$ by $\{x(s;\psi), s \in [-r,T]\}$. Then the corresponding $\mathbf{C}$-valued solution process $\{x_s(\psi), s \in [-r,T]\}$ of (4.3) is a strong Markov process with time-homogeneous probability transition function $p(\psi, s, B) \equiv p(0,\psi, s, B) = p(t,\psi, t+s, B)$ for all $s,t \geq 0$, $\psi \in \mathbf{C}$, and $B \in \mathcal{B}(\mathbf{C})$.

Assume L and Ψ are two $\|\cdot\|_2$-Lipschitz continuous real-valued functions on $[0,T] \times \mathbf{C}$ with at most polynomial growth in $L^2([-r,0]; \Re^n)$. In other words, we assume the following conditions.

Assumption 4.1.4   *1. (Lipschitz Continuity) There exist a constant $K_{lip} > 0$ such that*

$$|L(t,\psi) - L(s,\phi)| + |\Psi(t,\psi) - \Psi(s,\phi)|$$
$$\leq K_{lip}(\sqrt{|t-s|} + \|\psi - \phi\|_2), \quad \forall (t,\psi),(s,\phi) \in [0,T] \times \mathbf{C}.$$

2. (Polynomial Growth) There exist constants $K_p > 0$, and $k \geq 1$ such that

$$|L(t,\phi)| + |\Psi(t,\phi)| \leq \Lambda(1 + \|\phi\|_2)^k, \quad \forall (t,\phi) \in [0,T] \times \mathbf{C}.$$

3. (Integrability of L) For all initial $\psi \in \mathbf{C}$,

$$E^\psi \left[\int_t^T e^{-\alpha(s-t)} |L(\lambda, x_\lambda(\cdot; t, \psi))|\,dt \right] < \infty.$$

4. (Uniform Integrability of Ψ) The family $\{\Psi^-(x_\tau); \tau \in \mathcal{T}_t^T\}$ is uniformly integrable w.r.t. $P^{t,\psi}$ (the probability law of the $x(\cdot) = \{x(\cdot; t, \psi), t \in [0,T]\}$), given the initial datum $(t,\psi) \in [0,T] \times \mathbf{C}$, where $\Psi^- \equiv \max\{-\Psi, 0\}$ is the negative part of the function Ψ.

Since the L^2-norm $\|\cdot\|_2$ is weaker that the sup-norm $\|\cdot\|$, that is,

$$\|\phi\|_2 \leq \sqrt{r}\|\phi\|, \quad \forall \phi \in \mathbf{C},$$

2 of Assumption 4.1.4 implies that there exists a constant $K > 0$ such that

$$|L(t,\phi)| + |\Psi(t,\phi)| \le K(1 + \|\phi\|)^k, \quad \forall (t,\phi) \in [0,T] \times \mathbf{C}.$$

We state the optimal stopping problem (Problem (OSP1)) as follows.

Problem (OSP1). Given the initial datum $(t,\psi) \in [0,T] \times \mathbf{C}$, our objective is to find an optimal stopping time $\tau^* \in \mathcal{T}_t^T$ that maximizes the following expected performance index:

$$J(t,\psi;\tau) \equiv E\left[\int_t^\tau e^{-\alpha(s-t)} L(s,x_s)\, ds + e^{-\alpha(\tau-t)} \Psi(x_\tau)\right], \qquad (4.4)$$

where $\alpha > 0$ denotes a discount factor. In this case, the value function $V : [0,T] \times \mathbf{C} \to \Re$ is defined to be

$$V(t,\psi) \equiv \sup_{\tau \in \mathcal{T}_t^T} J(t,\psi;\tau). \qquad (4.5)$$

For the autonomous case (i.e., (4.3))

$$dx(s) = f(x_s)\, dt + g(x_s)\, dW(s), \quad s \in [0,T],$$

the following optimal stopping problem (Problem (OSP2)) is a special case of Problem (OSP1).

Problem (OSP2). Find an optimal stopping time $\tau^* \in \mathcal{T}_0^T$ that maximizes the following discounted objective functional:

$$J(\psi;\tau) \equiv E\left[\int_0^\tau e^{-\alpha s} L(x_s)\, ds + e^{-\alpha\tau} \Psi(x_\tau)\right]. \qquad (4.6)$$

In this case, the value function $V : \mathbf{C} \to \Re$ is defined to be

$$V(\psi) \equiv \sup_{\tau \in \mathcal{T}_0^T} J(\psi;\tau). \qquad (4.7)$$

4.2 Existence of Optimal Stopping

In this section we investigate the existence of optimal stopping for Problem (OSP1) via a construction of the least superharmonic majorant of an appropriate terminal reward functional.

4.2.1 The Infinitesimal Generator

We recall the following concepts, which were introduced earlier in Section 2.3 in Chapter 2. Let $\mathbf{C}^*$ and $\mathbf{C}^\dagger$ be the space of bounded linear functionals $\Phi : \mathbf{C} \to \Re$ and bounded bilinear functionals $\tilde{\Phi} : \mathbf{C} \times \mathbf{C} \to \Re$, of the space

C, respectively. They are equipped with the operator norms, which will be respectively denoted by $\|\cdot\|^*$ and $\|\cdot\|^\dagger$.

Let $\mathbf{B} = \{v\mathbf{1}_{\{0\}}, v \in \Re^n\}$, where $\mathbf{1}_{\{0\}} : [-r, 0] \to \Re$ is defined by

$$\mathbf{1}_{\{0\}}(\theta) = \begin{cases} 0 \text{ for } \theta \in [-r, 0) \\ 1 \text{ for } \theta = 0. \end{cases}$$

We form the direct sum

$$\mathbf{C} \oplus \mathbf{B} = \{\phi + v\mathbf{1}_{\{0\}} \mid \phi \in \mathbf{C}, v \in \Re^n\}$$

and equip it with the norm, also denoted by $\|\cdot\|$, defined by

$$\|\phi + v\mathbf{1}_{\{0\}}\| = \sup_{\theta \in [-r, 0]} |\phi(\theta)| + |v|, \quad \phi \in \mathbf{C}, v \in \Re^n.$$

Again, let $(\mathbf{C} \oplus \mathbf{B})^*$ and $(\mathbf{C} \oplus \mathbf{B})^\dagger$ be spaces of bounded linear and bilinear functionals of $\mathbf{C} \oplus \mathbf{B}$, respectively.

The following two results can be found in Lemma 2.2.3 and Lemma 2.2.4 in Chapter 2.

Lemma 4.2.1 *If* $\Gamma \in \mathbf{C}^*$, *then* Γ *has a unique (continuous) linear extension* $\bar{\Gamma} : \mathbf{C} \oplus \mathbf{B} \to \Re$ *satisfying the following weak continuity property:*
(W1). If $\{\xi^{(k)}\}_{k=1}^\infty$ *is a bounded sequence in* $\mathbf{C}$ *such that* $\xi^{(k)}(\theta) \to \xi(\theta)$ *as* $k \to \infty$ *for all* $\theta \in [-r, 0]$ *for some* $\xi \in \mathbf{C} \oplus \mathbf{B}$, *then* $\Gamma(\xi^{(k)}) \to \bar{\Gamma}(\xi)$ *as* $k \to \infty$. *The extension map* $\mathbf{C}^* \to (\mathbf{C} \oplus \mathbf{B})^*$, $\Gamma \mapsto \bar{\Gamma}$ *is a linear isometric injective map.*

Lemma 4.2.2 *If* $\Gamma \in \mathbf{C}^\dagger$, *then* Γ *has a unique (continuous) linear extension* $\bar{\Gamma} \in (\mathbf{C} \oplus \mathbf{B})^\dagger$ *satisfying the following weak continuity property:*
(W2). If $\{\xi^{(k)}\}_{k=1}^\infty$ *and* $\{\zeta^{(k)}\}_{k=1}^\infty$ *are bounded sequences in* $\mathbf{C}$ *such that* $\xi^{(k)}(\theta) \to \xi(\theta)$ *and* $\zeta^{(k)}(\theta) \to \zeta(\theta)$ *as* $k \to \infty$ *for all* $\theta \in [-r, 0]$ *for some* $\xi, \zeta \in \mathbf{C} \oplus \mathbf{B}$, *then* $\Gamma(\xi^{(k)}, \zeta^{(k)}) \to \bar{\Gamma}(\xi, \zeta)$ *as* $k \to \infty$.

For a sufficiently smooth functional $\Phi : \mathbf{C} \to \Re$, we can define its Fréchet derivatives with respect to $\phi \in \mathbf{C}$. It is clear from the results stated above that its first-order Fréchet derivative, $D\Phi(\phi) \in \mathbf{C}^*$, has a unique and (continuous) linear extension $\overline{D\Phi(\phi)} \in (\mathbf{C} \oplus \mathbf{B})^*$. Similarly, its second-order Fréchet derivative, $D^2\Phi(\phi) \in \mathbf{C}^\dagger$, has a unique and continuous linear extension $\overline{D^2\Phi(\phi)} \in (\mathbf{C} \oplus \mathbf{B})^\dagger$.

For a Borel measurable function $\Phi : \mathbf{C} \to \Re$, we also define

$$\mathcal{S}(\Phi)(\phi) \equiv \lim_{\epsilon \downarrow 0} \frac{1}{\epsilon} \left[\Phi(\tilde{\phi}_\epsilon) - \Phi(\phi) \right] \tag{4.8}$$

for all $\phi \in \mathbf{C}$, where $\tilde{\phi} : [-r, T] \to \Re^n$ is an extension of $\phi : [-r, 0] \to \Re^n$ defined by

$$\tilde{\phi}(t) = \begin{cases} \phi(t) \text{ if } t \in [-r, 0) \\ \phi(0) \text{ if } t \geq 0, \end{cases}$$

and again $\tilde{\phi}_t \in \mathbf{C}$ is defined by

$$\tilde{\phi}_t(\theta) = \tilde{\phi}(t+\theta), \quad \theta \in [-r, 0].$$

Let $\hat{\mathcal{D}}(\mathcal{S})$, the domain of the operator $\mathcal{S}$, be the set of $\Phi : \mathbf{C} \to \Re$ such that the above limit exists for each $\phi \in \mathbf{C}$. Define $\mathcal{D}(\mathcal{S})$ as the set of all functions $\Psi : [0, T] \times \mathbf{C} \to \Re$ such that $\Psi(t, \cdot) \in \hat{\mathcal{D}}(\mathcal{S})$, $\forall t \in [0, T]$.

Throughout, let $C_{lip}^{1,2}([0,T] \times \mathbf{C})$ be the space of functions $\Phi : [0,T] \times \mathbf{C} \to \mathbb{R}$ such that $\frac{\partial \Phi}{\partial t} : [0,T] \times \mathbf{C} \to \mathbb{R}$, $D\Phi : [0,T] \times \mathbf{C} \to \mathbf{C}^*$ and $D^2\Phi : [0,T] \times \mathbf{C} \to \mathbf{C}^\dagger$ exist and are continuous and its second-order Fréchet derivative $D^2\Phi$ satisfies the following global Lipschitz condition:

$$\|D^2\Phi(t,\phi) - D^2\Phi(t,\varphi)\|^\dagger \le K\|\phi - \varphi\| \quad \forall t \in [0,T], \ \phi, \varphi \in \mathbf{C}.$$

The following result (see Theorem 2.4.1 in Chapter 2) will be used later on in this chapter to derive the HJBVI for Problem (OSP1).

Theorem 4.2.3 *Suppose that $\Phi \in C_{lip}^{1,2}([0,T] \times \mathbf{C}) \cap \mathcal{D}(\mathcal{S})$ satisfies (i) and (ii). Let $\{x_s(\cdot; t, \psi), s \in [t-r, T]\}$ be the $\mathbf{C}$-valued Markov segment process for (4.1) through the initial data $(t, \varphi_t) \in [0,T] \times \mathbf{C}$. Then*

$$\lim_{\epsilon \downarrow 0} \frac{\mathbf{E}[\Phi(t+\epsilon, x_{t+\epsilon})] - \Phi(t, \psi)}{\epsilon} = \partial_t \Phi(t, \psi) + \mathbf{A}\Phi(t, \psi), \qquad (4.9)$$

where

$$\mathbf{A}\Phi(t, \psi) = \mathcal{S}(\Phi)(t, \psi) + \overline{D\Phi(t, \psi)}(f(t, \psi)\mathbf{1}_{\{0\}})$$
$$+ \frac{1}{2} \sum_{j=1}^{m} \overline{D^2\Phi(t, \psi)}(g(t, \psi)(\mathbf{e}_j)\mathbf{1}_{\{0\}}, g(t, \psi)(\mathbf{e}_j)\mathbf{1}_{\{0\}})$$

and $\mathbf{e}_j, j = 1, 2, \cdots, m$, is the jth unit vector of the standard basis in $\Re^m$.

4.2.2 An Alternate Formulation

For convenience, we propose an alternate formulation of the Problem (OSP1) by considering an optimal stopping problem with only terminal reward function. Problem (OSP1) can be reformulated into this format by absorbing the time variable as well as the integral reward term as follows.

First, we introduce the process $\{(s, x_s, y(s)), s \in [t, T]\}$ with values in $[0, T] \times \mathbf{C} \times \Re$, where

$$y(s) = y + \int_t^s e^{-\alpha(\lambda - t)} L(\lambda, x_\lambda) \, d\lambda, \quad \forall s \in [t, T].$$

The dynamics of the process $\{(s, x_s, y(s)), s \in [t, T]\}$ can be described by the following system of SHDEs:

$$dx(s) = f(s, x_s) \, ds + g(s, x_s) \, dW(s); \qquad (4.10)$$
$$dy(s) = e^{-\alpha(s-t)} L(s, x_s) \, ds, \quad s \in [t, T],$$

with the initial datum at $(t, x_t, y(t)) = (t, \psi, y)$ at the initial time $s = t$.

Lemma 4.2.4 *The process* $\{(s, x_s, y(s)), s \in [t, T]\}$ *is a strong Markov process with respect to the filtration* $\mathbf{G}(t)$.

Proof. It is clear from Theorem 1.5.2 in Chapter 1 that $\{(s, x_s), s \in [t, T]\}$ is a strong Markov $[0, T] \times \mathbf{C}$-valued process. We only need to note that the real-valued process $\{y(s), s \in [t, T]\}$ is strong Markov, because it is sample-path integral of a continuous functional of $\{(s, x_s), s \in [t, T]\}$ and is continuously dependent on its initial datum $y \in \Re$. $\qquad \square$

Let $C^{1,2,1}_{lip}([0, T] \times \mathbf{C} \times \Re)$ denote the space of real-valued functions G defined on $[0, T] \times \mathbf{C} \times \Re$, where $G(t, \phi, y)$ is continuously differentiable with respect to its first variable $t \in [0, T]$, twice continuously Fréchet differentiable with respect to its second variable $\phi \in \mathbf{C}$, continuously differentiable with respect to its third variable $y \in \Re$, and its second Fréchet derivative $D^2 G(t, \phi, y) \in \mathbf{C}^\dagger$ satisfies the following global Lipschitz condition in operator norm $\| \cdot \|^\dagger$: There exists a constant $K > 0$ such that for all $t \in [0, T]$, $y \in \Re$, and $\varphi, \phi \in \mathbf{C}$,

$$\|D^2 G(t, \varphi, y) - D^2 G(t, \phi, y)\|^\dagger \leq K \|\varphi - \phi\|.$$

The weak infinitesimal generator $\tilde{\mathcal{L}}$ for the strong Markov process $\{(s, x_s, y(s)), s \geq 0\}$ acting on any $G \in C^{1,2,1}_{lip}([0, T] \times \mathbf{C} \times \Re) \cap \mathcal{D}(\mathcal{S})$ can be written as

$$\tilde{\mathcal{L}}(G)(t, \psi, y) \equiv \lim_{\epsilon \downarrow 0} \frac{E[G(t + \epsilon, x_{t+\epsilon}, y(t + \epsilon)) - G(t, \psi, y)]}{\epsilon}$$

$$= \lim_{\epsilon \downarrow 0} \frac{E[G(t + \epsilon, x_{t+\epsilon}, y(t + \epsilon)) - G(t, \psi, y(t + \epsilon))]}{\epsilon}$$

$$+ \lim_{\epsilon \downarrow 0} \frac{E[G(t, \psi, y(t + \epsilon)) - G(t, \psi, y)]}{\epsilon}.$$

From (4.9), it is clear that

$$\lim_{\epsilon \downarrow 0} \frac{E[G(t + \epsilon, x_{t+\epsilon}, y(t + \epsilon)) - G(t, \psi, y)]}{\epsilon} = (\partial_t + \mathbf{A}) G(t, \psi, y),$$

since $\lim_{\epsilon \downarrow 0} y(t + \epsilon) = y$. Now, by the mean-valued theorem, the second limit in (4.11) becomes

$$\lim_{\epsilon \downarrow 0} \frac{E[G(t, \psi, y(t + \epsilon)) - G(t, \psi, y)]}{\epsilon}$$

$$= \lim_{\epsilon \downarrow 0} \partial_y G(t, \psi, y + (1 - \lambda)(y(t + \epsilon) - y)) \frac{E[y(t + \epsilon) - y]}{\epsilon} \quad \text{(by the chain rule)}$$

$$= L(t, \psi) \partial_y G(t, \psi, y).$$

Therefore,

$$\tilde{\mathcal{L}} G(t, \psi, y) = (\partial_t + \mathbf{A} + L(t, \psi) \partial_y) G(t, \psi, y), \tag{4.11}$$

where **A** is defined in (4.9).

Define the new terminal reward functional $\Phi : [0, T] \times \mathbf{C} \times \Re \to \Re$ by

$$\Phi(s, \phi, y) = e^{-\alpha(s-t)}\Psi(\phi) + y, \quad \forall (s, \phi, y) \in [0, T] \times \mathbf{C} \times \Re. \qquad (4.12)$$

Remark 4.2.5 *If Φ is the new terminal reward functional defined in (4.12) then*

$$\Phi \in C_{lip}^{1,2,1}([0, T] \times \mathbf{C} \times \Re) \cap \mathcal{D}(\mathcal{S})$$

and

$$\tilde{\mathcal{L}}(\Phi)(s, \psi, y) = \mathbf{A}\Psi(t, \psi) + L(t, \psi) - \alpha\Psi(t, \psi) \qquad (4.13)$$

provided that Ψ, the terminal reward functional of **Problem (OSP1)**, *is smooth enough, i.e., $\Psi \in C_{lip}^2([0, T] \times \mathbf{C}) \cap \mathcal{D}(\mathcal{S})$.*

Second, the reward functional (4.15) can be rewritten as

$$\tilde{J}(t, \psi; \tau) = E^{t,\psi,0}\left[\Phi(\tau, x_\tau, y(\tau))\right], \qquad (4.14)$$

where $\Phi : [0, T] \times \mathbf{C} \times \Re \to$ is the new terminal reward functional that is related to the original terminal reward functional $\Psi : \mathbf{C} \to \Re \to \Re$ through the relation described by (4.12).

We therefore have the following alternate optimal stopping problem with only terminal reward functional.

Problem (OSP3). Given the initial datum $(t, \psi, y) \in [0, T] \times \mathbf{C} \times \Re$, our objective is to find an optimal stopping time $\tau^* \in \mathcal{T}_t^T$ that maximizes the following expected performance index:

$$\tilde{J}(t, \psi, y; \tau) \equiv E^{t,\psi,y}\left[\Phi(\tau, x_\tau, y(\tau))\right] \qquad (4.15)$$

where Φ is the new terminal reward functional defined in (4.12). In this case, the value function $\tilde{V} : [0, T] \times \mathbf{C} \times \Re \to \Re$ is defined to be

$$\tilde{V}(t, \psi, y) \equiv \sup_{\tau \in \mathcal{T}_t^T} \tilde{J}(t, \psi, y; \tau) \qquad (4.16)$$

and the stopping time $\tau^* \in \mathcal{T}_t^T$ such that

$$\tilde{V}(t, \psi, y) = \tilde{J}(t, \psi, y; \tau^*)$$

is called an optimal stopping time.

Remark 4.2.6 *It is clear that Problem (OSP3) is equivalent to Problem (OSP1) in the sense that*

$$V(t, \psi) = \tilde{V}(t, \psi, 0), \quad \forall (t, \psi) \in [0, T] \times \mathbf{C},$$

and the optimal stopping times for both problems coincide. This is because when the initial datum for $\{y(s), s \in [t, T]\}$ is $y(t) = y = 0$ at time $s = t$,

$$\tilde{J}(t, \psi, 0; \tau) = E^{t,\psi,0}[\Phi(\tau, x_\tau, y(\tau)]$$

$$= E^{t,\psi,0}\left[e^{-\alpha(\tau-t)}\Psi(x_\tau) + y(\tau)\right]$$

$$= E^{t,\psi,0}\left[e^{-\alpha(\tau-t)}\Psi(x_\tau) + \int_t^\tau e^{-\alpha(s-t)}L(s, x_s)\,ds\right]$$

$$= J(t, \psi; \tau), \quad \forall \tau \in T_t^T.$$

In the following, we develop the existence of optimal stopping for Problem (OSP3) via excessive function, Snell envelop, and so forth, and parallel to those presented in Øksendal [Øks98] but with infinite-dimensional state space.

First, we introduce the concept of any semicontinuous function as follows. Let Ξ be a metric space and let $F : \Xi \to \Re$ be a Borel measurable function. Then the upper semicontinuous (USC) envelop $\bar{F} : \Xi \to \Re$ and the lower semicontinuous (LSC) envelop $\underline{F} : \Xi \to \Re$ of F are defined respectively by

$$\bar{F}(\mathbf{x}) = \limsup_{\mathbf{y}\to\mathbf{x}, \mathbf{y}\in\Xi} F(\mathbf{y}) \text{ and } \underline{F}(\mathbf{x}) = \liminf_{\mathbf{y}\to\mathbf{x}, \mathbf{y}\in\Xi} F(\mathbf{y}).$$

We let $USC(F)$ and $LSC(F)$ denote the set of USC and LSC functions on Ξ, respectively. Note that, in general, one has

$$\underline{F} \leq F \leq \bar{F},$$

and that F is USC if and only if $F = \bar{F}$, F is LSC if and only if $F = \underline{F}$. In particular, F is continuous if and only if

$$\underline{F} = F = \bar{F}$$

The following lemma will be useful later, but its proof can be found in Wheeden and Zygmound [WZ77] and will be omitted here.

Lemma 4.2.7 *If $\{F^{(k)}\}_{k=1}^\infty \subset LSC(\Xi)$ and $F^{(k)} \uparrow F$ for some $F : \Xi \to \Re$, then $F \in LSU(\Xi)$. If $\{F^{(k)}\}_{k=1}^\infty \subset USC(\Xi)$ and $F^{(k)} \downarrow F$ for some $F : \Xi \to \Re$, then $F \in USC(\Xi)$.*

In connection with Problem (OSP3), we now define the concept of supermeanvalued and superharmonic functions $\Theta : [0, T] \times \mathbf{C} \times \Re \to \Re$ as follows.

Definition 4.2.8 *A measurable function $\Theta : [0, T] \times \mathbf{C} \times \Re \to \Re$ is said to be supermeanvalued with respect to the process $\{(s, x_s, y(s)), s \in [t, T]\}$ if it satisfies*

$$\Theta(t, \psi, y) \geq E^{t,\psi,y}[\Theta(\tau, x_\tau, y(\tau))], \quad \forall(t, \psi, y) \in [0, T] \times \mathbf{C} \times \Re. \quad (4.17)$$

If, in addition, Θ is LSC, then $\Theta : [0, T] \times \mathbf{C} \times \Re \to \Re$ is said to be the following:

1. *a superharmonic function with respect to the process* $\{(s, x_s, y(s)), s \in [t, T]\}$ *if it satisfies*

$$\Theta(t, \psi, y) \geq E^{t,\psi,y}[\Theta(\tau, x_\tau, y(\tau))], \quad \forall(t, \psi, y) \in [0, T] \times \mathbf{C} \times \Re; \quad (4.18)$$

2. *an excessive function with respect to the process* $\{(s, x_s, y(s)), s \in [t, T]\}$ *if* $\forall(t, \psi, y) \in [0, T] \times \mathbf{C} \times \Re$, *and* $s \in [t, T]$, *and it satisfies*

$$\Theta(t, \psi, y) \geq E^{t,\psi,y}[\Theta(s, x_s, y(s))]. \quad (4.19)$$

Remark 4.2.9 *1. If* $\Theta : [0, T] \times \mathbf{C} \times \Re \to \Re$ *is a superharmonic function with respect to the process* $\{(s, x_s, y(s)), s \in [t, T]\}$, *then for any sequence of* $\mathbf{G}(t)$-*stopping times* $\{\tau_k\}_{k=1}^\infty$ *such that* $\lim_{K \to \infty} \tau_k = t$ *we have* $\forall(t, \psi, y) \in [0, T] \times \mathbf{C} \times \Re$

$$\Theta(t, \psi, y) = \lim_{k \to \infty} E^{t,\psi,y}[\Theta(\tau, x_\tau, y(\tau))]. \quad (4.20)$$

This is because of Fatou's lemma and the property of superharmonic functions.

$$\begin{aligned}
\Theta(t, \psi, y) &\leq E^{t,\psi,y}[\underline{\lim}_{k \to \infty} \Theta(\tau_k, x_{\tau_k}, y(\tau_k))] \\
&\leq \underline{\lim}_{k \to \infty} E^{t,\psi,y}[\Theta(\tau_k, x_{\tau_k}, y(\tau_k))] \\
&\leq \Theta(t, \psi, y).
\end{aligned}$$

2. *If* $\Theta \in C_{lip}^{1,2,1}([0, T] \times \mathbf{C} \times \Re) \cap \mathcal{D}(\mathcal{S})$, *then it follows from Dynkin's formula that it is superharmonic with respect to the process* $\{(s, x_s, y(s)), s \in [t, T]\}$ *if and only if*

$$\tilde{\mathcal{L}}\Theta \leq 0,$$

where $\tilde{\mathcal{L}}$ *is the infinitesimal generator given by (4.11).*

3. *It is clear that a superharmonic function is excessive. However, the converse is not as trivial.*

It should be understood in the following lemma that the domain of all functions involved is $[0, T] \times \mathbf{C} \times \Re$ and the words "supermeanvalued" and "superharmonic" are meant to be with respect to the process $\{(s, x_s, y(s)), s \in [t, T]\}$.

Lemma 4.2.10 *1. If* $\{\Theta^{(j)}\}_{j=1}^\infty$ *is a family of supermeanvalued functions, then* $\Theta(t, \psi, y) \equiv \inf_{j \in J}\{\Theta^{(j)}(t, \psi, y)\}$ *is supermeanvalued if it is measurable* $(J$ *is any set)*.

2. *If* $\Theta^{(1)}, \Theta^{(2)}, \dots$ *are superharmonic (supermeanvalued) functions and* $\Theta^{(k)} \uparrow \Theta$ *pointwise, then* Θ *is superharmonic (supermeanvalued)*.

3. *If* Θ *is supermeanvalued and* $\tau \leq \tilde{\tau}$ *are* $\mathbf{G}(t)$-*stopping times, then*

$$E^{t,\psi,y}[\Theta(\tilde{\tau}, x_{\tilde{\tau}}, y(\tilde{\tau}))] \geq E^{t,\psi,y}[\Theta(\tau, x_\tau, y(\tau)))].$$

4. *If Θ is supermeanvalued and $H \in \mathcal{B}([0,T] \times \mathbf{C} \times \Re)$, then $\tilde{\Theta}(t, \psi, y) \equiv$*
   $E^{t,\psi,y}[\Theta(\tau_H, x_{\tau_H}, y(\tau_H))]$ is supermeanvalued, where τ_H is the first exit
   time from H, that is,

$$\tau_H = \inf\{s \geq t \mid (s, x_s, y(s)) \notin H\}.$$

Proof.

1. Suppose $\Theta^{(j)}$ is supermeanvalued for all $j \in J$. Then

$$\Theta^{(j)}(t, \psi, y) \geq E^{t,\psi,y}[\Theta^{(j)}(\tau, x_\tau, y(\tau))]$$
$$\geq E^{t,\psi,y}[\Theta(\tau, x_\tau, y(\tau))], \quad \forall j \in J.$$

So,

$$\Theta(t, \psi, y) = \inf_{j \in J} \Theta^{(j)}(t, \psi, y)$$
$$\geq E^{t,\psi,y}[\Theta(\tau, x_\tau, y(\tau))], \text{ as required.}$$

Therefore Θ is supermeanvalued.

2. Suppose $\Theta^{(k)}$ is supermeanvalued and $\Theta^{(k)} \uparrow \Theta$ pointwise. Then

$$\Theta(t, \psi, y) \geq \Theta^{(k)}(t, \psi, y)$$
$$\geq E^{t,\psi,y}[\Theta^{(k)}(\tau, x_\tau, y(\tau))], \quad \forall k = 1, 2, \ldots.$$

Therefore,

$$\Theta(t, \psi, y) \geq \lim_{k \to \infty} E^{t,\psi,y}[\Theta^{(k)}(\tau, x_\tau, y(\tau))]$$
$$= E^{t,\psi,y}[\Theta(\tau, x_\tau, y(\tau))]$$

by monotone convergence. Therefore, Θ is supermeanvalued. If each $\Theta^{(k)}$ is also LSC, then

$$\Theta^{(k)}(t, \psi, y) = \lim_{j \to \infty} E^{t,\psi,y}[\Theta^{(k)}(\tau_j, x_{\tau_j}, y(\tau_j))]$$
$$\leq \underline{\lim}_{j \to \infty} E^{t,\psi,y}[\Theta(\tau_j, x_{\tau_j}, y(\tau_j))]$$

for any sequence of stopping times $\{\tau_j\}_{j=1}^{\infty}$ that converges to t. Therefore,

$$\Theta(t, \psi, y) \leq \underline{\lim}_{j \to \infty} E^{t,\psi,y}[\Theta(\tau_j, x_{\tau_j}, y(\tau_j))].$$

3. If Θ is supermeanvalued, we have by the Markov property of the process that $\{(s, x_s, y(s)), s \in [t, T]\}$; then for all $t \leq \tilde{s} \leq s \leq T$,

$$E^{t,\psi,y}[\Theta(s, x_s, y(s)) \mid \mathcal{F}(\tilde{s})] = E^{\tilde{s}, x_{\tilde{s}}, y(\tilde{s})}[\Theta(s - \tilde{s}, x_{s-\tilde{s}}, y(s - \tilde{s}))]$$
$$\leq \Theta(\tilde{s}, x_{\tilde{s}}, y(\tilde{s}))] \tag{4.21}$$

that is, the process $\{\Theta(s, x_s, y(s)), s \in [t, T]\}$ is a supermartingale with respect to the filtration $\mathbf{F} = \{\mathcal{F}(s), s \geq 0\}$ generated by the Brownian motion

$W(\cdot) = \{W(s), s \geq t\}$. Therefore, by Doob's optional sampling theorem (see Karatzas and Shreve [KS91, pp19 and 20])

$$E^{t,\psi,y}[\Theta(\tilde{\tau}, x_{\tilde{\tau}}, y(\tilde{\tau}))] \geq E^{t,\psi,y}[\Theta(\tau, x_\tau, y(\tau))]$$

for all stopping times $\tilde{\tau}, \tau \in \mathcal{T}_t^T$ with $\tilde{\tau} \leq \tau$ $P^{t,\psi,y}$-a.s.

4. Suppose Θ is supermeanvalued. By the strong Markov property, Theorem 1.5.2 in Chapter 1, we have for any stopping time $\sigma \in \mathcal{T}_t^T$,

$$\begin{aligned}
E^{t,\psi,y}[\tilde{\Theta}(\sigma, x_\sigma, y(\sigma))] &= E^{t,\psi,y}[E^{\sigma,x_\sigma,y(\sigma)}[\Theta(\tau_H, x_{\tau_H}, y(\tau_H))]] \\
&= E^{t,\psi,y}[E^{t,\psi,y}[\theta(\sigma)\Theta(\tau_H, x_{\tau_H}, y(\tau_H)) \mid \mathcal{G}(t,\sigma)]] \\
&= E^{t,\psi,y}[\theta(\sigma)\Theta(\tau_H, x_{\tau_H}, y(\tau_H))] \\
&= E^{t,\psi,y}[\theta(\sigma)\Theta(\tau_H, x_{\tau_H}, y(\tau_H))] \\
&= E^{t,\psi,y}[\Theta(\tau_H^\sigma, x_{\tau_H^\sigma}, y(\tau_H^\sigma))], \qquad (4.22)
\end{aligned}$$

where $\theta(\sigma)$ is the strong Markov shift operator defined by

$$\theta(\sigma)\Theta(s, x_s, y(s)) = \Theta(s + \sigma, x_{s+\sigma}, y(s + \sigma))$$

and $\tau_H^\sigma = \inf\{s > \sigma \mid (s, x_s, y(s)) \notin H\}$. Since $\tau_H^\sigma \geq \tau_H$, we have by item 3 of this lemma

$$\begin{aligned}
E^{t,\psi,y}[\tilde{\Theta}(\sigma, x_\sigma, y(\sigma))] &\leq E^{t,\psi,y}[\Theta(\tau_H, x_{\tau_H}, y(\tau_H))] \\
&= \tilde{\Theta}(t, \psi, y),
\end{aligned}$$

so $\tilde{\Theta}$ is supermeanvalued.   □

We have the following theorem.

Theorem 4.2.11 *Let $\Theta : [0,T] \times \mathbf{C} \times \Re \to \Re$. Then Θ is a superharmonic function with respect to the process $\{(s, x_s, y(s)), s \in [t,T]\}$ if and only if it is excessive with respect to the process $\{(s, x_s, y(s)), s \in [t,T]\}$.*

Proof in a Special Case. We only prove the theorem in the special case when $\Theta \in C_{lip}^{1,2,1}([0,T] \times \mathbf{C} \times \Re) \cap \mathcal{D}(\mathcal{S})$. By Dynkin's formula (see Theorem 2.4.1), we have

$$E^{t,\psi,y}[\Theta(s, x_s, y(s))] = \Theta(t, \psi, y) + E^{t,\psi,y}\left[\int_t^s \tilde{\mathcal{L}}(\Theta)(\lambda, x_\lambda, y(\lambda))d\lambda\right]$$

$$\forall (t, \psi, y) \in [0,T] \times \mathbf{C} \times \Re, \text{ and } s \in [t,T].$$

Therefore, if Θ is excessive, then $\tilde{\mathcal{L}}(\Theta) \leq 0$. Consequently, if $\tau \in \mathcal{T}_t^T$, we get $\forall (t, \psi, y) \in [0,T] \times \mathbf{C} \times \Re$, and $s \in [t,T]$,

$$E^{t,\psi,y}[\Theta(s \wedge \tau, x_{s\wedge\tau}, y(s \wedge \tau))] \leq \Theta(t, \psi, y).$$

Letting $s \uparrow T$, we see that Θ is superharmonic. A proof in the general case can be found in Dynkin [Dyn65] and is omitted here.

Definition 4.2.12 *If $\Theta : [0,T] \times \mathbf{C} \times \Re \to \Re$ is superharmonic (supermean-valued) and if $\tilde{\Theta} : [0,T] \times \mathbf{C} \times \Re \to \Re$ is also superharmonic (supermeanvalued) and $\tilde{\Theta} \geq \Theta$, then $\tilde{\Theta}$ is said to be a superharmonic (supermeanvalued) majorant of Θ. The function $\hat{\Theta} : [0,T] \times \mathbf{C} \times \Re \to \Re$ is said to be the least superharmonic (supermeanvalued) majorant of Θ if*

(i) $\hat{\Theta}$ is a superharmonic (supermeanvalued) majorant of Θ and
(ii) if $\check{\Theta}$ is another superharmonic (supermeanvalued) majorant of Θ, then $\check{\Theta} \geq \hat{\Theta}$.

Given a LSC function $\Theta : [0,T] \times \mathbf{C} \times \Re \to \Re$. The following result provides an iterative construction of the least superharmonic majorant of Θ.

Theorem 4.2.13 (Construction of the Least Superharmonic Majortant) *Let $\Theta = \Theta^{(0)}$ be a non-negative, LSC function on $[0,T] \times \mathbf{C} \times \Re$ and define inductively for $k = 1, 2, \dots$*

$$\Theta^{(k)}(t,\psi,y) = \sup_{s \in S^{(k)}} E^{t,\psi,y}[\Theta^{(k-1)}(s,x_s,y(s))], \qquad (4.23)$$

where $S^{(k)} = \{t + j \cdot 2^{-k} \mid 0 \leq j \leq 4^k\} \wedge T$, $k = 1, 2, \dots$. Then $\Theta^{(k)} \uparrow \hat{\Theta}$, where $\hat{\Theta}$ is the least superharmonic majorant of Θ and $\hat{\Theta} = \bar{\Theta}$, where $\bar{\Theta}$ is the least superexcessive majorant of Θ.

Proof. Note that $\{\Theta^{(k)}\}$ is increasing. Define $\check{\Theta}(t,\psi,y) = \lim_{k \to \infty} \Theta^{(k)}(t,\psi,y)$. Then

$$\check{\Theta}(t,\psi,y) \geq \Theta^{(k)}(t,\psi,y) \geq E^{t,\psi,y}[\Theta^{(k-1)}(s,x_s,y(s))], \quad \forall k \text{ and } \forall s \in S^{(k)}.$$

Hence, $\forall s \in S \equiv \cup_{k=1}^{\infty} S^{(k)}$,

$$\begin{aligned} \check{\Theta}(t,\psi,y) &\geq \lim_{k \to \infty} E^{t,\psi,y}[\Theta^{(k-1)}(s,x_s,y(s))] \\ &\geq E^{t,\psi,y}[\lim_{k \to \infty} \Theta^{(k-1)}(s,x_s,y(s))] \\ &= E^{t,\psi,y}[\check{\Theta}(s,x_s,y(s))]. \end{aligned} \qquad (4.24)$$

Since $\check{\Theta}$ is an increasing limit of LSC functions (see Lemma 4.2.7), $\check{\Theta}$ is LSC. Fix $s \in [t,T]$ and choose $t_k \in S$ such that $t_k \to s$. Then by (4.24), Fatou's lemma, and LSC,

$$\begin{aligned} \check{\Theta}(t,\psi,y) &\geq \underline{\lim}_{k \to \infty} E^{t,\psi,y}[\check{\Theta}(t_k,x_{t_k},y(t_k))] \\ &\geq E^{t,\psi,y}[\underline{\lim}_{k \to \infty} \check{\Theta}(t_k,x_{t_k},y(t_k))] \\ &\geq E^{t,\psi,y}[\check{\Theta}(s,x_s,y(s))]. \end{aligned}$$

So, $\check{\Theta}$ is excessive. Therefore, $\check{\Theta}$ is superharmonic by Theorem 4.2.11 and, hence, $\check{\Theta}$ is a superharmonic majorant of Θ. On the other hand, if $\tilde{\Theta}$ is any superharmonic majorant of Θ, then clearly, by induction,

$$\tilde{\Theta}(t,\psi,y) \geq \Theta^{(k)}(t,\psi,y), \quad \forall k = 1, 2, \dots,$$

and so $\tilde{\Theta}(t,\psi,y) \geq \check{\Theta}(t,\psi,y)$; this shows that $\check{\Theta}$ is the least supermeanvalued majorant $\bar{\Theta}$ of Θ. So, $\hat{\Theta} = \bar{\Theta}$.   □

4.2.3 Existence and Uniqueness

In this subsection, we establish the existence and uniqueness of the optimal stopping for Problem (OSP3). In the following, we let $\Phi : [0,T] \times \mathbf{C} \times \Re \to \Re$ be the terminal reward functional for Problem (OSP3) defined by (4.12). As a reminder,

$$\Phi(s,\phi,y) = e^{-\alpha(s-t)}\Psi(\phi) + y, \quad \forall(s,\phi,y) \in [0,T] \times \mathbf{C} \times \Re,$$

where $\Psi : \mathbf{C} \to \Re$ is the terminal reward functional of Problem (OSP1).

Theorem 4.2.14 (Existence Problem (OSP3)) *Let $\tilde{V} : [0,T] \times \mathbf{C} \times \Re \to \Re$ be the value function of Problem (OSP3) and $\hat{\Phi}$ be the least superharmonic majorant of a continuous terminal reward function $\Phi \geq 0$ of Problem (OSP3).*

1. *Then*

$$\tilde{V}(t,\psi,y) = \hat{\Phi}(t,\psi,y), \quad \forall(t,\psi,y) \in [0,T] \times \mathbf{C} \times \Re. \tag{4.25}$$

2. *For $\epsilon > 0$, let*

$$\mathbf{D}_\epsilon = \{(t,\psi,y) \in [0,T] \times \mathbf{C} \times \Re \mid \Phi(t,\psi,y) < \hat{\Phi}(t,\psi,y) - \epsilon\}. \tag{4.26}$$

Suppose Φ is bounded. Then stopping at the first time τ_ϵ of exit from $\mathbf{D}_\epsilon$, that is,

$$\tau_\epsilon = \inf\{s \geq t \mid (s, x_s, y(s)) \notin \mathbf{D}_\epsilon\},$$

is close to being optimal in the sense that $\forall(t,\psi,y) \in [0,T] \times \mathbf{C} \times \Re$,

$$|\tilde{V}(t,\psi,y) - E^{t,\psi,y}[\Phi(\tau_\epsilon, x_{\tau_\epsilon}, y(\tau_\epsilon))]| \leq 2\epsilon. \tag{4.27}$$

3. *For a continuous terminal reward functional $\Phi \geq 0$, let the continuation region $\mathbf{D}$ be defined as*

$$\mathbf{D} = \{(t,\psi,y) \in [0,T] \times \mathbf{C} \times \Re \mid \Phi(t,\psi,y) < \tilde{V}(t,\psi,y)\}. \tag{4.28}$$

For $N = 1, 2, \ldots$, define $\Phi_N = \Phi \wedge N$,

$$\mathbf{D}_N = \{(t,\psi,y) \in [0,T] \times \mathbf{C} \times \Re \mid \Phi_N(t,\psi,y) < \hat{\Phi}_N(t,\psi,y)\},$$

and $\tau_N = \tau_{\mathbf{D}_N}$. Then

$$\mathbf{D}_N \subset \mathbf{D}_{N+1}, \quad \mathbf{D}_N \subset \mathbf{D} \cap \Phi^{-1}([0,N]), \quad \mathbf{D} = \cup_N \mathbf{D}_N.$$

If $\tau_N < \infty$, $P^{t,\psi,y}$-a.s for all N, then

$$\tilde{V}(t,\psi,y) = \lim_{N\to\infty} E^{t,\psi,y}[\Phi(\tau_N, x_{\tau_N}, y(\tau_N))]. \tag{4.29}$$

4. *In particular, if $\tau_{\mathbf{D}} < \infty$, $P^{t,\psi,y}$-a.s. and the family $\Phi(\tau_N, x_{\tau_N}, y(\tau_N)))\}_{N=1}^{\infty}$ is uniformly integrable with respect to $P^{t,\psi,y}$, then*

$$\tilde{V}(t,\psi,y) = E^{t,\psi,y}[\Phi(\tau_{\mathbf{D}}, x_{\tau_{\mathbf{D}}}, y(\tau_{\mathbf{D}}))]$$

is the value function for Problem (OSP3) and $\tau^* = \tau_{\mathbf{D}}$ is an optimal stopping time for Problem (OSP3).

Proof. First, assume that Φ is bounded and define

$$\Phi_{\epsilon}(t,\psi,y) = E^{t,\psi,y}[\hat{\Phi}(\tau_{\epsilon}, x_{\tau_{\epsilon}}, y(\tau_{\epsilon}))], \quad \epsilon > 0. \tag{4.30}$$

Then Φ_{ϵ} is supermeanvalued by item 4 of Lemma 4.2.10. We claim that

$$\Phi(t,\psi,y) \leq \Phi_{\epsilon}(t,\psi,y) + \epsilon, \quad \forall(t,\psi,y). \tag{4.31}$$

We prove the claim by contradiction. Suppose

$$\beta \equiv \sup_{(t,\psi,y)\in[0,T]\times\mathbf{C}\times\Re} \{\Phi(t,\psi,y) - \Phi_{\epsilon}(t,\psi,y)\} > \epsilon. \tag{4.32}$$

Then, for all $\eta > 0$, we can find $(\bar{t}, \bar{\psi}, \bar{y})$ such that

$$\Phi(\bar{t}, \bar{\psi}, \bar{y}) - \Phi_{\epsilon}(\bar{t}, \bar{\psi}, \bar{y}) \geq \beta - \eta. \tag{4.33}$$

On the other hand, since $\Phi_{\epsilon} + \beta$ is a supermeanvalued majorant of Φ, we have

$$\hat{\Phi}(\bar{t}, \bar{\psi}, \bar{y}) \leq \Phi_{\epsilon}(\bar{t}, \bar{\psi}, \bar{y}) + \beta. \tag{4.34}$$

Combining (4.33) and (4.34), we get

$$\hat{\Phi}(\bar{t}, \bar{\psi}, \bar{y}) \leq \Phi(\bar{t}, \bar{\psi}, \bar{y}) + \eta. \tag{4.35}$$

Consider the following two possible cases.
Case 1. $\tau_{\epsilon} > 0$ $P^{\bar{t},\bar{\psi},\bar{y}}$-a.s. Then by (4.35) and the definition of $\mathbf{D}_{\epsilon}$,

$$\begin{aligned}
\Phi(\bar{t}, \bar{\psi}, \bar{y}) &\geq \hat{\Phi}(\bar{t}, \bar{\psi}, \bar{y}) \\
&\geq E^{\bar{t},\bar{\psi},\bar{y}}[\hat{\Phi}(s \wedge \tau_{\epsilon}, x_{s\wedge\tau_{\epsilon}}, y(s \wedge \tau_{\epsilon}))] \\
&\geq E^{\bar{t},\bar{\psi},\bar{y}}[(\Phi(s, x_s, y(s)) + \epsilon)\chi_{\{s<\tau_{\epsilon}\}}], \quad \forall s \in [t,T].
\end{aligned}$$

Hence, by the Fatou lemma and LSC of Φ,

$$\begin{aligned}
\Phi(\bar{t}, \bar{\psi}, \bar{y}) + \eta &\geq \underline{\lim}_{s\downarrow t} E^{\bar{t},\bar{\psi},\bar{y}}[(\Phi(s, x_s, y(s)) + \epsilon)\chi_{\{s<\tau_{\epsilon}\}}] \\
&\geq E^{\bar{t},\bar{\psi},\bar{y}}[\underline{\lim}_{s\downarrow t}(\Phi(s, x_s, y(s)) + \epsilon)\chi_{\{s<\tau_{\epsilon}\}}] \\
&\geq \Phi(\bar{t}, \bar{\psi}, \bar{y}) + \epsilon.
\end{aligned}$$

This is a contradiction, since $\eta < \epsilon$.
Case 2. $\tau_{\epsilon} = 0$ $P^{\bar{t},\bar{\psi},\bar{y}}$-a.s. Then

$$\Phi_\epsilon(\bar{t}, \bar{\psi}, \bar{y}) = \hat{\Phi}(\bar{t}, \bar{\psi}, \bar{y}),$$

so

$$\Phi(\bar{t}, \bar{\psi}, \bar{y}) \le \Phi_\epsilon(\bar{t}, \bar{\psi}, \bar{y}),$$

contradicting (4.33) for $\eta \le \beta$. Therefore, (4.32) leads to a contradiction. Thus, (4.31) is proved and we conclude that $\Phi_\epsilon + \epsilon$ is a supermeanvalued majorant of Φ. Therefore,

$$
\begin{aligned}
\hat{\Phi}(t, \psi, y) &\le \Phi_\epsilon(t, \psi, y) + \epsilon \\
&\le E^{t,\psi,y}[(\Phi + \epsilon)(\tau_\epsilon, x_{\tau_\epsilon}, y(\tau_\epsilon))] + \epsilon \\
&\le \tilde{V}(t, \psi, y) + 2\epsilon,
\end{aligned}
\tag{4.36}
$$

and since ϵ is arbitrary, we have, by the fact that $\hat{\Phi} \ge \tilde{V}$,

$$\hat{\Phi} = \tilde{V}.$$

If Φ is not bounded, let

$$\Phi_N = \Phi \wedge N, \quad N = 1, 2, \dots$$

and as before, let $\hat{\Phi}_N$ be the superharmonic majorant of Φ_N. Then,

$$\tilde{V} \ge \tilde{V}_N = \hat{\Phi}_N \uparrow h \quad \text{as } N \to \infty, \text{ where } h \ge \hat{\Phi}$$

since h is a superharmonic majorant of Φ. Thus, $h = \hat{\Phi} = \tilde{V}$ and this proves part 1, $\tilde{V} = \hat{g}$ for general Φ. From (4.36) and part 1, $\tilde{V} = \hat{g}$, we obtain part 2, that is,

$$|\tilde{V}(t, \psi, y) - E^{t,\psi,y}[\Phi(\tau_\epsilon, x_{\tau_\epsilon}, y(\tau_\epsilon))]| \le 2\epsilon, \quad \forall (t, \psi, y) \in [0, T] \times \mathbf{C} \times \Re.$$

To obtain parts 3 and 4, let us again first assume that Φ is bounded. Then, since

$$\tau_\epsilon \uparrow \tau_{\mathbf{D}} \quad \text{as} \quad \epsilon \downarrow 0$$

and $\tau_{\mathbf{D}} < \infty$ $P^{t,\psi,y}$-a.s., we have

$$\lim_{\epsilon \downarrow} E^{t,\psi,y}[\Phi(\tau_\epsilon, x_{\tau_\epsilon}, y(\tau_\epsilon))] = E^{t,\psi,y}[\Phi(\tau_{\mathbf{D}}, x_{\tau_{\mathbf{D}}}, y(\tau_{\mathbf{D}}))], \tag{4.37}$$

and, hence, by (4.36) and part 1,

$$\tilde{V}(t, \psi, y) = E^{t,\psi,y}[\Phi(\tau_{\mathbf{D}}, x_{\tau_{\mathbf{D}}}, y(\tau_{\mathbf{D}}))] \quad \text{if } \Phi \text{ is bounded.} \tag{4.38}$$

Finally, if Φ is not bounded, define

$$h = \lim_{N \to \infty} \hat{\Phi}.$$

Then h is superharmonic, and since $\hat{\Phi}_N \le \hat{\Phi}$ for all N, we have $h \le \hat{\Phi}$. On the other hand, $\Phi_N \le \hat{\Phi}_N \le h$ for all N and, therefore, $\Phi \le h$. Since $\hat{\Phi}$ is the least superharmonic majorant of Φ, we conclude that

$$h = \hat{\Phi}. \tag{4.39}$$

Hence, by (4.39), we obtain part 3:

$$\begin{aligned}
\tilde{V}(t, \psi, y) &= \lim_{N \to \infty} \hat{\Phi}_N(t, \psi, y) \\
&= \lim_{N \to \infty} E^{t, \psi, y}[\Phi_N(\tau_N, x_{\tau_N}, y(\tau_N))] \\
&= \lim_{N \to \infty} E^{t, \psi, y}[\Phi(\tau_N, x_{\tau_N}, y(\tau_N))] \\
&\le \tilde{V}(t, \psi, y).
\end{aligned}$$

Note that $\hat{\Phi}_N \le N$ everywhere, so if $\Phi_N(t, \psi, y) < \hat{\Phi}_N(t, \psi, y)$, then $\Phi_N(t, \psi, y) < N$ and, therefore,

$$\Phi(t, \psi, y) = \Phi_N(t, \psi, y) < \hat{\Phi}_N(t, \psi, y) \le \hat{\Phi}(t, \psi, y)$$

and

$$\Phi_{N+1}(t, \psi, y) = \Phi_N(t, \psi, y) < \hat{\Phi}_N(t, \psi, y) \le \hat{\Phi_{N+1}}(t, \psi, y).$$

Hence, $\forall N = 1, 2, \ldots,$

$$\mathbf{D}_N \subset \mathbf{D} \cap \{(t, \psi, y) \mid \Phi(t, \psi, y) < N\} \text{ and } \mathbf{D}_N \subset \mathbf{D}_{N+1}.$$

So, by (4.39), we conclude that $\mathbf{D}$ is the increasing union of the sets $\mathbf{D}_N$; $N = 1, 2, \ldots$. Therefore,

$$\tau_{\mathbf{D}} = \lim_{N \to \infty} \tau_N.$$

So, by part 3,

$$\tilde{V}(t, \psi, y) = \lim_{N \to \infty} E^{t, \psi, y}[\Phi(\tau_N, x_{\tau_N}, y(\tau_N))],$$

and uniform integrability, we have

$$\begin{aligned}
\tilde{V}(t, \psi, y) &= \lim_{N \to \infty} \hat{\Phi}_N(t, \psi, y) \\
&= \lim_{N \to \infty} E^{t, \psi, y}[\Phi_N(\tau_N, x_{\tau_N}, y(\tau_N))] \\
&= E^{t, \psi, y}[\lim_{N \to \infty} \Phi_N(\tau_N, x_{\tau_N}, y(\tau_N))] \\
&= E^{t, \psi, y}[\Phi(\tau_{\mathbf{D}}, x_{\tau_{\mathbf{D}}}, y(\tau_{\mathbf{D}}))],
\end{aligned}$$

and, therefore, the proof of Theorem 4.2.14 is complete.   □

Remark 4.2.15 *1. Note that the sets* $\mathbf{D}$, $\mathbf{D}_\epsilon$, *and* $\mathbf{D}_N$ *are open, since* $\hat{\Phi} = \tilde{V}$ *is LSC and* Φ *is continuous.*
2. By inspecting the proof, we see that part 1 of Theorem 4.2.14 holds under the weaker assumption that $\Phi \ge 0$ *is LSC.*

The following consequence of Theorem 4.2.14 is often useful.

Corollary 4.2.16 *Suppose there exists a $H \in \mathcal{B}([0,T] \times \mathbf{C} \times \Re)$ such that*

$$\Phi_H(t, \psi, y) \equiv E^{t,\psi,y}[\Phi(\tau_H, x_{\tau_H}, y(\tau_H))]$$

is a supermeanvalued majorant of Φ. Then

$$\tilde{V}(t, \psi, y) = \Phi_H(t, \psi, y), \quad so \ \tau^* = \tau_H \ is \ optimal.$$

Proof. If Φ_H is supermeanvalued majorant of Φ, then, clearly,

$$\bar{\Phi}(t, \psi, y) \leq \Phi_H(t, \psi, y),$$

where $\bar{\Phi}$ is the least supermeanvalued majorant of Φ. One the other hand, we have

$$
\begin{aligned}
\Phi_H(t, \psi, y) &\leq \sup_{\tau \in \mathcal{T}_t^T} E^{t,\psi,y}[\Phi(\tau, x_\tau, y(\tau))] \\
&= \sup_{\tau \in \mathcal{T}_t^T} \tilde{J}(t, \psi, y; \tau) \\
&= \tilde{V}(t, \psi, y),
\end{aligned}
$$

so $\tilde{V} = \Phi_H$ by Theorem 4.2.13 and part 1 of Theorem 4.2.14.   □

Corollary 4.2.17 *Let*

$$\mathbf{D} = \{(t, \psi, y) \mid \Phi(t, \psi, y) < \bar{\Phi}(t, \psi, y)\}$$

and put

$$\Phi_{\mathbf{D}}(t, \psi, y) \equiv E^{t,\psi,y}[\Phi(\tau_{\mathbf{D}}, x_{\tau_{\mathbf{D}}}, y(\tau_{\mathbf{D}}))].$$

If $\Phi_{\mathbf{D}} \geq \Phi$, then $\Phi_{\mathbf{D}} = \tilde{V}$.

Proof. Since $(\tau_{\mathbf{D}}, x_{\tau_{\mathbf{D}}}, y(\tau_{\mathbf{D}})) \notin \mathbf{D}$, we have

$$\Phi(\tau_{\mathbf{D}}, x_{\tau_{\mathbf{D}}}, y(\tau_{\mathbf{D}})) \geq \hat{\Phi}(\tau_{\mathbf{D}}, x_{\tau_{\mathbf{D}}}, y(\tau_{\mathbf{D}}))$$

and, therefore,

$$\Phi(\tau_{\mathbf{D}}, x_{\tau_{\mathbf{D}}}, y(\tau_{\mathbf{D}})) = \hat{\Phi}(\tau_{\mathbf{D}}, x_{\tau_{\mathbf{D}}}, y(\tau_{\mathbf{D}})), \quad P^{t,\psi,y}\text{-a.s.}$$

So,

$$\Phi_{\mathbf{D}}(t, \psi, y) = E^{t,\psi,y}[\hat{\Phi}(\tau_{\mathbf{D}}, x_{\tau_{\mathbf{D}}}, y(\tau_{\mathbf{D}}))]$$

is supermeanvalued since $\hat{\Phi}$ is, and the result follows from Corollary 4.2.16.
□

Theorem 4.2.18 (Uniqueness for Problem (OSP3)) *Define the continuation region $\mathbf{D}$ as earlier*

$$\mathbf{D} = \{(t, \psi, y) \in [0, T] \times \mathbf{C} \times \Re \mid \Phi(t, \psi, y) < \tilde{V}(t, \psi, y)\}.$$

Suppose there exists an optimal stopping time $\tau^* = \tau^*(t, \psi, y, \omega)$ for the Problem (OSP3) for all $(t, \psi, y) \in [0, T] \times \mathbf{C} \times \Re$. Then

$$\tau^* \geq \tau_{\mathbf{D}}, \quad \forall (t, \psi, y) \in [0, T] \times \mathbf{C} \times \Re, \qquad (4.40)$$

and

$$\tilde{V}(t, \psi, y) = E^{t,\psi,y}[\Phi(\tau_{\mathbf{D}}, x_{\tau_{\mathbf{D}}}, y(\tau_{\mathbf{D}}))], \quad \forall (t, \psi, y) \in [0, T] \times \mathbf{C} \times \Re. \quad (4.41)$$

Hence, $\tau_{\mathbf{D}}$ is an optimal stopping time for the Problem (OSP3).

Proof. Choose $(t, \psi, y) \in \mathbf{D}$. Let $\tau \in \mathcal{T}_t^T$ and assume $P^{t,\psi,y}[\tau < \tau_{\mathbf{D}}] > 0$. Since $\Phi(\tau, x_\tau, y(\tau)) < \tilde{V}(\tau, x_\tau, y(\tau))$ if $\tau < \tau_{\mathbf{D}}$ and $\Phi \leq V$ always, we have

$$
\begin{aligned}
E^{t,\psi,y}[\Phi(\tau, x_\tau, y(\tau))] &= \int_{\{\tau < \tau_{\mathbf{D}}\}} \Phi(\tau, x_\tau, y(\tau)) \, dP^{t,\psi,y} \\
&\quad + \int_{\{\tau \geq \tau_{\mathbf{D}}\}} \Phi(\tau, x_\tau, y(\tau)) \, dP^{t,\psi,y} \\
&< \int_{\{\tau < \tau_{\mathbf{D}}\}} \tilde{V}(\tau, x_\tau, y(\tau)) \, dP^{t,\psi,y} \\
&\quad \int_{\{\tau \geq \tau_{\mathbf{D}}\}} \tilde{V}(\tau, x_\tau, y(\tau)) \, dP^{t,\psi,y} \\
&= E^{t,\psi,y}[\tilde{V}(\tau, x_\tau, y(\tau))] \\
&\leq \tilde{V}(t, \psi, y),
\end{aligned}
$$

since $\tilde{V}$ is superharmonic. This proves (4.40).

To obtain (4.41), we first choose $(t, \psi, y) \in \mathbf{D}$. Since $\hat{\Phi}$ is superharmonic, we have by (4.40) and part 3 of Lemma 4.2.10,

$$
\begin{aligned}
\tilde{V}(t, \psi, y) &= E^{t,\psi,y}[\Phi(\tau^*, x_{\tau^*}, y(\tau^*))] \\
&\leq E^{t,\psi,y}[\hat{\Phi}(\tau^*, x_{\tau^*}, y(\tau^*))] \\
&\leq E^{t,\psi,y}[\hat{\Phi}(\tau_{\mathbf{D}}, x_{\tau_{\mathbf{D}}}, y(\tau_{\mathbf{D}}))] \\
&= E^{t,\psi,y}[\Phi(\tau^*, x_{\tau^*}, y(\tau^*))] \\
&\leq \tilde{V}(t, \psi, y),
\end{aligned}
$$

which proves (4.41) for $(t, \psi, y) \in \mathbf{D}$.

Next, choose $(t, \psi, y) \in \partial \mathbf{D}$ to be an *irregular* boundary point of $\mathbf{D}$. Then $\tau_{\mathbf{D}} > 0$, $P^{t,\psi,y}$-a.s.. Let $\{\tau_k\}_{k=1}^\infty$ be a sequence of stopping times such that $0 < \tau_k < \tau_{\mathbf{D}}$ and $\tau_k \to 0$ $P^{t,\psi}$-a.s., as $k \to \infty$. Then $(\tau_k, x_{\tau_k}, y(\tau_k)) \in \mathbf{D}$, so by (4.40), and the strong Markov Theorem 1.5.2, we have for all $k = 1, 2, \ldots$,

$$
\begin{aligned}
E^{t,\psi,y}[\Phi(\tau_{\mathbf{D}}, x_{\tau_{\mathbf{D}}}, y(\tau_{\mathbf{D}}))] &= E^{t,\psi,y}[\theta(\tau_k)\tilde{\Psi}(\tau_{\mathbf{D}}, x_{\tau_{\mathbf{D}}}, y(\tau_{\mathbf{D}}))] \\
&= E^{t,\psi,y}[E^{\tau_k, x_{\tau_k}, y(\tau_k)}[\Phi(\tau_{\mathbf{D}}, x_{\tau_{\mathbf{D}}}, y(\tau_{\mathbf{D}}))]] \\
&= E^{t,\psi,y}[\tilde{V}(\tau_{\mathbf{D}}, x_{\tau_{\mathbf{D}}}, y(\tau_{\mathbf{D}}))].
\end{aligned}
$$

Hence, by lower semicontinuity and the Fatou lemma,

$$\tilde{V}(t, \psi, y) \leq E^{t,\psi,y}[\underline{\lim}_{k\to\infty}\tilde{V}(\tau_k, x_{\tau_k}, y(\tau_k))]$$
$$\leq \underline{\lim}_{k\to\infty}E^{t,\psi,y}[\tilde{V}(\tau_k, x_{\tau_k}, y(\tau_k))]$$
$$= E^{t,\psi,y}[\Phi(\tau_{\mathbf{D}}, x_{\tau_{\mathbf{D}}}, y(\tau_{\mathbf{D}}))].$$

Finally, if $(t, \psi, y) \in \partial\mathbf{D}$ is a *regular* boundary point of $\mathbf{D}$ or if $(t, \psi, y) \notin \bar{\mathbf{D}}$, we have $\tau_{\mathbf{D}} = 0$, $P^{t,\psi}$-a.s. and, hence, $\tilde{V}(t, \psi, y) = E^{t,\psi,y}[\Phi(\tau_{\mathbf{D}}, x_{\tau_{\mathbf{D}}}, y(\tau_{\mathbf{D}}))]$.
□

Remark 4.2.19 *The following observation is sometimes useful: Assume* $\Phi \in C^{1,2,1}_{lip} \cap \mathcal{D}(\mathcal{S})$. *Define*

$$\mathbf{U} = \{(t, \psi, y) \in [0, T] \times \mathbf{C} \times \Re \mid \mathcal{A}\Phi(t, \psi, y) > 0\}. \tag{4.42}$$

Then, with $\mathbf{D}$ being defined in the previous theorem, we have

$$\mathbf{U} \subset \mathbf{D}. \tag{4.43}$$

Consequently, from (4.40) we conclude that it is never optimal to stop the process before it exits from $\mathbf{U}$.

Combining the above, Theorem 4.2.14, and the relationship between Problem (OSP1) and Problem (OSP3), we have the following existence and uniqueness result for the optimal stopping of Problem (OSP1).

Theorem 4.2.20 *Assume Assumption 4.1.2 holds and L and that Ψ are continuous functions on $[0, T] \times \mathbf{C}$ and $\mathbf{C}$, respectively. Then for each initial datum $(t, \psi) \in [0, T] \times \mathbf{C}$, the value function $V(t, \psi)$ and the optimal stopping time $\tau^*(t, \psi)$ exist and are unique.*

4.3 HJB Variational Inequality

Heuristic Derivation. We consider the following HJBVI:

$$\max\{\Psi - V, \ \partial_t V + \mathbf{A}V + L - \alpha V\} = 0 \tag{4.44}$$

on $[0, T] \times \mathbf{C}$, where

$$\mathbf{A}V(t, \psi) \equiv \mathcal{S}(V)(t, \psi) + \overline{DV(t, \psi)}(f(t, \psi)\mathbf{1}_{\{0\}})$$
$$+ \frac{1}{2}\sum_{i=1}^{m}\overline{D^2V(t, \psi)}(g(t, \psi)\mathbf{e}_i\mathbf{1}_{\{0\}}, g(t, \psi)\mathbf{e}_i\mathbf{1}_{\{0\}}). \tag{4.45}$$

The above inequality will be interpreted as follows:

$$\partial_t V + \mathbf{A}V + L - \alpha V \leq 0 \quad \text{and} \quad V \geq \Psi \tag{4.46}$$

and

$$(\partial_t V + \mathbf{A}V + L - \alpha V)(V - \Psi) = 0 \qquad (4.47)$$

on $[0,T] \times \mathbf{C}$, where $\mathbf{A}$ is defined by (4.45).

We heuristically derive the above variational inequality as follows. A rigorous derivation will be provided as a by-product when we prove that the value function is a viscosity solution of HJBVI (4.44) in the next section. First, we prove that $V(t,\psi) \geq \Psi(t,\psi)$ for all $(t,\psi) \in [0,T] \times \mathbf{C}$. Let $\tilde{\tau} \in \mathcal{T}_t^T$. If $\tilde{\tau} = t$, then by (4.15), we can get

$$J(t,\psi;\tilde{\tau}) = \Psi(t,\psi).$$

Therefore,

$$V(t,\psi) = \sup_{\tau \in \mathcal{T}_t^T} J(t,\psi;\tau) \geq J(t,\psi;\tilde{\tau}) = \Psi(t,\psi). \qquad (4.48)$$

The following dynamic programming principle (see Theorem 3.3.9 in Chapter 3) can and will be used to derive our HJBVI (4.44):

$$V(t,\psi) \geq E\left[\int_t^{t+\delta} e^{-\alpha(s-t)} L(s,x_s)\, ds + e^{-\alpha\delta} V(t+\delta, x_{t+\delta})\right], \ \forall \delta \geq 0.$$

From this principle, we have

$$\lim_{\delta \downarrow 0} \frac{E[e^{-\alpha\delta} V(t+\delta, x_{t+\delta}) - V(t,\psi)]}{\delta}$$

$$= \lim_{\delta \downarrow 0} \frac{E[e^{-\alpha\delta} V(t+\delta, x_{t+\delta}) - V(t+\delta, x_{t+\delta})]}{\delta}$$

$$+ \lim_{\delta \downarrow 0} \frac{E[V(t+\delta, x_{t+\delta}) - V(t,\psi)]}{\delta}$$

$$= -\alpha V(t,\psi) + \partial_t V(t,\psi) + \mathbf{A}V(t,\psi) + L(t,\psi)$$

$$\leq 0$$

for all $(t,\psi) \in [0,T] \times \mathbf{C}$ provided that $V \in C_{lip}^{1,2}([0,T] \times \mathbf{C}) \cap \mathcal{D}(\mathcal{S})$; that is,

$$\alpha V - \partial_t V - \mathbf{A}V - L \geq 0. \qquad (4.49)$$

From (4.48) and (4.49), it follows that

$$\min\{V - \Psi, \ \alpha V - \partial_t V - \mathbf{A}V - L\} \geq 0 \qquad (4.50)$$

on $[0,T] \times \mathbf{C}$. The derivation of the inequality

$$\min\{V - \Psi, \ \alpha V - \partial_t V - \mathbf{A}V - L\} \leq 0 \qquad (4.51)$$

can be found in the next section.

We therefore have the following result.

Theorem 4.3.1 *Suppose* $V : [0,T] \times \mathbf{C} \to \Re$ *is the value function of Problem (OSP1) and it satisfies the smoothness condition* $V \in C^{1,2}_{lip}([0,T] \times \mathbf{C}) \cap \mathcal{D}(\mathcal{S})$. *Then the value function* V *satisfies the following HJBVI:*

$$\max\{\Psi - V, \ \partial_t V + \mathbf{A}V + L - \alpha V\} = 0 \qquad (4.52)$$

on $[0,T] \times \mathbf{C}$, *and* $V(T,\psi) = \Psi(\psi)$, $\forall \psi \in \mathbf{C}$.

Note that it is not known that the value function V satisfies the smoothness conditions mentioned in the previous theorem. Therefore, we need to consider viscosity solutions instead of classical solutions for HJBVI (4.52). In fact, it will be shown that the value function is a unique viscosity solution of the HJBVI (4.52). These results will be given in Sections 4.5 and 4.6.

4.4 Verification Theorem

We have the following verification theorem for Problem (OSP1) but restricted to autonomous equation (4.3).

Theorem 4.4.1 (Verification Theorem) *Suppose we can find a function* $\Phi \in C^2_{lip}(\mathbf{C}) \cap \mathcal{D}(\mathcal{S})$ *such that the following hold:*
(i) $\Phi(\cdot) \geq \Psi(\cdot)$ *on* $\mathbf{C}$. *Define the open subset* $\mathbf{D}$ *of* $\mathbf{C}$ *by*

$$\mathbf{D} = \{\phi \in \mathbf{C} \mid \Phi(\phi) > \Psi(\phi)\}.$$

(ii) Assume that the $\mathbf{C}$-*valued segment process* $\{x_t(\psi), t \in [0,T]\}$ *for (4.3) spends zero time on the boundary* $\partial \mathbf{D}$ *of* $\mathbf{D}$ *a.s., that is,*

$$E\left[\int_0^T \chi_{\{\partial \mathbf{D}\}}(x_t(\psi))\,dt\right], \qquad \forall \psi \in \mathbf{C}.$$

(iii) $\mathbf{A}\Phi + \Psi \leq 0$ *on* $\mathbf{C} - \bar{\mathbf{D}}$.
(iv) $\mathbf{A}\Phi + \Psi = 0$ *on* $\mathbf{D}$.
(v) $\tau_{\mathbf{D}} \equiv \inf\{t > 0 \mid x_t(\psi) \notin \mathbf{D}\} < \infty$ *a.s. for all initial* $\psi \in \mathbf{C}$.
(vi) The family $\{\Phi(x_\tau(\psi)), \tau \leq \tau_{\mathbf{D}}\}$ *is uniformly integrable, for all initial* $\psi \in \mathbf{C}$.
Then, for all $\psi \in \mathbf{C}$,

$$\Phi(\psi) = V(\psi) = \sup_{\tau \in T_t^T} E\left[\int_0^\tau e^{\alpha t} L(x_t(\psi))\,dt + e^{\alpha \tau}\Psi(x_\tau(\psi))\right], \qquad (4.53)$$

and

$$\tau^* = \tau_{\mathbf{D}} \qquad (4.54)$$

is an optimal stopping time for Problem (OSP2).

Proof. For $R > 0$ and the initial $\psi \in \mathbf{C}$, put $T_R = R \wedge \inf\{t > 0 \mid \|x_t\| \geq R\}$ and let $\tau \in \mathcal{T}_t^T$. Then by Dynkin's formula in Subsection 2.4.1 of Chapter 2, and parts (i)-(iv),

$$\Phi(\psi) = E^\psi \left[-\int_0^{\tau \wedge T_R} e^{-\alpha t} \mathbf{A}\Phi(x_t)\, dt + \Phi(x_{\tau \wedge T_R}) \right]$$

$$\geq E^\psi \left[-\int_0^{\tau \wedge T_R} e^{-\alpha t} L(x_t)\, dt + e^{\tau \wedge T_R} \Psi(x_{\tau \wedge T_R}) \right]. \quad (4.55)$$

Hence, by the Fatou lemma,

$$\Phi(\psi) \geq \underline{\lim}_{R \to \infty} E^\psi \left[\int_0^{\tau \wedge T_R} e^{-\alpha t} L(x_t)\, dt + e^{\tau \wedge T_R} \Psi(x_{\tau \wedge T_R}) \right]$$

$$\geq E^\psi \left[\int_0^{\tau} e^{-\alpha t} L(x_t)\, dt + e^{-\alpha \tau} \Psi(x_\tau) \right].$$

Since $\tau \in \mathcal{T}_t^T$ is arbitrary, we conclude that

$$\Phi(\psi) \geq V(\psi), \quad \forall \psi \in \mathbf{C}. \quad (4.56)$$

We consider the following two cases: (i) $\psi \notin \mathbf{D}$ and (ii) $\psi \in \mathbf{D}$. If $\psi \notin \mathbf{D}$, then $\Phi(\psi) = \Psi(\psi) \leq V(\psi)$; so by (4.56), we have

$$\Phi(\psi) = V(\psi) \quad \text{and} \quad \hat{\tau} = \hat{\tau}(\psi, \omega) \equiv 0 \text{ is optimal.} \quad (4.57)$$

If $\psi \in \mathbf{D}$, we let $\{\mathbf{D}_k\}_{k=1}^\infty$ be an increasing sequence of open sets $\mathbf{D}_k$ such that $\bar{\mathbf{D}}_k \subset \mathbf{D}$, $\bar{\mathbf{D}}_k$ is compact and $\mathbf{D} = \cup_{k=1}^\infty \mathbf{D}_k$. Put $\tau_k = \inf\{t > 0 \mid x_t \notin \mathbf{D}_k\}$, $k = 1, 2, \ldots$. By Dynkin's formula we have, for $\psi \in \mathbf{D}_k$,

$$\Phi(\psi) = E^\psi \left[-\int_0^{\tau_k \wedge T_R} e^{-\alpha t} \mathbf{A}\Phi(x_t)\, dt + e^{-\alpha(\tau_k \wedge T_R)} (\Phi(x_{\tau_k \wedge T_R})) \right]$$

$$\geq E^\psi \left[-\int_0^{\tau_k \wedge T_R} e^{-\alpha t} L(x_t)\, dt + e^{-\alpha(\tau \wedge T_R)} \Phi(x_{\tau_k \wedge T_R}) \right].$$

So, by the uniform integrability assumption and parts (i), (iv), and (vi), we get

$$\Phi(\psi) = \lim_{R,k \to \infty} E^\psi \left[-\int_0^{\tau_k \wedge T_R} e^{-\alpha t} \mathbf{A}\Phi(x_t)\, dt + \Phi(x_{\tau_k \wedge T_R}) \right]$$

$$= E^\psi \left[\int_0^{\tau_\mathbf{D}} e^{-\alpha t} L(x_t)\, dt + e^{-\alpha \tau_\mathbf{D}} \Psi(x_{\tau_\mathbf{D}}) \right]$$

$$= J(\psi; \tau_\mathbf{D})$$

$$\leq V(\psi). \quad (4.58)$$

Combining (4.56) and (4.58), we get

$$\Phi(\psi) \geq V(\psi) \geq J(\psi; \tau_{\mathbf{D}} = \Phi(\psi).$$

Therefore,

$$\Phi(\psi) = V(\psi) \quad \text{and} \quad \hat{\tau}(\psi, \omega) \equiv \tau_{\mathbf{D}} \text{ is optimal.} \qquad (4.59)$$

From (4.57) and (4.59), we conclude that

$$\Phi(\psi) = V(\psi), \quad \forall \psi \in \mathbf{C}.$$

Moreover, the stopping time $\tau \in \mathcal{T}_t^T$ defined by

$$\hat{\tau}(\psi, \omega) = \begin{cases} 0 & \text{for } \psi \notin \mathbf{D} \\ \tau_{\mathbf{D}} & \text{for } \psi \in \mathbf{D} \end{cases}$$

is optimal. We therefore conclude that $\tau_{\mathbf{D}}$ is optimal as well. This proves the verification theorem.   □

4.5 Viscosity Solution

In this section, we will show that the value function V defined for Problem (OSP1) is actually the unique viscosity solution of the HJBVI (4.52). First, let us define the viscosity solution of (4.52) as follows.

Definition 4.5.1 *Let $w \in C([0,T] \times \mathbf{C})$. We say that w is a viscosity subsolution of (4.52) if for every $(t, \psi) \in [0,T] \times \mathbf{C}$ and for every $\Gamma : [0,T] \times \mathbf{C} \to \Re$ satisfying smoothness conditions (i) and (ii), $\Gamma \geq w$ on $[0,T] \times \mathbf{C}$, and $\Gamma(t, \psi) = w(t, \psi)$, we have*

$$\min\{\Gamma(t, \psi) - \Psi(t, \psi), \ \alpha\Gamma(t, \psi) - \partial_t\Gamma(t, \psi) - \mathbf{A}\Gamma(t, \psi) - L(t, \psi)\} \leq 0.$$

We say that w is a viscosity supersolution of (4.52) if, for every $(t, \psi) \in [0,T] \times \mathbf{C}$ and for every $\Gamma : [0,T] \times \mathbf{C} \to \Re$ satisfying Smoothness Conditions (i)-(ii), $\Gamma \leq w$ on $[0,T] \times \mathbf{C}$, and $\Gamma(t, \psi) = w(t, \psi)$, we have

$$\min\{\Gamma(t, \psi) - \Psi(t, \psi), \ \alpha\Gamma(t, \psi) - \partial_t\Gamma(t, \psi) - \mathbf{A}\Gamma(t, \psi) - L(t, \psi)\} \geq 0.$$

We say that w is a viscosity solution of (4.52) if it is a viscosity supersolution and a viscosity subsolution of (4.52).

As we can see in the definition, a viscosity solution must be continuous. So, first we will show that the value function V defined by (4.16) has this property. Actually, we have the following result.

Lemma 4.5.2 *The value function $V : [0,T] \times \mathbf{C} \to \Re$ is continuous and there exist constants $K > 0$ and $k \geq 1$ such that, for every $(t, \psi) \in [0,T] \times \mathbf{C}$, we have*

$$|V(t, \psi)| \leq K(1 + \|\psi\|_2)^k. \qquad (4.60)$$

Proof. It is clear that V has at most polynomial growth, since L and Φ have at most polynomial growth with the same $k \geq 1$ as in Assumption 4.1.4.

Let $\{x_s(t, \psi), s \in [t, T]\}$, be the **C**-valued segment process for (4.1) with initial data $(t, \psi) \in [0, T] \times \mathbf{C}$. It has been shown in Chapter 1 that the trajectory map $(t, \psi) \to x_s(t, \psi)$ from $[0, T] \times \mathbf{C}$ to $L^2(\Omega, \mathbf{C})$ is globally Lipschitz in ψ uniformly with respect to t on compact sets and continuous in t for fixed ψ. Therefore, given two **C**-valued segment processes

$$\Xi_1(s) = x_s(t, \psi_1) \text{ and } \Xi_2(s) = x_s(t, \psi_2), s \in [t, T]$$

of (4.1) with initial data (t, ψ_1) and (t, ψ_2), respectively, we have

$$\mathbf{E}\|\Xi_1(s) - \Xi_2(s)\| \leq K_{lip}\|\psi_1 - \psi_2\|_2, \tag{4.61}$$

where K is a positive constant that depends on the Lipschitz constant in Assumption 4.1.4 and T.

Using the Lipshitz continuity of $L, \Psi : [0, T] \times \mathbf{C} \to \Re$, there exists yet another constant $\Lambda > 0$ such that

$$|J(t, \psi_1; \tau) - J(t, \psi_2; \tau)| \leq \Lambda E[\|\Xi_1(\tau) - \Xi_2(\tau)\|]. \tag{4.62}$$

Therefore, using (4.62) and (4.61), we have

$$\begin{aligned}
|V(t, \psi_1) - V(t, \psi_2)| &\leq \sup_{\tau \in \mathcal{T}_t^T} |J(t, \psi_1; \tau) - J(t, \psi_2; \tau)| \\
&\leq \Lambda \sup_{\tau \in \mathcal{T}_t^T} E[\|\Xi_1(\tau) - \Xi_2(\tau)\|] \\
&\leq \Lambda \|\psi_1 - \psi_2\|_2. \tag{4.63}
\end{aligned}$$

This implies the (uniform) continuity of $V(t, \psi)$ with respect to ψ.

We next show the continuity of $V(t, \psi)$ with respect to t. Let $\Xi_1(s) = X_s(t_1, \psi)$, $s \in [t_1 T]$, and $\Xi_2(s) = X_s(t_2, \psi)$, $s \in [t_2, T]$, be two **C**-valued solutions of (4.1) with initial data (t_1, ψ) and (t_2, ψ), respectively.

Without lost of generality, we assume that $t_1 < t_2$. Then we can get

$$\begin{aligned}
J(t_1&, \psi; \tau) - J(t_2, \psi; \tau) \\
&= E\Bigg[\int_{t_1}^{t_2} e^{-\alpha(\xi - t_1)}[L(\xi, \Xi_1(\xi))]\, d\xi \\
&\quad + \int_{t_2}^{\tau} e^{-\alpha(\xi - t_2)}[L(\xi, \Xi_1(\xi)) - L(\xi, \Xi_2(\xi))]\, d\xi \\
&\quad + e^{-\alpha(\tau - t_1)}\Psi(\Xi_1(\tau)) - e^{-\alpha(\tau - t_2)}\Psi(\Xi_2(\tau))\Bigg]. \tag{4.64}
\end{aligned}$$

Therefore, there exists a constant $\Lambda > 0$ such that

$$\begin{aligned}
|J(t_1&, \psi; \tau) - J(t_2, \psi; \tau)| \\
&\leq \Lambda\bigg(|t_1 - t_2|E[\|\Xi_1(\tau)\|] + E[\|\Xi_1(\tau) - \Xi_2(\tau)\|]\bigg). \tag{4.65}
\end{aligned}$$

Let $\epsilon > 0$; using the compactness of $[0,T]$ and the uniform continuity of the trajectory map in t, there exists $\eta > 0$ such that if $|t_1 - t_2| < \eta$, then $E[\|\Xi_1(s) - \Xi_2(s)\|] \leq \frac{\epsilon}{2\Lambda}$. In addition, there exists a constant $K > 0$ such that

$$E\left[\sup_{s\in[t_1,T]} \|\Xi_1(s)\|\right] \leq K, \quad \forall t_1 \in [0,T].$$

Then, for $|t_1 - t_2| < \min\left\{\eta, \frac{\epsilon}{2\Lambda K}\right\}$, we have

$$|J(t_1,\psi;\tau) - J(t_2,\psi;\tau)| \leq \frac{\epsilon}{2} + \frac{\epsilon}{2} = \epsilon.$$

Consequently,

$$|V(t_1,\psi) - V(t_2,\psi)| \leq \epsilon.$$

This completes the proof.   □

Before we show that the value function is a viscosity solution of HJBVI (4.52), we need to prove some results related to the dynamic programming principle. The results are given next in Lemma 4.5.3 and Lemma 4.5.5.

Lemma 4.5.3 *Let $\underline{\tau}, \bar{\tau} \in T_t^T$ be $\mathbf{G}(t)$-stopping times and $t > 0$ such that $t \leq \underline{\tau} \leq \bar{\tau}$ a.s.. Then we have*

$$E\left[e^{-\alpha(\underline{\tau}-t)}V(\underline{\tau},x_{\underline{\tau}})\right] \geq E\left[\int_{\underline{\tau}}^{\bar{\tau}} e^{-\alpha(s-t)}L(s,x_s)\,ds\right]$$
$$+ E\left[e^{-\alpha(\bar{\tau}-t)}V(\bar{\tau},x_{\bar{\tau}})\right]. \quad (4.66)$$

Proof. It is known that

$$E\left[e^{-\alpha(\bar{\tau}-t)}V(\bar{\tau},x_{\bar{\tau}}) + \int_{\underline{\tau}}^{\bar{\tau}} e^{-\alpha(s-t)}L(s,x_s)\,ds\right]$$
$$= \sup_{\tau\in T_{\bar{\tau}}^T} E\left[\int_{\bar{\tau}}^{\tau} e^{-\alpha(s-\bar{\tau})}e^{-\alpha(\bar{\tau}-t)}L(s,x_s)\,ds + e^{-\alpha(\tau-\bar{\tau})}e^{-\alpha(\bar{\tau}-t)}\Psi(x_\tau)\right]$$
$$+ E\left[\int_t^{\bar{\tau}} e^{-\alpha(s-t)}L(s,x_s)\,ds\right]$$
$$= \sup_{\tau\in T_{\bar{\tau}}^T} E\left[\int_{\bar{\tau}}^{\tau} e^{-\alpha(s-t)}L(s,x_s)\,ds + e^{-\alpha(\tau-t)}\Psi(x_\tau)\right]$$
$$+ E\left[\int_{\underline{\tau}}^{\bar{\tau}} e^{-\alpha(s-t)}L(s,x_s)\,ds\right]$$
$$= \sup_{\tau\in T_{\bar{\tau}}^T} E\left[\int_{\underline{\tau}}^{\tau} e^{-\alpha(s-t)}L(s,x_s)\,ds + e^{-\alpha(\tau-t)}\Psi(x_\tau)\right]$$
$$\leq \sup_{\tau\in T_{\underline{\tau}}^T} E\left[\int_{\underline{\tau}}^{\tau} e^{-\alpha(\underline{\tau}-t)}e^{-\alpha(s-\underline{\tau})}L(s,x_s)\,ds + e^{-\alpha(\underline{\tau}-t)}e^{-\alpha(\tau-\underline{\tau})}\Psi(x_\tau)\right]$$
$$= E[e^{-\alpha(\underline{\tau}-t)}V(\underline{\tau},x_{\underline{\tau}})].$$

This completes the proof.   □

Now, let us give the definition of ϵ-optimal stopping time, which will be used in the next lemma.

Definition 4.5.4 *For each $\epsilon > 0$, a* $\mathbf{G}(t)$*-stopping time* $\tau_\epsilon \in T_t^T$ *is said to be ϵ-optimal if*

$$0 \leq V(t, \psi) - E\left[\int_t^{\tau_\epsilon} e^{-\alpha(s-t)} L(s, x_s)\, ds + e^{-\alpha(\tau_\epsilon - t)} V(\tau_\epsilon, x_{\tau_\epsilon})\right] \leq \epsilon.$$

Lemma 4.5.5 *Let θ be a stopping time such that $\theta \leq \tau_\epsilon$ a.s., for any $\epsilon > 0$, where $\tau_\epsilon \in T_t^T$ is ϵ-optimal. Then*

$$V(t, \psi) = E\left[\int_t^{\theta} e^{-\alpha(s-t)} L(s, x_s)\, ds + e^{-\alpha(\theta - t)} V(\theta, x_\theta)\right]. \tag{4.67}$$

Proof. Let θ be a stopping time such that $\theta \leq \tau_\epsilon$ a.s., for any ϵ-optimal $\tau_\epsilon \in T_t^T$. Using Lemma 4.5.3, we have

$$E[e^{-\alpha(\theta-t)} V(\theta, x_\theta)] \geq E\left[\int_\theta^{\tau_\epsilon} e^{-\alpha(s-t)} L(s, x_s)\, ds\right]$$
$$+ E[e^{-\alpha(\tau_\epsilon - t)} V(\tau_\epsilon, x_{\tau_\epsilon})].$$

This implies that

$$E[e^{-\alpha(\theta-t)} V(\theta, x_\theta)] + E\left[\int_t^{\theta} e^{-\alpha(s-t)} L(s, x_s)\, ds\right]$$
$$\geq E\left[\int_t^{\tau_\epsilon} e^{-\alpha(s-t)} L(s, x_s)\, ds\right] + E[V e^{-\alpha(\tau_\epsilon - t)}(\tau_\epsilon, x_{\tau_\epsilon})]. \tag{4.68}$$

Note that τ_ϵ is the ϵ-stopping time; then

$$0 \leq V(t, \psi) - E\left[\int_t^{\tau_\epsilon} e^{-\alpha(s-t)} L(s, x_s)\, ds + e^{-\alpha(\tau_\epsilon - t)} V(\tau_\epsilon, x_{\tau_\epsilon})\right] \leq \epsilon.$$

On the other hand, by virtue of (4.68), we can get

$$V(t, \psi) - E\left[e^{-\alpha(\theta-t)} V(\theta, x_\theta) + \int_t^{\theta} e^{-\alpha(s-t)} L(s, x_s)\, ds\right]$$
$$\leq V(t, \psi) - E\left[e^{-\alpha(\tau_\epsilon - t)} V(\tau_\epsilon, x_{\tau_\epsilon}) + \int_t^{\tau_\epsilon} e^{-\alpha(s-t)} L(s, x_s)\, ds\right]. \tag{4.69}$$

Thus, we can get

$$0 \leq V(t, \psi) - E\left[\int_t^{\theta} e^{-\alpha(s-t)} L(s, x_s)\, ds + e^{-\alpha(\theta - t)} V(\theta, x_\theta)\right] \leq \epsilon.$$

Now, we let $\epsilon \to 0$ in the above inequality and we can get

$$V(t,\psi) = E\left[\int_t^\theta e^{-\alpha(s-t)} L(s,x_s)\,ds\right] + E[e^{-\alpha(\theta-t)} V(\theta,x_\theta)].$$

This completes the proof.   □

Theorem 4.5.6 *The value function V is a viscosity solution of the HJBVI (4.52).*

Proof. We need to prove that the value function V is both a viscosity subsolution and a viscosity supersolution of (4.52).

First, we prove that V is a viscosity subsolution. Let $(t,\psi) \in [0,T] \times \mathbf{C}$ and $\Gamma \in C_{lip}^{1,2}([0,T] \times \mathbf{C}) \cap \mathcal{D}(\mathcal{S})$ satisfying $\Gamma \le V$ on $[0,T] \times \mathbf{C}$ and $\Gamma(t,\psi) = V(t,\psi)$. We want to prove the viscosity subsolution inequality, that is,

$$\min\left\{\Gamma(t,\psi) - \Psi(\psi),\ \alpha\Gamma(t,\psi) - \partial_t\Gamma(t,\psi) - \mathbf{A}\Gamma(t,\psi) - L(t,\psi)\right\} \ge 0. \quad (4.70)$$

We know that $V \ge \Psi$ and $\Gamma(t,\psi) = V(t,\psi)$, so we have

$$\Gamma(t,\psi) - \Psi(\psi) \ge 0.$$

Therefore, we just need to prove that

$$\alpha\Gamma(t,\psi) - \partial_t\Gamma(t,\psi) - \mathbf{A}\Gamma(t,\psi) - L(t,\psi) \ge 0.$$

Since $\Gamma \in C_{lip}^{1,2}([0,T] \times \mathbf{C}) \cap \mathcal{D}(\mathcal{S})$, by virtue of Theorem 4.2.3, for $t \le s \le T$, we have

$$E[e^{-\alpha(s-t)}\Gamma(s,x_s) - \Gamma(t,\psi)]$$
$$= E\left[\int_t^s e^{-\alpha(\xi-t)}\left(\partial_\xi\Gamma(\xi,x_\xi) + \mathbf{A}\Gamma(\xi,x_\xi) - \alpha\Gamma(\xi,x_\xi)\right)d\xi\right]. \quad (4.71)$$

For any $s \in [t,T]$, from Lemma 4.5.3, we can get

$$V(t,\psi) \ge E\left[\int_t^s e^{-\alpha(\xi-t)} L(\xi,x_\xi)\,d\xi\right] + E\left[e^{-\alpha(s-t)} V(s,x_s)\right].$$

By virtue of (4.71), $\Gamma \le V$, and $V(t,\psi) = \Gamma(t,\psi)$, we can get

$$0 \ge E\left[\int_t^s e^{-\alpha(\xi-t)} L(\xi,x_\xi)\,d\xi\right] + E\left[e^{-\alpha(s-t)} V(s,x_s)\right] - V(t,\psi)$$
$$\ge E\left[\int_t^s e^{-\alpha(\xi-t)} L(\xi,x_\xi)\,d\xi\right] + E\left[e^{-\alpha(s-t)}\Gamma(s,x_s)\right] - \Gamma(t,\psi)$$
$$\ge E\left[\int_t^s e^{-\alpha(\xi-t)}\left(L(\xi,x_\xi) + \partial_\xi\Gamma(\xi,x_\xi) + \mathbf{A}\Gamma(\xi,x_\xi)\right.\right.$$
$$\left.\left. - \alpha\Gamma(\xi,x_\xi)\right)d\xi\right]. \quad (4.72)$$

Dividing both sides of the above inequality by $(s - t)$, we have

$$0 \geq E\left[\frac{1}{s-t}\int_t^s e^{-\alpha(\xi-t)}\Big(L(\xi, x_\xi) + \partial_\xi \Gamma(\xi, x_\xi) + \mathbf{A}\Gamma(\xi, x_\xi)\right.$$
$$\left.-\alpha\Gamma(\xi, x_\xi)\Big)\, d\xi\right]. \tag{4.73}$$

Now, let $s \downarrow t$ in (4.73), and we obtain

$$\partial_t \Gamma(t, \psi) + \mathbf{A}\Gamma(t, \psi) + L(t, \psi) - \alpha\Gamma(t, \psi) \leq 0 \tag{4.74}$$

which proves the inequality (4.70).

Next, we want to prove that V is also a viscosity supersolution of (4.52). Let $(t, \psi) \in [0, T] \times \mathbf{C}$ and $\Gamma \in C_{lip}^{1,2}([0, T] \times \mathbf{C}) \cap \mathcal{D}(\mathcal{S})$ satisfying $\Gamma \geq V$ on $[0, T] \times \mathbf{C}$ and $\Gamma(t, \psi) = V(t, \psi)$; we want to prove that

$$\max\Big\{ \Psi(\psi) - \Gamma(t, \psi),\ \partial_t \Gamma(t, \psi) + \mathbf{A}\Gamma(t, \psi) + L(t, \psi) - \alpha\Gamma(t, \psi)\Big\} \geq 0. \tag{4.75}$$

Actually, it is sufficient to show that

$$\partial_t \Gamma(t, \psi) + \mathbf{A}\Gamma(t, \psi) + L(t, \psi) - \alpha\Gamma(t, \psi) \geq 0. \tag{4.76}$$

Let $\theta \in \mathcal{T}_t^T$ be a stopping time such that $\theta \leq \tau_\epsilon$ for every τ_ϵ, ϵ-optimal stopping time. Using Lemma 4.5.5, we can get

$$V(t, \psi) = E\left[\int_t^\theta e^{-\alpha(s-t)} L(s, x_s)\, ds\right] + E\left[e^{-\alpha(\theta-t)} V(\theta, x_\theta)\right]. \tag{4.77}$$

Using Dynkin's formula (see Theorem 2.4.1), we have

$$E\left[e^{-\alpha(\theta-t)}\Gamma(\theta, x_\theta)\right] - \Gamma(t, \psi)$$
$$= E\left[\int_t^\theta e^{-\alpha(s-t)}\Big(\partial_s\Gamma(s, x_s) + \mathbf{A}\Gamma(s, x_s) - \alpha\Gamma(s, x_s)\Big)ds\right].$$

Since $\Gamma \geq V$ and $\Gamma(t, \psi) = V(t, \psi)$, now we can get

$$E\left[e^{-\alpha(\theta-t)}V(\theta, x_\theta)\right] - V(t, \psi)$$
$$\leq E\left[\int_t^\theta e^{-\alpha(s-t)}\Big(\partial_s\Gamma(s, x_s) + \mathbf{A}\Gamma(s, x_s) - \alpha\Gamma(s, x_s)\Big)ds\right]$$

Combining this with (4.77), the above inequality implies

$$0 \leq E\left[\int_t^\theta e^{-\alpha(s-t)}\Big(L(s, x_s) + \partial_s\Gamma(s, x_s) + \mathbf{A}\Gamma(s, x_s) - \alpha\Gamma(s, x_s)\Big)ds\right]. \tag{4.78}$$

Dividing (4.78) by $E[(\theta - t)]$ and sending $E[\theta] \to t$, we deduce

$$\partial_t \Gamma(t, \psi) + \mathbf{A}\Gamma(t, \psi) + L(t, \psi) - \alpha\Gamma(t, \psi) \geq 0, \qquad (4.79)$$

which proves (4.75). Therefore, V is also a viscosity supersolution. This completes the proof of the theorem. $\square$

The following comparison principle is crucial for our uniqueness result. Its proof is rather lengthy but similar to the proof of the comparison principle provided in Chapter 3. We, therefore, provide only an outline.

Theorem 4.5.7 (Comparison Principle) *If $V_1(t, \psi)$ and $V_2(t, \psi)$ are both continuous with respect to the argument $(t, \psi) \in [0, T] \times \mathbf{C}$ and are respectively the viscosity subsolution and supersolution of (4.52) with at most a polynomial growth (i.e., there exist constants $\Lambda > 0$ and $k \geq 1$ such that*

$$|V_i(t, \psi)| \leq \Lambda(1 + \|\psi\|_2)^k, \ \ for \ \ (t, \psi) \in [0, T] \times \mathbf{C}, \ i = 1, 2). \qquad (4.80)$$

Then

$$V_1(t, \psi) \leq V_2(t, \psi) \quad for \ all \ (t, \psi) \in [0, T] \times \mathbf{C}. \qquad (4.81)$$

Since the value function $V : [0, T] \times \mathbf{C} \to \Re$ of Problem (OSP1) is a viscosity solution (and, hence, is both a subsolution and a supersolution) of HJBVI (4.52), the uniqueness result of the viscosity solution follows immediately from the above comparison principle.

We therefore have the following main result of this section.

Theorem 4.5.8 *The value function $V : [0, T] \times \mathbf{C} \to \Re$ for Problem (OSP1) is the unique viscosity solution of HJBVI (4.52).*

Proof. Suppose $V_1, V_2 : [0, T] \times \mathbf{C} \to \Re$ are two viscosity solutions of HJBVI (4.52). Then they are both subsolution and supersolution. From Theorem 4.5.7, we have

$$V_1(t, \psi) \leq V_2(t, \psi) \leq V_1(t, \psi), \quad \forall (t, \psi) \in [0, T] \times \mathbf{C}.$$

Therefore, $V_1 = V_2$. This shows that viscosity solution is unique. $\square$

4.6 A Sketch of a Proof of Theorem 4.5.7

With exception of a few minor differences, the proof of the comparison principle, Theorem 4.5.7, is very similar to that of Theorem 3.6.1 in Chapter 3. Instead of giving a detailed proof, a sketch of the proof is provided below for completeness.

Let V_1 and V_2 be respectively a viscosity subsolution and supersolution of (4.52). For any $0 < \delta, \gamma < 1$ and for all $\psi, \phi \in \mathbf{C}$ and $t, s \in [0, T]$, define

$$\Theta_{\delta\gamma}(t, s, \psi, \phi) \equiv \frac{1}{\delta}\left[\|\psi - \phi\|_2^2 + \|\psi^0 - \phi^0\|_2^2 + |t - s|^2\right] \qquad (4.82)$$

$$+\gamma\left[\exp(1 + \|\psi\|_2^2 + \|\psi^0\|_2^2) + \exp(1 + \|\phi\|_2^2 + \|\phi^0\|_2^2)\right]$$

and

$$\Phi_{\delta\gamma}(t, s, \psi, \phi) \equiv V_1(t, \psi) - V_2(s, \phi) - \Theta_{\delta\gamma}(t, s, \psi, \phi), \qquad (4.83)$$

where $\psi^0, \phi^0 \in \mathbf{C}$ with $\psi^0(\theta) = \frac{\theta}{-r}\psi(-r - \theta)$, $\phi^0(\theta) = \frac{\theta}{-r}\phi(-r - \theta)$ for $\theta \in [-r, 0]$. It is desirable in the following proof that $\Phi_{\delta\gamma} : [0, T] \times [0, T] \times \mathbf{C} \times \mathbf{C}$ has a global maximum. To achieve this goal, we first restrict its domain to the separable Hilbert space $[0, T] \times [0, T] \times W^{1,2}((-r, 0); \Re^n) \times W^{1,2}((-r, 0); \Re^n)$, where $W^{1,2}((-r, 0); \Re^n)$ is a real separable Sobolev space defined by

$$W^{1,2}((-r, 0); \Re^n) = \{\phi \in \mathbf{C} \mid \phi \text{ is absolutely continuous with}$$

$$\text{and } \|\phi\|_{1,2} < \infty\},$$

$$\|\phi\|_{1,2}^2 \equiv \|\phi\|_2^2 + \|\dot\phi\|_2^2,$$

and $\dot\phi$ is the derivative of ϕ in the distributional sense. Note that it can be shown that the Hilbertian norm $\|\cdot\|_{1,2}$ is weaker than the sup-norm $\|\cdot\|$; that is, there exists a constant $K > 0$ such that

$$\|\phi\|_{1,2} \leq K\|\phi\|, \quad \forall \phi \in W^{1,2}((-r, 0); \Re^n).$$

From the Sobolev embedding theorems, it is known that $W^{1,2}((-r, 0); \Re^n) \subset \mathbf{C}$ and that $W^{1,2}((-r, 0); \Re^n)$ is dense in $\mathbf{C}$.

We observe that, using the polynomial growth condition for V_1 and V_2, we have

$$\lim_{\|\psi\|_2 + \|\phi\|_2 \to \infty} \Phi_{\delta\gamma}(\psi, \phi) = -\infty. \qquad (4.84)$$

The function $\Phi_{\delta\gamma}$ is a real-valued function that is bounded above and continuous on $[0, T] \times [0, T] \times W^{1,2}((-r, 0); \Re^n) \times W^{1,2}((-r, 0); \Re^n)$ (since the Hilbertian norm $\|\cdot\|_{1,2}$ is weaker than the sup-norm $\|\cdot\|$). Therefore, from Lemma 3.6.2 in Chapter 3, for any $1 > \epsilon > 0$ there exists a continuous linear functional T_ϵ in the topological dual of $W^{1,2}((-r, 0); \Re^n) \times W^{1,2}((-r, 0); \Re^n)$, with norm at most ϵ, such that the function $\Phi_{\delta\gamma} + T_\epsilon$ attains it maximum in $[0, T] \times [0, T] \times W^{1,2}((-r, 0); \Re^n) \times W^{1,2}((-r, 0); \Re^n)$. Let us denote by

$$(t_{\delta\gamma\epsilon}, s_{\delta\gamma\epsilon}, \psi_{\delta\gamma\epsilon}, \phi_{\delta\gamma\epsilon})$$

the global maximum of $\Phi_{\delta\gamma} + T_\epsilon$ on $[0, T] \times [0, T] \times W^{1,2}((-r, 0); \Re^n) \times W^{1,2}((-r, 0); \Re^n)$.

Without loss of generality, we assume that for any given δ, γ, and ϵ, there exists a constant $M_{\delta\gamma\epsilon}$ such that the maximum value $\Phi_{\delta\gamma} + T_\epsilon + M_{\delta\gamma\epsilon}$ is zero. In other words, we have

$$\Phi_{\delta\gamma}(t_{\delta\gamma\epsilon}, s_{\delta\gamma\epsilon}, \psi_{\delta\gamma\epsilon}, \phi_{\delta\gamma\epsilon}) + T_\epsilon(\psi_{\delta\gamma\epsilon}, \phi_{\delta\gamma\epsilon}) + M_{\delta\gamma\epsilon} = 0. \qquad (4.85)$$

The lemmas stated in the remainder of this section are taken from Section 3.6. The readers are referred to that section for proofs.

Lemma 4.6.1 $(t_{\delta\gamma\epsilon}, s_{\delta\gamma\epsilon}, \psi_{\delta\gamma\epsilon}, \phi_{\delta\gamma\epsilon})$ *is the global maximum of* $\Phi_{\delta\gamma} + T_\epsilon$ *in* $[0,T] \times [0,T] \times \mathbf{C} \times \mathbf{C}$.

Lemma 4.6.2 *For each fixed* $\gamma > 0$, *we can find a constant* $\Lambda_\gamma > 0$ *such that*

$$\|\psi_{\delta\gamma\epsilon}\|_2 + \|\psi^0_{\delta\gamma\epsilon}\|_2 + \|\phi_{\delta\gamma\epsilon}\|_2 + \|\phi^0_{\delta\gamma\epsilon}\|_2 \le \Lambda_\gamma \qquad (4.86)$$

and

$$\lim_{\epsilon\downarrow 0, \delta\downarrow 0} \left(\|\psi_{\delta\gamma\epsilon} - \phi_{\delta\gamma\epsilon}\|_2^2 + \|\psi^0_{\delta\gamma\epsilon} - \phi^0_{\delta\gamma\epsilon}\|_2^2 + |t_{\delta\gamma\epsilon} - s_{\delta\gamma\epsilon}|^2 \right) = 0. \qquad (4.87)$$

Now, let us introduce a functional $F : \mathbf{C} \to \mathbb{R}$ defined by

$$F(\psi) \equiv \|\psi\|_2^2 \qquad (4.88)$$

and the linear map $H : \mathbf{C} \to \mathbf{C}$ defined by

$$H(\psi)(\theta) \equiv \frac{\theta}{-r}\psi(-r-\theta) = \psi^0(\theta), \qquad \theta \in [-r, 0]. \qquad (4.89)$$

Note that $H(\psi)(0) = \psi^0(0) = 0$ and $H(\psi)(-r) = \psi^0(-r) = -\psi(0)$. It is not hard to show that the map F is Fréchet differentiable and its derivative is given by $DF(u)h = 2(u|h)$. This comes from the fact that

$$\|\psi + h\|_2^2 - \|\psi\|_2^2 = 2(\psi|h) + \|h\|_2^2,$$

and we can always find a constant $\Lambda > 0$ such that

$$\frac{|\|\psi + h\|_2^2 - \|\psi\|_2^2 - 2(\psi|h)|}{\|h\|} = \frac{\|h\|_2^2}{\|h\|} \le \frac{\Lambda\|h\|^2}{\|h\|} = \Lambda\|h\|. \qquad (4.90)$$

Moreover, we have

$$2(\psi + h| \,\cdot\,) - 2(\psi| \,\cdot\,) = 2(h| \,\cdot\,).$$

We deduce that F is twice differentiable and $D^2F(u)(h,k) = 2(h|k)$.

In addition, the map H is linear, thus twice Fréchet differentiable. Therefore, $DH(\psi)(h) = H(h)$ and $D^2H(\psi)(h,k) = 0$, for all $\psi, h, k \in \mathbf{C}$.

From the definition of $\Theta_{\delta\gamma}$ and the definition of F, we can get that

$$\Theta_{\delta\gamma}(t, s, \psi, \phi) = \frac{1}{\delta}\left[F(\psi - \phi) + F(\psi^0 - \phi^0) + |t - s|^2 \right]$$
$$+ \gamma[e^{1+F(\psi)+F(H(\psi))} + e^{1+F(\phi)+F(H(\phi))}].$$

Given the above chain rule, we can say that $\Theta_{\delta\gamma}$ is Fréchet differentiable. Actually, for $h, k \in \mathbf{C}$, we can get

$$D_\psi\Theta_{\delta\gamma}(t, s, \psi, \phi)(h) = \frac{2}{\delta}\left[(\psi - \phi|h) + (H(\psi - \phi)|H(h)) \right]$$
$$+ 2\gamma e^{1+F(\psi)+F(H(\psi))}[(\psi|h) + (H(\psi)|H(h))]. \qquad (4.91)$$

Similarly,

$$D_\phi \Theta_{\delta\gamma}(t, s, \psi, \phi)(k) = \frac{2}{\delta}\Big[(\phi - \psi|k) + (H(\phi - \psi)|H(k))\Big]$$
$$+ 2\gamma e^{1 + F(\phi) + F(H(\phi))}[(\phi|k) + (H(\phi)|H(k))]. \quad (4.92)$$

Furthermore,

$$D_\psi^2 \Theta_{\delta\gamma}(t, s, \psi, \phi)(h, k)$$
$$= \frac{2}{\delta}\Big[(h|k) + (H(h)|H(k))\Big]$$
$$+ 2\gamma e^{1 + F(\psi) + F(H(\psi))}\Big[2((\psi|k) + (H(\psi)|H(k)))((\psi|h) + (H(\psi)|H(h)))$$
$$+ (k|h) + (H(k)|H(h))\Big]. \quad (4.93)$$

Similarly,

$$D_\phi^2 \Theta_{\delta\gamma}(t, s, \psi, \phi)(h, k)$$
$$= \frac{2}{\delta}\Big[(h|k) + (H(h)|H(k))\Big]$$
$$+ 2\gamma e^{1 + F(\phi) + F(H(\phi))}\Big[2((\phi|k) + (H(\phi)|H(k)))((\phi|h) + (H(\phi)|H(h)))$$
$$+ (k|h) + (H(k)|H(h))\Big]. \quad (4.94)$$

By the Hahn-Banach theorem (see, e.g., Siddiqi [Sid04]), we can extend the continuous linear functional T_ϵ to the space $\mathbf{C} \times \mathbf{C}$ and its norm is preserved. Thus, the first-order Fréchet derivatives of T_ϵ is just T_ϵ, that is,

$$D_\psi T_\epsilon(\psi, \phi)h = T_\epsilon(h, \phi),$$
$$D_\phi T_\epsilon(\psi, \phi)k = T_\epsilon(\psi, k) \qquad \text{for all } \psi, \phi, h, k \in \mathbf{C}.$$

Also, for the second derivative, we have

$$D_\psi^2 T_\epsilon(\psi, \phi)(h, k) = 0,$$
$$D_\phi^2 T_\epsilon(\psi, \phi)(h, k) = 0, \qquad \text{for all } \psi, \phi, h, k \in \mathbf{C}.$$

Observe that we can extend $D_\psi \Theta_{\delta\gamma}(t, s, \psi, \phi)$ and $D_\psi^2 \Theta_{\delta\gamma}(t, s, \psi, \phi)$, the first- and second-order Fréchet derivatives of $\Theta_{\delta\gamma}$ with respect to ψ, to the space $\mathbf{C} \oplus \mathbf{B}$ (see Lemma 2.2.3 and Lemma 2.2.4) by setting

$$\overline{D_\psi \Theta_{\delta\gamma}(t, s, \psi, \phi)}(h + v\mathbf{1}_{\{0\}})$$
$$= \frac{2}{\delta}\Big[(\psi - \phi|h + v\mathbf{1}_{\{0\}}) + (H(\psi - \phi)|H(h + v\mathbf{1}_{\{0\}}))\Big]$$
$$+ 2\gamma e^{1 + F(\psi) + F(H(\psi))}[(\psi|h + v\mathbf{1}_{\{0\}}) + (H(\psi)|H(h + v\mathbf{1}_{\{0\}}))] \quad (4.95)$$

and

$$\overline{D_\psi^2 \Theta_{\delta\gamma}}(t,s,\psi,\phi)(h+v\mathbf{1}_{\{0\}}, k+w\mathbf{1}_{\{0\}})$$

$$= \frac{2}{\delta}\Big[(h+v\mathbf{1}_{\{0\}}|k+w\mathbf{1}_{\{0\}}) + (H(h+v\mathbf{1}_{\{0\}})|H(k+w\mathbf{1}_{\{0\}}))\Big]$$

$$+2\gamma e^{1+F(\psi)+F(H(\psi))}\Big[2((\psi|k+w\mathbf{1}_{\{0\}}) + (H(\psi)|H(k+w\mathbf{1}_{\{0\}})))$$

$$\times ((\psi|h+v\mathbf{1}_{\{0\}}) + (H(\psi)|H(h+v\mathbf{1}_{\{0\}})))$$

$$+(k+w\mathbf{1}_{\{0\}}|h+v\mathbf{1}_{\{0\}}) + (H(k+w\mathbf{1}_{\{0\}})|H(h+v\mathbf{1}_{\{0\}}))\Big] \quad (4.96)$$

for $v, w \in \mathbf{R}^n$ and $h, k \in \mathbf{C}$.

Moreover, it is easy to see that these extensions are continuous in that there exists a constant $\Lambda > 0$ such that

$$|(\psi - \phi|h + v\mathbf{1}_{\{0\}})| \leq \|\psi - \phi\|_2 \cdot \|h + v\mathbf{1}_{\{0\}}\|_2$$
$$\leq \Lambda\|\psi - \phi\|_2(\|h\| + |v|) \quad (4.97)$$

$$|(\psi|h + v\mathbf{1}_{\{0\}})| \leq \|\psi\|_2 \cdot \|h + v\mathbf{1}_{\{0\}}\|_2$$
$$\leq \Lambda\|\psi\|_2(\|h\| + |v|) \quad (4.98)$$

$$|(\psi|k + w\mathbf{1}_{\{0\}})| \leq \|\psi\|_2 \cdot \|k + w\mathbf{1}_{\{0\}}\|_2$$
$$\leq \Lambda\|\psi\|_2(\|k\| + |w|) \quad (4.99)$$

and

$$|(k + w\mathbf{1}_{\{0\}}|h + v\mathbf{1}_{\{0\}})| \leq \|k + w\mathbf{1}_{\{0\}}\|_2\|h + v\mathbf{1}_{\{0\}}\|_2$$
$$\leq \Lambda(\|k\| + |w|)(\|h\| + |v|). \quad (4.100)$$

Similarly, we can extend the first- and second-order Fréchet derivatives of $\Theta_{\delta\gamma}$ with respect to ϕ to the space $\mathbf{C} \oplus \mathbf{B}$ and obtain similar expressions for $\overline{D_\phi \Theta_{\delta\gamma}}(t,s,\psi,\phi)(k+w\mathbf{1}_{\{0\}})$ and $\overline{D_\phi^2 \Theta_{\delta\gamma}}(t,s,\psi,\phi)(h+v\mathbf{1}_{\{0\}}, k+w\mathbf{1}_{\{0\}})$. The same is also true for the bounded linear functional T_ϵ whose extension is still written as T_ϵ.

In addition, it is easy to verify that for any $\phi \in \mathbf{C}$ and $v, w \in \Re^n$, we have

$$(\phi|v\mathbf{1}_{\{0\}}) = \int_{-r}^0 \langle \phi(s), v\mathbf{1}_{\{0\}}(s) \rangle ds = 0, \quad (4.101)$$

$$(w\mathbf{1}_{\{0\}}|v\mathbf{1}_{\{0\}}) = \int_{-r}^0 \langle w\mathbf{1}_{\{0\}}(s), v\mathbf{1}_{\{0\}}(s) \rangle ds = 0, \quad (4.102)$$

$$H(v\mathbf{1}_{\{0\}}) = v\mathbf{1}_{\{-r\}}, \quad (4.103)$$

$$(H(\psi)|H(v\mathbf{1}_{\{0\}})) = 0, \quad (H(w\mathbf{1}_{\{0\}})|H(v\mathbf{1}_{\{0\}})) = 0. \quad (4.104)$$

These observations will be used later.

Next, we need several lemmas about the operator $\mathcal{S}$.

Lemma 4.6.3 *Given $\phi \in C$, we have*

$$S(F)(\phi) = |\phi(0)|^2 - |\phi(-r)|^2, \tag{4.105}$$

$$S(F)(\phi^0) = -|\phi(0)|^2, \tag{4.106}$$

where F is the functional defined in (4.88) and S is the operator defined in (4.8).

Let S_ψ and S_ϕ denote the operator S applied to ψ and ϕ, respectively. We have the following lemma.

Lemma 4.6.4 *Given $\phi, \psi \in C$,*

$$S_\psi(F)(\phi - \psi) + S_\phi(F)(\phi - \psi) = |\psi(0) - \phi(0)|^2 - |\psi(-r) - \phi(-r)|^2 \tag{4.107}$$

and

$$S_\psi(F)(\phi^0 - \psi^0) + S_\phi(F)(\phi^0 - \psi^0) = -|\psi(0) - \phi(0)|^2, \tag{4.108}$$

where F is the functional defined in (4.88) and S is the operator defined in (4.8).

Lemma 4.6.5 *Given $\phi \in C$, we define a new operator G as follows:*

$$G(\phi) = e^{1+F(\phi)+F(\phi^0)}. \tag{4.109}$$

We have

$$S(G)(\phi) = (-|\phi(-r)|^2)e^{1+F(\phi)+F(\phi^0)}, \tag{4.110}$$

where F is the functional defined in (4.88) and S is the operator defined in (4.8).

Lemma 4.6.6 *For any $\psi, \phi \in C$, we have*

$$\lim_{\epsilon\downarrow 0} |S_\psi(T_\epsilon)(\psi, \phi)| = 0 \quad and \quad \lim_{\epsilon\downarrow 0} |S_\phi(T_\epsilon)(\psi, \phi)| = 0. \tag{4.111}$$

Given all of the above results, now we are ready to prove Theorem 4.5.7.

Proof of Theorem 4.5.7.

Define

$$\Gamma_1(t, \psi) \equiv V_2(s_{\delta\gamma\epsilon}, \phi_{\delta\gamma\epsilon}) + \Theta_{\delta\gamma}(t, s_{\delta\gamma\epsilon}, \psi, \phi_{\delta\gamma\epsilon}) - T_\epsilon(\psi, \phi_{\delta\gamma\epsilon}) - M_{\delta\gamma\epsilon} \tag{4.112}$$

and

$$\Gamma_2(s, \phi) \equiv V_1(t_{\delta\gamma\epsilon}, \psi_{\delta\gamma\epsilon}) - \Theta_{\delta\gamma}(t_{\delta\gamma\epsilon}, s, \psi_{\delta\gamma\epsilon}, \phi) + T_\epsilon(\psi_{\delta\gamma\epsilon}, \phi) + M_{\delta\gamma\epsilon} \tag{4.113}$$

for all $s, t \in [0, T]$ and $\psi, \phi \in C$. Recall that

$$\Phi_{\delta\gamma}(t,s,\psi,\phi) = V_1(t,\psi) - V_2(s,\phi) - \Theta_{\delta\gamma}(t,s,\psi,\phi),$$

and $\Phi_{\delta\gamma} + T_\epsilon + M_{\delta\gamma\epsilon}$ reaches its maximum value zero at $(t_{\delta\gamma\epsilon}, s_{\delta\gamma\epsilon}, \psi_{\delta\gamma\epsilon}, \phi_{\delta\gamma\epsilon})$ in $[0,T] \times [0,T] \times \mathbf{C} \times \mathbf{C}$. By the definition of Γ_1 and Γ_2, it is easy to verify that, for all ϕ and ψ, we have

$$\Gamma_1(t,\psi) \geq V_1(t,\psi), \quad \Gamma_2(s,\phi) \leq V_2(s,\phi), \quad \forall t,s \in [0,T] \text{ and } \phi, \psi \in \mathbf{C},$$

and

$$V_1(t_{\delta\gamma\epsilon}, \psi_{\delta\gamma\epsilon}) = \Gamma_1(t_{\delta\gamma\epsilon}, \psi_{\delta\gamma\epsilon}) \text{ and } V_2(s_{\delta\gamma\epsilon}, \phi_{\delta\gamma\epsilon}) = \Gamma_2(s_{\delta\gamma\epsilon}, \phi_{\delta\gamma\epsilon}).$$

Using the definitions of the viscosity subsolution of V_1 and Γ_1, we have

$$\min\Big\{ V_1(t_{\delta\gamma\epsilon}, \psi_{\delta\gamma\epsilon}) - \Psi(t, \psi_{\delta\gamma\epsilon}),$$
$$\alpha V_1(t_{\delta\gamma\epsilon}, \psi_{\delta\gamma\epsilon}) - \partial_t \Gamma_1(t_{\delta\gamma\epsilon}, \psi_{\delta\gamma\epsilon})$$
$$-\mathbf{A}(\Gamma_1)(t_{\delta\gamma\epsilon}, \psi_{\delta\gamma\epsilon}) - L(t_{\delta\gamma\epsilon}, \psi_{\delta\gamma\epsilon}) \Big\} \leq 0. \qquad (4.114)$$

By the definitions of the operator $\mathbf{A}$ and Γ_1 and the fact that the second-order Fréchet derivatives of $T_\epsilon = 0$, we have, by combining (4.91), (4.92), (4.93), (4.94), (4.95), (4.96), (4.101), (4.102), and (4.104),

$$\mathbf{A}(\Gamma_1)(t_{\delta\gamma\epsilon}, \psi_{\delta\gamma\epsilon})$$
$$= \mathcal{S}(\Gamma_1)(t_{\delta\gamma\epsilon}, \psi_{\delta\gamma\epsilon}) + \overline{D_\psi \Theta_{\delta\gamma}(\cdots)}(f(t_{\delta\gamma\epsilon}, \psi_{\delta\gamma\epsilon})\mathbf{1}_{\{0\}})$$
$$+ \frac{1}{2}\sum_{j=1}^{m} \overline{D_\psi^2 \Theta_{\delta\gamma}(\cdots)}\Big(g(t_{\delta\gamma\epsilon}, \psi_{\delta\gamma\epsilon})(\mathbf{e}_j)\mathbf{1}_{\{0\}}, g(t_{\delta\gamma\epsilon}, \psi_{\delta\gamma\epsilon})(\mathbf{e}_j)\mathbf{1}_{\{0\}} \Big)$$
$$- \overline{D_\psi T_\epsilon(\psi_{\delta\gamma\epsilon}, \phi_{\delta\gamma\epsilon})}(f(t_{\delta\gamma\epsilon}, \psi_{\delta\gamma\epsilon})\mathbf{1}_{\{0\}})$$
$$= \mathcal{S}(\Gamma_1)(t_{\delta\gamma\epsilon}, \psi_{\delta\gamma\epsilon}) - \overline{T_\epsilon(f(t_{\delta\gamma\epsilon}, \psi_{\delta\gamma\epsilon})\mathbf{1}_{\{0\}}, \phi_{\delta\gamma\epsilon})}.$$

Note that $\Theta_{\delta\gamma}(\cdots)$ is an abbreviation for $\Theta_{\delta\gamma}(t_{\delta\gamma\epsilon}, s_{\delta\gamma\epsilon}, \psi_{\delta\gamma\epsilon}, \phi_{\delta\gamma\epsilon})$ in the above equation and the following.

It follows from (4.114) and the above inequality together yield that

$$\min\Big\{ V_1(t_{\delta\gamma\epsilon}, \psi_{\delta\gamma\epsilon}) - \Psi(t_{\delta\gamma\epsilon}, \psi_{\delta\gamma\epsilon}),$$
$$\alpha V_1(t_{\delta\gamma\epsilon}, \psi_{\delta\gamma\epsilon}) - \mathcal{S}(\Gamma_1)(t_{\delta\gamma\epsilon}, \psi_{\delta\gamma\epsilon}) - \partial_t \Gamma_1(t_{\delta\gamma\epsilon}, \psi_{\delta\gamma\epsilon})$$
$$- \Big[-\overline{T_\epsilon(f(t_{\delta\gamma\epsilon}, \psi_{\delta\gamma\epsilon})\mathbf{1}_{\{0\}}, \phi_{\delta\gamma\epsilon})} + L(t_{\delta\gamma\epsilon}, \psi_{\delta\gamma\epsilon}) \Big] \Big\} \leq 0. \quad (4.115)$$

Similarly, using the definitions of the viscosity supersolution of V_2 and Γ_2 and by the virtue of the same techniques similar to (4.115), we have

$$\min\Big\{ V_2(s_{\delta\gamma\epsilon}, \phi_{\delta\gamma\epsilon}) - \Psi(s_{\delta\gamma\epsilon}, \phi_{\delta\gamma\epsilon}),$$
$$\alpha V_2(s_{\delta\gamma\epsilon}, \phi_{\delta\gamma\epsilon}) - \mathcal{S}(\Gamma_2)(s_{\delta\gamma\epsilon}, \phi_{\delta\gamma\epsilon}) - \partial_s \Gamma_2(s_{\delta\gamma\epsilon}, \phi_{\delta\gamma\epsilon})$$
$$- \Big[\overline{T_\epsilon(\psi_{\delta\gamma\epsilon}, f(s_{\delta\gamma\epsilon}, \phi_{\delta\gamma\epsilon})\mathbf{1}_{\{0\}})} + L(s_{\delta\gamma\epsilon}, \phi_{\delta\gamma\epsilon}) \Big] \Big\} \geq 0. \quad (4.116)$$

Inequality (4.115) is equivalent to

$$V_1(t_{\delta\gamma\epsilon}, \psi_{\delta\gamma\epsilon}) - \Psi(t_{\delta\gamma\epsilon}, \psi_{\delta\gamma\epsilon}) \leq 0, \tag{4.117}$$

or

$$\alpha V_1(t_{\delta\gamma\epsilon}, \psi_{\delta\gamma\epsilon}) - \mathcal{S}(\Gamma_1)(t_{\delta\gamma\epsilon}, \psi_{\delta\gamma\epsilon}) - 2(t_{\delta\gamma\epsilon} - s_{\delta\gamma\epsilon})$$
$$- \left[- \overline{T_\epsilon(f(t_{\delta\gamma\epsilon}, \psi_{\delta\gamma\epsilon}) \mathbf{1}_{\{0\}}, \phi_{\delta\gamma\epsilon})} + L(t_{\delta\gamma\epsilon}, \psi_{\delta\gamma\epsilon}) \right] \leq 0. \tag{4.118}$$

Similarly, Inequality (4.116) is equivalent to

$$V_2(s_{\delta\gamma\epsilon}, \phi_{\delta\gamma\epsilon}) - \Psi(s_{\delta\gamma\epsilon}, \phi_{\delta\gamma\epsilon}) \geq 0, \tag{4.119}$$

and

$$\alpha V_2(s_{\delta\gamma\epsilon}, \phi_{\delta\gamma\epsilon}) - \mathcal{S}(\Gamma_2)(s_{\delta\gamma\epsilon}, \phi_{\delta\gamma\epsilon}) - 2(s_{\delta\gamma\epsilon} - t_{\delta\gamma\epsilon})$$
$$- \left[\overline{T_\epsilon(\psi_{\delta\gamma\epsilon}, f(s_{\delta\gamma\epsilon}, \phi_{\delta\gamma\epsilon}) \mathbf{1}_{\{0\}})} + L(s_{\delta\gamma\epsilon}, \phi_{\delta\gamma\epsilon}) \right] \geq 0. \tag{4.120}$$

If we have (4.117), using (4.119), we can get that there exists a constant $\Lambda > 0$ such that

$$V_1(t_{\delta\gamma\epsilon}, \psi_{\delta\gamma\epsilon}) - V_2(s_{\delta\gamma\epsilon}, \phi_{\delta\gamma\epsilon}) \leq \Psi(t_{\delta\gamma\epsilon}, \psi_{\delta\gamma\epsilon}) - \Psi(s_{\delta\gamma\epsilon}, \phi_{\delta\gamma\epsilon})$$
$$\leq \Lambda \left(|t_{\delta\gamma\epsilon} - s_{\delta\gamma\epsilon}| + \|\psi_{\delta\gamma\epsilon} - \phi_{\delta\gamma\epsilon}\|_2 \right). \tag{4.121}$$

Thus, applying Lemma 4.6.2 to (4.121), we have

$$\limsup_{\delta\downarrow 0, \epsilon\downarrow 0} (V_1(t_{\delta\gamma\epsilon}, \psi_{\delta\gamma\epsilon}) - V_2(s_{\delta\gamma\epsilon}, \phi_{\delta\gamma\epsilon})) \leq 0. \tag{4.122}$$

On the other hand, by virtue of (4.118) and (4.120), we obtain

$$\alpha(V_1(t_{\delta\gamma\epsilon}, \psi_{\delta\gamma\epsilon}) - V_2(s_{\delta\gamma\epsilon}, \phi_{\delta\gamma\epsilon}))$$
$$\leq \mathcal{S}(\Gamma_1)(t_{\delta\gamma\epsilon}, \psi_{\delta\gamma\epsilon}) - \mathcal{S}(\Gamma_2)(s_{\delta\gamma\epsilon}, \phi_{\delta\gamma\epsilon}) + 4(t_{\delta\gamma\epsilon} - s_{\delta\gamma\epsilon})$$
$$+ \left[L(t_{\delta\gamma\epsilon}, \psi_{\delta\gamma\epsilon}) - \overline{T_\epsilon(f(t_{\delta\gamma\epsilon}, \psi_{\delta\gamma\epsilon}) \mathbf{1}_{\{0\}}, \phi_{\delta\gamma\epsilon})} \right]$$
$$- \left[L(s_{\delta\gamma\epsilon}, \phi_{\delta\gamma\epsilon}) + \overline{T_\epsilon(\psi_{\delta\gamma\epsilon}, f(s_{\delta\gamma\epsilon}, \phi_{\delta\gamma\epsilon}) \mathbf{1}_{\{0\}})} \right]. \tag{4.123}$$

From definition (4.8) of $\mathcal{S}$, it is clear that $\mathcal{S}$ is linear and takes zero on constants. Recall that

$$\Gamma_1(t, \psi) = V_2(s_{\delta\gamma\epsilon}, \phi_{\delta\gamma\epsilon}) + \Theta_{\delta\gamma}(t, s_{\delta\gamma\epsilon}, \psi, \phi_{\delta\gamma\epsilon}) - T_\epsilon(\psi, \phi_{\delta\gamma\epsilon}) - M_{\delta\gamma\epsilon} \tag{4.124}$$

and

$$\Gamma_2(s, \phi) = V_1(t_{\delta\gamma\epsilon}, \psi_{\delta\gamma\epsilon}) - \Theta_{\delta\gamma}(t_{\delta\gamma\epsilon}, s, \psi_{\delta\gamma\epsilon}, \phi) + T_\epsilon(\psi_{\delta\gamma\epsilon}, \phi) + M_{\delta\gamma\epsilon}. \tag{4.125}$$

Thus we have,

$$\mathcal{S}(\Gamma_1)(t_{\delta\gamma\epsilon}, \psi_{\delta\gamma\epsilon}) = \mathcal{S}_\psi(\Theta_{\delta\gamma})(t_{\delta\gamma\epsilon}, s_{\delta\gamma\epsilon}, \psi_{\delta\gamma\epsilon}, \phi_{\delta\gamma\epsilon}) - \mathcal{S}_\psi(T_\epsilon)(\psi_{\delta\gamma\epsilon}, \phi_{\delta\gamma\epsilon}),$$
(4.126)

and

$$\mathcal{S}(\Gamma_2)(s_{\delta\gamma\epsilon}, \phi_{\delta\gamma\epsilon}) = -\mathcal{S}_\phi(\Theta_{\delta\gamma})(t_{\delta\gamma\epsilon}, s_{\delta\gamma\epsilon}, \psi_{\delta\gamma\epsilon}, \phi_{\delta\gamma\epsilon}) + \mathcal{S}_\phi(T_\epsilon)(\psi_{\delta\gamma\epsilon}, \phi_{\delta\gamma\epsilon}).$$
(4.127)

Therefore,

$$\begin{aligned}
&\mathcal{S}(\Gamma_1)(t_{\delta\gamma\epsilon}, \psi_{\delta\gamma\epsilon}) - \mathcal{S}(\Gamma_2)(s_{\delta\gamma\epsilon}, \phi_{\delta\gamma\epsilon}) \\
&= \mathcal{S}_\psi(\Theta_{\delta\gamma})(t_{\delta\gamma\epsilon}, s_{\delta\gamma\epsilon}, \psi_{\delta\gamma\epsilon}, \phi_{\delta\gamma\epsilon}) + \mathcal{S}_\phi(\Theta_{\delta\gamma})(t_{\delta\gamma\epsilon}, s_{\delta\gamma\epsilon}, \psi_{\delta\gamma\epsilon}, \phi_{\delta\gamma\epsilon}) \\
&\quad - [\mathcal{S}_\psi(T_\epsilon)(\psi_{\delta\gamma\epsilon}, \phi_{\delta\gamma\epsilon}) + \mathcal{S}_\phi(T_\epsilon)(\psi_{\delta\gamma\epsilon}, \phi_{\delta\gamma\epsilon})].
\end{aligned}$$
(4.128)

Recall that

$$\Theta_{\delta\gamma}(t, s, \psi, \phi) = \frac{1}{\delta}\left[F(\psi - \phi) + F(\psi^0 - \phi^0) + |t - s|^2\right] + \gamma(G(\psi) + G(\phi)).$$

Therefore, we have

$$\begin{aligned}
&\left(\mathcal{S}_\psi(\Theta_{\delta\gamma}) + \mathcal{S}_\phi(\Theta_{\delta\gamma})\right)(t_{\delta\gamma\epsilon}, s_{\delta\gamma\epsilon}, \psi_{\delta\gamma\epsilon}, \phi_{\delta\gamma\epsilon}) \\
&\equiv \mathcal{S}_\psi(\Theta_{\delta\gamma})(t_{\delta\gamma\epsilon}, s_{\delta\gamma\epsilon}, \psi_{\delta\gamma\epsilon}, \phi_{\delta\gamma\epsilon}) \\
&\quad + \mathcal{S}_\phi(\Theta_{\delta\gamma})(t_{\delta\gamma\epsilon}, s_{\delta\gamma\epsilon}, \psi_{\delta\gamma\epsilon}, \phi_{\delta\gamma\epsilon}) \\
&= \frac{1}{\delta}[\mathcal{S}_\psi(F)(\psi_{\delta\gamma\epsilon} - \phi_{\delta\gamma\epsilon}) + \mathcal{S}_\phi(F)(\psi_{\delta\gamma\epsilon} - \phi_{\delta\gamma\epsilon}) \\
&\quad + \mathcal{S}_\psi(F)(\psi_{\delta\gamma\epsilon}^0 - \phi_{\delta\gamma\epsilon}^0) + \mathcal{S}_\phi(F)(\psi_{\delta\gamma\epsilon}^0 - \phi_{\delta\gamma\epsilon}^0)] \\
&\quad + \gamma[\mathcal{S}_\psi(G)(\psi_{\delta\gamma\epsilon}) + \mathcal{S}_\phi(G)(\phi_{\delta\gamma\epsilon})].
\end{aligned}$$
(4.129)

Using Lemma 4.6.4 and Lemma 4.6.5, we deduce

$$\begin{aligned}
&\left(\mathcal{S}_\psi(\Theta_{\delta\gamma}) + \mathcal{S}_\phi(\Theta_{\delta\gamma})\right)(t_{\delta\gamma\epsilon}, s_{\delta\gamma\epsilon}, \psi_{\delta\gamma\epsilon}, \phi_{\delta\gamma\epsilon}) \\
&= \frac{1}{\delta}\left[-|\psi_{\delta\gamma\epsilon}(-r) - \phi_{\delta\gamma\epsilon}(-r)|^2\right] \\
&\quad - \gamma\left(|\psi_{\delta\gamma\epsilon}(-r)|^2 e^{1+F(\psi_{\delta\gamma\epsilon})+F(\psi_{\delta\gamma\epsilon}^0)}\right. \\
&\quad \left. + |\phi_{\delta\gamma\epsilon}(-r)|^2 e^{1+F(\phi_{\delta\gamma\epsilon})+F(\phi_{\delta\gamma\epsilon}^0)}\right) \\
&\leq 0.
\end{aligned}$$
(4.130)

Thus, by virtue of (4.128) and Lemma 4.6.6, we have

$$\limsup_{\delta\downarrow 0, \epsilon\downarrow 0}\left[\mathcal{S}(\Gamma_1)(t_{\delta\gamma\epsilon}, \psi_{\delta\gamma\epsilon}) - \mathcal{S}(\Gamma_2)(s_{\delta\gamma\epsilon}, \phi_{\delta\gamma\epsilon})\right] \leq 0.$$
(4.131)

Moreover, we know that the norm of T_ϵ is less than ϵ; thus, for any $\gamma > 0$, using (4.128) and taking the lim sup on both sides of (4.123) as δ and ϵ go to 0, we obtain

$$\limsup_{\epsilon \downarrow 0, \delta \downarrow 0} \alpha(V_1(t_{\delta\gamma\epsilon}, \psi_{\delta\gamma\epsilon}) - V_2(s_{\delta\gamma\epsilon}, \phi_{\delta\gamma\epsilon}))$$

$$\leq \limsup_{\epsilon \downarrow 0, \delta \downarrow 0} \left\{ \mathcal{S}(\Gamma_1)(t_{\delta\gamma\epsilon}, \psi_{\delta\gamma\epsilon}) - \mathcal{S}(\Gamma_2)(s_{\delta\gamma\epsilon}, \phi_{\delta\gamma\epsilon}) \right.$$

$$+ [L(t_{\delta\gamma\epsilon}, \psi_{\delta\gamma\epsilon}) - \overline{T_\epsilon(f(t, \psi_{\delta\gamma\epsilon})\mathbf{1}_{\{0\}}, \phi_{\delta\gamma\epsilon})}]$$

$$\left. - [L(s_{\delta\gamma\epsilon}, \phi_{\delta\gamma\epsilon}) + \overline{T_\epsilon(\psi_{\delta\gamma\epsilon}, f(t, \phi_{\delta\gamma\epsilon})\mathbf{1}_{\{0\}})}] \right\}$$

$$\leq \limsup_{\epsilon \downarrow 0, \delta \downarrow 0} \left\{ \left| [L(t_{\delta\gamma\epsilon}, \psi_{\delta\gamma\epsilon}) - L(s_{\delta\gamma\epsilon}, \phi_{\delta\gamma\epsilon})] \right| \right\}. \tag{4.132}$$

Using the Lipschitz continuity of L and Lemma 4.6.2, we see that

$$\limsup_{\delta \downarrow 0, \epsilon \downarrow 0} |L(t_{\delta\gamma\epsilon}, \psi_{\delta\gamma\epsilon}) - L(s_{\delta\gamma\epsilon}, \phi_{\delta\gamma\epsilon})|$$

$$\leq \limsup_{\delta \downarrow 0, \epsilon \downarrow 0} C\left(|t_{\delta\gamma\epsilon} - s_{\delta\gamma\epsilon}| + \|\psi_{\delta\gamma\epsilon} - \phi_{\delta\gamma\epsilon}\|_2 \right)$$

$$= 0; \tag{4.133}$$

moreover, by virtue of (4.133), we get

$$\limsup_{\epsilon \downarrow 0, \delta \downarrow 0} \alpha(V_1(t_{\delta\gamma\epsilon}, \psi_{\delta\gamma\epsilon}) - V_2(s_{\delta\gamma\epsilon}, \phi_{\delta\gamma\epsilon})) \leq 0. \tag{4.134}$$

Since $(t_{\delta\gamma\epsilon}, s_{\delta\gamma\epsilon}, \psi_{\delta\gamma\epsilon}, \phi_{\delta\gamma\epsilon})$ is the maximum of $\Phi_{\delta\gamma} + T_\epsilon$ in $[0, T] \times [0, T] \times \mathbf{C} \times \mathbf{C}$, then, for all $(t, \psi) \in [0, T] \times \mathbf{C}$, we have

$$\Phi_{\delta\gamma}(t, t, \psi, \psi) + T_\epsilon(\psi, \psi) \leq \Phi_{\delta\gamma}(t_{\delta\gamma\epsilon}, s_{\delta\gamma\epsilon}, \psi_{\delta\gamma\epsilon}, \phi_{\delta\gamma\epsilon}) + T_\epsilon(\psi_{\delta\gamma\epsilon}, \phi_{\delta\gamma\epsilon}). \tag{4.135}$$

Then we get

$$V_1(t, \psi) - V_2(t, \psi)$$

$$\leq V_1(t_{\delta\gamma\epsilon}, \psi_{\delta\gamma\epsilon}) - V_2(s_{\delta\gamma\epsilon}, \phi_{\delta\gamma\epsilon})$$

$$- \frac{1}{\delta} \left[\|\psi_{\delta\gamma\epsilon} - \phi_{\delta\gamma\epsilon}\|_2^2 + \|\psi_{\delta\gamma\epsilon}^0 - \phi_{\delta\gamma\epsilon}^0\|_2^2 + |t_{\delta\gamma\epsilon} - s_{\delta\gamma\epsilon}|^2 \right]$$

$$+ 2\gamma \exp(1 + \|\psi\|_2^2 + \|\psi^0\|_2^2)$$

$$- \gamma(\exp(1 + \|\psi_{\delta\gamma\epsilon}\|_2^2 + \|\psi_{\delta\gamma\epsilon}^0\|_2^2) + \exp(1 + \|\phi_{\delta\gamma\epsilon}\|_2^2 + \|\phi_{\delta\gamma\epsilon}^0\|_2^2))$$

$$+ T_\epsilon(\psi_{\delta\gamma\epsilon}, \phi_{\delta\gamma\epsilon}) - T_\epsilon(\psi, \psi)$$

$$\leq V_1(t_{\delta\gamma\epsilon}, \psi_{\delta\gamma\epsilon}) - V_2(s_{\delta\gamma\epsilon}, \phi_{\delta\gamma\epsilon})$$

$$+ 2\gamma \exp(1 + \|\psi\|_2^2 + \|\psi^0\|_2^2) + T_\epsilon(\psi_{\delta\gamma\epsilon}, \phi_{\delta\gamma\epsilon}) - T_\epsilon(\psi, \psi), \tag{4.136}$$

where the last inequality comes from the fact that $\delta > 0$ and $\gamma > 0$. By virtue of (4.122) and (4.134), when we take the lim sup on (4.136) as δ, ϵ, and γ go to zero, we can obtain

$$V_1(t, \psi) - V_2(t, \psi) \leq \limsup_{\gamma \downarrow 0, \epsilon \downarrow 0, \delta \downarrow 0} \Big(V_1(t_{\delta\gamma\epsilon}, \psi_{\delta\gamma\epsilon}) - V_2(s_{\delta\gamma\epsilon}, \phi_{\delta\gamma\epsilon})$$

$$+ 2\gamma \exp(1 + \|\psi\|_2^2 + \|\psi^0\|_2^2) + T_\epsilon(\psi_{\delta\gamma\epsilon}, \phi_{\delta\gamma\epsilon}) - T_\epsilon(\psi, \psi) \Big)$$

$$\leq 0. \tag{4.137}$$

Therefore, we have

$$V_1(t, \psi) \leq V_2(t, \psi), \quad \forall (t, \psi) \in [0, T] \times \mathbf{C}. \tag{4.138}$$

This completes the proof of Theorem 4.5.7.   □

4.7 Conclusions and Remarks

This chapter investigates an optimal stopping problem for a general system of SHDEs with a bounded delay. Using the concept and a construction of the least superharmonic majorant (see Shireyaev [Shi78] and Øksendal [Øks00], the existence and uniqueness results are extended to the optimal stopping problem Problem (OSP01). An infinite-dimensional HJBVI is derived using a Bellman-type dynamic programming principle. It is shown that the value function is the unique viscosity solution of the HJBVI. Due to the length of the current chapter, computational issues of the HJBVI are not addressed in this chapter.

5

Discrete Approximations

In this chapter we address some computational issues and propose various discrete approximations for the optimal classical control considered in Chapter 3. Although it is feasible that these approximations may be applicable to the optimal stopping problems outlined in Chapter 4, we, however, do not attempt to deal with them for the sake of space.

The methods of discrete approximation for the optimal classical control problem include (1) a two-step semidiscretization scheme; (2) a Markov chain approximation; and (3) a finite difference approximation. These three different methods of discrete approximation will be described in Sections 5.2, 5.3, and 5.4, respectively.

Roughly speaking, the two-step semidiscretization scheme only discretizes the time variable of the solution process but not the spatial variable and, therefore, is not effective in providing explicit a numerically approximating solution to the problem. However, the scheme provides the error bound or rate of convergence that the other two methods fail to provide. The Markov chain approximation for controlled diffusion (without delay) started from a series of work done by Kushner and his collaborators (see [Kus77]) and was extended to various cases and summarized in Kushner and Dupuis [KD01]. The ideas behind this is to approximate the original optimal control problem by a sequence of discrete controlled Markov chain problems in which the mean and covariance of the controlled Markov chains satisfy the so-called *local consistency condition*. As an extension to the controlled diffusion, the Markov chain approximation for a certain class of optimal control problems that involved stochastic differential equations with delays have recently been done by Kushner [Kus05, Kus06] and Fischer and Reiss [FR06]. The basic idea behind the two-step semidiscretization scheme and the Markov chain approximation is to approximate the original control problem by a control problem with a suitable discretization approximation, solve the Bellman equation for the approximation, and then prove the convergence of the value functions to that of the original control problem when the discretization mesh goes to zero.

The main ideas behind the two-step semidiscretization scheme and presented in Section 5.2 is derived from Fischer and Nappo [FN07]. The materials in Section 5.3 are mainly due to [Kus05], [Kus06], and [FR06]. The finite-difference scheme presented in Section 5.4, deviates from that of Sections 5.2 and 5.3. The result presented here is an extension of that obtained by Barles and Songanidis [BS91], which approximates the viscosity solution of the infinite-dimensional HJBE directly. The convergence result of the approximation is given in and the computational algorithm, based on the result obtained, is also summarized.

5.1 Preliminaries

The optimal classical problem treated in Chapter 3 is briefly restated for convenience of the readers.

Optimal Classical Control Problem. Given any initial data $(t, \psi) \in [0, T] \times \mathbf{C}$, find an admissible control $u^*(\cdot) \in \mathcal{U}[0, T]$ that maximizes the objective functional $J(t, \psi; u(\cdot))$ defined by

$$
J(t, \psi; u(\cdot)) = E\left[\int_t^T e^{-\alpha(s-t)} L(s, x_s(\cdot; t, \psi, u(\cdot)), u(s)) \, ds \right.
$$

$$
\left. + e^{-\alpha(T-t)} \Psi(x_T(\cdot; t, \psi, u(\cdot)))\right] \tag{5.1}
$$

and subject to the following controlled SHDE with a bounded memory $0 < r < \infty$:

$$
dx(s) = f(s, x_s, u(s)) \, ds + g(s, x_s, u(s)) \, dW(s), \quad s \in [t, T]. \tag{5.2}
$$

Again, the value function $V : [0, T] \times \mathbf{C} \to \Re$ is defined by

$$
V(t, \psi) = \sup_{u(\cdot) \in \mathcal{U}[t,T]} J(t, \psi; u(\cdot)). \tag{5.3}
$$

To obtain more explicit results in error bound and/or rate of convergence for the semidiscretization scheme in Section 5.2, we often use the following set of conditions instead of Assumption 3.1.1 in Chapter 3.

Assumption 5.1.1 *The functions f, g, L, and Ψ satisfy the following conditions:*
(A5.1.1). (Measurability) $f : [0, T] \times \mathbf{C} \times U \to \Re^n$, $g : [0, T] \times \mathbf{C} \times U \to \Re^{n \times m}$, $L : [0, T] \times \mathbf{C} \times U \to \Re$, and $\Psi : \mathbf{C} \to \Re$ are Borel measurable.
(A5.1.2). (Boundedness) The functions f, g, L and Ψ are uniformly bounded by a constant $K_b > 0$, that is,

$$
|f(t, \phi, u)| + |g(t, \phi, u)| + |L(t, \phi, u)| + |\Psi(\psi)| \leq K_b,
$$

$$\forall (t, \phi, u) \in [0, T] \times \mathbf{C} \times U.$$

(A5.1.3). (Uniform Lipschitz and Hölder condition) There is a constant $K_{lip} > 0$ such that for all $\phi, \varphi \in \mathbf{C}$, $t, s \in [0, T]$, and $u \in U$,

$$|f(s, \phi, u) - f(t, \varphi, u)| + |g(s, \phi, u) - g(t, \varphi, u)| \leq K_{lip}(\|\phi - \varphi\| + \sqrt{|t - s|}),$$

$$|L(s, \phi, u) - L(t, \varphi, u)| + |\Psi(\phi) - \Psi(\varphi)| \leq K_{lip}(\|\phi - \varphi\| + \sqrt{|t - s|}).$$

(A5.1.4). (Continuity in the Control) $f(t, \phi, \cdot)$, $g(t, \phi, \cdot)$, and $L(t, \phi, \cdot)$ are continuous functions on U for any $(t, \phi) \in [0, T] \times \mathbf{C}$.

Note that the linear growth condition in the previous chapters has been replaced by uniform boundedness Assumption (A5.1.2) for convenience of convergence analysis.

5.1.1 Temporal and Spatial Discretizations

Let $N \in \aleph \equiv 1, 2, \ldots$ (the set of all positive integers). In order to construct the Nth approximation of the optimal classical control problem, we set $h^{(N)} := \frac{r}{N}$ and define $\lfloor \cdot \rfloor_N$ by $\lfloor t \rfloor_N := h^{(N)} \lfloor \frac{t}{h}^{(N)} \rfloor$, where $\lfloor a \rfloor$ is the integer part of the real number $a \in \Re$. We also set $T^{(N)} = \lfloor \frac{T}{h^{(N)}} \rfloor$ and $\lfloor T \rfloor_N = h^{(N)} \lfloor \frac{T}{h^{(N)}} \rfloor$ and $\mathbf{I}^{(N)} := \{kh^{(N)} \mid k = 0, 1, 2, \ldots\} \cap [0, T^{(N)}]$. As T is the time horizon for the original control problem, $T^{(N)}$ will be the time horizon for the Nth approximating problem. It is clear that $\lfloor T \rfloor_N \rightarrow T$ and $\lfloor t \rfloor_N \rightarrow t$ for any $t \in [0, T]$. The set $\mathbf{I}^{(N)}$ is the time grid of discretization degree N.

Let $\pi^{(N)}$ be the partition of the interval $[-r, 0]$, that is,

$$\pi^{(N)} : r = -Nh^{(N)} < (-N + 1)h^{(N)} < \cdots < -h^{(N)} < 0.$$

Define $\tilde{\pi}^{(N)} : \mathbf{C} \rightarrow (\Re^n)^{N+1}$ as the $(N+1)$-point-mass projection of a continuous function $\phi \in \mathbf{C}$ based on the partition $\pi^{(N)}$, that is,

$$\tilde{\pi}^{(N)} \phi = (\phi(-Nh^{(N)}), \phi((-N + 1)h^{(N)}), \ldots, \phi(-h^{(N)}), \phi(0)).$$

Define $\Pi^{(N)} : (\Re^n)^{N+1} \rightarrow \mathbf{C}$ by $\Pi^{(N)} \mathbf{s} = \tilde{\mathbf{x}}$ for each

$$\mathbf{x} = (x(-Nh^{(N)}), x(-N + 1)h^{(N)}), \ldots, x(-h^{(N)}), x(0)) \in (\Re^n)^{N+1}$$

and $\tilde{\mathbf{x}} \in \mathbf{C}$ by making the linear interpolation between the two (consecutive) time-space points $(kh, x(kh))$ and $((k+1)h, x((k+1)h))$. Therefore, if $\theta \in [-r, 0]$ and $\theta = kh$ for some $k = -N, -N + 1, \ldots, -1, 0$, then $\tilde{\mathbf{x}}(kh) = x(kh)$. If $\theta \in [-r, 0]$ is such that $kh < \theta < (k + 1)h$ for some $k = -N, -N + 1, \ldots, -1$, then

$$\tilde{\mathbf{x}}(\theta) = x(kh) + \frac{(x((k + 1)h) - x(kh))(\theta + kh)}{h}.$$

With a little abuse of notation, we also denote by $\Pi^{(N)} : \mathbf{C} \rightarrow \mathbf{C}$ the operator that maps a function $\varphi \in \mathbf{C}$ to its piecewise linear interpolation $\Pi^{(N)} \varphi$ on the grid $\pi^{(N)}$; that is, if $\theta \in [-(k + 1)h^{(N)}, -kh^{(N)}]$ for some $k = 0, 1, 2, \ldots, N$,

$$(\Pi^{(N)}\varphi)(\theta) = \varphi(-(k+1)h^{(N)})$$
$$+ \frac{(\varphi(-kh^{(N)}) - \varphi(-(k+1)h^{(N)}))(\theta + (k+1)h^{(N)})}{h^{(N)}}.$$

In addition to temporal discretization introduced earlier, we define spatial discretization of the $\Re^n$:

$$\mathbf{S}^{(N)} = \{(k_1, k_2, \ldots, k_n)\sqrt{h^{(N)}} \mid k_i = 0, \pm 1, \pm 2, \ldots, \text{ for } i = 1, 2, \ldots, n\}.$$

When $n = 1$, we simply write $\mathbf{S}^{(N)} = \{kh^{(N)} \mid k = 0, \pm 1, \pm 2, \ldots\}$. Given N, let $(\mathbf{S})^{N+1} = \mathbf{S} \times \cdots \times \mathbf{S}$ be the $(N+1)$-folds Cartesian product of $\mathbf{S}$.

The semidiscretization scheme presented in Section 5.2 involves temporal discretization of the controlled state process. The Markov chain and finite difference approximations presented in Sections 5.3 and 5.4 involve both temporal and spatial discretization.

For simplicity of the notation, we sometime omit N in the superscript and write $h = h^{(N)}$, $\pi = \pi^{(N)}$, $\Pi^{(N)} = \Pi$, $\tilde{\pi} = \tilde{\pi}^{(N)}$, and so forth, whenever there is no danger of ambiguity. This is particularly true when we are working in the context of a fixed N. However, we will carry the full superscripts and/or subscripts when we are working with the quantities that involve different Ns such as the limiting quantity when $N \to \infty$.

5.1.2 Some Lemmas

To prepare what follows, we first recall the Gronwall lemma in Chapter 1 and the two results for the value function $V : [0, T] \times \mathbf{C} \to \Re$ as follows.

Lemma 5.1.2 *Suppose that $h \in L^1([t, T]; \Re)$ and $\alpha \in L^\infty([t, T]; \Re)$ satisfy, for some $\beta \geq 0$,*

$$0 \leq h(s) \leq \alpha(s) + \beta \int_t^s h(\lambda)\,d\lambda \quad \text{for a.e. } s \in [t, T]. \tag{5.4}$$

Then

$$h(s) \leq \alpha(s) + \beta \int_t^s \alpha(\lambda)e^{\beta(\lambda-t)}\,d\lambda \quad \text{for a.e. } s \in [t, T]. \tag{5.5}$$

If, in addition, α is nondecreasing, then

$$h(s) \leq \alpha(s)e^{\beta(s-t)} \quad \text{for a.e. } s \in [t, T]. \tag{5.6}$$

Proposition 5.1.3 *Assume Assumptions (A5.1.1)-(A5.1.3) hold. The value function V satisfies the following properties: There is a constant $K_V \geq 0$ not greater than $3K_{lip}(T+1)e^{3T(T+4m)K_{lip}^2}$ such that for all $t \in [0, T]$ and $\phi, \varphi \in \mathbf{C}$, we have*

$$|V(t, \phi)| \leq K_b(T+1) \quad \text{and} \quad |V(t, \phi) - V(t, \varphi)| \leq K_V\|\phi - \varphi\|.$$

Proof. The first inequality is due to the global boundedness of $L : [0,T] \times \mathbf{C} \times U \to \Re$ and $\Psi : \mathbf{C} \to \Re$ and the second inequality is proven in Lemma 3.3.7 in Chapter 3. $\square$

Proposition 5.1.4 *Assume Assumptions (A5.1.1)-(A5.1.3) hold. Let the initial segment $\psi \in \mathbf{C}$. If ψ is γ-Hölder continuous with $0 < \gamma \le K_H < \infty$, then the function $V(\cdot,\psi) : [0,T] \to \Re$ is Hölder continuous; that is, there is a constant $\tilde{K}_V > 0$ depending only on K_H, K_{lip}, T, and the dimensions such that for all $t, \tilde{t} \in [0,T]$,*

$$|V(t,\psi) - V(\tilde{t},\psi)| \le \tilde{K}_V \left(|t - \tilde{t}|^\gamma \vee \sqrt{|t - \tilde{t}|} \right).$$

Proof. See Lemma 3.3.11 in Chapter 3. We need the following moments of the modulus of continuity of the Itô diffusion, due to Lemma A.4 of Slominski [Slo01].

Lemma 5.1.5 (Slominski's Lemma) *Let $(\Omega, \mathcal{F}, P, \mathbf{F}, W(\cdot))$ be a given m-dimensional Brownian basis. Let $y(\cdot) = \{y(s), s \in [0,T]\}$ be an Itô diffusion of the form*

$$y(s) = y(0) + \int_0^s \tilde{f}(t)\,dt + \int_0^s \tilde{g}(t)\,dW(t), \quad s \in [0,T],$$

where $y(0) \in \Re^n$ and $\tilde{f}(\cdot)$ and $\tilde{g}(\cdot)$ are $\mathbf{F}$-adapted processes with values in $\Re^n$ and $\Re^{n \times m}$, respectively. If $|\tilde{f}|(\cdot)$ and $|\tilde{g}|(\cdot)$ are bounded by a constant $K_b > 0$, then for every $k > 0$ and every $T > 0$, there exists a constant $K_{k,T}$ depending only on K_b, the dimensions n and m, k, and T such that

$$E\left[\sup_{t,s\in[0,T],|t-s|\le h} |y(s) - y(t)|^k\right] \le K_{k,T}\left(h \ln\left(\frac{1}{h}\right)\right)^{\frac{k}{2}}, \quad \forall h \in \left(0, \frac{1}{2}\right].$$

Sketch of the Proof. To save space, we will omit the details of the proof but to point out here that Lemma 5.1.5 can be proved for the special case $m = 1$. The full statement is then derived by a componentwise estimate and a time-change argument; see Theorem 3.4.6 in [KS91], for example. One way of proving the assertion for Brownian motion is to follow the derivation of Lévy's exact modulus of continuity given in Exercise 2.4.8 of [SV79]. The main ingredient there is an inequality due to Garsia, Rademich, and Rumsey; see Theorem 2.1.3, of [SV79, p.47]. $\square$

5.2 Semidiscretization Scheme

In this section we study a semidiscretization scheme for Problem (OCCP) described in Section 3.1 of Chapter 3 and restated below. The discretization scheme consists of two steps: (1) piecewise linear interpolation of the

C-valued segment process for the solution of the controlled SHDE and then (2) piecewise constant control process. By discretizing time in two steps, we construct a sequence of approximating finite-dimensional Markovian optimal control problems. It is shown that the sequence of the value functions for the finite-dimensional Markovian optimal control problems converge to the value function of the original problem. An upper bound on the discretization error or, equivalently, an estimate for the rate of convergence is also given. Much of the material presented in this section can be found in Fischer and Nappo [FN06].

5.2.1 First Approximation Step: Piecewise Constant Segments

Throughout the end of this section, the segment space will be extended from $[-r, 0]$ to $[-r - h^{(N)}, 0]$ and the domain of the admissible control $u(\cdot) \in \mathcal{U}[t, T]$ will be enlarged from $[t, T]$ to $[t - h^{(N)}, T]$ by setting $u(s) = u(t)$ for all $s \in [t - h^{(N)}, t]$ according to the discretization degree N if the initial time $t \in [0, T]$ is such that $t \neq \lfloor t \rfloor_N$. Denote by $\mathbf{C}(N)$ the space $C([-r - h^{(N)}, 0]; \Re^n)$ of n-dimensional continuous functions defined on $[-r - h^{(N)}, 0]$ and the class of admissible control with an enlarged time domain $[t - h^{(N)}, T]$ defined above as $\mathcal{U}[t - h^{(N)}, T]$. For a continuous function or process $z(\cdot)$ defined on the interval $[-r - h^{(N)}, \infty)$, let $\tilde{z}_t^{(N)}$ denote the segment of $z(\cdot)$ at time $t \geq 0$ of the length $r + h^{(N)}$, that is,

$$\tilde{z}_t^{(N)}(\theta) = z(t + \theta), \quad \theta \in [-r - h^{(N)}, 0].$$

Given the initial datum $(t, \psi) \in [0, T] \times \mathbf{C}(N)$ and $u(\cdot) \in \mathcal{U}[t - h^{(N)}, T]$, we define the Euler-Maruyama approximation $z(\cdot) = \{z^{(N)}(s; t, \psi, u(\cdot)), s \geq t\}$ of degree N of the solution of $x(\cdot) = \{x(s; t, \psi, u(\cdot)), s \geq t\}$ to (5.2) under the control process $u(\cdot)$ with the initial datum (t, ψ) as the solution to

$$z(s) = \psi(0) + \int_{\lfloor t \rfloor_N}^s f^{(N)}(\lambda, \tilde{z}_\lambda^{(N)}, u(\lambda)) \, d\lambda$$

$$+ \int_{\lfloor t \rfloor_N}^s g^{(N)}(\lambda, \tilde{z}_\lambda^{(N)}, u(\lambda)) \, dW(\lambda), \quad s \in [t, T] \qquad (5.7)$$

with the initial datum $(t, \psi) \in [0, T] \times \mathbf{C}(N)$, where

$$f^{(N)}(t, \phi, u) := f(\lfloor t \rfloor_N, \Pi^{(N)} \phi_{\lfloor t \rfloor_N - t}, u),$$
$$g^{(N)}(t, \phi, u) := g(\lfloor t \rfloor_N, \Pi^{(N)} \phi_{\lfloor t \rfloor_N - t}, u), \quad (t, \psi, u) \in [0, T] \times \mathbf{C}(N) \times U.$$

Similarly to $f^{(N)}$ and $g^{(N)}$, we also defined the functions $L^{(N)}$ and $\Psi^{(N)}$ as

$$L^{(N)}(t, \phi, u) := L(\lfloor t \rfloor_N, \Pi^{(N)} \phi_{\lfloor t \rfloor_N - t}, u),$$
$$\Psi^{(N)}(\phi) := \Psi(\Pi^{(N)} \phi_{\lfloor t \rfloor_N - t}), \quad (t, \phi, u) \in [0, T] \times \mathbf{C}(N) \times U.$$

Thus, $f^{(N)}(t, \phi, u)$, $g^{(N)}(t, \phi, u)$, $L^{(N)}(t, \phi, u)$, and $\Psi^{(N)}(\phi)$ are calculated by evaluating the functions f, g, and L at $(t, \hat{\phi}, u)$ and Ψ at $\hat{\phi}$, where $\hat{\phi}$ is the segment in $\mathbf{C}$ that arises from the piecewise linear interpolation with mesh size $\frac{r}{N}$ of the restriction of ϕ to the interval $[\lfloor t \rfloor_N - t - r, \lfloor t \rfloor_N - t]$. Notice that the control action $u \in U$ remains unchanged.

Note that for each N, the functions $f^{(N)}(t, \phi, u)$ and $g^{(N)}(t, \phi, u)$ satisfy Assumption 5.1.1. This guarantees that given any admissible control $u(\cdot) \in \mathcal{U}[t - h^{(N)}, T]$, (5.7) has a unique solution for each initial datum $(t, \psi) \in [0, T] \times \mathbf{C}(N)$. Thus, the process $z(\cdot) = \{z^{(N)}(s; t, \psi, u(\cdot)), s \in [t, T]\}$ of degree N is well defined.

Define the objective functional $J^{(N)} : [0, T^{(N)}] \times \mathbf{C}(N) \times \mathcal{U}[t - h^{(N)}, T] \to \Re$ of discretization degree N by

$$J^{(N)}(t, \psi, u(\cdot)) = E\left[\int_{\lfloor t \rfloor_N}^{T^{(N)}} L^{(N)}(s, \tilde{z}_s^{(N)}, u(s)) \, ds + \Psi^{(N)}(\tilde{z}_{T^{(N)}}^{(N)}) \right]. \tag{5.8}$$

As $f^{(N)}$, $g^{(N)}$, $L^{(N)}$, and $\Psi^{(N)}$ are Lipschitz continuous in the segment variable ϕ (uniformly in $(t, u) \in [0, T] \times U$) under the sup-norm on $\mathbf{C}(N)$, the value function $V^{(N)} : [0, T^{(N)}] \times \mathbf{C}(N) \to \Re$ is determined by

$$V^{(N)}(t, \psi) = \sup_{u(\cdot) \in \mathcal{U}[t - h^{(N)}, T]} J^{(N)}(t, \psi; u(\cdot)). \tag{5.9}$$

When the initial time $t \in [0, T]$ is such that $t = \lfloor t \rfloor_N$, then (5.7), (5.8), and (5.9) reduce to the following three equations, respectively.

$$z(s) = \psi(0) + \int_t^s f(\lambda, \Pi^{(N)} z_{\lfloor \lambda \rfloor_N}, u(\lambda)) \, d\lambda$$
$$+ \int_t^s g(\lambda, \Pi^{(N)} z_{\lfloor \lambda \rfloor_N}, u(\lambda)) \, dW(\lambda), \quad s \in [t, \lfloor T \rfloor_N], \tag{5.10}$$

$$J^{(N)}(t, \psi, u(\cdot)) = E\left[\int_t^{T^{(N)}} L(s, \Pi^{(N)} z_{\lfloor s \rfloor_N}, u(s)) \, ds + \Psi^{(N)}(\Pi^{(N)} z_{T^{(N)}}) \right];$$
$$\tag{5.11}$$

and

$$V^{(N)}(t, \psi) = \sup_{u(\cdot) \in \mathcal{U}[t, T]} J^{(N)}(t, \psi; u(\cdot)). \tag{5.12}$$

The following proposition estimates the difference between the solution $x(\cdot)$ of (5.2) and the solution $z^{(N)}(\cdot)$ of (5.7).

Proposition 5.2.1 *Assume Assumptions (A5.1.1)-(A5.1.3) hold. Let $x(\cdot)$ be the solution to (5.2) under admissible $u(\cdot) \in \mathcal{U}[t, T]$ and with initial datum (t, ϕ) and $z^{(N)}(\cdot)$ is the solution to (5.7) of discretization degree N under the same control process $u(\cdot)$ and with initial datum $(t, \psi) \in [0, T] \times \mathbf{C}(N)$ being such that $\psi|_{[-r,0]} = \phi$. Let the initial segment $\psi \in \mathbf{C}$ be γ-Hölder continuous with $0 \leq \gamma \leq K_H < \infty$. Then there is a positive constant $\tilde{K}$ depending only on*

γ, K_H, K_{lip}, K_b, T, and the dimensions n and m such that for all $N = 1, 2, \ldots$ with $N \geq 2r$, all $t \in \mathbf{I}^{(N)}$, $u(\cdot) \in \mathcal{U}[t, T]$ such that

$$E\left[\sup_{s \in [-r, T]} |x(s) - z^{(N)}(s)|\right] \leq \tilde{K}\left((h^{(N)})^{\gamma} \vee \sqrt{h^{(N)} ln(\frac{1}{h^{(N)}})}\right).$$

Proof. Notice that $h^{(N)} \leq \frac{1}{2}$ since $N \geq 2r$ and observe that $z(\cdot) := z^{(N)}(\cdot)$ depends on the initial segment ψ only through $\psi|_{[-r, 0]} = \phi$.

Using Hölder inequality , we find that

$$E\left[\sup_{s \in [t-r, T]} |x(s) - z(s)|^2\right]$$

$$= E\left[\sup_{s \in [t, T]} |x(s) - z(s)|^2\right] \quad (\text{since } \psi|_{[-r, 0]} = \phi)$$

$$\leq 2E\left[\left|\int_t^T f(s, x_s, u(s)) - f(\lfloor s \rfloor_N, \Pi^{(N)}(z_{\lfloor s \rfloor_N}), u(s))\, ds\right|^2\right]$$

$$+ 2E\left[\sup_{s \in [t, T]} \left|\int_t^s (g(\lambda, x_\lambda, u(\lambda)) - g(\lfloor \lambda \rfloor_N, \Pi^{(N)}(z_{\lfloor \lambda \rfloor_N}), u(\lambda)))\, dW(\lambda)\right|^2\right].$$

Now, by Hölder inequality, Assumption (A5.1.3), and the Fubini theorem, we have

$$E\left[\left|\int_t^T f(s, x_s, u(s)) - f(\lfloor s \rfloor_N, \Pi^{(N)}(z_{\lfloor s \rfloor_N}), u(s))\, ds\right|^2\right]$$

$$\leq 2T E\left[\int_t^T |f(s, x_s, u(s)) - f(\lfloor s \rfloor_N, \Pi^{(N)}(z_{\lfloor s \rfloor_N}), u(s))|^2\, ds\right]$$

$$\leq 4T K_{lip}^2 \left(\int_t^T |s - \lfloor s \rfloor_N|\, ds + E\left[\int_t^T \|x_s - \Pi^{(N)}(z_{\lfloor s \rfloor_N})\|^2\, ds\right]\right)$$

$$\leq 4T K_{lip}^2 \left(T h^{(N)} + \int_t^T E[\|x_s - \Pi^{(N)}(z_{\lfloor s \rfloor_N})\|^2]\, ds\right).$$

Using Doob's maximal inequality (1.18)

$$E\left[\sup_{s \in [0, T]} \left|\int_0^s g(t)\, dW(t)\right|^2\right] \leq K \int_0^T E[|g(s)|^2]\, ds,$$

$$\forall T > 0, \text{ and } \forall g(\cdot) \in L_{\mathbf{F}}^{2, loc}(0, \infty; \Re^{n \times m}),$$

Itô isometry, Assumption (A5.1.3), and the Fubini theorem, we have

$$E\left[\sup_{s\in[t,T]}\left|\int_t^s(g(\lambda,x_\lambda,u(\lambda))-g(\lfloor\lambda\rfloor_N,\Pi^{(N)}(z_{\lfloor\lambda\rfloor_N}),u(\lambda)))\,dW(\lambda)\right|^2\right]$$

$$\leq 4E\left[\left|\int_t^T(g(\lambda,x_\lambda,u(\lambda))-g(\lfloor\lambda\rfloor_N,\Pi^{(N)}(z_{\lfloor\lambda\rfloor_N}),u(\lambda)))\,dW(\lambda)\right|^2\right]$$

$$\leq 8mE\left[\int_t^T\left|g(s,x_s,u(s))-g(\lfloor s\rfloor_N,\Pi^{(N)}(z_{\lfloor s\rfloor_N}).u(s)))\right|^2\,ds\right]$$

Therefore,

$$E\left[\sup_{s\in[t,T]}\left|\int_t^s(g(\lambda,x_\lambda,u(\lambda))-g(\lfloor\lambda\rfloor_N,\Pi^{(N)}(z_{\lfloor\lambda\rfloor_N}),u(\lambda)))\,dW(\lambda)\right|^2\right]$$

$$\leq 16mTK_{lip}^2\left(\int_t^T|s-\lfloor s\rfloor_N|\,ds+E\left[\int_t^T\|x_s-\Pi^{(N)}(z_{\lfloor s\rfloor_N})\|^2\,ds\right]\right)$$

$$\leq 16mTK_{lip}^2(Th^{(N)}+\int_t^TE[\|x_s-\Pi^{(N)}(z_{\lfloor s\rfloor_N})\|^2]\,ds.$$

By triangular inequality, we have

$$\int_t^TE[\|x_s-\Pi^{(N)}(z_{\lfloor s\rfloor_N})\|^2]\,ds=3\int_t^TE[\|x_s-x_{\lfloor s\rfloor_N}\|^2]\,ds$$

$$+3\int_t^TE[\|x_{\lfloor s\rfloor_N}-z_{\lfloor s\rfloor_N}\|^2]\,ds$$

$$+3\int_t^TE[\|z_{\lfloor s\rfloor_N}-\Pi^{(N)}z_{\lfloor s\rfloor_N}\|^2]\,ds.$$

Therefore,

$$E\left[\sup_{s\in[t-r,T]}|x(s)-z(s)|^2\right]\leq 4(T+4m)K_{lip}^2\left(Th^{(N)}+3\int_t^TE[\|x_s-x_{\lfloor s\rfloor_N}\|^2]\,ds\right.$$

$$\left.+\int_t^TE[\|z_{\lfloor s\rfloor_N}-\Pi^{(N)}z_{\lfloor s\rfloor_N}\|^2]\,ds\right)$$

$$+12(T+4m)K_{lip}^2\int_t^TE[\|x_{\lfloor s\rfloor_N}-z_{\lfloor s\rfloor_N}\|^2]\,ds.$$

In the last step of the above estimate, we use the following two preliminary steps:

First, for all $s\in[0,T]$,

$$E[\|x_s - x_{\lfloor s \rfloor_N}\|^2] \leq 2E \left[\sup_{\theta, \tilde{\theta} \in [-r,0], |\theta - \tilde{\theta}| \leq h^{(N)}} |\phi(\theta) - \phi(\tilde{\theta})|^2 \right]$$

$$+2E \left[\sup_{s, \tilde{s} \in [t,T], |s - \tilde{s}| \leq h^{(N)}} |x(s) - x(\tilde{s})|^2 \right]$$

$$\leq 2K^2(H)(h^{(N)})^{2\gamma} + 2K_{2,T}h^{(N)}ln(\frac{1}{h^{(N)}}), \quad \forall s \in [t, T].$$

Second,

$$E[\|z_{\lfloor s \rfloor_N} - \Pi^{(N)}z_{\lfloor s \rfloor_N}\|^2]$$

$$= E \left[\sup_{\tilde{s} \in [\lfloor s \rfloor_N - r, \lfloor s \rfloor_N]} \|z_{\lfloor s \rfloor_N} - \tilde{z}_{\lfloor s \rfloor_N}^{(N)}\|^2 \right]$$

$$\leq 2E \left[\sup_{\theta \in [-r,0)} (|\phi(\theta) - \phi(\lfloor \theta \rfloor_N)|^2 + |\phi(\theta - \phi(\lfloor \theta \rfloor_N)|^2) \right]$$

$$+2E \left[\sup_{\tilde{s} \in [0,s)} (|z(\tilde{s}) - z(\lfloor \tilde{s} \rfloor_N)|^2 + |z(\tilde{s}) - z(\lfloor \tilde{s} \rfloor_N) + h^{(N)}|^2) \right]$$

$$\leq 4K^2(H)(h^{(N)})^{2\gamma} + 4E \left[\sup_{t, \tilde{t} \in [0,s], |t - \tilde{t}| \leq h^{(N)}} |z(t) - z(\tilde{t})|^2 \right]$$

$$\leq 4K^2(H)(h^{(N)})^{2\gamma} + K_{2,T}h^{(N)}ln(\frac{1}{h^{(N)}}).$$

Therefore,

$$E \left[\sup_{s \in [t-r,T]} |x(s) - z(s)|^2 \right]$$

$$= E \left[\sup_{s \in [t,T]} |x(s) - z(s)|^2 \right]$$

$$\leq 4T(T + 4m)K_{lip}^2 \left(h^{(N)} + 18K^2(H)(h^{(N)})^{2\gamma} + 18K_{2,T}h^{(N)}ln\left(\frac{1}{h^{(N)}}\right) \right)$$

$$+12(T + 4m)K_{lip}^2 \int_t^T E \left[\sup_{\lambda \in [t,s]} |x(\lambda) - z(\lambda)|^2 \right] ds.$$

Applying Gronwall's lemma (see Lemma 5.1.2), we have the assertion that

$$E \left[\sup_{s \in [-r,T]} |x(s) - z^{(N)}(s)| \right] \leq \tilde{K} \left((h^{(N)})^{\gamma} \vee \sqrt{h^{(N)}ln(\frac{1}{h^{(N)}})} \right)$$

for some constant $\tilde{K} > 0$. $\quad \square$

The order of approximation error or rate of convergence obtained in Proposition 5.2.1 for the underlying dynamics carries over to the approximation of the corresponding value function. This works due to the Lipschitz continuity of L and Ψ in the segment variable, the bound on the moments of the modulus of continuity from Lemma 5.1.5, and the fact that the error bound in Proposition 5.2.1 is uniform over all $u(\cdot) \in \mathcal{U}[t, T]$. By Proposition 5.1.4, it is known that the original value function V is Hölder continuous in time provided that the initial segment is Hölder continuous. It is therefore enough to compare V and $V^{(N)}$ on the grid $\mathbf{I}^{(N)} \times \mathbf{C}$. This is the content of the next proposition. Again, the order of the error will be uniform only over those initial segments that are γ-Hölder continuous of some $\gamma > 0$; the constant in the error bound also depends on the Hölder constant of the initial segment. We start with comparing solutions to (5.2) and (5.10) for initial times $t \in \mathbf{I}^{(N)}$.

Theorem 5.2.2 *Assume Assumption 5.1.1 holds. Let $\phi \in \mathbf{C}$ be γ-Hölder continuous with $0 < \gamma \leq K_H < \infty$. Then there is a constant $\tilde{K}$ depending only on γ, K_H, K_{lip}, K_b, T, and the dimensions (n and m) such that for all $N = 1, 2, \ldots$ with $N \geq 2r$, all initial times $t \in \mathbf{I}^{(N)}$, we have*

$$|V(t, \phi) - V^{(N)}(t, \psi)| \leq \sup_{u(\cdot) \in \mathcal{U}[t,T]} |J(t, \phi; u(\cdot)) - J^{(N)}(t, \psi; u(\cdot))|$$

$$\leq \tilde{K} \left((h^{(N)})^\gamma \vee \sqrt{h^{(N)} ln(\frac{1}{h^{(N)}})} \right),$$

where $\psi \in \mathbf{C}(N)$ is such that $\psi|_{[-r,0]} = \phi$, and $\mathbf{C}(N)$ is the space of n-dimensional continuous functions defined on the extended interval $[\cdot, \cdot]$.

Proof. The proof of the first inequality is straightforward.

$$|V(t, \phi) - V^{(N)}(t, \psi)| \leq \left| \sup_{u(\cdot) \in \mathcal{U}[t,T]} J(t, \phi; u(\cdot)) - \sup_{u(\cdot) \in \mathcal{U}[t,T]} J^{(N)}(t, \psi; u(\cdot)) \right|$$

$$\leq \left| \sup_{u(\cdot) \in \mathcal{U}[t,T]} (J(t, \phi; u(\cdot)) - J^{(N)}(t, \psi; u(\cdot))) \right|$$

$$\leq \sup_{u(\cdot) \in \mathcal{U}[t,T]} |J(t, \phi; u(\cdot)) - J^{(N)}(t, \psi; u(\cdot))|.$$

Now, let $u(\cdot) \in \mathcal{U}[t, T]$ be any admissible control. Let $x(\cdot) = \{x(s) = x(s; t, \phi, u(\cdot)), s \in [t, T]\}$ be the (strong) solution to (5.2) under $u(\cdot)$ and with the initial datum $(t, \phi) \in [0, T] \times \mathbf{C}$ and let $z(\cdot) = \{z(s) = z(s; t, \psi, u(\cdot)), s \in [t, T]\}$ be the (strong) solution to (5.7) under $u(\cdot)$ and with the initial datum $(t, \psi) \in [0, T] \times \mathbf{C}(N)$. Using Assumption (A5.1.2) and the hypothesis that $t \in \mathbf{I}^{(N)}$, the difference

$$|J(t, \phi; u(\cdot)) - J^{(N)}(t, \psi; u(\cdot))|$$

can be estimated as follows:

$$|J(t, \phi; u(\cdot)) - J^{(N)}(t, \psi; u(\cdot))|$$

$$= E\left[\left|\int_t^T L(s, x_s, u(s))\, ds - \int_t^{\lfloor T \rfloor_N} L(\lfloor s \rfloor_N, \Pi^{(N)}(z_{\lfloor s \rfloor_N}, u(s))\, ds\right|\right]$$

$$+ E\left[\Psi(x_T) + \Psi(\Pi^{(N)}(z_{\lfloor T \rfloor_N})\right].$$

Therefore,

$$|J(t, \phi; u(\cdot)) - J^{(N)}(t, \psi; u(\cdot))|$$

$$\leq E\left[\int_t^T |L(s, x_s, u(s))\, ds - L(\lfloor s \rfloor_N, \Pi^{(N)}(z_{\lfloor s \rfloor_N}, u(s))|ds\right]$$

$$= E\left[\int_{\lfloor T \rfloor_N}^T |L(\lfloor s \rfloor_N, \Pi^{(N)}(z_{\lfloor s \rfloor_N}, u(s))|ds\right]$$

$$+ E\left[|\Psi(x_T) - \Psi(\Pi^{(N)}(z_{\lfloor T \rfloor_N})|\right]$$

$$\leq K_b|T - T^{(N)}| + E[|\Psi(\Pi^{(N)}(z_{T^{(N)}})) - \Psi(x_T)|]$$

$$+ E\left[\int_t^{T^{(N)}} |L(\lfloor s \rfloor_N, \Pi^{(N)}(z_{\lfloor s \rfloor_N}), u(s)) - L(s, x_s, u(s))|ds\right].$$

Recall that $|T - \lfloor T \rfloor_N| = T - \lfloor T \rfloor_N \leq h^{(N)}$. Hence, $K_b|T - \lfloor T \rfloor_N| \leq K_b h^{(N)}$. Now, using Assumption (A5.1.3), we see that

$$E[|\Psi(\Pi^{(N)}(z_{\lfloor T \rfloor_N})) - \Psi(x_T)|]$$

$$\leq K_{lip}\big(E[\|z_{\lfloor T \rfloor_N} - x_{\lfloor T \rfloor_N}\|] + E[\|\Pi^{(N)}(z_{\lfloor T \rfloor_N}) - z_{\lfloor T \rfloor_N}\|]$$

$$+ E[\|x_{\lfloor T \rfloor_N} - x_T\|]\big)$$

$$\leq K_{lip}\bigg(\tilde{K}((h^{(N)})^\gamma \vee \sqrt{h^{(N)} ln\left(\frac{1}{h^{(N)}}\right)} + 3K_H(h^{(N)})^\gamma$$

$$+ 3K_{1,T}\sqrt{h^{(N)} ln\left(\frac{1}{h^{(N)}}\right)}\bigg),$$

where $\tilde{K}$ is a constant as in Proposition 5.2.1 and $K_{1,T}$ is a constant as in Lemma 5.1.5. Notice that $x(\cdot)$ as well as $z(\cdot)$ are Itô diffusions with coefficients bounded by the constant K_b from Assumption (A5.1.2). In the same way, also using the Hölder continuity of L in time and recalling that $|s - \lfloor s \rfloor_N| \leq h^{(N)}$ for all $s \in [t, T]$, an estimate of

$$E\left[\int_t^{\lfloor T \rfloor_N} |L(\lfloor s \rfloor_N, \Pi^{(N)}(z_{\lfloor s \rfloor_N}), u(s)) - L(s, x_s, u(s))|\, ds\right]$$

is given by

$$E\left[\int_t^{\lfloor T\rfloor_N}|L(\lfloor s\rfloor_N,\Pi^{(N)}(z_{\lfloor s\rfloor_N}),u(s))-L(s,x_s,u(s))|\,ds\right]$$

$$\leq K_{lip}(\lfloor T\rfloor_N-t)\left(\sqrt{h^{(N)}}+3K_{1,T}\sqrt{h^{(N)}ln\left(\frac{1}{h^{(N)}}\right)}\right)$$

$$+(\tilde{K}+3K_H)((h^{(N)})^\gamma\vee\sqrt{h^{(N)}ln\left(\frac{1}{h^{(N)}}\right)}).$$

Combining the above three estimates, we obtain the assertion.   □

By virtue of the above theorem, we can replace the original classical control problem with the sequence of approximating control problems defined above. The error at approximation degree N in terms of the difference between the corresponding value functions V and $V^{(N)}$ is not greater than a multiple of $(\frac{r}{N})^\gamma$ for γ-Hölder continuous initial segments if $\gamma\in(0,\frac{1}{2})$, where the proportionality factor is affine in the Hölder constant, and is less than a multiple of $\sqrt{\frac{ln(N)}{N}}$ if $\gamma\geq\frac{1}{2}$.

Although we obtain an error bound for the approximation of V by sequence of value functions $\{V^{(N)},N=1,2,\ldots\}$ only for Hölder continuous initial segments, the proofs of Proposition 5.2.1 and Theorem 5.2.2 show that pointwise convergence of the value functions holds true for all initial segments $\phi\in\mathbf{C}$. Recall that a function $\phi:[-r,0]\to\Re^n$ is continuous if and only if $\sup_{t,s\in[-r,0],|t-s|\leq h}|\phi(t)-\phi(s)|$ tends to zero as $h\downarrow0$. We, therefore, have the following result.

Corollary 5.2.3 *Assume Assumptions (A5.1.1)-(A5.1.3). Then for all* $(t,\psi)\in[0,T]\times\mathbf{C}$,

$$\lim_{N\to\infty}|V(t,\psi)-V^{(N)}(\lfloor t\rfloor_N,\psi)|=0.$$

Similarly to the value function of the original optimal classical control problem, we can also show that the function $V^{(N)}(t,\cdot):\mathbf{C}(N)\to\Re$ is Lipschitz continuous uniformly in $t\in I^{(N)}$ with the Lipschitz constant not depending on the discretization degree N. Since $t\in I^{(N)}$, we may interpret $V^{(N)})(t,\cdot)$ as a function defined on $\mathbf{C}$.

Proposition 5.2.4 *Assume Assumptions (A5.1.1)-(A5.1.3) hold. Let* $V^{(N)}$ *be the value function of discretization degree* N. *Then* $|V^{(N)}|$ *is bounded by* $K_b(T+1)$. *Moreover, if* $t\in\mathbf{I}^{(N)}$, *then* $V^{(N)}(t,\cdot)$ *as a function of* $\mathbf{C}$ *satisfies the following Lipschitz condition:*

$$|V^{(N)}(t,\phi)-V^{(N)}(t,\varphi)|\leq3K_{lip}(T+1)e^{3T(T+4m)K_{lip}^2}\|\phi-\varphi\|,\quad\forall\phi,\varphi\in\mathbf{C}.$$

5.2.2 Second Approximation Step: Piecewise Constant Strategies

Notice in the previous subsection that we only discretize in the time variable as well as the segment space in time of the controlled SHDE and the objective function without discretizing the control process $u(\cdot)$. In this subsection, we will also discretize the time variable of the admissible controls $u(\cdot) \in \mathcal{U}[t,T]$. To this end, for $M = 1, 2, \ldots$, set

$$\mathcal{U}^{(M)}[t,T] = \{u(\cdot) \in \mathcal{U}[t,T] \mid u(s) \text{ is } \mathcal{F}(\lfloor s \rfloor_M) \text{ measurable and}$$

$$u(s) = u(\lfloor s \rfloor_M) \text{ for each } s \in [t,T]\}. \tag{5.13}$$

Recall $\lfloor s \rfloor_M = \frac{r}{M}\lfloor \frac{M}{r}s \rfloor$. Hence, $\mathcal{U}^{(M)}[t,T]$ is the set of all U-valued $\mathbf{F}(t)$-progressively measurable processes that are right-continuous and piecewise constant in time relative to the grid $\{k\frac{r}{M} \mid k = 0, 1, 2, \ldots\}$ and, in addition, $\{\mathcal{F}(\lfloor s \rfloor_M), s \in [t,T]\}$-adapted. For the purpose of approximating the control problem of degree N, we will use strategies in $\mathcal{U}^{(NM)}[t,T]$. We write $\mathcal{U}^{(N,M)}[t,T]$ for $\mathcal{U}^{(NM)}[t,T]$. Note that $u(\cdot) \in \mathcal{U}^{(N,M)}[t,T]$ has M-times finer discretization of that of the discretization of degree N.

With the same dynamics and the same performance criterion as in the previous subsection, for each $N = 1, 2, \ldots$, we introduce a family of value functions $\{V^{(N,M)}, M = 1, 2, \ldots\}$ defined on $[t, T^{(N)}] \times \mathbf{C}(N)$ by setting

$$V^{(N,M)}(t,\psi) := \sup_{u(\cdot) \in \mathcal{U}^{(N,M)}[t,T]} J^{(N)}(t,\psi; u(\cdot)). \tag{5.14}$$

We will refer to $V^{(N,M)}$ as the value function of degree (N,M). Note that, by construction, we have $V^{(N,M)}(t,\psi) \leq V^{(N)}(t,\psi)$ for all $(t,\psi) \in [0, T^{(N)}] \times \mathbf{C}(N)$, since $\mathcal{U}^{(M)}[t,T] \subset \mathcal{U}[t,T]$. Hence, in estimating the approximation error, we only need an upper bound for $V^{(N)} - V^{(N,M)}$.

As with $V^{(N)}$, if the initial time $t \in \mathbf{I}^{(N)}$, then $V^{(N,M)}(t,\psi)$ depends on ψ only through its restriction $\psi|_{[-r,0]} \in \mathbf{C}$ to the interval $[-r,0]$. We write $V^{(N,M)}(t,\psi)$ for this function. The dynamics and costs, in this case, can again be represented by (5.7) and (5.8), respectively. Again, if $t \in \mathbf{I}^{(N)}$, we have $V^{(N,M)}(t,\phi) = V^{(N,M)}(t, \Pi^{(N)}(\phi))$ for all $\phi \in \mathbf{C}$.

The following two propositions state Bellman's DDP for the value functions $V^{(N)}$ and $V^{(N,M)}$, respectively. The proofs of of the following two discrete versions of DDP can be reproduced from that of Theorem 3.3.9 in Chapter 3 and are, therefore, omitted here.

Proposition 5.2.5 *Assume Assumptions (A5.1.1)-(A5.1.3) hold. Let $(t,\psi) \in [0, T^{(N)}] \times \mathbf{C}(N)$. Then for each $s \in \mathbf{I}^{(N)}$,*

$$V^{(N)}(t,\psi) = \sup_{u(\cdot) \in \mathcal{U}[t,T]} E\left[\int_t^s L^{(N)}(\lambda, \Pi^{(N)}(z_\lambda), u(\lambda))\, ds + V^{(N)}(s, \Pi^{(N)}(z_s))\right],$$

where $z(\cdot)$ is the solution to (5.7) of degree N under $u(\cdot) \in \mathcal{U}[t,T]$ and with initial condition (t,ψ). If $t \in \mathbf{I}^{(N)}$ and $s \in \mathbf{I}^{(N)} \cap [0,T]$, then

$$V^{(N)}(t,\psi) = \sup_{u(\cdot) \in \mathcal{U}[t,T]} E\left[\int_t^s L(\lfloor \lambda \rfloor_N, \Pi^{(N)}(z_{\lfloor \lambda \rfloor_N}), u(\lambda))\, d\lambda\right.$$

$$\left. + V^{(N)}(s, \Pi^{(N)}(z_s))\right],$$

where $V^{(N)}(t, \cdot)$ and $V^{(N)}(s, \cdot)$ are defined as functionals on $\mathbf{C}$ and ϕ is the restriction of ψ to the interval $[-r, 0]$.

Proposition 5.2.6 *Assume Assumptions (A5.1.1)-(A5.1.3) hold. Let $(t, \psi) \in [0, T^{(N)}] \times \mathbf{C}(N)$. Then for $s \in \mathbf{I}^{(NM)} \cap [0, T]$,*

$$V^{(N,M)}(t, \psi) = \sup_{u(\cdot) \in \mathcal{U}^{(N,M)}[t,T]} E\left[\int_t^s L^{(N)}(\lfloor \lambda \rfloor_N, \Pi^{(N)}(z_\lambda), u(\lambda)) \, d\lambda \right.$$
$$\left. + V^{(N)}(s, \Pi^{(N)}(z_s)) \right],$$

where $z(\cdot)$ is the solution to (5.7) of degree N under control process $u(\cdot)$ and with the initial datum (t, ψ). If $t \in \mathbf{I}^{(N)}$ and $s \in \mathbf{I}^{(N)} \cap [t, T]$, then

$$V^{(N,M)}(t, \phi) = \sup_{u(\cdot) \in \mathcal{U}^{(N,M)}[t,T]} E\left[\int_t^s L(\lfloor \lambda \rfloor_N, \Pi^{(N)}(z_{\lfloor \lambda \rfloor_N}), u(\lambda)) \, d\lambda \right.$$
$$\left. + V^{(N,M)}(s, \Pi^{(N)}(z_s)) \right],$$

where $V^{(N,M)}(t, \cdot)$ and $V^{(N,M)}(s, \cdot)$ are defined as functions on $\mathbf{C}$ and ϕ is the restriction of ψ to the interval $[-r, 0]$.

The next result gives a bound on the order of the global approximation error between the value functions of degree N and (N, M) provided that the local approximation error is of order greater than 1 in the discretization step.

Theorem 5.2.7 *Assume Assumptions (A5.1.1)-(A5.1.3) hold. Let $N, M = 1, 2, \ldots$. Suppose that for some positive constants $\hat{K}, \delta > 0$, the following holds: For any $(t, \phi) \in \mathbf{I}^{(N)} \times \mathbf{C}$, $u(\cdot) \in \mathcal{U}[t, T]$, there is $\bar{u}(\cdot) \in \mathcal{U}^{(N,M)}[t, T]$ such that*

$$E\left[\int_t^{t+h^{(N)}} L(s, \Pi^{(N)}\phi, \bar{u}(s)) \, ds + V^{(N)}(t + h^{(N)}, \bar{z}_{t+h^{(N)}}) \right]$$
$$\leq E\left[\int_t^{t+h^{(N)}} L(s, \Pi^{(N)}\phi, u(s)) \, ds + V^{(N)}(t + h^{(N)}, z_{t+h^{(N)}}) \right]$$
$$+ \hat{K}(h^{(N)})^{1+\delta}, \tag{5.15}$$

where $z(\cdot)$ is the solution to (5.7) of degree N under control process $u(\cdot)$ and $\bar{z}(\cdot)$ is the solution to (5.7) of degree N under control process $\bar{u}(\cdot)$, both with initial datum $(t, \psi) \in \mathbf{I}^{(N)} \times \mathbf{C}(N)$ and $\psi|_{[-r,0]} = \phi$. Then

$$\left| V^{(N,M)}(t, \phi) - V^{(N)}(t, \phi) \right| \leq T\hat{K}(h^{(N)})^\delta, \quad \forall (t, \phi) \in \mathbf{I}^{(N)} \times \mathbf{C}.$$

Proof. Let $N, M = 1, 2, \ldots$. Recall that $V^{(N,M)} \leq V^{(N)}$ by construction. It is therefore sufficient to prove the upper bound for $V^{(N)} - V^{(N,M)}$. Suppose

Condition (5.15) is satisfied for N, M, and some constants $\hat{K}, \delta > 0$. Observe that $V^{(N,M)}(\lfloor T \rfloor_N, \cdot) = V^{(N)}(\lfloor T \rfloor_N, \cdot) = \Psi(\Pi^{(N)} \cdot)$.

Let the initial datum (t, ϕ) be such that $t \in \mathbf{I}^{(N)} - \{T^{(N)}\}$ and $\phi \in \mathbf{C}$. Choose any $\psi \in \mathbf{C}(N)$ such that $\psi|_{[-r,0]} = \phi$. Given $\epsilon > 0$, by virtue of Proposition 5.2.5, we can find an $u(\cdot) \in \mathcal{U}[t, T]$ such that

$$V^{(N)}(t, \phi) - \epsilon$$
$$\leq E\left[\int_t^{t+h^{(N)}} L(s, \Pi^{(N)} \phi, u(s))\, ds + V^{(N)}(t + h^{(N)}, \Pi^{(N)}(z_{h^{(N)}})) \right],$$

where $z(\cdot)$ is the solution to (5.7) of degree N under control process $u(\cdot)$ with initial datum (t, ψ). For this $u(\cdot)$, choose $\bar{u}(\cdot) \in \mathcal{U}^{(N,M)}[t, T^{(N)}]$ according to Condition (5.15) and let $\bar{z}(\cdot)$ be the solution to (5.7) of degree N under control process $\bar{u}(\cdot)$ with the same initial datum as for $z(\cdot)$. Then, using the above inequality and Proposition 5.2.6, we have

$$V^{(N)}(t, \phi) - V^{(N,M)}(t, \phi)$$
$$\leq E\left[\int_t^{t+h^{(N)}} L(s, \Pi^{(N)}(\phi), u(s))\, ds + V^{(N)}(t + h^{(N)}, \Pi^{(N)}(z_{h^{(N)}})) \right] + \epsilon$$
$$- E\left[\int_t^{t+h^{(N)}} L(s, \Pi^{(N)}(\phi), \bar{u}(s))\, ds + V^{(N,M)}(t + h^{(N)}, \Pi^{(N)}(\bar{z}_{h^{(N)}})) \right]$$
$$= E\left[\int_t^{t+h^{(N)}} L(s, \Pi^{(N)}(\phi), u(s))\, ds + V^{(N)}(t + h^{(N)}, \Pi^{(N)}(z_{h^{(N)}})) \right] + \epsilon$$
$$- E\left[\int_t^{t+h^{(N)}} L(s, \Pi^{(N)}(\phi), \bar{u}(s))\, ds + V^{(N)}(t + h^{(N)}, \Pi^{(N)}(\bar{z}_{h^{(N)}})) \right]$$
$$+ V^{(N)}(t + h^{(N)}, \Pi^{(N)}(\bar{z}_{h^{(N)}})) - V^{(N,M)}(t + h^{(N)}, \Pi^{(N)}(\bar{z}_{h^{(N)}}))$$
$$\leq \hat{K}((h^{(N)})^{1+\delta} + \sup_{\tilde{\phi} \in \mathbf{C}}\{V^{(N)}(t + h^{(N)}, \tilde{\phi}) - V^{(N,M)}(t + h^{(N)}, \tilde{\phi})\} + \epsilon,$$

where in the last line, Condition (5.15) has been exploited. Since $\epsilon > 0$ is arbitrary and neither the first nor the last line of the above inequalities depend on $u(\cdot)$ or $\bar{u}(\cdot)$, it follows that for all $t \in \mathbf{I}^{(N)} - \{T^{(N)}\}$,

$$\sup_{\phi \in \mathbf{C}}\{V^{(N)}(t, \phi) - V^{(N,M)}(t, \phi)\}$$
$$\leq \sup_{\phi \in \mathbf{C}}\{V^{(N)}(t + h^{(N)}, \phi) - V^{(N,M)}(t + h^{(N)}, \phi)\} + \hat{K}(h^{(N)})^{1+\delta}.$$

Recalling the equality $V^{(N)}(T^{(N)}, \cdot) = V^{(N,M)}(T^{(N)}, \cdot)$, we conclude that for all $t \in \mathbf{I}^{(N)}$,

$$\sup_{\phi \in \mathbf{C}}\{V^{(N)}(t, \phi) - V^{(N,M)}(t, \phi)\} \leq \frac{1}{h^{(N)}}(T^{(N)} \hat{K}(h^{(N)}))^{1+\delta} \leq T\hat{K}(h^{(N)})^{\delta}.$$

This proves the assertion.   □

Let the initial datum (t, ϕ) be such that $t \in \mathbf{I}^{(N)} - \{T^{(N)}\}$ and $\phi \in \mathbf{C}$. Choose any $\psi \in \mathbf{C}(N)$ such that $\psi|_{[-r,0]} = \phi$. Given $\epsilon > 0$, by virtue of Proposition 5.2.5, we can find an $u(\cdot) \in \mathcal{U}[t, T]$ such that

$$V^{(N)}(t, \phi) - \epsilon$$
$$\leq E\left[\int_t^{t+h^{(N)}} L(s, \Pi^{(N)}\phi, u(s))\, ds + V^{(N)}(t + h^{(N)}, \Pi^{(N)} z_{h^{(N)}})\right],$$

where $z(\cdot)$ is the solution to (5.7) of degree N under control process $u(\cdot)$ with initial datum (t, ψ). For this $u(\cdot)$, choose $\bar{u}(\cdot) \in \mathcal{U}^{(N,M)}[t, T^{(N)}]$ according to Condition (5.15) and let $\bar{z}(\cdot)$ be the solution to (5.7) of degree N under control process $\bar{u}(\cdot)$ with the same initial datum as for $z(\cdot)$. Then, using the above inequality and Proposition 5.2.6, we have

$$V^{(N)}(t, \phi) - V^{(N,M)}(t, \phi)$$
$$\leq E\left[\int_t^{t+h^{(N)}} L(s, \Pi^{(N)}\phi, u(s))\, ds + V^{(N)}(t + h^{(N)}, \Pi^{(N)} z_{h^{(N)}})\right] + \epsilon$$
$$- E\left[\int_t^{t+h^{(N)}} L(s, \Pi^{(N)}\phi, \bar{u}(s))\, ds + V^{(N,M)}(t + h^{(N)}, \Pi^{(N)} \bar{z}_{h^{(N)}})\right]$$
$$= E\left[\int_t^{t+h^{(N)}} L(s, \Pi^{(N)}\phi, u(s))\, ds + V^{(N)}(t + h^{(N)}, \Pi^{(N)} z_{h^{(N)}})\right] + \epsilon$$
$$- E\left[\int_t^{t+h^{(N)}} L(s, \Pi^{(N)}\phi, \bar{u}(s))\, ds + V^{(N)}(t + h^{(N)}, \Pi^{(N)} \bar{z}_{h^{(N)}})\right]$$
$$+ V^{(N)}(t + h^{(N)}, \Pi^{(N)} \bar{z}_{h^{(N)}}) - V^{(N,M)}(t + h^{(N)}, \Pi^{(N)} \bar{z}_{h^{(N)}})$$
$$\leq \hat{K}((h^{(N)})^{1+\delta} + \sup_{\tilde{\phi} \in \mathbf{C}} \{V^{(N)}(t + h^{(N)}, \tilde{\phi}) - V^{(N,M)}(t + h^{(N)}, \tilde{\phi})\} + \epsilon,$$

where in the last line, Condition (5.15) has been exploited. Since $\epsilon > 0$ is arbitrary and neither the first nor the last line of the above inequalities depend on $u(\cdot)$ or $\bar{u}(\cdot)$, it follows that for all $t \in \mathbf{I}^{(N)} - \{T^{(N)}\}$,

$$\sup_{\phi \in \mathbf{C}} \{V^{(N)}(t, \phi) - V^{(N,M)}(t, \phi)\}$$
$$\leq \sup_{\phi \in \mathbf{C}} \{V^{(N)}(t + h^{(N)}, \phi) - V^{(N,M)}(t + h^{(N)}, \phi)\} + \hat{K}(h^{(N)})^{1+\delta}.$$

Recalling the equality $V^{(N)}(T^{(N)}, \cdot) = V^{(N,M)}(T^{(N)}, \cdot)$, we conclude that for all $t \in \mathbf{I}^{(N)}$,

$$\sup_{\phi \in \mathbf{C}} \{V^{(N)}(t, \phi) - V^{(N,M)}(t, \phi)\} \leq \frac{1}{h^{(N)}}(T^{(N)} \hat{K}(h^{(N)})^{1+\delta}) \leq T\hat{K}(h^{(N)})^{\delta}.$$

This proves the assertion.   □

In order to apply Theorem 5.2.7, we must check if Condition (5.15) is satisfied. Given a time grid of width $\frac{r}{N}$ for the discretization in time and segment space, we would expect the condition to be satisfied provided we choose the subgrid for the piecewise constant controls fine enough; that is, the time discretization of the control processes should be of degree M, with M being sufficiently large in comparison to N. This is the content of the next result.

Theorem 5.2.8 *Assume Assumptions (A5.1.1)-(A5.1.3) hold. Let $\beta > 3$. Then there exists a constant $\hat{K} > 0$ depending only on K_b, r, K_{lip}, the dimensions n and m, and β such that Condition (5.15) in Theorem 5.2.7 is satisfied with constants $\hat{K}$ and $\delta = \frac{\beta-3}{4}$ for all $N, M = 1, 2, \ldots$ such that $N \geq r$ and $M \geq N^\beta$.*

Proof. Let $N, M = 1, 2, \ldots$ be such that $N \geq r$ and $M \geq N^\beta$. Let the initial datum $(t, \phi) \in \mathbf{I}^{(N)} \times \mathbf{C}$. Define the following functions:

$$
\begin{aligned}
\tilde{f} &: U \to \Re^n, & \tilde{f} &= f(t, \Pi^{(N)}\phi, u), \\
\tilde{g} &: U \to \Re^{n \times m}, & \tilde{g} &= g(t, \Pi^{(N)}\phi, u), \\
\tilde{L} &: U \to \Re, & \tilde{L} &= L(t, \Pi^{(N)}\phi, u), \\
\tilde{V} &: \Re^n \to \Re^n, & \tilde{V} &= V^{(N)}(t + h^{(N)}, \Pi^{(N)}S(\phi, x)),
\end{aligned}
$$

where for each $\phi \in \mathbf{C}$ and $x \in \Re^n$, $S(\phi, x) \in \mathbf{C}$ is defined for each $\theta \in [-r, 0]$ by

$$
S(\phi, x)(\theta) = \begin{cases} \phi(\theta + h^{(N)}) & \text{if } \theta \in [-r, -h^{(N)}] \\ \phi(0) + \frac{\theta + h^{(N)}}{h^{(N)}}x & \text{if } \theta \in (-h^{(N)}, 0]. \end{cases}
$$

As a consequence of Assumption (A5.1.4), the functions $\tilde{f}$, $\tilde{g}$, and $\tilde{L}$ defined above are continuous functions on U. By Assumption (A5.1.2), $|\tilde{f}|$, $|\tilde{g}|$, and $|\tilde{L}|$ are all bounded by K_b. As a consequence of Proposition 5.2.1, the function $\tilde{V}$ is Lipschitz continuous, and for the Lipschitz constant K_{lip}, we have

$$
\sup_{x, y \in \Re^n, x \neq y} \frac{|\tilde{V}(x) - \tilde{V}(y)|}{|x - y|} \leq 3K_{lip}(T + 1)e^{3T(T+4m)K_{lip}^2}.
$$

Let $u(\cdot) \in \mathcal{U}[t, T]$ and $z(\cdot; u(\cdot))$ be the solution to (5.7) of degree N under control process $u(\cdot)$ with the initial datum $(t, \psi) \in \mathbf{I}^{(N)} \times \mathbf{C}(N)$ with $\psi|[-r, 0] = \phi$. As $z(\cdot)$ also satisfies (5.10), we see that

$$
z(s; u(\cdot)) - \phi(0) = \int_t^s \tilde{f}(u(\lambda)) \, d\lambda + \int_t^s \tilde{g}(u(\lambda)) \, dW(\lambda), \quad \forall s \in [t, t + h^{(N)}].
$$

We need the following Krylov stochastic mean value theorem (see Theorem 2.7 in [Kry01]), which is restated as follows.

Theorem. (Krylov) Let $\bar{T} > 0$. There is a constant $\bar{K} > 0$ depending only K_b and the dimensions n and m such that the following holds: For any $N = 1, 2, \ldots$ such that $N \geq r$, any bounded continuous function $\tilde{L} : U \to \Re$, any bounded Lipschitz continuous function $\tilde{V} : \Re^n \to \Re$, and any $u(\cdot) \in \mathcal{U}[t, T]$, there exists $u^{(N)}(\cdot) \in \mathcal{U}^{(N)}[t, T]$ such that

$$E\left[\int_t^{\bar{T}} \tilde{L}(u^{(N)}(s))\, ds + \tilde{V}(z_{\bar{T}}(u^{(N)}(\cdot)))\right] - E\left[\int_t^{\bar{T}} \tilde{L}(u(s))\, ds + \tilde{V}(z_{\bar{T}}(u(\cdot)))\right]$$

$$\leq \bar{K}(1 + \bar{T})(h^{(N)})^{\frac{1}{4}}\left[(h^{(N)})^{\frac{1}{4}} \sup_{u \in U} |\tilde{L}(u)| + \sup_{x, y \in \Re^n, x \neq y} \frac{|\tilde{V}(x) - \tilde{V}(y)|}{|x - y|}\right].$$

(Note that in the above theorem, the difference between the two expectations may be inverted, since we can take $-\tilde{L}$ in place of $\tilde{L}$ and $-\tilde{V}$ in place of $\tilde{V}$.)

By the above Krylov theorem, we find $\bar{u}(\cdot) \in \mathcal{U}^{(N,M)}[t, T]$ such that

$$E\left[\int_t^{t+h^{(N)}} \tilde{L}(\bar{u}(s))\, ds + \tilde{V}(x(t + h^{(N)}; \bar{u}(\cdot)))\right]$$

$$- E\left[\int_t^{t+h^{(N)}} \tilde{L}(u(s))\, ds + \tilde{V}(z(t + h^{(N)}) - \phi(0); u(\cdot)))\right]$$

$$\leq \bar{K}(1 + h^{(N)})(h^{(N)}/M)^{\frac{1}{4}}$$

$$\times \left[(h^{(N)}/M)^{\frac{1}{4}} \sup_{u \in U} |\tilde{L}(u)| + \sup_{x, y \in \Re^n, x \neq y} \frac{|\tilde{V}(x) - \tilde{V}(y)|}{|x - y|}\right],$$

where $x(\cdot; \bar{u}(\cdot))$ satisfies

$$x(s; \bar{u}(\cdot)) = \int_t^s \tilde{f}(\bar{u}(\lambda))\, d\lambda + \int_t^s \tilde{g}(\bar{u}(\lambda))\, dW(\lambda), \quad \forall s \geq t.$$

Notice that the constant $\bar{K}$ above only depends on K_b and the dimensions n and m. Let $z(\cdot; \bar{u}(\cdot))$ be the solution to (5.7) of degree N under control process $\bar{u}(\cdot)$ with initial datum $(t, \psi) \in \mathbf{I}^{(N)} \times \mathbf{C}(N)$, where $\psi|_{[-r,0]} = \phi$ as above. Then, by construction, $z(s; \bar{u}(\cdot)) - \phi(0) = x(s; \bar{u}(\cdot))$ for all $s \in [t, t + h^{(N)}]$. Set

$$\hat{K} := 2\bar{K}r^{-\frac{\beta}{4}}(K_b + 3K_{lip}(T + 1)e^{3T(T+4m)K_{lip}^2}).$$

Since $M \geq N^\beta$ by hypothesis, $\frac{1+\beta}{4} = 1 + \delta > 1$, and $h^{(N)} = \frac{r}{N}$, we have

$$r^{\frac{1}{4}}(NM)^{-\frac{1}{4}} \leq r^{\frac{1}{4}}N^{-\frac{1+\beta}{4}} = r^{-\frac{\beta}{4}}(h^{(N)})^{1+\delta}.$$

Recalling the definitions of $\tilde{f}$, $\tilde{g}$, $\tilde{L}$, and $\tilde{V}$, we have thus found a piecewise constant strategy $\bar{u}(\cdot) \in \mathcal{U}^{(N,M)}[t, T]$ such that

$$E\left[\int_t^{t+h^{(N)}} L(s, \Pi^{(N)}\phi, \bar{u}(s))\, ds + V^{(N)}(t + h^{(N)}, \bar{z}_{t+h^{(N)}})\right]$$

$$\leq E\left[\int_t^{t+h^{(N)}} L(s, \Pi^{(N)}\phi, u(s))\, ds + V^{(N)}(t + h^{(N)}, z_{t+h^{(N)}})\right]$$

$$+ \hat{K}(h^{(N)})^{1+\delta}.$$

This proves the theorem.   □

Corollary 5.2.9 *Assume Assumptions (A5.1.1)-(A5.1.3) hold. Then there is a positive constant $\bar{K}$ depending only on K_b, K_{lip}, and the dimensions n and m such that for all $\beta > 3$, $N = 1, 2, \ldots$ with $N \geq r$, $M = 1, 2, \ldots$ with $M \geq N^\beta$, and all initial datum $(t, \phi) \in \mathbf{I}^{(N)} \times \mathbf{C}$, it holds that*

$$|V^{(N)}(t, \phi) - V^{(N,M)}(t, \phi)| \leq \bar{K}Tr^{-\frac{\beta}{1+\beta}}\left(\frac{r}{N^{1+\beta}}\right)^{\frac{\beta-3}{4(1+\beta)}}.$$

In particular, with $M = \lceil N^\beta \rceil$, where $\lceil a \rceil$ is the least integer not smaller than a for all $a \in \mathfrak{R}$, the upper bound on the discretization error can be rewritten as

$$|V^{(N)}(t, \phi) - V^{(N,\lceil N^\beta \rceil)}(t, \phi)| \leq \bar{K}Tr^{-\frac{\beta}{1+\beta}}\left(\frac{r}{N^{1+\beta}}\right)^{\frac{\beta-3}{4(1+\beta)}}.$$

From the above corollary we see that, in terms of the total number of the time steps $N\lceil N^\beta \rceil$, we can achieve any rate of convergence smaller than $\frac{1}{4}$ by choosing the subdiscretization order β sufficiently large.

5.2.3 Overall Discretization Error

Theorem 5.2.10 *Assume Assumptions (A5.1.1)-(A5.1.3) hold. Let $0 < \gamma \leq K_H < \infty$. Then there is a constant $\bar{K}$ depending on γ, K_H, K_{lip}, K_b, T, and dimensions n and m such that for all $\beta > 3$, $N = 1, 2, \ldots$ with $N \geq 2r$, and all initial datum $(t, \phi) \in \mathbf{I}^{(N)} \times \mathbf{C}$ with ϕ being γ-Hölder continuous, it holds that, with $h = \frac{r}{N^{1+\beta}}$,*

$$|V^{(N)}(t, \phi) - V^{(N,\lceil N^\beta \rceil)}(t, \phi)|$$

$$\leq \bar{K}\left(r^{\frac{\gamma\beta}{1+\beta}}h^{\frac{\gamma}{1+\beta}} \vee r^{\frac{\beta}{2(1+\beta)}}\sqrt{\ln\left(\frac{1}{h}\right)}h^{\frac{1}{2(1+\beta)}} + r^{-\frac{\beta}{1+\beta}}h^{\frac{\beta-3}{4(1+\beta)}}\right).$$

In particular, with $\beta = 5$ and $h = \frac{r}{N^6}$, it holds that

$$|V^{(N)}(t, \phi) - V^{(N,N^5)}(t, \phi)| \leq \bar{K}\left(r^{\frac{5\gamma}{6}}h^{\frac{2\gamma-1}{12}} \vee r^{\frac{5}{12}}\sqrt{\ln\left(\frac{1}{h}\right)} + r^{-\frac{5}{6}}\right)h^{\frac{1}{12}}.$$

Proof. Clearly, $|v - V^{(N,\lceil N^\beta \rceil)}| \leq |V - V^{(N)}| + |V^{(N)} - V^{(N,\lceil N^\beta \rceil)}|$. The assertion now follows from Corollary 5.2.9 and Theorem 5.2.7, where we assume $\ln\left(\frac{1}{h^{(N)}}\right) = \ln\left(\frac{r}{N}\right)$ by $\ln\left(\frac{N^{1+\beta}}{r}\right) = \ln\left(\frac{1}{h}\right)$.   □

Theorem 5.2.11 *Assume Assumptions (A5.1.1)-(A5.1.3) hold. Let $0 < \gamma \leq K_H$. Then there is a constant $\bar{K}(r)$ depending on γ, K_H, K_{lip}, K_b, T, dimensions n and m, and delay duration r such that for all $\beta > 3$, $N, M = 1, 2, \ldots$ with $N \geq 2r$ and $M \geq N^\beta$, and all initial datum $(t, \phi) \in \mathbf{I}^{(N)} \times \mathbf{C}$ with ϕ being γ-Hölder continuous, the following holds: If $\bar{u}(\cdot) \in \mathcal{U}^{(N,M)}[t, T]$ is such that*

$$V^{(N,M)}(t, \phi) - J^{(N)}(t, \phi; \bar{u}(\cdot)) \leq \epsilon,$$

then with $h = \frac{r}{N^{1+\beta}}$,

$$V(t, \phi) - J(t, \phi; \bar{u}(\cdot)) \leq \bar{K}(r) \left(h^{\frac{\gamma}{1+\beta}} \vee \sqrt{\ln\left(\frac{1}{h}\right)} h^{\frac{1}{2(1+\beta)}} + h^{\frac{\beta-3}{4(1+\beta)}} \right) + \epsilon.$$

Proof. Let $\bar{u}(\cdot) \in \mathcal{U}^{(N,M)}[t, T]$ be such that $V^{(N,M)}(t, \phi) - J^{(N)}(t, \phi; \bar{u}(\cdot)) \leq \epsilon$. Then

$$V(t, \phi) - J(t, \phi; \bar{u}(\cdot)) \leq V(t, \phi) - V^{(N,M)}(t, \phi) + V^{(N,M)}(t, \phi) - J^{(N)}(t, \phi; \bar{u}(\cdot))$$
$$+ J^{(N)}(t, \phi; \bar{u}(\cdot)) - J(t, \phi; \bar{u}(\cdot))$$

The assertion is now a consequence of Theorem 5.2.10 and Theorem 5.2.11.

$\square$

5.3 Markov Chain Approximation

To avoid complications arising from the discretization of the space $\Re^n$ and the degeneracy of the diffusion, we will assume without loss of generality that the controlled SHDE (5.2) is autonomous and one dimensional (i.e., $n = m = 1$). Specifically, we consider the following autonomous one-dimensional controlled equation:

$$dx(t) = f(x_s, u(s)) \, ds + g(x_s) \, dW(s), \quad s \in [0, T], \tag{5.16}$$

with the initial segment $\psi \in \mathbf{C}$ (here, $\mathbf{C} = C([-r, 0]; \Re)$ throughout this section) at the initial time $t = 0$.

Assumption 5.3.1 *In this section, we assume that the real-valued functions f, g, L, and Ψ defined on $\mathbf{C} \times U$, $\mathbf{C}$, $\mathbf{C} \times U$, and $\mathbf{C}$, respectively, satisfy the following conditions:*
(A5.3.1). (Global Boundedness) The functions $|f|$ and $|g|$ are bounded by a constant $K_b > 0$, that is,

$$|f(\phi, u)| + |g(\phi)| \leq K_b, \quad \forall (\phi, u) \in \mathbf{C} \times U.$$

(A5.3.2). (Uniform Lipschitz Condition) There is a constant $K_{lip} > 0$ such that for all $\phi, \varphi \in \mathbf{C}$ and $u \in U$,

$$|f(\phi, u) - f(\varphi, u)| + |g(\phi) - g(\varphi)|$$
$$+ |L(\phi, u) - L(\varphi, u)| + |\Psi(\phi) - \Psi(\varphi)| \leq K_{lip} \|\phi - \varphi\|.$$

(A5.3.3). (Ellipticity of the Diffusion Coefficient) $g(\phi) \geq \sigma$ for all $\phi \in \mathbf{C}$, *where $\sigma > 0$ is a given positive constant.*

The objective of the optimal classical control problem is to find an admissible control $u(\cdot) \in \mathcal{U}[0, T]$ so that the following expected objective functional is maximized:

$$
\begin{aligned}
J(\psi; u(\cdot)) = E\Bigg[\int_0^T & e^{-\alpha s} L(s, x_s(\cdot; \psi, u(\cdot)), u(s))\, ds \\
& + e^{-\alpha T} \Psi(x_T(\cdot; \psi, u(\cdot))) \Bigg],
\end{aligned}
\tag{5.17}
$$

Again, the value function of the optimal control problem is defined by

$$
V(\psi) = \sup_{u(\cdot) \in \mathcal{U}[0, T]} J(\psi; u(\cdot)), \quad \psi \in \mathbf{C}.
\tag{5.18}
$$

We also recall, in the following, the corresponding optimal relaxed control problem in which the controlled state equation is described by

$$
dx(s) = \int_U f(x_s, u)\dot{\mu}(s, du)\, ds + g(x_s)\, dW(s), \quad s \in [0, T],
\tag{5.19}
$$

and the objective function is defined by

$$
\begin{aligned}
J(\psi; \mu(\cdot, \cdot)) = E\Bigg[\int_U \int_0^T & e^{-\alpha s} L(s, x_s(\cdot; \psi, u), u)\dot{\mu}(s, du)\, ds \\
& + e^{-\alpha T} \Psi(x_T(\cdot; \psi, u(\cdot))) \Bigg].
\end{aligned}
\tag{5.20}
$$

The objective of the optimal relaxed control problem is to find an optimal relaxed control $\mu^*(\cdot, \cdot) \in \mathcal{F}[0, T]$ that maximizes $J(\psi; \mu(\cdot, \cdot))$. Again, the value function (using a slightly different notation from those of (5.18)) is described by

$$
\hat{V}(\psi) = \sup_{\mu(\cdot, \cdot) \in \mathcal{R}[0, T]} J(\psi; \mu(\cdot, \cdot)), \quad \psi \in \mathbf{C}
\tag{5.21}
$$

(using a slightly different notation from those of (5.18)), where $\mathcal{R}[0, T]$ is the class of admissible relaxed controls (see Section 3.2.2 in Chapter 3).

Note that every $u(\cdot) \in \mathcal{U}[0, T]$ can be represented as an admissible relaxed control $\mu(\cdot, \cdot)$ via

$$
\mu(A) = \int_0^T \int_U \mathbf{1}_{\{(t, u) \in B\}} \delta_{u(t)}(du)\, dt, \quad A \in \mathcal{A}([0, T] \times U),
\tag{5.22}
$$

where δ_u is the Dirac measure at $u \in U$ and $\mathbf{1}_{\{(t, u) \in B\}}$ is the indicator function of the set $\{(t, u) \in B\}$. Recall in Section 3.2.2 that $V(\psi) = \hat{V}(\psi), \forall \psi \in \mathbf{C}$.

In the following subsections, we formulate optimal control problems for appropriate approximating Markov chains. It is shown when the approximating parameter $N \to \infty$ (or equivalently $h^{(N)} \to 0$) that the value functions of the approximating optimal control problems converges to that of the optimal relaxed control problem and, hence, to the value function of the original optimal control problem. The difference between the Markov chain approximation and the semidiscretization scheme presented in Section 5.2 is that the Markov chain approximation requires discretion of the spatial variable of the state process to the nearest multiple of $\sqrt{h^{(N)}}$.

5.3.1 Controlled Markov Chains

Let $N = 1, 2, \ldots$ be given and recall $T^{(N)} = h^{(N)} \lfloor \frac{T}{h^{(N)}} \rfloor$. Hence, we often suppress in the following the superscript (N) to avoid the complication of notation when there is no danger of ambiguity. We call a one-dimensional discrete-time process

$$\{\zeta(kh), k = -N, -N+1, \ldots, 0, 1, 2, \ldots, T^{(N)}\}$$

defined on a complete filtered probability space $(\Omega, \mathcal{F}, P, \mathbf{F})$ a *discrete chain* of degree N if it takes values in $\mathbf{S} \equiv \{k\sqrt{h} \mid k = 0, \pm 1, \pm 2, \ldots\}$ and $\zeta(kh)$ is $\mathcal{F}(kh)$-measurable for all $k = 0, 1, 2, \ldots, T^{(N)}$. We define the $\Re^{N+1}$-valued discrete process $\{\zeta_{kh}, k = 0, 1, 2, \cdots, T^{(N)}\}$ by setting for each $k = 0, 1, 2, \cdots, T^{(N)}$,

$$\zeta_{kh} = (\zeta((k-N)h), \zeta((k-N+1)h), \ldots, \zeta((k-1)h), \zeta(kh)).$$

Note that the one-dimensional *discrete chain* needs not be Markovian.

With a little abuse of notation, we also define $\Pi(\zeta_{kh}) \in \mathbf{C}$ and $\Pi(\phi) \in \mathbf{C}$ as the linear interpolation of $\zeta_{kh} \in \Re^{N+1}$ and $\phi \in \mathbf{C}$. A sequence

$$u^{(N)}(\cdot) = \{u^{(N)}(kh), k = 0, 1, 2, \ldots, T^{(N)}\}$$

is said to be a discrete admissible control if $u^{(N)}(kh)$ is $\mathcal{F}(kh)$-measurable for each $k = 0, 1, 2, \ldots, T^{(N)}$ and

$$E\left[\sum_{k=0}^{T^{(N)}} |u^{(N)}(kh)|^2\right] < \infty.$$

As in Section 5.1.2 but with $M = 1$ for simplicity, we also let $\mathcal{U}^{(N)}[0, T]$ be the class of a continuous-time admissible control process $\bar{u}(\cdot) = \{\bar{u}(s), s \in [0, T]\}$, where for each $s \in [0, T]$, $\bar{u}(s) = \bar{u}(\lfloor s \rfloor_N)$ is $\mathcal{F}(\lfloor s \rfloor_N)$-measurable and takes only finite different values in U.

Remark 5.3.2 *We observe the following facts or convention:*

1. *Without loss of generality, we can and will assume that all real-valued discrete processes*

$$\zeta(\cdot) = \{\zeta(kh), k = -N, -N + 1, \ldots, 0, 1, 2, \cdots, T^{(N)}\}$$

discussed in this section are actually taking discrete values in the set **S**. *This can be done by truncation down to the nearest multiple of* $\sqrt{h}$. *Implicitly, we are also discretizing the value of the chain* $\zeta(\cdot)$, *which is not required in Section 5.2 for a semidiscretization scheme.*

2. *Similar to the fact that the real-valued solution* $x(\cdot)$ *to (5.16) is not a Markov process and yet its corresponding* **C**-*valued segment process* $\{x_s, s \in [0, T]\}$ *is. We note here that the* **S**-*valued process* $\zeta(\cdot)$ *is not Markovian but it is desirable under appropriate conditions that its corresponding* $(\mathbf{S})^{N+1}$-*valued segment process* $\{\zeta_{kh}, k = 0, 1, 2 \ldots, T^{(N)}\}$ *is Markovian with respect to the discrete filtration* $\{\mathcal{F}(kh), k = 0, 1, 2, \ldots\}$.

Given a one-step Markov transition function $p^{(N)} : (\mathbf{S})^{N+1} \times U \times (\mathbf{S})^{N+1} \to [0, 1]$, where $p^{(N)}(\mathbf{x}, u; \mathbf{y})$ will be interpreted as the probability that the $\zeta_{(k+1)h} = \mathbf{y} \in (\mathbf{S})^{N+1}$ given that $\zeta_{kh} = \mathbf{x}$ and $u(kh) = u$. We define a sequence of *controlled Markov chains* associated with the initial segment ψ and $\bar{u}(\cdot) \in \mathcal{U}^{(N)}[0, T]$ as a family $\{\zeta^{(N)}(\cdot), N = 1, 2, \ldots\}$ of processes such that $\zeta^{(N)}(\cdot)$ is a $\mathbf{S}^{(N)}$-valued discrete chain of degree N defined on the same stochastic basis as $u^{(N)}(\cdot)$, provided the following conditions are satisfied:

Assumption 5.3.3 *(i) Initial Condition:* $\zeta(-kh) = \psi(-kh) = \mathbf{S}$ *for* $k = -N$, $-N + 1, \cdots, 0, 1, \ldots, T^{(N)}$.
(ii). Extended Markov Property: For any $k = 1, 2, \ldots$, *and*

$$\mathbf{y} = (y(-Nh), y((-N + 1)h), \cdots, y(0)) \in (\mathbf{S})^{N+1},$$

$$
\begin{aligned}
&P\{\zeta_{(k+1)h} = \mathbf{y} \mid \zeta(ih), u(ih), i \leq k\} \\
&= \begin{cases} p^{(N)}(\zeta_{kh}, u(kh); \mathbf{y}), & \text{if } y(-ih) = \zeta((-i+1)h) \text{ for } 1 \leq i \leq N \\ 0, & \text{otherwise.} \end{cases}
\end{aligned}
$$

(iii) Local Consistency with the Drift Coefficient :

$$
\begin{aligned}
b(kh) &\equiv E_\psi^{(N,u)}[\zeta((k+1)h) - \zeta(kh)] \\
&= hf(\Pi(\zeta_{kh}), u(kh)) + o(h) \\
&\equiv hf^{(N)}(\zeta_{kh}, u(kh)),
\end{aligned}
$$

where $f^{(N)} : \Re^{N+1} \times U \to \Re$ *is defined by*

$$f^{(N)}(\mathbf{x}, u) = f(\Pi^{(N)}(\mathbf{x}), u), \quad \forall (\mathbf{x}, u) \in \Re^{N+1} \times U,$$

$E_\psi^{N,u(\cdot)}$ *is the conditional expectation given the discrete admissible control* $u^{(N)}(\cdot) = \{u(kh), k = 0, 1, 2, \cdots, T^{(N)}\}$, *and the initial function*

$$\pi^{(N)}\psi \equiv (\psi(-Nh), \psi((-N+1)h), \cdots, \psi(-h), \psi(0)).$$

(iv) Local Consistency with the Diffusion Coefficient:

$$E_\psi^{N,u(\cdot)}[(\zeta((k+1)h) - \zeta(kh) - b(kh))^2] = hg^2(\zeta_{kh}, u(kh)) + o(h)$$
$$\equiv h(g^{(N)})^2(\zeta_{kh}, u(kh)),$$

where $g^{(N)} : \Re^{N+1} \to \Re$ *is defined by*

$$g^{(N)}(\mathbf{x}) = g(\Pi^{(N)}(\mathbf{x})), \quad \forall \mathbf{x} \in \Re^{N+1}.$$

(v) Jump Heights: There is a positive number $\tilde{K}$ *independent of* N, *such that*

$$\sup_k |\zeta((k+1)h) - \zeta(kh)| \le \tilde{K}\sqrt{h} \quad \text{for some } \tilde{K} > 0.$$

It is straightforward, under Assumption 5.3.3, that a sequence of extended transition functions with the jump height and the local consistency conditions can be constructed. The following is a construction example of the extended Markov transition functions that satisfy the above conditions.

Examples for the Markov Transition Function

(A) We can define the Markov probability transition function

$$p^{(N)} : (\mathbf{S})^{N+1} \times U \times (\mathbf{S})^{N+1} \to [0,1]$$

as follows:

$$p^{(N)}(\mathbf{x}, u; \mathbf{x} \oplus h) = \frac{g^2(\Pi(\mathbf{x})) + hf(\Pi(\mathbf{x}), u)}{2g^2(\Pi(\mathbf{x}))}$$

and

$$p^{(N)}(\mathbf{x}, u; \mathbf{x} \ominus h) = \frac{g^2(\Pi(\mathbf{x})) - hf(\Pi(\mathbf{x}), u)}{2g^2(\Pi(\mathbf{x}))},$$

for all $\mathbf{x} = (x(-Nh), x((-N+1)h), \ldots, x(-h), x(0)) \in (\mathbf{S})^{N+1}$,

$$\mathbf{x} \oplus h = (x(-N), x((-N+1)h), \ldots, x(-h), x(0) + h),$$

and

$$\mathbf{x} \ominus h = (x(-Nh), x(-N+1), \ldots, x(-1), x(0) - h),$$

where $\Pi(\mathbf{x}) \in \mathbf{C}$ is the linear interpolation of $\mathbf{x}$.
(B) If the diffusion coefficient $g : \mathbf{C} \to \Re$ is such that $g(\phi) \equiv 1$ for all $\phi \in \mathbf{C}$, then the controlled SHDE (5.16) reduces to the following equation:

$$dx(s) = f(x_s, u(s))ds + dW(s), \quad s \in [0,T].$$

The discrete chain

$$\zeta(\cdot) = \{\zeta(kh), k = -N, -N+1, \ldots, -1, 0, 1 \cdots, T^{(N)}\}$$

corresponding to the above controlled SHDE can be written as

$$\zeta((k+1)h) = \zeta(kh) + hf(\Pi^{(N)}(\zeta_{kh}), u(kh)) + \Delta W(kh), \quad k = 0, 1, 2, \ldots, T^{(N)},$$

and

$$(\zeta(-Nh), \zeta((-N+1)h), \ldots, \zeta(-h), \zeta(0))$$
$$= (\psi(-Nh), \psi((-N+1)h), \ldots, \psi(-h), \psi(0)).$$

Since $\Delta W(kh) \equiv W((k+1)h) - W(kh)$ is Gaussian with $E[\Delta W(kh)] = 0$ and $Var[\Delta W(kh)] = kh$, it is clear that the Markov transition function $p^{(N)}$ in Assumption 5.3.3 can be similaryly defined for this special case by taking $g(\phi) \equiv 1$.

5.3.2 Optimal Control of Markov Chains

Let $N = 1, 2, \ldots$ be given. Using the notation and concept developed in the previous subsection, we assume that the $(\mathbf{S})^{N+1}$-valued process $\{\zeta_{kh}, k = 0, 1, 2, \ldots\}$ is a controlled $(\mathbf{S})^{N+1}$-valued Markov chain with initial datum $\zeta_0 = \pi^{(N)}\psi \in (\mathbf{S})^{N+1}$ and the Markov probability transition function $p^{(N)}$: $(\mathbf{S})^{N+1} \times U \times (\mathbf{S})^{N+1} \to [0, 1]$ that satisfies Assumption 5.3.3.

In the following, we will define the objective function $J^{(N)}$: $(\mathbf{S})^{N+1} \times \mathcal{U}^{(N)}[0, T] \to \Re$ and the value function $V^{(N)}$: $(\mathbf{S})^{N+1} \to \Re$ of the approximating optimal control problem as follows.

Define the discrete objective functional of degree N by

$$J^{(N)}(\pi^{(N)}; u^{(N)}(\cdot)) = E\left[\sum_{k=0}^{T^{(N)}-1} e^{-\alpha kh} L(\Pi^{(N)}(\zeta_{kh}), u(kh))h + \Psi(\Psi(\zeta_{\lfloor \frac{T}{h} \rfloor_N}^{(N)}))\right],$$

$$(5.23)$$

where

$$\psi^{(N)} = (\psi(-Nh), \psi((-N+1)h), \ldots, \psi(-h), \psi(0)) \in (\mathbf{S})^{N+1}, \quad \text{for } \psi \in \mathbf{C},$$

and $\{u^{(N)}(\cdot) = \{u(kh), k = 0, 1, 2, \ldots\}$ is adapted to $\{\mathcal{F}^{(N)}(kh), k = 0, 1, 2, \ldots\}$ and taking values in U.

$$V^{(N)}(\psi^{(N)}) = \sup_{u^{(N)}(\cdot)} J(\psi^{(N)}; u^{(N)}(\cdot)), \quad \psi \in \mathbf{C}, \quad (5.24)$$

with the terminal condition $V^{(N)}(\zeta_{\lfloor \frac{T}{h} \rfloor h}^{(N)}) = \Psi(\zeta_{\lfloor \frac{T}{h} \rfloor h}^{(N)})$.

For each $N = 1, 2, \ldots$, we state without proof, the following dynamic programming principle (DDP) for the controlled Markov chain $\{\zeta_{kh}^{(N)}, k = 0, 1, \ldots, T^{(N)}\}$ and discrete admissible control process $u(\cdot) \in \mathcal{U}^{(N)}[0, T]$ as follows.

Proposition 5.3.4 *Assume Assumption 5.3.3 holds. Let $\psi \in \mathbf{C}$ and let $\{\zeta_{kh}, k = 0, 1, 2, \ldots, T^{(N)}\}$ be an $(\mathbf{S})^{N+1}$-valued Markov chain determined by the probability transition function $p^{(N)}$. Then for each $k = 0, 1, 2, \ldots, T^{(N)} - 1$,*

$$V^{(N)}(\psi^{(N)}) = \sup_{u^{(N)}(\cdot)} E\left[V^{(N)}(\zeta^{(N)}_{(k+1)h}) + \sum_{i=0}^{k} hL(\Pi(\zeta_{ih}), u^{(N)}(ih))\right].$$

The optimal control of Markov chain problem (OCMCP) is stated below.

Problem (OCMCP). Given $N = 1, 2, \ldots$ and $\psi \in \mathbf{C}$, find an admissible discrete control process $u^{(N)}(\cdot)$ that maximizes the objective functional $J^{(N)}(\psi^{(N)}; u^{(N)}(\cdot))$ in (5.23).

The detail of optimal control of Markov chain is an entire subject of study by itself and the space will not be well spent if we attempt to reproduce the theory here. The readers are referred to monographs by Dynkin and Yushkevich [DY79] and Berzekas [Ber76] for details.

Given $N = 1, 2, \ldots$ and initial segment $\psi \in \mathbf{C}$, we outline a backward-in-time computational algorithm for Problem (OCMCP), based on Proposition 5.3.4, as follows.

A Computational Algorithm

Step 1. For $M = T^{(N)}$, set $V_M^{(N)}(\psi) = \Psi(\psi)$ for all $\phi \in \mathbf{C}$. Compute

$$J_M^{(N)}(x, u) \equiv E\left[V_M^{(N)}(x_M) \mid x_{M-1} = x, u((M-1)h) = u\right]$$
$$= \Psi(x \oplus \sqrt{h})p^{(N)}(x, u, x \oplus \sqrt{h})$$
$$+ \Psi(x \ominus \sqrt{h}, u)p^{(N)}(x, u, x \ominus \sqrt{h}),$$

where $x = (x(-N), x(-N+1), \ldots, x(-1), x(0)) \in (\mathbf{S})^{N+1}$,

$$x \oplus \sqrt{h} = (x(-N), x(-N+1), \ldots, x(-1), x(0) + \sqrt{h}),$$

and

$$x \ominus \sqrt{h} = (x(-N), x(-N+1), \ldots, x(-1), x(0) - \sqrt{h}).$$

Step 2. Find $u \in U$ that maximizes the following function by setting

$$u^* = \arg\max_{u \in U}\{L(x, u) + J_M^{(N)}(x, u)\}$$

Let $\bar{u}^*(s) = u*$ for all $(M-1)h \leq s \leq Mh$, and

$$J_{M-1}^{(N)}(x, u) \equiv E\Big[L(x_{M-1}, u((M-1)h))$$
$$+ J_M^{(N)}(x_M, u*) \mid x_{M-1} = x, u((M-1)h) = u\Big]$$
$$= L(x, u) + J_{M-1}^{(N)}(x \oplus \sqrt{h}, u)p^{(N)}(x, u, x \oplus \sqrt{h})$$
$$+ J_{M-1}^{(N)}(x \ominus \sqrt{h}, u)p^{(N)}(x, u, x \ominus \sqrt{h}),$$

where, again, $x = (x(-N), x(-N+1), \ldots, x(-1), x(0)) \in (\mathbf{S})^{N+1}$,

$$x \oplus \sqrt{h} = (x(-N), x(-N+1), \ldots, x(-1), x(0) + \sqrt{h}),$$

and

$$x \ominus \sqrt{h} = (x(-N), x(-N+1), \ldots, x(-1), x(0) - \sqrt{h}).$$

Step k. For $k = 1, 2, \ldots, M$, find $u \in U$ that maximizes the function $L(x, u) + J_{M-k}^{(N)}(x, u)$ by setting

$$u^* = \arg\max_{u \in U} \{L(x, u) + J_{M-k}^{(N)}(x, u)\}.$$

Let $\bar{u}^*(s) = u*$ for all $(M - k - 1)h \leq s \leq (M - k)h$, and

$$\begin{aligned}
J_{M-k-1}^{(N)}(x, u) &\equiv E\Big[L(x_{M-k-1}, u((M-k-1)h)) \\
&\quad + J_{M-k}^{(N)}(x_M, u*) \mid x_{M-k-1} = x, u((M-k-1)h) = u\Big] \\
&= L(x, u) + J_{M-k}^{(N)}(x \oplus \sqrt{h}, u^*)p^{(N)}(x, u, x \oplus \sqrt{h}) \\
&\quad + J_{M-k}^{(N)}(x \ominus \sqrt{h}, u^*)p^{(N)}(x, u, x \ominus \sqrt{h}).
\end{aligned}$$

Step M. Given the initial segment $\psi \in \mathbf{C}$, set

$$\begin{aligned}
J_1^{(N)}(x, u) &\equiv E\Big[L(x_1, u(h)) \\
&\quad + J_1^{(N)}(x_M, u*) \mid x_{M-k-1} = x, u((M-k-1)h) = u\Big] \\
&= L(x, u) + J_{M-k}^{(N)}(x \oplus \sqrt{h}, u^*)p^{(N)}(x, u, x \oplus \sqrt{h}) \\
&\quad + J_{M-k}^{(N)}(x \ominus \sqrt{h}, u^*)p^{(N)}(x, u, x \ominus \sqrt{h}).
\end{aligned}$$

Let $u \in U$ that maximizes the function $L(x, u) + J_1^{(N)}(x, u)$ by setting

$$u^* = \arg\max_{u \in U} \{L(x, u) + J_1^{(N)}(x, u)\}.$$

Let $\bar{u}^*(s) = u*$ for all $0 \leq s \leq h$. In this case, the value function of the optimal Markov chain control problem can be set as $V^{(N)}(\pi\psi) = J_1^{(N)}(\pi\psi, u^*)$ and the optimal piecewise constant control $\bar{u}^*(\cdot) \in \mathcal{U}^{(N)}[0, T]$ can be constructed.

5.3.3 Embedding the Controlled Markov Chain

Given $N = 1, 2, \ldots$ (hence the superscript (N) will be suppressed when appropriate), we will represent the discrete chain

$$\zeta(\cdot) = \zeta(kh), k = -N, -N+1, \ldots, -1, 0, 1, 2, \ldots, T^{(N)}$$

as a solution to a discretized equation

$$\zeta(kh) = \psi(0) + \sum_{i=0}^{k-1} h \cdot f^{(N)}(\zeta_{ih}, u(ih)) + \xi(kh), \quad k = 1, 2, \ldots, T^{(N)}, \quad (5.25)$$

corresponding to (5.16) with admissible constant control process $u(\cdot) = u^{(N)}(\cdot) \in \mathcal{U}^{(N)}[0, T]$ and the initial datum

$$\psi^{(N)} \equiv (\psi(-Nh), \psi((-N+1)h), \ldots, \psi(-h), \psi(0))$$
$$= (\zeta(-Nh), \zeta((-N+1)h), \ldots, \zeta(-h), \zeta(0)) \in \mathbf{S}^{N+1}.$$

Define the discrete process $\{\xi(kh), k = 0, 1, 2, \ldots, T^{(N)}\}$ by $\xi(0) = 0$ and

$$\xi(kh) = \zeta(kh) - \psi(0) - \sum_{i=0}^{k-1} h \cdot f^{(N)}(\Pi\zeta_{ih}, u(ih)), \quad k = 1, 2, \ldots, T^{(N)}. \quad (5.26)$$

We have the following two lemmas regarding $\zeta(\cdot)$ and $\xi(\cdot)$.

Lemma 5.3.5 *The one-step Markov transition function* $p^{(N)} : \mathbf{S}^{N+1} \times U \times \mathbf{S}^{N+1} \to [0, 1]$ *for the controlled discrete chain* $\zeta(\cdot)$ *described by (5.25) satisfies Assumption 5.3.3.*

Lemma 5.3.6 *The process* $\xi(\cdot) = \{\xi(kh), k = 0, 1, 2, \cdots\}$ *defined by (5.26) is a discrete-time martingale with respect to the filtration* $\{\mathcal{F}^{(N)}(kh), k = 0, 1, 2, \ldots\}$.

Proof. From the definition of $\xi^{(N)}(\cdot)$, we have

$$E[\xi((k+1)h) \mid \mathcal{F}^{(N)}(kh)]$$
$$= E\left[\zeta((k+1)h) - \psi(0) - \sum_{i=0}^{k} h \cdot f^{(N)}(\Pi(\zeta_{ih}), u(ih)) \mid \mathcal{F}^{(N)}(kh)\right]$$
$$= E\left[\zeta((k+1)h) - \zeta(kh) + \sum_{i=0}^{k}(\zeta((i+1)h) - \zeta(ih)) \mid\mid \mathcal{F}^{(N)}(kh)\right]$$
$$- \sum_{i=0}^{k} E\left[h \cdot f^{(N)}(\Pi(\zeta_{ih}), u(ih)) \mid \mathcal{F}^{(N)}(kh)\right]$$
$$= \xi(kh).$$

Therefore, the process $\xi(\cdot)$ is a discrete-time martingale with respect to the filtration $\{\mathcal{F}^{(N)}(kh), k = 0, 1, 2, \ldots\}$. $\square$

Set

$$\epsilon_1^{(N)}(s) := \sum_{i=0}^{\lfloor \frac{s}{h} \rfloor - 1} h \cdot f^{(N)}(\Pi\zeta_{ih}^{(N)}, \bar{u}^{(N)}(ih)) - \int_0^s f(\Pi\zeta_{\lfloor t \rfloor_N}^{(N)}, \bar{u}^{(N)}(t)) dt, \quad s \geq 0,$$

With $T > 0$, for the error term we have

$$E^{(N)} \left[\sup_{s \in [0,T]} \left| \epsilon_1^{(N)}(s) \right| \right]$$

$$\leq \sum_{i=0}^{\lfloor \frac{T}{h} \rfloor - 1} h E^{(N)} \left[f^{(N)}(\Pi \zeta_{ih}^{(N)}, u^{(N)}(ih)) - f(\Pi \zeta_{ih}^{(N)}, u^{(N)}(ih)) | \right] + K_b h$$

$$\times \int_0^{h \lfloor \frac{T}{h} \rfloor} E^{(N)} \left[f(\Pi \zeta_{\lfloor s \rfloor_N}^{(N)}, \bar{u}^{(N)}(s)) - f(\Pi \zeta_s^{(N)}, \bar{u}^{(N)}(s)) | \right] ds,$$

which tends to zero as N goes to infinity by Assumption 5.3.3, dominated convergence, and the defining properties of $\{\zeta^{(N)}(kh), k = -N, -N + 1, \ldots, -1, 0, 1, 2, \ldots\}$. Moreover, $|\epsilon_1^{(N)}(s)|$ is bounded by $2K \cdot T$ for all $s \in [0, T]$ and all N large enough. Therefore,

$$\lim_{N \to \infty} E^{(N)} \left[\sup_{s \in [0,T]} \left| \epsilon_1^{(N)}(s) \right|^2 \right] = 0.$$

The discrete-time martingale $\{\xi^{(N)}(kh), k = 0, 1, 2, \ldots\}$ can be rewritten as discrete stochastic integral.

Define $\{W^{(N)}(kh), k = 0, 1, 2, \cdots\}$ by setting $W^{(N)}(0) = 0$ and

$$W^{(N)}(kh) = \sum_{i=0}^{k-1} \frac{1}{g(\Pi \zeta_{ih}^{(N)})} (\xi^{(N)}((i+1)h) - \xi^{(N)}(ih)), \quad k = 1, 2, \ldots.$$

Using the piecewise constant interpolation $\bar{W}^{(N)}(\cdot)$ of $W^{(N)}(\cdot)$, the linearly interpolated process $\tilde{\zeta}^{(N)}(\cdot)$ can be expressed as a solution to

$$\tilde{\zeta}^{(N)}(s) = \psi^{(N)}(0) + \int_0^s f(\Pi \zeta_{\lfloor s \rfloor_N}^{(N)}, \bar{u}^{(N)}(s)) \, ds$$

$$+ \int_0^s g(\Pi \zeta_{\lfloor s \rfloor_N}^{(N)}) \, d\bar{W}^{(N)}(s) + \epsilon_2^{(N)}(s)$$

$$\equiv C^{(N)}(s) + \int_0^s g(\Pi \zeta_{\lfloor s \rfloor_N}^{(N)}) \, d\bar{W}^{(N)}(s), \quad s \in [0, T], \quad (5.27)$$

where the error terms $\epsilon_2^{(N)}(\cdot)$ converges to $\epsilon_1^{(N)}(\cdot)$ in the following sense:

$$\lim_{N \to \infty} E^{(N)} \left[\sup_{s \in [0,T]} \left| \epsilon_2^{(N)}(s) - \epsilon_1^{(N)}(s) \right|^2 \right] = 0.$$

5.3.4 Convergence of Approximations

In order to present convergence results, we will retain the superscript (N) in each of the applicable quantities in order to highlight its dependence on the discretization parameter N.

Let $\{\bar{u}^{(N)}(\cdot), N = 1, 2, \ldots\}$ be the sequence of maximizing piecewise constant controls; that is, for each $N = 1, 2, \ldots$,

$$\bar{u}^{(N)}(s) = u^{(N)}(kh) \quad \text{for } kh \leq s < (k+1)h, \text{ and } k = 0, 1, 2, \ldots, T^{(N)} - 1,$$

for the optimal control of Problem (OCMCP) and let $\{\mu^{(N)}(\cdot, \cdot), N = 1, 2, \ldots\}$ be the corresponding sequence of admissible relax controls, where

$$\mu^{(N)}(A) = \int_0^T \int_U \mathbf{1}_{\{(t,u) \in A\}} \delta_{\bar{u}(t)}(du)\, dt, \quad A \in \mathcal{B}([0, T] \times U),$$

with $\delta_{\bar{u}(t)}$ being the Dirac measure at $\bar{u}^{(N)}(t)$ and $\mathbf{1}_{\{(t,u) \in A\}}$ being the indicator function of the set $\{(t, u) \in A\}$.

We have the following result, which looks very similar to Proposition 3.2.16 in Chapter 3. However, their proofs differ slightly and, therefore, are included for completeness.

Proposition 5.3.7 *Assume Assumption 5.3.3 holds. Suppose the sequence* $\{\psi^{(N)}, n = 1, 2, \ldots\}$ *in* **C** *converges to the initial function* $\psi \in$ **C**. *Then the sequence of processes*

$$\{(\bar{\zeta}^{(N)}(\cdot), \bar{u}^{(N)}(\cdot), \bar{W}^{(N)}(\cdot)), N = 1, 2, \ldots\}$$

is tight. Let $(x(\cdot), \mu(\cdot, \cdot), W(\cdot))$ *be a limit point of the above sequence and set* $\mathcal{F}(s) = \sigma((x(t), \mu(t, A), W(t)), A \in \mathcal{B}(U), t \leq s)$ *for* $s \in [0, T]$. *Then we have the following.*
(i) $W(\cdot)$ *is an* $\{\mathcal{F}(s), s \in [0, T]\}$*-adapted Brownian motion.*
(ii) $\mu(\cdot, \cdot)$ *is an admissible relaxed control.*
(iii) $x(\cdot)$ *is a solution to (5.19) under* $(\mu(\cdot, \cdot), W(\cdot))$ *and with initial condition* $\psi \in$ **C**.

We note here that since $\psi^{(N)}$ is taken to be $\pi^{(N)}\psi$ for the initial segment $\psi \in$ **C**,

$$\lim_{N \to \infty} \|\psi^{(N)} - \psi\| \equiv \lim_{N \to \infty} \left[\sup_{\theta \in [-r, 0]} |\psi^{(N)}(\theta) - \psi(\theta)\| \right] = 0,$$

from the classical Weierstrauss theorem.

We need the following lemmas for a proof of Proposition 5.3.7.

Lemma 5.3.8 *The family of relaxed controls* $\{\mu^{(N)}(\cdot, \cdot), N = 1, 2, \ldots\}$ *corresponding to the family of maximizing piecewise constant controls* $\{\bar{u}^{(N)}(\cdot), N = 1, 2, \ldots\}$ *is tight and converges to an admissible relaxed control* $\mu(\cdot, \cdot)$ *of the optimal relaxed control problem.*

Proof. Since the space of admissible relaxed controls $\mathcal{R}[0, T]$ is compact under the weak convergence topology, the sequence $\{\mu^{(N)}(\cdot, \cdot), N = 1, 2, \ldots\}$ has a subsequence (which is still denoted by $\{\mu^{(N)}(\cdot, \cdot), N = 1, 2, \cdots\}$ with a little abuse of notation) that converges weakly to $\mu(\cdot, \cdot) \in \mathcal{R}[0, T]$.   □

Lemma 5.3.9 *The family of piecewise constant processes* $\{\bar{\zeta}^{(N)}(\cdot)), N = 1, 2, \ldots\}$ *is tight. Its limiting process* $x(\cdot) = \{x(t), t \in [0, T]\}$ *is a strong solution of (5.19) with the initial segment* $\psi \in \mathbf{C}$ *and under the admissible relaxed control* $\mu(\cdot, \cdot) \in \mathcal{R}[0, T]$.

Proof. It is clear that $\psi^{(N)}$ converges to the initial datum $\psi \in \mathbf{C}$, since $\psi^{(N)} = \Pi^{(N)}\psi$ for each $N = 1, 2, \ldots$. Tightness of the sequence $\{\bar{\zeta}^{(N)}(\cdot)), N = 1, 2, \ldots\}$ follows from the Aldous criterion (see Theorem 3.2.6 in Chapter 2): For any $\mathbf{F}^{(N)}$-stopping time τ and $\epsilon > 0$, we have

$$E^{(N)}\left[|\bar{\zeta}^{(N)}(\tau + \epsilon) - \bar{\zeta}^{(N)}(\tau)|^2 \mid \mathcal{F}^{(N)}(\tau)\right] \leq 2K_b^2\epsilon(\epsilon + 1) \qquad \square$$

Lemma 5.3.10 *The family of one dimensional piecewise constant processes* $\{\bar{W}^{(N)}(\cdot)), N = 1, 2, \ldots\}$ *is tight and it converges weakly to the one-dimensional standard Brownian motion* $W(\cdot) = \{W(s), s \in [0, T]\}$.

Proof. In this proof we will (i) establish the tightness of $\bar{W}^{(N)}(\cdot)$; and (ii) the identify the limit points

$$\langle W^{(N)}\rangle(kh) \equiv \sum_{i=0}^{k-1} E[|W^{(N)}((i + 1)h) - W^{(N)}(ih)|^2 \mid \mathcal{F}^{(N)}(ih)]$$

$$= kh + o(h)\sum_{i=0}^{k-1}\frac{1}{g^2(\bar{\zeta}_{ih}^{(N)})}, \quad \forall N = 1, 2, \ldots; k = 0, 1, \ldots, T(N),$$

for all $N = 1, 2, \ldots$ and $k = 0, 1, 2, \ldots$. Taking into account Assumption (A5.3.1) and the definition of the continuous-time piecewise constant processes $\{\bar{W}^{(N)}(\cdot), N = 1, 2, \ldots\}$, we see that $\langle W^{(N)}(\cdot)\rangle$ converges to the constant process 1 in probability uniformly on $[0, T]$. Since the sequence $\{\bar{W}^{(N)}(\cdot), N = 1, 2, \ldots\}$ is of *uniformly controlled variations*, hence a *good* sequence of integrators in the sense of Kurtz and Protter [KP91], because its jump heights are uniformly bounded and it is a martingale for each $N = 1, 2, \ldots$, we conclude without going into the detail of proof that $\{\bar{W}^{(N)}(\cdot), N = 1, 2, \ldots\}$ converges weakly in $C([0, T]; \Re)$ to a standard Brownian motion $W(\cdot)$. We also note that $W(\cdot)$ has independent increments with respect to the filtration $\{\mathcal{F}(s), s \in [0, T]\}$. This can be seen by considering the first and second conditional moments of the increments of $W^{(N)}(\cdot)$ for each N and applying the conditions on local consistency and the jump heights of $\{\zeta^{(N)}(\cdot), N = 1, 2, \ldots\}$. In addition, the results of [KP91] guarantee weak convergence of the corresponding adapted quadratic variation processes; that is, the sequence $\{[\bar{W}^{(N)}, \bar{W}^{(N)}], N = 1, 2, \ldots\}$ converges weakly to $[W, W]$ in $D[0, T]$, the space of right-continuous with finite left-hand limits under Skorokhod topology, where the square brackets indicate the adapted quadratic variation. Convergence also holds for the sequence of process pairs $(W^{(N)}, [W^{(N)}, W^{(N)}])$ in $D([0, T]; \Re^2)$ (see Kurtz and Protter [KP04]). $\qquad \square$

Proof of Proposition 5.3.7. The proof of this proposition is a slight modification of Proposition 3.2.16 in Chapter 3, where we prove the existence
of an optimal classical control. With a little abuse of notation, suppose
$\{(\bar{\zeta}^{(N)}(\cdot), \bar{u}^{(N)}(\cdot), \bar{W}^{(N)}(\cdot)), N = 1, 2, \ldots\}$ is a weakly convergent subsequence
with limit point $(x(\cdot), \mu(\cdot, \cdot), W(\cdot))$. The remaining part that needs to be veri-
fied is the identification of $x(\cdot)$ as a solution to (5.19) under the relaxed control
$\mu(\cdot, \cdot)$ with the initial condition ψ.

Define cádlág and bounded (due to Assumption (A5.2.2)) processes

$$C^{(N)}(\cdot) = \{C^{(N)}(s), s \in [0, T]\} \quad \text{and} \quad C(\cdot) = \{C(s), s \in [0, T]\}$$

by

$$C^{(N)}(s) = \psi^{(N)}(0) + \int_0^s f(\Pi^{(N)}(\zeta_{\lfloor t \rfloor N}), u^{(N)}(t)) \, dt + \epsilon_2^{(N)}(s), \quad s \in [0, T],$$

$$C(s) = \psi(0) + \int_0^s \int_U f(x_s, u) \mu(du, s) \, ds, \quad s \in [0, T].$$

Invoking Skorohod's representation theorem in Section 3.2 of Chapter 3, one
can establish weak convergence of $C^{(N)}(\cdot)$ to $C(\cdot)$.

We now know that each of the sequences

$$\bar{\zeta}^{(N)}(\cdot), \ C^{(N)}(\cdot), \bar{W}^{(N)}(\cdot), \text{ and } [\bar{W}^{(N)}, \bar{W}^{(N)}](\cdot), \quad N = 1, 2, \ldots$$

is weakly convergent in $D([0, T])$. Actually, we have weak convergence for the
sequence of quadruples

$$(\bar{\zeta}^{(N)}(\cdot), C^{(N)}(\cdot), \bar{W}^{(N)}(\cdot), [\bar{W}^{(N)}, \bar{W}^{(N)}](\cdot)), \quad N = 1, 2, \ldots,$$

which converges in $D([0, T]; \Re^4)$, since each of these four components converges weakly in $D[0, T]$. To see this, we notice that each of the four sequences

$$(\bar{\zeta}^{(N)} + C^{(N)})(\cdot), \quad (\bar{\zeta}^{(N)} + \bar{W}^{(N)})(\cdot),$$

$$(\bar{\zeta}^{(N)} + [\bar{W}^{(N)}, \bar{W}^{(N)}])(\cdot), \quad \text{and} \quad (\bar{W}^{(N)} + [\bar{W}^{(N)}, \bar{W}^{(N)}])(\cdot), \quad N = 1, 2, \ldots,$$

is tight in $D[0, T]$, because the limit processes $C(\cdot)$, $x(\cdot)$, $W(\cdot)$, and $[W, W] = 1$
are all continuous on $[0, T]$. This implies tightness of the sequence of quadruples in $D([0, T]; \Re^4)$. Consequently, it has a unique limit point, namely
$(x(\cdot), C(\cdot), W(\cdot), [W, W])$. By virtue of Skorohod's theorem, the convergence
is with probability 1. This proves the proposition. □

Proposition 5.3.11 *Assume Assumptions (A5.4.1)-(A5.4.3) hold. If the sequence*

$$(\bar{\zeta}^{(N)}(\cdot), \bar{u}^{(N)}(\cdot), \bar{W}^{(N)}(\cdot)), \quad N = 1, 2, \ldots,$$

*of the interpolated processes converges weakly to a limit point $(x(\cdot), \mu(\cdot), W(\cdot))$,
then $x(\cdot)$ is a solution to (5.19) under the relaxed control $\mu(\cdot, \cdot)$ with the initial
condition $\psi \in \mathbf{C}$ and*

$$\lim_{N \to \infty} J^{(N)}(\psi^{(N)}; u^{(N)}(\cdot)) = J(\psi; \mu(\cdot, \cdot)).$$

Proof. The convergence assertion for the objective functionals is a consequence of Proposition 5.3.7, Assumption (A5.4.3), and the definition of $J^{(N)}$ and J.   □

Theorem 5.3.12 *Assume Assumption (A5.4.1)-(A5.4.3) hold. Then we have*

$$\lim_{N\to\infty} V^{(N)}(\psi^{(N)}) \to \hat{V}(\psi).$$

Proof. We first note that

$$\underline{\lim}_{N\to\infty} V^{(N)}(\psi^{(N)}) \geq \hat{V}(\psi)$$

as a consequence of Propositions 5.3.7 and 5.3.11. In order to show that

$$\overline{\lim}_{N\to\infty} V^{(N)}(\psi^{(N)}) \leq \hat{V}(\psi),$$

we choose a relaxed control $\mu(\cdot,\cdot)$ so that $J(\psi;\mu(\cdot,\cdot)) = \hat{V}(\psi)$ according to Proposition 5.3.7. Given $\epsilon > 0$, one can construct a sequence of discrete admissible controls $\{u^{(N)}(\cdot), N = 1,2,\ldots\}$ such that

$$\{(\bar{\zeta}^{(N)}(\cdot), \bar{u}^{(N)}(\cdot), \bar{W}^{(N)}(\cdot)), N = 1,2,\ldots\}$$

is weakly convergent, where each of $\bar{\zeta}^{(N)}(\cdot)$ and $\bar{W}^{(N)}(\cdot))$ is constructed in Proposition 5.3.7 and

$$\overline{\lim}_{N\to\infty}|J^{(N)}(\psi^{(N)}; u^{(N)}(\cdot)) - J(\psi;\mu(\cdot,\cdot))| \leq \epsilon.$$

The existence of such a sequence of discrete admissible controls is guaranteed. By definition, there exists an admissible control $u^{(N)}(\cdot)$ such that $V^{(N)}(\psi^{(N)}) - \epsilon \leq J^{(N)}(\psi^{(N)}; u^{(N)}(\cdot))$ Using Proposition 5.3.11, we find that

$$\overline{\lim}_{N\to\infty} V^{(N)}(\psi^{(N)}) - \epsilon \leq \overline{\lim}_{N\to\infty} J^{(N)}(\psi^{(N)}; u^{(N)}(\cdot) \leq \hat{V}(\psi).$$

Since ϵ is arbitrary, the assertion follows.   □

5.4 Finite Difference Approximation

In this section we consider an explicit *finite difference scheme* and show that it converges to the unique viscosity solution of the HJBE

$$\alpha V(t,\psi) - \partial_t V(t,\psi) - \max_{v\in U} [\mathbf{A}^v V(t,\psi) + L(t,\psi,v)] = 0 \qquad (5.28)$$

on $[0,T] \times \mathbf{C}$, and the terminal condition $V(T,\psi) = \Psi(\psi)$, $\forall \psi \in \mathbf{C}$.

In the above, we recall from Section 3.4.1 for the convenience of the readers that

$$\mathbf{A}^v \Phi(t, \psi) \equiv \mathcal{S}\Phi(t, \psi) + \overline{D\Phi(t, \psi)}(f(t, \psi, v)\mathbf{1}_{\{0\}}) \qquad (5.29)$$

$$+ \frac{1}{2} \sum_{j=1}^{m} \overline{D^2\Phi(t, \psi)}(g(t, \psi, v)\mathbf{e}_j\mathbf{1}_{\{0\}}, g(t, \psi, v)\mathbf{e}_j\mathbf{1}_{\{0\}}),$$

for any $\Phi \in C_{lip}^{1,2}([0,T] \times \mathbf{C}) \cap \mathcal{D}(\mathcal{S})$, where $\mathbf{e}_j$ is the jth vector of the standard basis in $\Re^m$, the function $\mathbf{1}_{\{0\}} : [-r, 0] \to \Re$ is defined by

$$\mathbf{1}_{\{0\}}(\theta) = \begin{cases} 0 \text{ for } \theta \in [-r, 0) \\ 1 \quad \text{for } \theta = 0, \end{cases}$$

and $\mathcal{S}(\Phi)(t, \psi)$ is defined as

$$\mathcal{S}\Phi(t, \psi) = \lim_{\epsilon \downarrow 0} \frac{\Phi(t, \tilde{\psi}_\epsilon) - \Phi(t, \psi)}{\epsilon}, \qquad (5.30)$$

and $\tilde{\psi} : [-r, T] \to \Re^n$ is the extension of $\psi \in \mathbf{C}$ from $[-r, 0]$ to $[-r, T]$ and is defined by

$$\tilde{\psi}(t) = \begin{cases} \psi(0) \quad \text{for } t \geq 0, \\ \psi(t) \text{ for } t \in [-r, 0). \end{cases}$$

Note that $D\Phi(t, \psi) \in \mathbf{C}^*$ and $D^2\Phi(t, \psi) \in \mathbf{C}^\dagger$ are the first- and second-order Fréchet derivatives of Φ with respect to its second argument $\psi \in \mathbf{C}$. In addition, $\overline{D\Phi(t, \psi)} \in (\mathbf{C} \oplus \mathbf{B})^*$ is the extension of of $D\Phi(t, \psi)$ from $\mathbf{C}^*$ to $(\mathbf{C} \oplus \mathbf{B})^*$ and $\overline{D^2V(t, \psi)} \in (\mathbf{C} \oplus \mathbf{B})^\dagger$ is the extension of $D^2V(t, \psi)$ from $\mathbf{C}^\dagger$ to $(\mathbf{C} \oplus \mathbf{B})^\dagger$. These results can be found in Lemma 2.2.3 and Lemma 2.2.4 in Chapter 2.

In this section, we assume the following conditions, which are the same as those of Assumption 3.1.1 in Chapter 3 and are repeated here for convenience.

Assumption 5.4.1 *The functions f, g, L, and Ψ satisfy the following conditions:*
(A5.4.1) (Lipschitz Continuity) There exists a constant $K_{lip} > 0$ such that

$$|f(t, \phi, u) - f(s, \varphi, v)| + |g(t, \phi, u) - g(s, \varphi, v)|$$
$$+ |L(t, \phi, u) - L(s, \varphi, v)| + |\Psi(\phi) - \Psi(\varphi)|$$
$$\leq K_{lip}(\sqrt{|t - s|} + \|\phi - \varphi\| + |u - v|),$$

$$\forall s, t \in [0, T], u, v \in U, \text{ and } \phi, \varphi \in \mathbf{C}.$$

(A5.4.2) (Linear and Polynomial Growth) There exists a constant $K_{grow} > 0$ such that

$$|f(t, \phi, u)| + |g(t, \phi, u)| \leq K_{grow}(1 + \|\phi\|)$$

and

$$|L(t, \phi, u)| + |\Psi(\phi)| \leq K_{grow}(1 + \|\phi\|_2)^k, \quad \forall(t, \phi) \in [0, T] \times \mathbf{C} \text{ and } u \in U.$$

(A5.4.3) The initial function ψ belongs to the space $L^2(\Omega, \mathbf{C}, \mathcal{F}(t))$ of $\mathcal{F}(t)$-measurable elements in $L^2(\Omega : \mathbf{C})$ such that

$$\|\psi\|^2_{L^2(\Omega;\mathbf{C})} \equiv E[\|\psi\|^2] < \infty.$$

Based on Assumptions (A5.4.1)-(A5.4.3), we will extend a method introduced by Barles and Souganidis [BS91] to an infinite-dimensional setting that is suitable for the HJBE described in (5.28). The finite difference scheme and convergence results will be described in Section 5.3.1. A computational algorithm based on the finite difference scheme will be summarized in Section 5.4.2.

5.4.1 Finite Difference Scheme

Given a positive integer M, we consider the following truncated optimal control problem with value function $V_M : [0, T] \times \mathbf{C} \to \Re$ defined by

$$V_M(t, \psi) = \sup_{u(\cdot) \in \mathcal{U}[t,T]} E\left[\int_t^T e^{-\alpha(s-t)}(L(s, x_s, u(s)) \wedge M)\, ds \right.$$
$$\left. + e^{-\alpha(T-t)}(\Psi(x_T) \wedge M) \right], \tag{5.31}$$

where $a \wedge b$ is defined by $a \wedge b = \min\{a, b\}$ for all $a, b \in \Re$.

The corresponding truncated HJBE is given by

$$\alpha V_M(t, \psi) - \partial_t V_M(t, \psi) - \max_{u \in U} \left[\mathbf{A}^u V_M(t, \psi) + (L(t, \psi, u) \wedge M)\right] = 0 \tag{5.32}$$

on $[0, T] \times \mathbf{C}$, and $V_M(T, \psi) = \Psi(\psi) \wedge M$, $\forall \psi \in \mathbf{C}$. The corresponding truncated Hamiltonian is

$$\mathcal{H}_M(t, \psi, V_M(t, \psi), \partial_t V_M(t, \psi), \overline{DV_M(t, \psi)}, \overline{D^2 V_M(t, \psi)})$$
$$= \mathcal{S}(V_M)(t, \psi) + \partial_t V_M(t, \psi)$$
$$+ \sup_{u \in U}\left[\overline{DV_M(t, \psi)}(f(t, \psi, u)\mathbf{1}_{\{0\}}) + (L(t, \psi, u) \wedge M) \right.$$
$$\left. + \frac{1}{2}\sum_{i=1}^m \overline{D^2 V_M(t, \psi)}(g(t, \psi, u)\mathbf{e}_i \mathbf{1}_{\{0\}}, g(t, \psi, u)\mathbf{e}_i \mathbf{1}_{\{0\}}) \right]. \tag{5.33}$$

Similar to what had been presented in Sections 3.5 and 3.6, it can be shown that the value function $V_M : [0, T] \times \mathbf{C} \to \Re$ is the unique viscosity solution (see Section 3.5 of Chapter 3 for a definition of a viscosity solution) of the HJBE

$$0 = \alpha V_M(t, \psi) - \mathcal{H}_M(t, \psi, V_M(t, \psi), \partial_t V_M(t, \psi), \overline{DV_M(t, \psi)}, \overline{D^2 V_M(t, \psi)}),$$
$$\text{on } [0, T] \times \mathbf{C} \text{ with } V(T, \psi) = \Psi(\psi) \wedge M, \quad \forall \psi \in \mathbf{C}. \tag{5.34}$$

Moreover, it is easy to see that $V_M(t, \psi) \to V(t, \psi)$ for each $(t, \psi) \in [0, T] \times \mathbf{C}$ as $M \to \infty$. In view of these, we need only to find the numerical solution for $V_M(t, \psi)$ for each $(t, \psi) \in [0, T] \times \mathbf{C}$.

Let ϵ with $0 < \epsilon < 1$ be the step size for variable ψ and η, with $0 < \eta < 1$ be the step size for t. We define the finite difference operators $\Delta_\eta \Phi$, $D_\epsilon \Phi$ and $\Delta_\epsilon^2 \Phi$ for each Borel measurable function $\Phi : [0, T] \times \mathbf{C} \to \Re$ by

$$\Delta_\eta \Phi(t, \psi) = \frac{\Phi(t + \eta, \psi) - \Phi(t, \psi)}{\eta},$$

$$D_\epsilon \Phi(t, \psi)(\phi + v\mathbf{1}_{\{0\}}) = \frac{\Phi(t, \psi + \epsilon(\phi + v\mathbf{1}_{\{0\}})) - \Phi(t, \psi)}{\epsilon},$$

$$D_\epsilon^2 \Phi(t, \psi)(\phi + v\mathbf{1}_{\{0\}}, \varphi + w\mathbf{1}_{\{0\}}) = \frac{\Phi(t, \psi + \epsilon(\phi + v\mathbf{1}_{\{0\}})) - \Phi(t, \psi)}{\epsilon^2}$$
$$+ \frac{\Phi(t, \psi - \epsilon(\varphi + w\mathbf{1}_{\{0\}})) - \Phi(t, \psi)}{\epsilon^2},$$

where $\phi, \varphi \in \mathbf{C}$ and $v, w \in \Re^n$. Recall that

$$S\Phi(t, \psi) = \lim_{\epsilon \downarrow 0} \frac{1}{\epsilon} \left[\Phi(t, \tilde{\psi}_\epsilon) - \Phi(t, \psi) \right].$$

Therefore, we define

$$S_\epsilon \Phi(t, \psi) = \frac{1}{\epsilon} \left[\Phi(t, \tilde{\psi}_\epsilon) - \Phi(t, \psi) \right].$$

It is clear that

$$\lim_{\epsilon \downarrow 0} S_\epsilon \Phi(t, \psi) = S\Phi(t, \psi), \quad \forall (t, \psi) \in [0, T] \times \mathbf{C}.$$

We have the following lemma.

Lemma 5.4.2 *For any $\Phi \in C_{lip}^{1,2}([0, T] \times \mathbf{C})$ such that Φ can be smoothly extended on $[0, T] \times (\mathbf{C} \oplus \mathbf{B})$, we have for every $\phi, \varphi \in \mathbf{C}$, $v, w \in \Re^n$, and $t \in [0, T]$,*

$$\lim_{\epsilon \to 0} D_\epsilon \Phi(t, \psi)(\phi + v\mathbf{1}_{\{0\}}) = \overline{D\Phi(t, \psi)}(\phi + v\mathbf{1}_{\{0\}}) \qquad (5.35)$$

and

$$\lim_{\epsilon \to 0} D_\epsilon^2 \Phi(t, \psi)(\phi + v\mathbf{1}_{\{0\}}, \varphi + w\mathbf{1}_{\{0\}}) = \overline{D^2\Phi(t, \psi)}(\phi + v\mathbf{1}_{\{0\}}, \varphi + w\mathbf{1}_{\{0\}}). \quad (5.36)$$

Proof. Note that the function Φ can be extended from $[0, T] \times \mathbf{C}$ to $[0, T] \times (\mathbf{C} \oplus \mathbf{B})$. Let us denote by $\bar{\Phi}$ the smooth extension of Φ on $[0, T] \times (\mathbf{C} \oplus \mathbf{B})$. It is clear that

$$\lim_{\epsilon \to 0} \Delta_\epsilon \Phi(t, \psi)(\phi + v\mathbf{1}_{\{0\}}) = \bar{\Phi}^{(1)}(t, \psi)(\phi + v\mathbf{1}_{\{0\}})$$

$$= \frac{d}{d\epsilon} \bar{\Phi}^{(1)}(t, \psi + \epsilon(\phi + v\mathbf{1}_{\{0\}})) \bigg|_{\epsilon=0},$$

where $\bar{\Phi}^{(1)}$ denotes the first-order Gâteau derivative of $\bar{\Phi}$ with respect to its second variable $\psi \in \mathbf{C}$. Since Φ is smooth, then the Gâteau derivative and the Fréchet derivative of $\bar{\Phi}$ coincide and are a continuous extension of the $D\Phi$, the Fréchet derivative of Φ. In other words, we have

$$D\bar{\Phi}(t, \psi)(\phi + v\mathbf{1}_{\{0\}}) = \overline{D\Phi(t, \psi)}(\phi + v\mathbf{1}_{\{0\}}).$$

By the uniqueness of the linear continuous extension, we have

$$\lim_{\epsilon \to 0} D_\epsilon \Phi(t, \psi)(\phi + v\mathbf{1}_{\{0\}}) \tag{5.37}$$

$$= \lim_{\epsilon \to 0} D_\epsilon \bar{\Phi}(t, \psi)(\phi + v\mathbf{1}_{\{0\}})$$

$$= \overline{D\Phi(t, \psi)}(\phi + v\mathbf{1}_{\{0\}}). \tag{5.38}$$

Similarly, the same argument can be used for the second-order finite difference approximation

$$D_\epsilon^2 \Phi(t, \psi)(\phi + v\mathbf{1}_{\{0\}}, \varphi + w\mathbf{1}_{\{0\}}) = \frac{\Phi(t, \psi + \epsilon(\phi + v\mathbf{1}_{\{0\}})) - \Phi(t, \psi)}{\epsilon^2}$$

$$+ \frac{\Phi(t, \psi - \epsilon(\varphi + w\mathbf{1}_{\{0\}})) - \Phi(t, \psi)}{\epsilon^2}.$$

Therefore,

$$\lim_{\epsilon \to 0} D_\epsilon^2 \Phi(t, \psi)(\phi + v\mathbf{1}_{\{0\}}, \varphi + w\mathbf{1}_{\{0\}}) = \overline{D^2\Phi(t, \psi)}(\phi + v\mathbf{1}_{\{0\}}, \varphi + w\mathbf{1}_{\{0\}}).$$

This proves the lemma.   □

Let $\epsilon, \eta > 0$. The corresponding discrete version of (5.32) is given by

$$0 = \alpha V_M(t, \psi) - \Delta_\eta V_M(t, \psi)$$

$$- \sup_{u \in U} \left[\mathcal{S}_\epsilon(V_M)(t, \psi) + D_\epsilon V_M(t, \psi)(\phi + (f(t, \psi, u)\mathbf{1}_{\{0\}})) \right.$$

$$= \frac{1}{2} \sum_{i=1}^m D_\epsilon^2 V_M(t, \psi)(g(t, \psi, u)\mathbf{e}_j \mathbf{1}_{\{0\}}), g(t, \psi, u)\mathbf{e}_j \mathbf{1}_{\{0\}})$$

$$\left. + (L(t, \psi, u) \wedge M) \right].$$

Substituting the terms

$$\Delta_\eta V_M(t, \psi), \quad \mathcal{S}_\epsilon(V_M)(t, \psi),$$

$$D_\epsilon V_M(t, \psi)(\phi + (f(t, \psi, u)\mathbf{1}_{\{0\}})),$$

and

$$D_\epsilon^2 V_M(t, \psi)(g(t, \psi, u)\mathbf{e}_j \mathbf{1}_{\{0\}}), g(t, \psi, u)\mathbf{e}_j \mathbf{1}_{\{0\}})$$

into the equation above, we have equivalently,

$$
\alpha V_M(t, \psi) = \frac{V_M(t, \tilde{\psi}_\epsilon) - V_M(t, \psi)}{\epsilon} + \frac{V_M(t + \eta, \psi) - V_M(t, \psi)}{\eta}
$$
$$
+ \sup_{u \in U} \left[\frac{V_M(t, \psi + \epsilon(f(t, \psi, u)\mathbf{1}_{\{0\}})) - V_M(t, \psi)}{\epsilon} \right.
$$
$$
+ \frac{1}{2} \sum_{i=1}^{m} \left(\frac{V_M(t, \psi + \epsilon(g(t, \psi, u)\mathbf{e}_i \mathbf{1}_{\{0\}})) - V_M(t, \psi)}{\epsilon^2} \right.
$$
$$
\left. + \frac{V_M(t, \psi - \epsilon(g(t, \psi, u)\mathbf{e}_i \mathbf{1}_{\{0\}})) - V_M(t, \psi)}{\epsilon^2} \right)
$$
$$
\left. + (L(t, \psi, u) \wedge M) \right]. \tag{5.39}
$$

Rearranging terms, we obtain

$$
\sup_{u \in U} \left[\frac{V_M(t, \tilde{\psi}_\epsilon)}{\epsilon} + \frac{V_M(t + \eta, \psi)}{\eta} + \frac{V_M(t, \psi + \epsilon(f(t, \psi, u)\mathbf{1}_{\{0\}}))}{\epsilon} \right.
$$
$$
+ \frac{1}{2} \sum_{i=1}^{m} \frac{V_M(t, \psi + \epsilon(g(t, \psi, u)\mathbf{e}_i \mathbf{1}_{\{0\}})) + V_M(t, \psi - \epsilon(g(t, \psi, u)\mathbf{e}_i \mathbf{1}_{\{0\}}))}{\epsilon^2}
$$
$$
\left. + (L(t, \psi, u) \wedge M) - \left(\frac{2}{\epsilon} + \frac{1}{\eta} + \frac{m}{\epsilon^2} + \alpha \right) V_M(t, \psi) \right] = 0. \tag{5.40}
$$

Since the term $\left(\frac{2}{\epsilon} + \frac{1}{\eta} + \frac{m}{\epsilon^2} + \alpha \right)$ is always positive, (5.39) is equivalent to

$$
\sup_{u \in U} \left[\frac{1}{\frac{2}{\epsilon} + \frac{1}{\eta} + \frac{m}{\epsilon^2} + \alpha} \left(\frac{V_M(t, \tilde{\psi}_\epsilon)}{\epsilon} + \frac{V_M(t, \psi + \epsilon(f(t, \psi, u)\mathbf{1}_{\{0\}}))}{\epsilon} \right. \right.
$$
$$
+ \frac{1}{2} \sum_{i=1}^{m} \frac{V_M(t, \psi + \epsilon(g(t, \psi, u)\mathbf{e}_i \mathbf{1}_{\{0\}})) + V_M(t, \psi - \epsilon(g(t, \psi, u)\mathbf{e}_i \mathbf{1}_{\{0\}}))}{\epsilon^2}
$$
$$
\left. \left. + \frac{V_M(t + \eta, \psi)}{\eta} + (L(t, \psi, u) \wedge M) \right) - V_M(t, \psi) \right] = 0. \tag{5.41}
$$

Let $C_b([0, T] \times (\mathbf{C} \oplus \mathbf{B}))$ denote the space of bounded continuous functions Φ from $[0, T] \times (\mathbf{C} \oplus \mathbf{B})$ to $\Re$ equipped with the sup-norm $\| \cdot \|_{C_b}$ defined by

$$
\| \Phi \|_{C_b} = \sup_{(t, \phi \oplus v \mathbf{1}_{\{0\}})} |\Phi(t, \phi \oplus v \mathbf{1}_{\{0\}})|,
$$
$$
\forall \Phi \in C_b([0, T] \times (\mathbf{C} \oplus \mathbf{B})).
$$

In the following we make necessary preparations for establishing the fixed point for an appropriate mapping as the limiting point of finite difference approximation. Define a mapping $S_M : (0, 1)^2 \times [0, T] \times \mathbf{C} \times \Re \times C_b([0, T] \times (\mathbf{C} \oplus \mathbf{B})) \to \Re$ as follows:

$$S_M(\epsilon, \eta, t, \psi, x, \Phi)$$

$$= \epsilon \sup_{u \in U} \left[\frac{\Phi(t, \tilde{\psi}_\epsilon)}{\epsilon} + \frac{\Phi(t + \eta, \psi)}{\eta} \right.$$

$$+ \frac{\Phi(t, \psi + \epsilon(f(t, \psi, u)\mathbf{1}_{\{0\}}))}{\epsilon} + (L(t, \psi, u) \wedge M)$$

$$+ \frac{1}{2} \sum_{i=1}^m \frac{\Phi(t, \psi + \epsilon(g(t, \psi, u)\mathbf{e}_i\mathbf{1}_{\{0\}})) + \Phi(t, \psi - \epsilon(g(t, \psi, u)\mathbf{e}_i\mathbf{1}_{\{0\}}))}{\epsilon^2} \right]$$

$$- \epsilon \left(\frac{2}{\epsilon} + \frac{1}{\eta} + \frac{m}{\epsilon^2} + \alpha \right) x. \tag{5.42}$$

Then, (5.39) is equivalent to $S_M(\epsilon, \eta, t, \psi, V_M(t, \psi), V_M) = 0$. Moreover, note that the coefficient of x in S_M is negative. This implies that S_M is monotone: that is, for all $x, y \in \Re$, $\epsilon, \eta \in (0, 1)$, $t \in [0, T]$, $\psi \in \mathbf{C}$, and $\Phi \in C_b([0, T] \times (\mathbf{C} \oplus \mathbf{B}))$

$$S_M(\epsilon, \eta, t, \psi, x, \Phi) \le S_M(\epsilon, \eta, t, \psi, y, \Phi) \quad \text{whenever } x \ge y.$$

Definition 5.4.3 *The scheme S_M is said to be consistent if for every $t \in [0, T]$, $\psi \in \mathbf{C} \oplus \mathbf{B}$, and for every test function $\Phi \in [0, T] \times (\mathbf{C} \oplus \mathbf{B})$ such that $\Phi \in C_{lip}^{1,2}([0, T] \times (\mathbf{C} \oplus \mathbf{B})) \cap \mathcal{D}(\mathcal{S})$,*

$$\alpha\Phi(t, \psi) - \mathcal{H}_M(t, \psi, \Phi(t, \psi), \partial_t \Phi(t, \psi), \overline{D\Phi(t, \psi)}, \overline{D^2\Phi(t, \psi)})$$

$$= \lim_{(\tau, \phi) \to (t, \psi), \, \epsilon, \eta \downarrow 0, \, \xi \to 0} \frac{S_M(\epsilon, \eta, \tau, \phi, \Phi(\tau, \phi) + \xi, \Phi + \xi)}{\epsilon}.$$

In the above definition, $\Phi + \xi$ should be understood to be a real-valued function defined on $[0, T] \times \mathbf{C}$ such that

$$(\Phi + \xi)(t, \psi) = \Phi(t, \psi) + \xi, \quad \forall (t, \xi) \in [0, T] \times \mathbf{C}.$$

Lemma 5.4.4 *The scheme S_M is consistent.*

Proof. Let $\Phi \in C_{lip}^{1,2}([0, T] \times (\mathbf{C} \oplus \mathbf{B})) \cap \mathcal{D}(\mathcal{S})$. We write

$$\frac{S_M(\epsilon, \eta, \tau, \phi, \Phi(\tau, \phi) + \xi, \Phi + \xi)}{\epsilon}$$

$$= \sup_{u \in U} \left[\left(\frac{\Phi(\tau, \tilde{\psi}_\epsilon) + \xi}{\epsilon} + \frac{\Phi(\tau, \psi + \epsilon(f(\tau, \psi, u)\mathbf{1}_{\{0\}})) + \xi}{\epsilon} \right. \right.$$

$$+ (L(\tau, \psi, u) \wedge M) + \frac{\Phi(t + \eta, \psi) + \xi}{\eta}$$

$$\left. \left. + \frac{1}{2} \sum_{i=1}^m \frac{\Phi(\tau, \psi + \epsilon(g(t, \psi, u)\mathbf{e}_i\mathbf{1}_{\{0\}})) + 2\xi + \Phi(\tau, \psi - \epsilon(g(t, \psi, u)\mathbf{e}_i\mathbf{1}_{\{0\}}))}{\epsilon^2} \right) \right]$$

$$- (\frac{2}{\epsilon} + \frac{1}{\eta} + \frac{m}{\epsilon^2} + \alpha)(\Phi(\tau, \phi) + \xi).$$

Sending $\xi \to 0$, $\tau \to t$, $\phi \to \psi$, $\epsilon, \eta \to 0$, we have

$$\alpha \Phi(t, \psi) - \mathcal{H}_M(t, \psi, \Phi(t, \psi), \partial_t \Phi(t, \psi), \overline{D\Phi(t, \psi)}, \overline{D^2\Phi(t, \psi)})$$

$$= \lim_{(\tau, \phi) \to (t, \psi), \epsilon, \eta, \downarrow 0, \xi \to 0} \frac{S_M(\epsilon, \eta, \tau, \phi, \Phi(\tau, \phi) + \xi, \Phi + \xi)}{\epsilon}.$$

This completes the proof.   □

Using (5.41), we see that the equation $S_M(\epsilon, \eta, t, \psi, \Phi(t, \psi), \Phi) = 0$ is equivalent to the equation

$$\Phi(t, \psi) = \sup_{u \in U} \left[\frac{1}{\frac{2}{\epsilon} + \frac{1}{\eta} + \frac{m}{\epsilon^2} + \alpha} \left(\frac{\Phi(t, \tilde{\psi}_\epsilon)}{\epsilon} + \frac{\Phi(t, \psi + \epsilon(f(t, \psi, u)\mathbf{1}_{\{0\}}))}{\epsilon} \right. \right.$$

$$+ \frac{1}{2} \sum_{i=1}^{m} \frac{\Phi(t, \psi + \epsilon(g(t, \psi, u)\mathbf{e}_i \mathbf{1}_{\{0\}})) + \Phi(t, \psi - \epsilon(g(t, \psi, u)\mathbf{e}_i \mathbf{1}_{\{0\}}))}{\epsilon^2}$$

$$\left. \left. + \frac{\Phi(t + \eta, \psi)}{\eta} + (L(t, \psi, u) \wedge M) \right) \right]. \tag{5.43}$$

For each $\epsilon > 0$ and $\eta > 0$, we define an operator $\mathcal{T}_{\epsilon, \eta}$ on $C_b([0, T] \times (\mathbf{C} \oplus \mathbf{B}))$ as follows:

$$\mathcal{T}_{\epsilon, \eta} \Phi(t, \psi)$$

$$\equiv \sup_{u \in U} \left[\frac{1}{\frac{2}{\epsilon} + \frac{1}{\eta} + \frac{m}{\epsilon^2} + \alpha} \left(\frac{\Phi(t, \tilde{\psi}_\epsilon)}{\epsilon} + \frac{\Phi(t, \psi + \epsilon(f(t, \psi, u)\mathbf{1}_{\{0\}}))}{\epsilon} \right. \right.$$

$$+ \frac{1}{2} \sum_{i=1}^{m} \frac{W(t, \psi + \epsilon(g(t, \psi, u)\mathbf{e}_i \mathbf{1}_{\{0\}})) + \Phi(t, \psi - \epsilon(g(t, \psi, u)\mathbf{e}_i \mathbf{1}_{\{0\}}))}{\epsilon^2}$$

$$\left. \left. + \frac{\Phi(t + \eta, \psi)}{\eta} + L(t, \psi, u) \wedge M \right) \right]. \tag{5.44}$$

Note that to find a $\Phi \in C_b([0, T] \times (\mathbf{C} \oplus \mathbf{B})$ that satisfies

$$S_M(\epsilon, \eta, t, \psi, \Phi(t, \psi), \Phi) = 0 \quad \text{for each } \epsilon, \eta > 0 \text{ and } (t, \psi) \in [0, T] \times \mathbf{C}.$$

is equivalent to finding a fixed point of the map $\mathcal{T}_{\epsilon, \eta}$.

Lemma 5.4.5 *For each $\epsilon > 0$ and $\eta > 0$, $\mathcal{T}_{\epsilon, \eta}$ is a contraction map.*

Proof. To prove that $\mathcal{T}_{\epsilon, \eta}$ is a contraction, we need to show that there exists $0 < \beta < 1$ such that

$$\|\mathcal{T}_{\epsilon, \eta} \Phi_1 - \mathcal{T}_{\epsilon, \eta} \Phi_2\|_{C_b} \leq \beta \|\Phi_1 - \Phi_2\|_{C_b}, \quad \forall \Phi_1, \Phi_2 \in C_b([0, T] \times (\mathbf{C} \oplus \mathbf{B})),$$

where $\|\cdot\|_{C_b}$ is the sup-norm for the space $C_b([0, T] \times (\mathbf{C} \oplus \mathbf{B}))$. Let us define $c_{\epsilon, \eta}$ by

$$c_{\epsilon,\eta} = \frac{2}{\epsilon} + \frac{1}{\eta} + \frac{m}{\epsilon^2} + \alpha.$$

Note that

$$|\mathcal{T}_{\epsilon,\eta}\Phi_1(t,\psi) - \mathcal{T}_{\epsilon,\eta}\Phi_2(t,\psi)|$$

$$\leq \sup_{u \in U}\left[\frac{1}{c_{\epsilon,\eta}}\left|\left(\frac{\Phi_1(t,\tilde{\psi}_\epsilon)}{\epsilon} + \frac{\Phi_1(t,\psi + \epsilon(f(t,\psi,u)\mathbf{1}_{\{0\}}))}{\epsilon} + \frac{\Phi_1(t+\eta,\psi)}{\eta}\right.\right.\right.$$

$$\left. + \frac{1}{2}\sum_{i=1}^{m}\frac{\Phi_1(t,\psi + \epsilon(g(t,\psi,u)\mathbf{e}_i\mathbf{1}_{\{0\}})) + \Phi_1(t,\psi - \epsilon(g(t,\psi,u)\mathbf{e}_i\mathbf{1}_{\{0\}}))}{\epsilon^2}\right)$$

$$-\left(\frac{1}{\epsilon}\Phi_2(t,\tilde{\psi}_\epsilon) + \frac{\Phi_2(t,\psi + \epsilon(f(t,\psi,u)\mathbf{1}_{\{0\}}))}{\epsilon} + \frac{\Phi_2(t+\eta,\psi)}{\eta}\right.$$

$$\left.\left.\left. + \frac{1}{2}\sum_{i=1}^{m}\frac{\Phi_2(t,\psi + \epsilon(g(t,\psi,u)\mathbf{e}_i\mathbf{1}_{\{0\}})) + \Phi_2(t,\psi - \epsilon(g(t,\psi,u)\mathbf{e}_i\mathbf{1}_{\{0\}}))}{\epsilon^2}\right)\right|\right].$$

This implies that for all $(t,\psi) \in [0,T] \times \mathbf{C}$,

$$|\mathcal{T}_{\epsilon,\eta}\Phi_1(t,\psi) - \mathcal{T}_{\epsilon,\eta}\Phi_2(t,\psi)| \leq \left[\frac{\frac{2}{\epsilon} + \frac{1}{\eta} + \frac{m}{\epsilon^2}}{c_{\epsilon,\eta}}\right]\|\Phi_1 - \Phi_2\|_{C_b}. \tag{5.45}$$

In addition, note that

$$\frac{\frac{2}{\epsilon} + \frac{1}{\eta} + \frac{m}{\epsilon^2}}{c_{\epsilon,\eta}} = \frac{\frac{2}{\epsilon} + \frac{1}{\eta} + \frac{m}{\epsilon^2}}{\frac{2}{\epsilon} + \frac{1}{\eta} + \frac{m}{\epsilon^2} + \alpha} < 1.$$

Let

$$\beta_{\epsilon,\eta} = \frac{\frac{2}{\epsilon} + \frac{1}{\eta} + \frac{m}{\epsilon^2}}{c_{\epsilon,\eta}}.$$

Therefore,

$$\|\mathcal{T}_{\epsilon,\eta}\Phi_1 - \mathcal{T}_{\epsilon,\eta}\Phi_2\|_{C_b} \leq \beta_{\epsilon,\eta}\|\Phi_1 - \Phi_2\|_{C_b}.$$

This proves that the operator $\mathcal{T}_{\epsilon,\eta}$ is a contraction map.   □

Definition 5.4.6 *The scheme S_M is said to be stable if for every $\epsilon, \eta \in (0,1)$, there exists a bounded solution $\Phi_{\epsilon,\eta} \in C_b([0,T] \times (\mathbf{C} \oplus \mathbf{B}))$ to the equation*

$$S_M(\epsilon, \eta, t, \psi, \Phi(t,\psi), \Phi) = 0, \tag{5.46}$$

with the bound independent of ϵ and η.

Lemma 5.4.7 *The scheme S_M is said to be stable.*

Proof. By the Banach fixed-point theorem, the strict contraction $\mathcal{T}_{\epsilon,\eta}$ has a unique fixed point that we denote by $\Phi_{\epsilon,\eta}^M$. Given any function $\Phi_0 \in C_b([0,T] \times (\mathbf{C} \oplus \mathbf{B}))_b$, we construct a sequence as follows: $\Phi_{k+1} = \mathcal{T}_{\epsilon,\eta}\Phi_k$ for $k \geq 0$. It is clear that

$$\lim_{k \to \infty} \Phi_k = \Phi_{\epsilon,\eta}^M.$$

Moreover, note that

$$\Phi_{k+1}(t, \psi)$$

$$= \sup_{u \in U} \left[\frac{1}{\frac{2}{\epsilon} + \frac{1}{\eta} + \frac{m}{\epsilon^2} + \alpha} \left(\frac{\Phi_k(t, \tilde{\psi}_\epsilon)}{\epsilon} + \frac{\Phi_k(t, \psi + \epsilon(f(t, \psi, u)\mathbf{1}_{\{0\}}))}{\epsilon} \right. \right.$$

$$+ \frac{1}{2} \sum_{i=1}^{m} \frac{\Phi_k(t, \psi + \epsilon(g(t, \psi, u)e_i \mathbf{1}_{\{0\}})) + \Phi_k(t, \psi - \epsilon(g(t, \psi, u)e_i \mathbf{1}_{\{0\}}))}{\epsilon^2}$$

$$\left. \left. + \frac{\Phi_k(t + \eta, \psi)}{\eta} + (L(t, \psi, u) \wedge M) \right) \right]$$

$$\leq \beta_{\epsilon,\eta} \|\Phi_k\|_{C_b} + \frac{1}{c_{\epsilon,\eta}} M. \tag{5.47}$$

In addition, we have

$$\beta_{\epsilon,\eta} = \frac{c_{\epsilon,\eta} - \rho}{c_{\epsilon,\eta}} < 1.$$

This implies that

$$\|\Phi_{k+1}\|_{C_b} \leq \frac{c_{\epsilon,\eta} - \rho}{c_{\epsilon,\eta}} \|\Phi_k\|_{C_b} + \frac{1}{c_{\epsilon,\eta}} M. \tag{5.48}$$

From (5.48), we deduce that

$$\|\Phi_{k+1}\|_{C_b} \leq \left(\frac{c_{\epsilon,\eta} - \rho}{c_{\epsilon,\eta}} \right)^{k+1} \|\Phi_0\|_{C_b} + \frac{M}{c_{\epsilon,\eta}} \sum_{i=0}^{k} \left(\frac{c_{\epsilon,\eta} - \rho}{c_{\epsilon,\eta}} \right)^{i}.$$

Taking the limit as $k \to \infty$, we obtain

$$\|\Phi_{\epsilon,\eta}^M\|_{C_b} \leq \frac{M}{c_{\epsilon,\eta}} \cdot \frac{1}{1 - \frac{c_{\epsilon,\eta} - \alpha}{c_{\epsilon,\eta}}} = \frac{M}{\alpha}.$$

This implies the stability of the scheme S_M. □

Theorem 5.4.8 *Let $\Phi_{\epsilon,\eta}^M$ denote the solution to (5.46). Then, as $\epsilon, \eta \downarrow 0$, the sequence $\Phi_{\epsilon,\eta}^M$ converges uniformly on $[0, T] \times \mathbf{C}$ to the unique viscosity solution V_M of (5.32).*

Proof. Define

$$\Phi_M^*(t, \psi) = \limsup_{(\tau,\phi) \to (t,\psi), \epsilon,\eta \downarrow 0} \Phi_{\epsilon,\eta}^M(\tau, \phi),$$

$$\Phi_{*M}(t, \psi) = \liminf_{(\tau,\phi) \to (t,\psi), \epsilon,\eta \downarrow 0} \Phi_{\epsilon,\eta}^M(\tau, \phi).$$

We claim that Φ_M^* and Φ_{*M} are the viscosity subsolution and supersolution of (5.32), respectively. To prove this claim, we only consider the case for Φ_M^*. The argument for that of Φ_{*M} is similar. We want to show that

$$\alpha \Gamma(t,\psi) - \mathcal{H}_M(t,\psi,\Gamma(t,\psi),\partial_t\Gamma(t,\psi),\overline{D\Gamma(t,\psi)},\overline{D^2\Gamma(t,\psi)}) \le 0$$

for any test function $\Gamma \in C_{lip}^{1,2}([0,T] \times (\mathbf{C} \oplus \mathbf{B})) \cap \mathcal{D}(\mathcal{S})$ such that (t,ψ) is a strictly local maximum of $\Phi_M^* - \Gamma$. Without loss of generality, we may assume that $\Phi_M^* \le \Gamma$ and $\Phi_M^*(t,\psi) = \Gamma(t,\psi)$, and because of the stability of our scheme, we can also assume that $\Gamma \ge 2\sup_{\epsilon,\eta} \|\Phi_{\epsilon,\eta}^M\|$ outside of the ball $B((t,\psi),l)$ centered at (t,ψ) with radius l, where $l > 0$ is such that

$$\Phi_M^*(\tau,\phi) - \Gamma(\tau,\phi) \le 0 = \Phi_M^*(t,\psi) - \Gamma(t,\psi) \quad \text{for } (\tau,\phi) \in B((t,\psi),l).$$

This implies that there exist sequences $\epsilon_k > 0$, $\eta_k > 0$, and $(\tau_k,\phi_k) \in [0,T] \times (\mathbf{C} \oplus \mathbf{B})$ such that as $k \to \infty$, we have

$$\epsilon_k \to 0, \ \eta_k \to 0, \ \tau_k \to t, \ \phi_k \to \psi, \ \Phi_{\epsilon_k,\eta_k}^M(\tau_k,\phi_k) \to \Phi_M^*(t,\psi), \qquad (5.49)$$

and (τ_k,ϕ_k) is a global maximum $\Phi_{\epsilon_k,\eta_k}^M - \Gamma$. Denote $\gamma_k = \Phi_{\epsilon_k,\eta_k}^M(\tau_k,\phi_k) - \Gamma(\tau_k,\phi_k)$. Obviously, $\gamma_k \to 0$ and

$$\Phi_{\epsilon_k,\eta_k}^M(\tau,\phi) \le \Gamma(\tau,\phi) + \gamma_k, \quad \forall(\tau,\phi) \in [0,T] \times (\mathbf{C} \oplus \mathbf{B}). \qquad (5.50)$$

We know that

$$S_M(\epsilon_k,\eta_k,\tau_k,\phi_k,\Phi_{\epsilon_k,\eta_k}^M(\tau_k,\phi_k),\Phi_{\epsilon_k,\eta_k}^M) = 0.$$

The monotonicity of S_M implies

$$\begin{aligned}
&S_M(\epsilon_k,\eta_k,\tau_k,\phi_k,\Gamma(\tau_k,\phi_k)+\gamma_k,\Gamma(\tau_k,\phi_k)+\gamma_k)\\
&\le S_M(\epsilon_k,\eta_k,\tau_k,\phi_k,\Phi_{\epsilon_k,\eta_k}^M(\tau_k,\phi_k),\Phi_{\epsilon_k,\eta_k}^M)\\
&= 0.
\end{aligned} \qquad (5.51)$$

Therefore,

$$\lim_{k\to\infty} \frac{S_M(\epsilon_k,\eta_k,\tau_k,\phi_k,\Phi_{\epsilon_k,\eta_k}^M(\tau_k,\phi_k),\Phi_{\epsilon_k,\eta_k}^M)}{\epsilon_k} \le 0,$$

so

$$\begin{aligned}
&\alpha\Phi_M^*(t,\psi) - \mathcal{H}_M(t,\psi,\Phi_M^*(t,\psi),\mathrm{D}_t\Gamma(t,\psi),\overline{D\Gamma(t,\psi)},\overline{D^2\Gamma(t,\psi)})\\
&= \lim_{k\to\infty} \frac{S_M(\epsilon_k,\eta_k,\tau_k,\phi_k,\Phi_{\epsilon_k,\eta_k}^M(\tau_k,\phi_k),\Phi_{\epsilon_k,\eta_k}^M)}{\epsilon_k}\\
&\le 0.
\end{aligned}$$

This proves that $\Phi_M^* : [0,T] \times \mathbf{C} \to \Re$ is a viscosity subsolution of (5.32) and, similarly, we can prove that Φ_{*M} is a viscosity supersolution. By the comparison principle (see Theorem 3.6.1 in Chapter 3), we can get that

$$\Phi_{*M}(t,\psi) \ge \Phi_M^*(t,\psi), \quad \forall(t,\psi) \in [0,T] \times \mathbf{C}. \qquad (5.52)$$

On the other hand, by the definition, of Φ_{*M} and Φ_M^*, it is easy to see that

$$\Phi_{*M}(t,\psi) \le \Phi_M^*(t,\psi), \quad \forall (t,\psi) \in [0,T] \times \mathbf{C}.$$

Combined with (5.52), the above implies

$$\Phi_{*M}(t,\psi) = \Phi_M^*(t,\psi), \quad \forall (t,\psi) \in [0,T] \times \mathbf{C}.$$

Since Φ_{*M} is a viscosity supersolution and Φ_M^* is a viscosity subsolution, they are also viscosity solutions of (5.32). Now, using the uniqueness of the viscosity solution (5.32), we see that $V_M = \Phi_M^* = \Phi_{*M}$. Therefore, we conclude that the sequence $(\Phi_{\epsilon,\eta}^M)_{\epsilon,\eta}$ converges locally uniformly to V_M as desired.   □

5.4.2 Discretization of Segment Functions

Although we are able to prove convergence in theory of the finite difference scheme described in the previous subsection, it is clear that we still will not be able to extract any useful computational algorithm for the value function $V : [0,T] \times \mathbf{C} \to \Re$ from it. To overcome this shortcoming, we propose to further discretize segment functions in $\mathbf{C}$ as follows.

We again use the same discretization notation in Section 5.1 and define for each $N = 1, 2, \cdots$ and any

$$\mathbf{x} = (x(-Nh), x((-N+1)h), \ldots, x(-h), x(0)) \in (\Re^n)^{N+1}$$

a continuous-time function $\tilde{\mathbf{x}} \in \mathbf{C} = C([-r,0]; \Re^n)$ by making the linear interpolation between the time-space points $(kh, x(kh))$ and $((k+1)h, x((k+1)h)$. Therefore, if $\theta \in [-r,0]$ and $\theta = kh$ for some $k = -N, -N+1, \ldots, -1, 0$, $\tilde{\mathbf{x}}(kh) = x(kh)$. If $\theta \in [-r,0]$ is such that $kh < \theta < (k+1)h$ for some $k = -N, -N+1, \ldots, -1$, then

$$\tilde{\mathbf{x}}(\theta) = x(kh) + \frac{x((k+1)h) - x(kh)(\theta + kh)}{h}.$$

For the functions $f : [0,T] \times \mathbf{C} \times U \to \Re^n$, $g : [0,T] \times \mathbf{C} \times U \to \Re^{n \times m}$, $L : [0,T] \times \mathbf{C} \times U \to \Re$, and $\Psi : \mathbf{C} \to \Re$, we define respectively the functions $\hat{f} : \mathbf{I} \times \mathbf{S}^{N+1} \times U \to \Re^n$, $\hat{g} : \mathbf{I} \times \mathbf{S}^{N+1} \times U \to \Re^{n \times m}$, $\hat{L} : \mathbf{I} \times \mathbf{S}^{N+1} \times U \to \Re$, and $\hat{\Psi} : \mathbf{S}^{N+1} \to \Re$ as follows:

$$\hat{f}(\lfloor t \rfloor_N, \pi\phi, u) = f(t, \phi, u), \quad \hat{g}(\lfloor t \rfloor_N, \pi\phi, u) = g(t, \phi, u),$$

$$\hat{L}(\lfloor t \rfloor_N, \pi\phi, u) = L(t, \phi, u), \quad \text{and} \quad \hat{\Psi}(\pi\phi) = \Psi(\phi), \quad \forall (t, \phi, u) \in [0,T] \times \mathbf{C} \times U,$$

where $\pi : \mathbf{C} \to \mathbf{S}^{(N+1}$ is the truncated point-mass projection function defined by

$$\pi\phi = (\phi(-Nh), \phi((-N+1)h), \ldots, \phi(-h), \phi(0)).$$

Using these notations, we define the discrete version of $S_\epsilon \Phi$, $\partial_{t,\eta}\Phi$, $D_\epsilon \Phi$, and $D_\epsilon^2 \Phi$ in Section 5.4.1 by

$$\partial_{t,\eta}\tilde{\hat{\Phi}}(\lfloor t \rfloor_N, \pi\psi) = \frac{\tilde{\Phi}(\lfloor t \rfloor_N + \eta, \pi\psi) - \tilde{\Phi}(\lfloor t \rfloor_N, \pi\psi)}{\eta},$$

$$D_\epsilon\tilde{\Phi}(\lfloor t \rfloor_N, \pi\psi)(\pi\phi \oplus v) = \frac{\tilde{\Phi}(\lfloor t \rfloor_N, \pi\psi + \epsilon(\pi\phi \oplus v)) - \tilde{\Phi}(\lfloor t \rfloor_N, \pi\psi)}{\epsilon},$$

$$D_\epsilon^2\tilde{\Phi}(\lfloor t \rfloor_N, \pi\psi)(\pi\phi \oplus v, \pi\varphi \oplus w) = \frac{\hat{\Phi}(\lfloor t \rfloor_N, \pi\psi + \epsilon(\pi\phi \oplus v)) - \hat{\Phi}(\lfloor t \rfloor_N, \pi\psi)}{\epsilon^2}$$
$$+ \frac{\hat{\Phi}(\lfloor t \rfloor_N, \pi\psi - \epsilon(\pi\phi \oplus v)) - \hat{\Phi}(\lfloor t \rfloor_N, \pi\psi)}{\epsilon^2},$$

and

$$\mathcal{S}_\epsilon(\tilde{\Phi})(\lfloor t \rfloor_N, \pi\phi) = \frac{\hat{\Phi}(\lfloor t \rfloor_N, \tilde{\phi}_\epsilon) - \hat{\Phi}(\lfloor t \rfloor_N, \pi\phi)}{\epsilon},$$

where, $\phi, \varphi \in \mathbf{C}$ and $v, w \in \Re^n$,

$$\pi\phi \oplus v = (\phi(-Nh), \phi((-N+1)h), \ldots, \phi(-h), \phi(0) + v).$$

Therefore, we define

$$\mathcal{S}_\epsilon(\Phi)(t, \psi) = \frac{1}{\epsilon}\left[\Phi(t, \tilde{\psi}_\epsilon) - \Phi(t, \psi)\right].$$

The discrete version of the mappings S_M and $\mathcal{T}_{\epsilon,\eta}$ are given as $S_M^{(N)}$: $(0,1)^2 \times [0, \lfloor T \rfloor_N] \times (\mathbf{S})^{N+1} \times \Re \times B([0, \lfloor t \rfloor_N] \times (\mathbf{S})^{N+1} \oplus \Re^n)) \to \Re$ as follows:

$$S_M^{(N)}(\epsilon, \eta, \lfloor t \rfloor_N, \pi\psi, x, \hat{\Phi})$$
$$= \epsilon \max_{u \in U}\left[\frac{\hat{\Phi}(\lfloor t \rfloor_N, \pi\tilde{\psi}_\epsilon)}{\epsilon} + \frac{\hat{\Phi}(\lfloor t \rfloor_N + \eta, \pi\psi)}{\eta}\right.$$
$$+ \frac{\hat{\Phi}(\lfloor t \rfloor_N, \pi\psi \oplus \epsilon f(\lfloor t \rfloor_N, \pi\psi, u))}{\epsilon} + (L(\lfloor t \rfloor_N, \pi\psi, u) \wedge M)$$
$$\left. + \frac{1}{2}\sum_{i=1}^m \frac{\hat{\Phi}(\lfloor t \rfloor_N, \pi\psi \oplus \epsilon(g(\lfloor t \rfloor_N, \pi\psi, u))) + \hat{\Phi}(\lfloor t \rfloor_N, \pi\psi \ominus \epsilon(g(\lfloor t \rfloor_N, \pi\psi, u)))}{\epsilon^2}\right]$$
$$- \epsilon\left(\frac{2}{\epsilon} + \frac{1}{\eta} + \frac{m}{\epsilon^2} + \alpha\right)x. \tag{5.53}$$

and

$$\mathcal{T}_{\epsilon,\eta}^{(N)}\hat{\Phi}(\lfloor t \rfloor_N, \pi\psi)$$
$$\equiv \max_{u \in U}\left[\frac{1}{\frac{2}{\epsilon} + \frac{1}{\eta} + \frac{m}{\epsilon^2} + \alpha}\left(\frac{\hat{\Phi}(\lfloor t \rfloor_N, \pi\tilde{\psi}_\epsilon)}{\epsilon} + \frac{\hat{\Phi}(\lfloor t \rfloor_N, \pi\psi \oplus \epsilon(\hat{f}(\lfloor t \rfloor_N, \pi\psi, u)))}{\epsilon}\right.\right.$$
$$+ \frac{1}{2}\sum_{i=1}^m \frac{\hat{\Phi}(\lfloor t \rfloor_N, \pi\psi \oplus \epsilon(\hat{g}(\lfloor t \rfloor_N, \pi\psi, u))) + \hat{\Phi}(\lfloor t \rfloor_N, \pi\psi \ominus \epsilon(\hat{g}(\lfloor t \rfloor_N, \pi\psi, u)))}{\epsilon^2}$$
$$\left.\left. + \frac{\hat{\Phi}(\lfloor t \rfloor_N + \eta, \pi\psi)}{\eta} + \hat{L}(\lfloor t \rfloor_N, \pi\psi, u) \wedge M\right)\right]. \tag{5.54}$$

The following two lemmas can be proven in ways similar to that of Lemma 5.4.5 and Lemma 5.4.7. The proofs are omitted here.

Lemma 5.4.9 *For each $\epsilon > 0$ and $\eta > 0$, $T_{\epsilon,\eta}^{(N)}$ is a contraction map.*

Lemma 5.4.10 *The scheme $S_M^{(N)}$ is monotone, consistent, and stable.*

We have the following convergence result.

Theorem 5.4.11 *For each $(t, \psi) \in [0, T] \times \mathbf{C}$,*

$$\lim_{N \to \infty} V_M^{(N)}(\lfloor t \rfloor_N, \pi^{(N)}\psi) = V_M(t, \psi).$$

Proof. Since $\lfloor t \rfloor_N \to t$ and $\pi^{(N)}\psi \to \psi$ (in $\mathbf{C}$) as $N \to \infty$, by the definition of $V_M : [0, T] \times \mathbf{C} \to \Re$, the convergence follows very easily.   □

5.4.3 A Computational Algorithm

Based on the results obtained in the last two subsections, we can construct a computational algorithm for each $N \in \aleph$ to obtain a numerical solution. For example, one algorithm can be as follows:
Step 0. Let $(t, \psi) \in [0, T] \times \mathbf{C}$.
(i) Compute $\lfloor t \rfloor_N$ and $\pi\psi$.
(ii) Choose any function $\Phi^{(0)} \in C_b([0, T] \times \mathbf{C} \oplus \mathbf{B})$.
(iii) Compute $\hat{\Phi}^{(0,N)}(\lfloor t \rfloor_N, \pi^{(N)}\psi) = \Phi^{(0)}(\pi^{(N)}\psi)$.

Step 1. Pick the starting values for $\epsilon(1), \eta(1)$. For example, we can choose $\epsilon(1) = 10^{-2}, \eta(1) = 10^{-3}$.

Step 2. For the given $\epsilon, \eta > 0$, compute the function

$$\hat{\Phi}_{\epsilon(1),\eta(1)}^{(1,N)} \in C_b([0, T] \times (\mathbf{S}^{(N)})^{N+1})$$

by the formula

$$\hat{\Phi}_{\epsilon(1),\eta(1)}^{(1,N)} = T_{\epsilon(1),\eta(1)}^{(N)} \Phi^{(0,N)},$$

where $T_{\epsilon(1),\eta(1)}^{(N)}$, which is defined on $C_b([0, T] \times (\mathbf{S}^{(N)})^{N+1})$, is given by (5.54).
Step 3. Repeat Step 2 for $i = 2, 3, \ldots$ using

$$\hat{\Phi}_{\epsilon(1),\eta(1)}^{(i,N)}(\lfloor t \rfloor_N, \pi\psi) = T_{\epsilon(1),\eta(1)}^{(N)} \Phi_{\epsilon(1),\eta(1)}^{(i-1,N)}(\lfloor t \rfloor_N, \pi\psi).$$

Stop the iteration when

$$|\hat{\Phi}_{\epsilon(1),\eta(1)}^{(i+1,N)}(t, \psi) - \hat{\Phi}_{\epsilon(1),\eta(1)}^{(i,N)}(t, \psi)| \leq \delta_1,$$

where δ_1 is a preselected number small enough to achieve the accuracy we want. Denote the final solution by $\hat{\Phi}_{\epsilon(1),\eta(1)}(\lfloor t \rfloor_N, \pi\psi)$.

Step 4. Choose two sequences of $\epsilon(k)$ and $\eta(k)$, such that

$$\lim_{k\to\infty} \epsilon(k) = \lim_{k\to\infty} \eta(k) = 0.$$

For example, we may choose $\epsilon(k) = \eta(k) = 10^{-(2+k)}$. Now, repeat Step 2 and Step 3 for each $\epsilon(k), \eta(k)$ until

$$|\hat{\Phi}^{(i,N)}_{\epsilon(k+1),\eta(k+1)}(\lfloor t \rfloor_N, \pi\psi) - \hat{\Phi}^{(i,N)}_{\epsilon(k),\eta(k)}(\lfloor t \rfloor_N, \pi\psi)| \le \delta_2,$$

where δ_2 is chosen to obtain the expected accuracy.

5.5 Conclusions and Remarks

This chapter presents three different discrete approximation schemes: semi-discrete scheme, Markov chain approximation, and finite difference approximation for the viscosity solution of the infinite-dimensional HJBE. The basic idea behind the semidiscrete scheme and Markov chain approximation is to discrete-time variable and/or space variable for each of the controlled SHDEs, the objective functional. Using the discrete version of the Bellman's dynamic programming principle, the value functions of the discrete optimal control problems can be obtained and are shown to converge to the value function of the original optimal classical control problem. These two discrete approximations do not deal with the infinite-dimensional HJBE at all. However, the finite difference method provides a discretization scheme for approximating the infinite-dimensional HJBE. A contraction map is set up so that the value function of the optimal classical control problem can be approximated by the fixed point of the contraction map. The finite difference method can be further discretized temporally and spatially for computational convenience. A computational algorithm is provided based on the finite difference scheme obtained.

One important area of SHDEs not addressed in this monograph and requires more research effort is the theory and applications of nonlinear filtering. However, we mention here some related computational results on this subject by Chang [Cha87] and, more recently, by Calzolari et al. [CFN06, CFN07].

6

Option Pricing

A contingent claim (or option) is a contract giving the buyer of the contract (or simply the buyer) the right to buy from or sell to the contract writer (or simply writer) a share of an underlying stock at a predetermined price $q > 0$ (called the strike price) and at or prior to a prespecified time $T > 0$ (called the expiration date) in the future. The right to buy (respectively, to sell) a share of the stock is called a *call* (respectively, a *put*) option. The European (call or put) option can only be excised at the expiration date T, but the American (call or put) option can be excised at any time prior to or at the expiration date. The pricing problem of a contingent claim is, briefly, to determine the fee (called the *rational price*) that the writer should receive from the buyer for the rights of the contract and also to determine the trading strategy the writer should use to invest this fee in a (B, S)-market in such a way as to ensure that the writer will be able to cover the option if and when it is exercised. The fee should be large enough that the writer can, with riskless investing, cover the option but be small enough that the writer does not make an unfair (i.e., riskless) profit.

Throughout this chapter, any financial market consisting of one (riskless) bank account and one (risky) stock account will be referred to as a (B, S)-market, although the specifics of each (B, S)-market may depend on the characteristics of the bank account and the underlying stock along with their associated parameters. The characterizations of rational pricing function of the contingent claim therefore depend on the specifics of the (B, S)-market considered.

The pricing of European or American options in the continuous-time (B, S)-market has been a subject of extensive research in recent years. Explicit results for the European call option obtained (see, e.g., Black and Scholes [BS73], Harrison and Kreps [HK79], Harrison and Pliska [HP81], Merton [Mer73, Mer90], Shiryaev et al. [SKKM94]) for the idealized Black-Scholes market often consisting of a riskless bank account $\{B(t), t \geq 0\}$ that grows continuously with a constant interest rate $\lambda > 0$, that is,

$$dB(t) = \lambda B(t)\, dt, \ \ B(0) = x,$$

and a stock whose price process $\{S(t), t \geq 0\}$ satisfies the following linear stochastic differential equation:

$$dS(t) = \mu S(t)dt + \sigma S(t)dW(t), \ \ S(0) = y,$$

or, equivalently, the geometric Brownian motion,

$$S(t) = S(0) \exp\left\{ (\mu - \frac{1}{2}\sigma^2)\, t + \sigma W(t) \right\}, \ \ \forall t \geq 0.$$

In the above, $W = \{W(t), t \geq 0\}$ is a one-dimensional standard Brownian motion defined on a complete filtered probability space $(\Omega, \mathcal{F}, P, \mathbf{F})$, μ and σ are positive constants that represent respectively the the stock appreciation rate and the stock volatility rate.

The objective of the standard European call option under the Black-Scholes market is to determine a rational price $Q(t, y)$ (i.e., the fee the writer should receive) when the contract is sold to the buyer at time $t \in [0, T]$ given the stock price $S(t) = y$. It is now well known that the pricing function $Q(t, y)$, $(t, y) \in [0, T] \times [0, \infty)$, can be described by the following expression:

$$Q(t,y) = \tilde{E}\left[e^{-\lambda(T-t)}(S(T) - q)^+ \mid S(t) = y \right]$$
$$= \tilde{E}\left[e^{-\lambda(T-t)}(S(T-t) - q)^+ \mid S(0) = y \right],$$

where q is the strike price, $\tilde{E}$ is the expectation corresponding to the probability measure $\tilde{P}$ under which the process $\{\tilde{W}(t), t \geq 0\}$ described by

$$\tilde{W}(t) = W(t) + \frac{\mu - \lambda}{\sigma}t, \ \ t \geq 0,$$

is a new standard Brownian motion, and

$$(S(T) - q)^+ \equiv \max\{S(T) - q, 0\}$$

is the payoff function that represents the profit $S(T) - q$ for the buyer at the expiration time T. This payoff function can be interpreted as follows. If $S(T) > q$, then the buyer shall exercise the option by buying (from the writer) a share of the stock at the strike price q and immediately selling it at the open market at the price $S(T)$, thus making an instant profit of $S(T) - q$. On the other hand, if $S(T) \leq q$, then the buyer shall not exercise the option, and therefore, the payoff will be zero.

It has also been shown (see, e.g., [SKKM94]) that the pricing function $Q(t, y)$ may be expressed via the following Black-Scholes formula:

$$Q(t,y) = \frac{1}{\sqrt{2\pi}} \int_{-\infty}^{\infty} \Lambda\left(y \exp\left\{ \sigma u \sqrt{T-t} + \left(\lambda - \frac{\sigma^2}{2}\right)(T-t) \right\} \right) e^{-u^2/2}\, du$$
$$= \frac{1}{y} \int_{-\infty}^{\infty} \Lambda(u)\varphi\left(T - t, \frac{u}{y}, \lambda - \frac{\sigma^2}{2}, \sigma \right) du,$$

where

$$\Lambda(y) = (y - q)^+$$

is the payoff function, with

$$\varphi(t, y, a, b) = \frac{1}{by\sqrt{2\pi t}} \exp\left\{ -\frac{\log(y - at)^2}{2b^2 t} \right\}.$$

Equivalently, if the rational pricing function for the European option Q : $[0, T] \times \Re \to \Re$ is sufficiently smooth, then it satisfies the following well-known Black-Scholes equation:

$$\lambda Q(t, y) = \partial_t Q(t, y) + \lambda y \partial_y Q(t, y)$$
$$+ \frac{1}{2}\sigma^2 y^2 \partial_y^2 Q(t, y), \quad (t, y) \in [0, T] \times [0, \infty),$$

with the terminal condition $Q(T, y) = \Lambda(y) = (y - q)^+$.

On the other hand, based on the same Black-Scholes financial market described above, it has been shown by many authors (see, e.g., Myneni [Myn92] and Karatzas [Kar88]) that the rational pricing function $Q(t, y)$ for the standard American call option is the value function of the following optimal stopping problem:

$$Q(t, y) = \sup_{\tau \in \mathcal{T}_t^T} \tilde{E}\left[e^{-\lambda(\tau - t)}(S(\tau) - q)^+ \mid S(t) = y \right]$$
$$= \sup_{\tau \in \mathcal{T}_0^{T-t}} \tilde{E}\left[e^{-\lambda(\tau - t)}(S(\tau - t) - q)^+ \mid S(0) = y \right],$$

where $\tau \in \mathcal{T}_t^T$ is a certain stopping time.

For both of the European and American call options, the pricing function $Q(t, y)$ for the Black-Scholes market given above is explicit and computable. However, its assumption of constant stock growth and volatility rates are not always satisfied by the real-life financial markets, as the probability distribution of an equity often has a fatter left tail and thinner right tail than the geometric Brownian motion model of the stock prices (see [Bat96]). In order to better model real-life options, various generalizations of the model have been made. These generalizations include the popular stochastic volatility models (see e.g. Hull [Hul00] and Fouque et al [FPS00] and references given therein) and a stock market that has hereditary structure or delayed responses (see, e.g., Chang and Youree [CY99, CY07], Arriojas et al. [AHMP07], Kazmerchuk et al. [KSW04a, KSW04b, KSW04c]) in which various versions of stochastic hereditary differential equations (SHDEs) were introduced to describe the dynamics of the bank account and the stock price.

This chapter treats both European and American call options for a (B, S)-financial market in which the bank account follows a linear functional differential equation (see (6.1)) and the stock price follows a general nonlinear SHDE with bounded memory (see (6.4)). For the European call option, the material presented here is based on results obtained in a recent paper of Chang and Youree [CY07]. For the American call option, the results presented here are consequences of optimal stopping obtained in Chapter 4. Under the (B, S)-market described by (6.1) and (6.4), and with a very general path-dependent payoff function $\Psi(S_T)$ for the European option and $\Psi(S_{\tau \wedge T})$ for the American option at expiration time T or execution time $\tau \wedge T$, we derive an infinite-dimensional Black-Scholes equation (see (6.28) and (6.29)) for the pricing function for the European call option and an infinite-dimensional HJB variational inequality (HJBVI) (see (6.46) and (6.47)) for the pricing function of the American call option. Both the infinite-dimensional Black-Scholes equation and HJBVI involve extended Fréchet derivatives of functions on the function space $\mathbf{C}$ (see Section 2.2 in Chapter 2 for definitions). Under certain smoothness conditions, it can be shown that the pricing function is a classical solution of the infinite-dimensional Black-Scholes equation and the pricing function of the American call option is the classical solution of the infinite-dimensional HJBVI in the sense that all Fréchet derivatives and infinitesimal generator of the shift operator of these two pricing functions exist, are continuous, and satisfy respectively the infinite-dimensional Black-Scholes equation and the HJBVI pointwise in $[0, T] \times \mathbf{C}$. Unfortunately, for a general payoff function, it is not known whether the pricing functions are smooth enough to be a solution of these two infinite-dimensional equations and inequalities in the classical sense. The main results contained in this chapter are that we have shown that the pricing function for the European call option is the unique viscosity solution of the infinite-dimensional Black-Scholes equation and the pricing function for the American call option is the unique viscosity solution of the infinite-dimensional HJBVI as a consequence of the optimal stopping from Chapter 4. For computational purposes, we also present an algorithm for computing the viscosity solution of the infinite-dimensional Black-Scholes equation. The computational algorithm is adopted from Chang and Youree [CY07].

Although other types of infinite-dimensional Black-Scholes equation and HJBVI for optimal stopping problems and their applications to pricing of the American option have been studied very recently by a few researchers, they either considered a stochastic delay equation of special form (see, e.g., Gapeev and Reiss [GR05a] and [GR05b]) or stochastic equations in Hilbert spaces (see e.g. Gatarek and Święch [GS99] and Barbu and Marinelli [BM06]). The material presented in this chapter differs from the aforementioned papers in the following significant ways: (i) The segmented solution process $\{S_t, t \in [0, T]\}$ for the stock price dynamics is a strong Markov process in the Banach space $\mathbf{C}$ with the sup-norm that is not differentiable and is therefore more difficult to handle than any Hilbert space considered in [GS99] and [BM06]; (ii) the infinite-dimensional Black-Scoles equation and HJBVI that characterize the pricing function of the European and American call options uniquely involve the extensions $\overline{DV(t, \psi)}$ and $\overline{D^2V(t, \psi)}$ of first- and second-order Fréchet derivatives $DV(t, \psi)$ and $D^2V(t, \psi)$ from $\mathbf{C}^*$ and $\mathbf{C}^\dagger$ to $(\mathbf{C}\oplus\mathbf{B})^*$ and $(\mathbf{C}\oplus\mathbf{B})^\dagger$ (see Sections 2.2 and 2.3 of Chapter 2 for definitions of these spaces), respectively; and (iii) the infinite-dimensional Black-Scholes and HJBVI also involve the infinitesimal generator $SV(t, \psi)$ of the semigroup of shift operators of the value functions that does not appear in the special class of equations in the aforementioned papers.

This chapter is organized as follows. Section 6.1 describes the (B, S)-market that possesses hereditary structure and gives definitions for the contingent claims with a general path-dependent reward function. Examples are given to justify consideration of the path-dependent reward function $\Psi : \mathbf{C} \to [0, \infty)$. In Section 6.2, we introduce the concept of the self-financing trading strategy in the market. In Section 6.3, the concept of risk-neutral martingale measure is constructed for the (B, S) and the derivation of pricing functions for European and American call options are given. The infinite-dimensional Black-Scholes equation is derived in Section 6.4. In there, it is shown that the pricing function for the European call option is the unique viscosity solution of the infinite-dimensional Black-Scholes equation. As a consequence of the optimal stopping in Chapter 4, the infinite-dimensional HJBVI is derived in Section 6.5 once it is recognized that the pricing function is actually the value function of an optimal stopping problem treated in Chapter 4. In there it is also stated without proof that the pricing function is the unique viscosity solution of the infinite-dimensional HJBVI. Finally, in Section 6.6, we present an approximation algorithm for the solution of the infinite-dimensional Black-Scholes equation by considering the Taylor series solution of the equation. Section 6.7 contains conclusions and supplementary remarks for the chapter.

6.1 Pricing with Hereditary Structure

6.1.1 The Financial Market

To describe the (B, S)-market with hereditary structures, we start by defining appropriate function spaces as follows. Let $0 < r < \infty$ be the duration of bounded memory. Throughout this chapter, we let $n = 1$ and we consider $\mathbf{C} = C[-r, 0]$, the space of real-valued functions that are continuous on $[-r, 0]$, and

$$\mathbf{C}_+ = \{\phi \in \mathbf{C} \mid \phi(\theta) \geq 0 \ \forall \theta \in [-r, 0]\}$$

for simplicity.

We again use the following convention that has been adopted throughout the monograph: If $\psi \in C[-r, T]$ and $t \in [0, T]$, let $\psi_t \in \mathbf{C}$ be defined by $\psi_t(\theta) = \psi(t + \theta)$, $\theta \in [-r, 0]$.

The new model for the (B, S)-market considered herein was first introduced in Chang and Youree [CY99,CY07]. The market is said to have a hereditary structure in the sense that the rate of change of stock prices and the bank account depend not only on the current price but also on the entire historical prices of time duration $r > 0$. Specifically, we assume that the bank (or savings) account $B(\cdot)$ and the stock prices $S(\cdot)$ evolve according to the following two equations.

(A) The Bank (Savings) Account

To describe the bank account holdings $B(\cdot) = \{B(t), t \in [-r, T]\}$, we assume that it satisfies the following linear (deterministic) functional differential equation:

$$dB(t) = L(B_t)dt, \quad t \in [0, T], \tag{6.1}$$

with an initial function $\phi \in C_+$, where L is a bounded linear functional on C that can and will be represented as the following Lebesque-Stieltjes integral

$$L(\phi) = \int_{-r}^{0} \phi(\theta) \, d\eta(\theta)$$

for some function $\eta : [-r, 0] \to \Re$ of bounded variation. Equivalently, (6.1) can be written as

$$dB(t) = \int_{-r}^{0} B(t + \theta) \, d\eta(\theta) \, dt, \quad t \geq 0.$$

It is assumed that the function $\eta : [-r, 0] \to \Re$ satisfies the following assumption.

Assumption 6.1.1 *The function $\eta : [-r, 0] \to \Re$ is nondecreasing (and hence of bounded variation) with $\eta(0) - \eta(-r) > 0$.*

We can and will extend the domain of the above function to the entire $\Re$ by defining $\eta(\theta) = \eta(-r)$ for $\theta \leq -r$ and $\eta(\theta) = \eta(0)$ for $\theta \geq 0$.

For the purpose of analyzing the discount rate for the bank account, let us assume that the solution process $B(\phi) = \{B(t), t \in [-r, T]\}$ of (6.1) with the initial function $\phi \in C_+$ takes the following form:

$$B(t) = \phi(0)e^{\lambda t}, \quad t \in [0, T], \tag{6.2}$$

and $B_0 = \phi \in C_+$. Then the constant $\lambda > 0$ satisfies the following equation:

$$\lambda = \int_{-r}^{0} e^{\lambda \theta} \, d\eta(\theta). \tag{6.3}$$

Lemma 6.1.2 *Under Assumption 6.1.1, there exists a unique $\lambda > 0$ that satisfies (6.3).*

Proof. Since $\eta : [-r, 0] \to \Re$ is nondecreasing with $\eta(0) - \eta(-r) > 0$, it is clear for all $\lambda > 0$ that

$$\int_{-r}^{0} \theta e^{\lambda\theta} \, d\eta(\theta) < 0.$$

To prove the lemma, we let $\gamma : [0, \infty) \to [0, \infty)$ by

$$\gamma(\lambda) = \int_{-r}^{0} e^{\lambda\theta} \, d\eta(\theta).$$

Then $\gamma(0) = \eta(0) - \eta(-r) > 0$ and

$$\dot{\gamma}(\lambda) = \int_{-r}^{0} \theta e^{\lambda\theta} \, d\eta(\theta) < 0, \quad \forall \lambda > 0.$$

Therefore, (6.3) has a unique solution $\lambda > 0$.   □

Remark 6.1.3 *The constant $\lambda > 0$ specified in Lemma 6.1.2 will be referred to as the effective interest rate of the bank account.*

(B) Price Dynamics of the Stock

Let $(\Omega, \mathcal{F}, P, \mathbf{F}, W(\cdot))$ be a one-dimensional standard Brownian motion, where $\mathbf{F} = \{\mathcal{F}(t), t \geq 0\}$ is the P-augmented natural filtration of $W(\cdot)$; that is,

$$\mathcal{F}(t) = \sigma(W(s), 0 \leq s \leq t) \vee \mathcal{N}, \quad t \geq 0,$$

where

$$\mathcal{N} = \{A \subset \Omega \mid \exists B \in \mathcal{F} \text{ such that } A \subset B \text{ and } P(B) = 0\}.$$

Assume that the stock price process $\{S(t), t \in [-r, T]\}$ follows the following nonlinear SHDE:

$$\frac{dS(t)}{S(t)} = f(S_t) \, dt + g(S_t) \, dW(t), \quad t \in [0, T], \tag{6.4}$$

with initial price function $\psi \in \mathbf{C}_+$, where f, and g are some continuous real-valued functions defined on the real Banach space $\mathbf{C}$. At any time $t \in [0, T]$, the terms $f(S_t)$ and $g(S_t)$ (referred to respectively as the stock appreciation rate and the stock volatility rate at time t) in (6.4) are random functions that depend not only on the current stock price $S(t)$ but also on the stock prices, S_t, over the time interval $[t - r, t]$.

We make the following assumption on the functions $f, g : \mathbf{C} \to \Re$.

Assumption 6.1.4 *The functions $f, g : \mathbf{C} \to \Re$ are continuous and satisfy the following local Lipschitz and linear growth conditions: There exist positive constants $K_{1,N}$ and K_2 for all $N = 1, 2, \ldots$ such that*

$$|\phi(0)f(\phi) - \varphi(0)f(\varphi)| + |\phi(0)g(\phi) - \varphi(0)g(\varphi)|$$
$$\leq K_{1,N}\|\phi - \varphi\|, \quad \forall \phi, \varphi \in \mathbf{C} \text{ with } \|\phi\|, \|\varphi\| \leq N,$$

and

$$0 \leq |\phi(0)f(\phi)| + |\phi(0)g(\phi)| \leq K_2(1 + \|\phi\|), \quad \forall \phi \in \mathbf{C}.$$

Note that Assumption 6.1.4 is standard for the existence and uniqueness of a strong solution of (6.4) (see Theorem 1.3.12 in Chapter 1).

As a reminder, we recall the meaning of the *strong* solution $\{S(t), t \in [-r, T]\}$ of (6.4) as follows.

Definition 6.1.5 *A process* $\{S(t), t \in [-r, T]\}$ *in* $\Re$ *is said to be a strong solution of (6.4) through the initial function* $\psi \in \mathbf{C}_+$ *if it satisfies the following conditions:*

1. $S(t) = \psi(t), \; \forall t \in [-r, 0]$.
2. $\{S(t), t \in [0, T]\}$ *is* **F**-*adapted; that is,* $S(t)$ *is* $\mathcal{F}(t)$-*measurable for each* $t \in [0, T]$.
3. *The process* $\{S(t), t \in [0, T]\}$ *is continuous and satisfies the following stochastic integral equation* P-*a.s. for all* $t \in [0, T]$:

$$S(t) = \psi(0) + \int_0^t S(s)f(S_s)\,ds + \int_0^t S(s)g(S_s)\,dW(s) \quad t \in [0, T].$$

4. $\int_0^T |S(t)f(S_t)|\,dt < \infty$ *and* $\int_0^T |S(t)g(S_t)|^2\,dt < \infty$ P − *a.s.*

Definition 6.1.6 *The strong solution* $\{S(t), t \in [-r, T]\}$ *of (6.4) through the initial datum* $\psi \in \mathbf{C}_+$ *and on the interval* $[-r, T]$ *is said to be strongly (or pathwise) unique if* $\{\tilde{S}(t), t \in [-r, T]\}$ *is another strong solution of (6.4) through the same initial function* ψ *and on the same time interval* $[0, T]$. *Then*

$$P\{S(t) = \tilde{S}(t), \quad \forall t \in [0, T]\} = 1.$$

Under Assumption 6.1.4, it can be shown that for each initial historical price function $\psi \in \mathbf{C}_+$, the price process $\{S(t), t \in [-r, T]\}$ exists and is non-negative, continuous, and **F**-adapted.

Theorem 6.1.7 *Assume Assumption 6.1.4 holds. Then for each* $\psi \in \mathbf{C}_+$, *there exists a unique non-negative strong solution process* $\{S(t), t \in [-r, T]\}$ *through the initial datum* ψ *and on the time interval* $[0, T]$.

Proof. The existence and uniqueness of the strong solution process $\{S(t), t \in [-r, T]\}$ follows from Theorem 1.3.12 in Chapter 1. Therefore, we only need to show that for each initial $S_0 = \psi \in \mathbf{C}_+$, then $S(t) \geq 0$ for each $t \in [0, T]$. To show this, we note that $\{t \in [0, T] \mid S(t) \geq 0\} \neq \emptyset$ by sample-path continuity of the solution process and nonnegativity of the initial data $\psi \in \mathbf{C}_+$. Now, let

$$\tau = \inf\{t \in [0, T] \mid S(t) \geq 0\}.$$

If $\tau < T$, then $dS(t) = 0$ for all $t \geq \tau$, with $S(\tau) = 0$. This implies that $S(t) = 0$ for all $t \in [\tau, T]$ and, hence, $S(t) \geq 0$ for all $t \in [0, T]$. The same conclusion holds if $\tau = T$. $\quad\square$

Using the convention that for each $t \in [0, T]$, $S_t(\theta) = S(t+\theta)$, $\forall \theta \in [-r, 0]$, we also consider the associated **C**-valued segment process $\{S_t, t \in [0, T]\}$.

Let $\mathbf{G} = \{\mathcal{G}(t), t \in [0,T]\}$ be the subfiltration of $\mathbf{F} = \{\mathcal{F}(t), t \in [0,T]\}$ generated by $\{S(t), t \in [-h,T]\}$; that is

$$\mathcal{G}(t) = \sigma(S(s), 0 \leq s \leq t) \vee \mathcal{N}, \quad t \geq 0.$$

From Lemma 1.3.4 in Chapter 1, it can be shown that

$$\mathcal{G}(t) = \sigma(S_s, 0 \leq s \leq t) \vee \mathcal{N}, \quad t \geq 0.$$

If $\tau : \Omega \to [0,T]$ is a $\mathbf{G}$-stopping time, (i.e., the event $\{\tau \leq t\} \in \mathcal{G}(t)$ for each $t \in [0,T]$), denote the sub-σ-algebra $\mathcal{G}(\tau)$ by

$$\mathcal{G}(\tau) = \{A \in \mathcal{F} \mid A \cap \{\tau \leq t\} \in \mathcal{G}(t), \forall t \leq \tau\}.$$

The following strong Markov property of the $\mathbf{C}$-valued process $\{S_t, t \in [0,T]\}$ also holds under Assumption 6.1.4.

Theorem 6.1.8 *Under Assumption 6.1.4, the $\mathbf{C}$-valued process $\{S_t, t \in [0,T]\}$ satisfies the following strong Markov property: For each $t \in [0,T]$ and each $\mathbf{G}$-stopping time τ with $P\{\tau \leq t\} = 1$, we have*

$$P\{S_t \in A \mid \mathcal{G}(\tau)\} = P\{S_t \in A \mid S_\tau\}, \quad \forall A \in \mathcal{B}(\mathbf{C}).$$

It is clear that when $r = 0$, (6.4) reduces to the nonlinear stochastic ordinary differential equation

$$\frac{dS(t)}{S(t)} = f(S(t)) \, dt + g(S(t)) \, dW(t), \quad t \in [0,T], \tag{6.5}$$

of which the Black-Scholes (B,S)-market is a special case by considering the case in which $f(x) \equiv \mu$ and $g(x) \equiv \sigma$. In addition, (6.4) is general enough to include the following linear model considered by Chang and Youree [CY99] and pure discrete delay models considered by Arriojas et al. [AHMP07], and Kazmerchuk et al. [KSW04a, KSW04b, KSW04c]:

$$dS(t) = M(S_t) \, dt + N(S_t) \, dW(t), \text{ (see [CY99])}$$
$$= \int_{-r}^{0} S(t+\theta) \, d\xi(\theta)dt + \int_{-r}^{0} S(t+\theta) \, d\zeta(\theta)dW(t), \quad t \geq 0,$$

where $\xi, \zeta : [-r,0] \to \Re$ are certain functions of bounded variation.

$$dS(t) = f(S_t) \, dt + g(S(t-b))S(t) \, dW(t), \quad t \geq 0 \text{ (see [AHMP07])}, \tag{6.6}$$

where $0 < b \leq r$ and $f(S_t) = \mu S(t-a)S(t)$ or $f(S_t) = \mu S(t-a)$ and

$$\frac{dS(t)}{S(t)} = \mu S(t-a) \, dt + \sigma(S(t-b)) \, dW(t), \ t \geq 0 \ \text{ (see [KSW04a])}.$$

6.1.2 Contingent Claims

As briefly mentioned earlier, a *contingent claim* or *option* is a contract conferred by the contract *writer* on the contract *holder* giving the *holder* the right (but not the obligation) to buy from or to sell to the *writer* a share of the stock at a prespecified price prior to or at the contract expiry time $T > 0$. The right for the *holder* to buy from the *writer* a share of the stock will be called a *call option* and the right to sell to the *writer* a share of the stock will be called a *put option*. If an *option* is purchased at time $t \geq 0$ and is exercised by the *holder* at time $\tau \in [t, T]$, then he will receive a payoff of the amount $\Psi(S_\tau)$ from the *writer*, where $\Psi : \mathbf{C} \to [0, \infty)$ is the payoff function and $\{S(s), s \in [-r, T]\}$ is the price of the underlying stock. To secure such a contract, the contract *holder*, however, has to pay the *writer* at contract purchase time t a fee that is mutually agreeable to both parties. The determination of such a fee is called the pricing of the *contingent claim*.

In determining a fair price for the *contingent claim*, the *writer* of the contingent claim seeks to invest in the (B, S)-market the fee x received from the *holder* and trades over the time interval $[t, T]$ between the *bank* account and the *stock* account in an optimal and prudent manner so that his total wealth will replicate or exceed that of the payoff $\Psi(S_\tau)$ he has to pay to the *holder* if and when the contingent claim is exercised. The smallest such x is called the fair price of the *contingent claim*.

The *contingent claims* can be classified according to the their allowable exercise times τ. If the *contingent claim* can be exercised at any time between the option purchase time t and option expiry time T (i.e., $\tau \in [t, T]$), it is called a *contingent claim* of the American type and it will be denoted by $ACC(\Psi)$. In this case, the exercise time $\tau \in [t, T]$ of the $ACC(\Psi)$ should satisfy the fundamental assumption that it is an **F**-stopping time (i.e., $\tau \in \mathcal{M}_t^T$), since the information of the stock prices up to the exercise time τ are available to both parties and neither the *writer* nor the *holder* of the contingent claim can anticipate the future price of the underlying stock. If the *contingent claim* can only be exercised at the expiry time T (i.e., $\tau \equiv T$), then the *contingent claim* is called a European type and it will be denoted by $ECC(\Psi)$.

We make the following assumption on the payoff function.

Assumption 6.1.9 *The payoff function* $\Psi : \mathbf{C} \to [0, \infty)$ *is globally convex; that is, for all* $\phi, \varphi \in \mathbf{C}$,

$$\Psi(\alpha\phi + (1 - \alpha)\varphi) \leq \alpha\Psi(\phi) + (1 - \alpha)\Psi(\varphi), \quad \alpha \in [0, 1].$$

We state the following lemma without proof.

Lemma 6.1.10 *Assuming the global convexity of the reward function* $\Psi : \mathbf{C} \to [0, \infty)$, *then* Ψ *is uniformly Lipschitz on* $\mathbf{C}$; *that is, there exists a constant* $K > 0$ *such that*

$$|\Psi(\phi) - \Psi(\varphi)| \leq K\|\phi - \varphi\|, \quad \forall\phi, \varphi \in \mathbf{C}.$$

The following are some examples of the *contingent claims* or *options* that have been traded in option exchanges around the world.

Example 6.1 (Standard Call Option). The payoff function for the standard call option is given by

$$\Psi(S_s) = \max\{S(s) - q, 0\},$$

where $S(s)$ is the stock price at time s and $q > 0$ is the strike price of the standard call option. If the standard American option is exercised by the *holder* at the **F**-stopping time $\tau \in T_t^T$, the *holder* will receive the payoff of the amount $\Psi(S_\tau)$. Of course, the option will be called the standard European call option if $\tau \equiv T$. In this case, the amount of the reward will be $\Psi(S_T) = \max\{S(T) - q, 0\}$.

We offer a financial interpretation for the standard American (or European) call option as follows. If the option has not been exercised and the current stock price is higher than the strike price, then the *holder* can exercise the option and buy a share of the stock from the *writer* at the strike price q and immediately sells it at the open market and make an instant profit of the amount $S(\tau) - q > 0$. It the strike price is higher than the current stock price, then the option is worthless to the *holder*. In this case, the *holder* will not exercise the option and therefore the payoff will be zero dollar.

Example 6.2 (Standard Put Option). The payoff function for the standard put option is given by

$$\Psi(S_s) = \max\{q - S(t), 0\}.$$

This is similar to Example 6.1, except now the *holder* of the contract has the option to sell to the *writer* a share of the underlying *stock* at the prespecified price $q > 0$ and the *writer* of the contract has the obligation to buy at this price if and when the contract is exercised at $\tau \in T_t^T$. Clearly, the payoff for the standard American put option is $\Psi(S_\tau) = \max\{q - S(\tau), 0\}$ and that of the European put option is $\Psi(S(T)) = \max\{q - S(T), 0\}$.

One characteristic of a *path-dependent* option, whether it is of American type or European type, is that the payoff depends explicitly on the historical prices when the option is exercised. The following are some examples of the *path-dependent* option.

Example 6.3 (Modified Russian Option). The payoff process of a Russian call option can be expressed as

$$\Psi(S_s) = \max\{ \sup_{s \in [0 \wedge (t-r), t]} S(s) - q, 0\}$$

for some strike price $q > 0$. The payoff for a Russian put option can easily be written also. Note that the payoff $\Psi(S_\tau)$ of a Russian option depends on the highest price on the time interval $[\tau - r, \tau]$ if the option is exercised at the time τ.

Example 6.4 (Modified Barrier Options). The payoff for the barrier call option is given by

$$\Psi(S_t) = \max\{S(t) - q, 0\}\chi_{\{\tau(a) \leq t\}}$$

for some $a > q > 0$, $a > S(0)$ and with $\tau(a) = \inf\{s \in R_+ \mid S(s) \geq a\}$.

This is similar to the call option of Example 6.1, except now the stock price has to reach a certain "barrier" level $a > q \vee S(0)$ for the option to become activated.

Example 6.5 (Modified Asian Option). The payoff for the Asian call option is given by

$$\Psi(S_t) = \max\left\{\frac{1}{r}\int_{(t-r)\vee 0}^{t} S(s)\,ds - q, 0\right\}.$$

This is similar to a European call option with strike price $q > 0$ except that now the "moving-average stock price" $\frac{1}{r}\int_{(t-r)\vee 0}^{t} S(s)\,ds$, over the interval $[t - r, t]$, is used in place of the "terminal stock price" $S(t)$.

All of the last three examples (Examples 6.3-6.5) demonstrate that the payoff $\Psi(S_t)$ depends not only on the stock price $S(t)$ at time $t \in [0, T]$ but also explicitly on the prices of the underlying stock over the time interval $[t - r, t]$. In addition, the payoff function $\Psi : \mathbf{C} \to [0, \infty)$ is globally convex in all five examples.

6.2 Admissible Trading Strategies

Based on the (B, S)-market described by (6.1) and (6.4), the option *holder* will have the control over the trading between his *bank* account and the *stock*. It is assumed throughout this chapter that there is neither cost nor tax incurred in each of the transaction made by the *holder*.

Definition 6.2.1 *A trading strategy in the (B, S)-market is an $\mathbf{F}$-progressively measurable two-dimensional vector process*

$$\pi(\cdot) = \{\pi(t) = (\pi_1(t), \pi_2(t)), t \in [0, T]\}$$

defined on $(\Omega, \mathcal{F}, P, \mathbf{F})$ such that

$$\int_0^T E[|\pi(t)|^2]\,dt = \int_0^T E\left[\sum_{i=1}^{2}(\pi_i(t))^2\right]dt < \infty,$$

where $\pi_1(t)$ and $\pi_2(t)$ represent respectively the number of units of the bank account and the stock owned by the investor at time $t \in [0, T]$.

With a trading strategy $\pi(\cdot)$ described above, the *holder's* total asset in the (B, S)-market is described by the wealth process $\{X^\pi(t), t \in [0, T]\}$ defined by

$$X^\pi(t) = X_1^\pi(t) + X_2^\pi(t)$$
$$= \pi_1(t)B(t) + \pi_2(t)S(t), \quad t \in [0, T], \tag{6.7}$$

where, since there is no transaction cost or taxes, $X_1^\pi(t) = \pi_1(t)B(t)$ and $X_2^\pi(t) = \pi_2(t)S(t)$ denote the investor's holdings at time t in the bank account and the stock, respectively, and the pair $(X_1^\pi(t), X_2^\pi(t))$ will be called the *investor's* portfolio in the (B, S)-market at time t. Once again, $\{B(t), t \in [-r, T]\}$ and $\{S(t), t \in [-r, T]\}$ satisfy (6.1) and (6.4) with initial functions ϕ and $\psi \in \mathbf{C}_+$, respectively.

Definition 6.2.2 *Given the initial wealth $X^\pi(0) = x$, trading strategy $\pi(\cdot)$ is said to be admissible at x if*

$$X^\pi(t) \geq -\zeta, \quad P\text{-a.s.}, \forall t \in [0, T], \tag{6.8}$$

for some non-negative random variable ζ with $E[\zeta^{1+\delta}] < \infty$, where $\delta > 0$. Denote the collection of trading strategies $\pi(\cdot)$ that are admissible by $\mathcal{A}(x)$.

In a reasonable market, it is important that the constraint described by (6.8) be satisfied, for otherwise it is possible to create a *doubling strategy* (i.e., a trading strategy that attain arbitrarily large values of wealth with probability one at $t = T$, starting with zero initial capital $X^\pi(0) = 0$ at $t = 0$). The following example due to Shreve is taken from Karatzas [Kar96].

Example 6.6. In this example, we consider the continuous (B, S)-market described by
$$dB(t) = 0, \quad t \in [0, T] \text{ with } B(0) = 0,$$
and
$$dS(t) = dW(t), \quad t \in [0, T] \text{ with } S(0) = 0.$$

Let $\mathbf{F} = \{\mathcal{F}(t), t \in [0, T]\}$, where $\mathcal{F}(t) = \sigma\{W(s), 0 \leq s \leq t\}$. Consider the $\mathbf{F}$-martingale $\{M(t), t \in [0, T]\}$ described by

$$M(t) \equiv \int_0^t \frac{dW(s)}{\sqrt{T-s}}, \quad t \in [0, T],$$

with $E[M^2(t)] = \int_0^t \frac{ds}{T-s} = \log(\frac{T}{T-t})$ and $E[M(t)] = 0$ for all $t \in [0, T]$. From the Girsanov theorem (see Theorem 1.2.16 in Chapter 1), we have $\tilde{M}(t) = \log(\frac{T}{T-t})$, where

$$\tilde{M}(\cdot) \equiv \{M(T - Te^{-u}), u \in [0, \infty)\}$$

is a Brownian motion. Therefore, for b sufficiently large, the **F**-stopping time

$$\tau(b) \equiv \inf\{t \in [0, T] \mid M(t) = b\} \wedge T$$

satisfies $P\{0 < \tau(b) < T\} = 1$; we choose the trading strategy $\pi(\cdot) = \{\pi(t) = (\pi_1(t), \pi_2(t)), t \in [0, T]\}$ as follows:

$$\pi_1(t) = 0 \quad \text{and} \quad \pi_2(t) = \frac{1}{\sqrt{T - t}} \chi_{\{t \le \tau(b)\}}.$$

In this case, the holding in the stock becomes $X_2^\pi(t) \equiv (1/\sqrt{T-t}) \chi_{\{t \le \tau(b)\}}$ and the total wealth corresponding to this strategy becomes

$$X^\pi(t) = \int_0^t \frac{1}{\sqrt{T - s}} \chi_{\{s \le \tau(b)\}} \, dW(s) = \tilde{W}(t \wedge \tau(b)), \quad t \in [0, T].$$

This shows that $X^\pi(T) = \tilde{M}(\tau(b)) = b$. Notice that admissibility condition of π stated in Definition 6.2.2 fails, for otherwise by Fatou's lemma, we should have

$$X^\pi(T) \le \underline{\lim}_{t \uparrow T} X^\pi(t) = 0.$$

This is a contradiction.

For the option pricing problems under the (B, S)-market described by (6.1) and (6.4) and without transaction costs or taxes, we will make the following basic assumption throughout the end of this chapter.

Assumption 6.2.3 *(Self-Financing Condition) In the general (B, S)-market described by (6.1) and (6.4), the trading-consumption strategy $\pi(\cdot) \in \mathcal{A}(x)$ is said to satisfy the following self-financing condition if the wealth process $\{X^\pi(t), t \in [0, T]\}$ satisfies the following equation:*

$$dX^\pi(t) = dX_1^\pi(t) + dX_2^\pi(t) \tag{6.9}$$
$$= \pi^1(t) \, dB(t) + \pi^2(s) \, dS(s), \quad P\text{-a.s. } t \in [0, T]. \tag{6.10}$$

Financially speaking, the trading strategy $\pi(\cdot) \in \mathcal{A}(x)$ is said to satisfy the *self-financing condition* if there is no net inflow or outflow of wealth in and from the portfolio and the increments of the *investor's* holdings are due only to the increments of the bond price and stock price and consumptions. The financial meaning of the *self-financing* can be made clearer if we consider the following discrete-time version:

$$X^\pi(t + \Delta t) - X^\pi(t) = \pi_1(t)(B(t + \Delta t) - B(t)) + \pi_2(t)(S(t + \Delta t) - S(t))$$

for sufficiently small $\Delta t > 0$.

Denote the set of all *self-financing* trading strategies $\pi(\cdot) \in \mathcal{A}(x)$ by $SF(x)$.

Remark 6.2.4 *To have a more precise description of $X_1^\pi(\cdot)$, $X_2^\pi(\cdot)$, and $X^\pi(\cdot)$ when $\pi(\cdot) \in SF(x)$, let $\alpha(t)$ and $\beta(t)$ be respectively the cumulative number of shares of stock purchased and sold over the time interval $[0, t]$. Assuming that proceeds for selling stock are to be deposited in the bank account and purchases of stock are to be paid for from the bank account, it is clear that the following relations are true:*

$$\pi_2(t) = \pi_2(0) + \alpha(t) - \beta(t), \quad t \in [0, T],$$

$$\pi_1(t) = \pi_1(0) + \int_0^t \frac{S(s)}{B(s)} \, d(\beta(s) - \alpha(s)), \quad t \in [0, T].$$

In this case,

$$\begin{aligned} dX_1^\pi(t) &= d(\pi_1(t)B(t)) \\ &= \pi_1(t) \, dB(t) + B(t) \, d\pi_1(t) \\ &= \pi_1(t) \, dB(t) + S(t) \, d(\beta(s) - \alpha(t)), \end{aligned}$$

$$\begin{aligned} dX_2^\pi(t) &= d(\pi_2(t)S(t)) \\ &= \pi_2(t) \, dS(t) + S(t) \, d\pi_2(t) \\ &= \pi_2(t) \, dS(t) + S(t) \, d(\alpha(t) - \beta(t)) \end{aligned}$$

and

$$\begin{aligned} dX^\pi(t) &= dX_1^\pi(t) + dX_2^\pi(t) \\ &= \pi_1(t) \, dB(t) + \pi_2(t) \, dS(t), \quad t \in [0, t]. \end{aligned}$$

Definition 6.2.5 *The trading strategy $\pi(\cdot) \in SF(x)$ for the European contingent claim $ECC(\Psi)$ is called an arbitrage or realizing an arbitrage possibility if*

$$X^\pi(0) \leq 0 \Rightarrow X^\pi(T) \geq 0 \ P\text{-a.s. and } P\{X^\pi(T) > 0\} > 0.$$

Definition 6.2.6 *The trading strategy $\pi(\cdot) \in SF(x)$ for the American contingent claim $ACC(\Psi)$ is called an arbitrage or realizing an arbitrage possibility, if there exists $\tau \in \mathcal{T}_0^T$ such that*

$$X^\pi(0) \leq 0 \Rightarrow X^\pi(\tau) \geq 0 \ P\text{-a.s. and } P\{X^\pi(\tau) > 0\} > 0.$$

Financially speaking, a self-financing trading strategy is an arbitrage (or free lunch) if it provides opportunity for making profit without risk.

6.3 Risk-Neutral Martingale Measures

In this section, we study the existence and uniqueness (or nonuniqueness) question of a risk-neutral martingale measure for the (B, S)-market described by (6.1) and (6.4).

Assuming $\pi(\cdot) \in SF(x)$, we first study the concept of risk-neutral martingale measures. Let $\mathcal{P}$ be the collection of probability measures $\tilde{P}$ defined on the measurable space $(\Omega, \mathcal{F})$ that satisfy the following two conditions:
(i) $\tilde{P}$ is equivalent to the underlying probability measure P; that is, $\tilde{P}$ and P are mutually absolutely continuous.
(ii) The discounted wealth process $\{Y^\pi(t), t \in [0, T]\}$ is an **F**-martingale with respect to $\tilde{P}$, where $Y^\pi(t) = \frac{X^\pi(t)}{B(t)}$.

In the following, it will be shown that $\mathcal{P} \neq \emptyset$ for the (B, S)-market described by (6.1) and (6.4).

For the unit price of the bank account $B(\cdot) = \{B(t), t \in [0, T]\}$ and the stock $S(\cdot) = \{S(t), t \in [0, T]\}$ described in (6.1) and (6.4), define

$$\tilde{W}(t) = W(t) + \int_0^t \gamma(B_s, S_s)\, ds, \quad t \in [0, T], \tag{6.11}$$

where $\gamma : \mathbf{C}_+ \times \mathbf{C}_+ \to \Re$ is defined by

$$\gamma(\phi, \psi) = \frac{\phi(0)f(\psi) - L(\phi)}{\phi(0)g(\psi)}. \tag{6.12}$$

Define the process $Z(\cdot) = \{Z(t), t \in [0, T]\}$ by

$$Z(t) = \exp\left\{ \int_0^t \gamma(B_s, S_s)\, dW(s) - \frac{1}{2} \int_0^t |\gamma(B_s, S_s)|^2\, ds \right\}. \tag{6.13}$$

Lemma 6.3.1 *The process $Z(\cdot)$ defined by (6.13) is a martingale defined on $(\Omega, \mathcal{F}, P; \mathbf{F})$.*

Proof. It is clear from the assumptions on f, g, and L that we have

$$|\gamma(B_t, S_t)| \leq \frac{|B(t)f(S_t)| + |L(B_t)|}{|B(t)g(S_t)|}$$

$$\leq \frac{\phi(0)e^{\lambda t}K_b\|S_t\| + \lambda\phi(0)e^{\lambda t}}{\phi(0)e^{\lambda t}\sigma}$$

$$\leq c\|S_t\| \text{ for some positive constant } c.$$

Since $E[\int_0^a \|S_t\|^2 dt] < \infty$ for each $a \geq 0$, it follows that

$$E\left[\exp\left\{ \frac{1}{2} \int_0^a |\gamma(B_t, S_t)|^2\, ds \right\} \right] < \infty, \quad 0 \leq a < \infty.$$

Therefore, from the martingale characterization theorem in Section 1.2 of Chapter 1, the process is a martingale. □

Lemma 6.3.2 *There exists a unique probability measure $\tilde{P}$ defined on the canonical measurable space $(\Omega, \mathcal{F})$ such that*

$$\tilde{P}(A) = E[1_A Z(T)], \quad \forall A \in \mathcal{F}(T),$$

where 1_A is the indicator function of $A \in \mathcal{F}(T)$.

Proof. This follows from Lemma 6.3.1, since $Z(\cdot)$ is a martingale defined on $(\Omega, \mathcal{F}, P; \mathbf{F})$ and $Z(0) = 1$.   □

Lemma 6.3.3 *The process $\tilde{W}(\cdot)$ defined by (6.11) is a standard Brownian motion defined on the filtered probability space $(\Omega, \mathcal{F}, \tilde{P}; \mathbf{F})$.*

Proof. This follows from Lemmas 6.3.1 and 6.3.2 and the Girsanov transformation Theorem 1.2.16.   □

Consider the following nonlinear SHDE:

$$\frac{dS(t)}{S(t)} = \lambda \, dt + g(S_t) \, d\tilde{W}(t), \quad t \in [0, T], \tag{6.14}$$

where $\lambda > 0$ is the effective interest rate of the bank account described in (6.1.2).

Theorem 6.3.4 *The strong solution $S(\cdot) = \{S(t), t \in [-r, T]\}$ for (6.4) under the probability measure P is also a unique strong solution of (6.14) with $S_0 = \psi \in \mathbf{C}_+$ under the new probability measure $\tilde{P}$ in the sense that they have the same distribution.*

Proof. Since $\tilde{W}(\cdot)$ is a standard Brownian motion defined on the new filtered probability space $(\Omega, \mathcal{F}, \tilde{P}; \mathbf{F})$, we have

$$\frac{dS(t)}{S(t)} = f(S_t) \, dt + g(S_t) \, dW(t)$$

$$= f(S_t) \, dt + g(S_t) \left(d\tilde{W}(t) - \frac{B(t)f(S_t) - L(B_t)}{B(t)g(S_t)} \, dt \right)$$

$$= f(S_t) \, dt + g(S_t) \, d\tilde{W}(t) - f(S_t) \, dt + \frac{L(B_t)}{B(t)} \, dt$$

$$= \frac{L(B_t)}{B(t)} \, dt + g(S_t) \, d\tilde{W}(t)$$

$$= \lambda \, dt + g(S_t) \, d\tilde{W}(t). \quad □$$

6.4 Pricing of Contingent Claims

An option pricing problem in the (B, S)-market is, briefly, to determine the *rational price* $\Upsilon(\Psi)$ for the European contingent claim $ECC(\Psi)$ and the American contingent claim $ACC(\Psi)$ described in Section 6.1. It is the price that the writer of the contract should receive from the buyer for the rights of the contract and also to determine the trading strategy $\pi(\cdot) = \{(\pi_1(s), \pi_2(s)), s \in [0, T]\}$ the writer should use to invest this fee in the (B, S)-market in such a way as to ensure that the writer will be able to cover the option if and when it is exercised. The fee should be large enough that the writer can, with riskless investing, cover the option but be small enough that the writer does not make an unfair (i.e., riskless) profit.

Definition 6.4.1 *A self financing trading strategy $\pi(\cdot)$ is a $ECC(\Psi)$-hedge if*

$$X^\pi(0) = \pi_1(0)\phi(0) + \pi_2(0)\psi(0) = x$$

and P-a.s.

$$X^\pi(T) \geq \Psi(S_T).$$

We say that a $ECC(\Psi)$-hedge self-financing trading strategy $\pi^*(\cdot)$ is minimal for the European contingent claim if

$$X^\pi(T) \geq X^{\pi^*}(T)$$

for any $ECC(\Psi)$-hedge self-financing strategy $\pi(\cdot)$ for the European contingent claim.

Definition 6.4.2 *Let $\Pi_{ECC(\Psi)}(x)$ be the set of $ECC(\Psi)$-hedge strategies from $SF(x)$ for the European contingent claim. Define*

$$\Upsilon(ECC(\Psi)) = \inf\{x \geq 0 : \Pi_{ECC(\Psi)}(x) \neq \emptyset\}. \tag{6.15}$$

Definition 6.4.3 *A trading strategy $\pi(\cdot) \in SF(x)$ is a $ACC(\Psi)$-hedge of the American contingent claim $ACC(\Psi)$ if*

$$X^\pi(0) = \pi_1(0)\phi(0) + \pi_2(0)\psi(0) = x$$

and P-a.s.

$$X^\pi(t) \geq \Psi(S_t), \quad \forall t \in [0,T].$$

Definition 6.4.4 *Let $\Pi_{ACC(\Psi)}(x)$ be the set of $ACC(\Psi)$-hedge strategies from $SF(x)$ for the American contingent claim. Define*

$$\Upsilon(ACC(\Psi)) = \inf\{x \geq 0 : \Pi_{ACC(\Psi)}(x) \neq \emptyset\}. \tag{6.16}$$

The value $\Upsilon(ECC(\Psi))$ and $\Upsilon(ACC(\Psi))$ defined above are called the *rational price* of the European contingent claim and the American contingent claim, respectively. If the infimum in (6.15) (respectively, (6.16)) is achieved, then $\Upsilon(ECC(\Psi))$ (respectively, $\Upsilon(ACC(\Psi))$) is the *minimal* possible initial capital x for which there exists a trading strategy $\pi(\cdot) \in SF^+(x)$ possessing the property that $\tilde{P}$-a.s. $X^\pi(T) \geq \Psi(S_T)$ (respectively, $X^\pi(t) \geq \Psi(S_t), t \in [0,T]$).

The main purpose of this section is twofold:
(a) Determine the *rational prices* $\Upsilon(ECC(\Psi))$ and $\Upsilon(ACC(\Psi))$.
(b) Find the $ECC(\Psi)$-hedging and $ACC(\Psi)$-hedging strategies $\pi(\cdot) \in SF(x)$ for $\Upsilon(ECC(\Psi))$ and $\Upsilon(ACC(\Psi))$.

We will treat the European option and the American option separately in the following two subsections.

6.4.1 The European Contingent Claims

Let $Y(\cdot) = \{Y(t), t \in [0,T]\}$ be defined by

$$Y(t) = \tilde{E}\left[\left.\frac{\Psi(S_T)}{B(T)}\right| \mathcal{F}(t)\right], \quad 0 \le t \le T.$$

Then the process $Y(\cdot) = \{Y(t), t \in [0,T]\}$ is a martingale defined on $(\Omega, \mathcal{F}, \tilde{P}; \mathbf{F})$, since $B(T) = e^{\lambda T}\phi(0) > 0$ for $\phi(0) > 0$ and, consequently,

$$\tilde{E}\left[\left|\frac{\Psi(S_T)}{B(T)}\right|\right] = \frac{1}{e^{\lambda T}\phi(0)}\tilde{E}[\Psi(S_T)] < \infty.$$

From the well-known Ito-Clark theorem on martingale representation in Section 1.3 of Chapter 1, there exists a process $\beta(\cdot) = \{\beta(t), t \in [0,T]\}$ that is $\mathbf{F}$-adapted and $\int_0^T \beta^2(t)\,dt < \infty$ (P-a.s.) such that

$$Y(t) = Y(0) + \int_0^t \beta(s)\,d\tilde{W}(s), \quad t \in [0,T]. \tag{6.17}$$

Furthermore, since $\mathcal{F}(t) = \sigma(S_s, 0 \le s \le t)$, there exists a nonanticipating functional $\beta^* : [0,T] \times C_+[0,T] \to \Re$ (where $C_+[0,T]$ is the set of all non-negative continuous functions defined on $[0,T]$) such that $\beta(t) = \beta^*(t, S(\cdot))$, $t \in [0,T]$.

Let $\pi^*(\cdot) = \{(\pi_1^*(t), \pi_2^*(t)), t \in [0,T]\}$ be a trading strategy, where

$$\pi_2^*(t) = \frac{\beta(t)B(t)}{g(S_t)} \tag{6.18}$$

and

$$\pi_1^*(t) = Y(t) - \frac{S(t)}{B(t)}\pi_2^*(t), \quad t \in [0,T]. \tag{6.19}$$

We have the following lemma concerning the trading strategy $\pi^*(\cdot)$ and the process $Y(\cdot)$.

Lemma 6.4.5 $\pi^*(\cdot) \in SF(x)$ and for each $t \in [0,T]$, $Y(t) = Y^{\pi^*}(t)$, where, again, $Y^{\pi^*}(\cdot)$ is the process defined in (6.17), with the minimal strategy $\pi^*(\cdot)$ defined in (6.19) and (6.23).

Proof. It is clear that

$$X^{\pi^*}(t) = \pi_1^*(t)B(t) + \pi_2^*(t)S(t)$$
$$= \left[Y(t) - \pi_2^*(t)\frac{S(t)}{B(t)}\right]B(t) + \pi_2^*(t)S(t)$$
$$= Y(t)B(t).$$

This shows that $Y(t) = \frac{X^{\pi^*}(t)}{B(t)}$, $0 \le t \le T$. In this case,

$$dX^{\pi^*}(t) = B(t)\,dY(t) + Y(t)\,dB(t)$$
$$= B(t)\beta(t)\,d\tilde{W}(t) + Y(t)\,dB(t)$$
$$= \left[Y(t) - \frac{\beta(t)S(t)}{N(S_t)} \right]\,dB(t)$$
$$+ \frac{B(t)\beta(t)}{N(S_t)}(L(S_t)\,dt + N(S_t)\,d\tilde{W}(t))$$
$$= \pi_1^*(t)\,dB(t) + \pi_2^*(t)\,dS(t).$$

This shows that the trading strategy $\pi^*(\cdot)$ is *self-financing*.   □

In the following, we assume that a contingent claim is written at time $t \in [0, T]$ instead at time zero. Since the (B, S)-market are described by the two autonomous equations (6.1) and (6.4), all of the results obtained thus far can be restated with appropriate translation of time by t.

Theorem 6.4.6 *Let* $\Psi : \mathbf{C} \to [0, \infty)$ *be the (convex) payoff function. Then the rational price* $\Upsilon(\Psi; t, \psi)$ *for* $ECC(\Psi)$ *given* $S_t = \psi \in \mathbf{C}$ *at time* $t \in [0, T]$ *is given by*

$$\Upsilon(\Psi; t, \psi) = \tilde{E}[e^{-\lambda(T-t)}\Psi(S_T) \mid S_t = \psi]$$
$$= \tilde{E}[e^{-\lambda(T-t)}\Psi(S_{T-t}) \mid S_0 = \psi],$$

where $\lambda > 0$ *is the effective interest rate described by Lemma 6.1.2. Furthermore, there exists a minimal hedge* $\pi^*(\cdot) = \{(\pi_1^*(s), \pi_2^*(s)), s \in [t, T]\}$, *where*

$$\pi_2^*(s) = \frac{\beta(s)B(s)}{S(s)g(S_s)},$$

$$\pi_1^*(s) = Y^{\pi^*}(s) - \pi_2^*(s)\frac{S(s)}{B(s)}, \quad s \in [t, T], \tag{6.20}$$

and the process $\beta(\cdot) = \{\beta(t), t \in [0, T]\}$ *is given in (6.17).*

Proof. Evidently

$$X^{\pi^*}(t) = \tilde{E}[e^{-\lambda(T-t)}\Psi(S_T) \mid S_t = \psi],$$

$$X^{\pi^*}(T) = \Psi(\psi) \text{ (P-a.s.)}.$$

Therefore, π^* is a (Ψ, x)-hedge with initial capital $x = \tilde{E}[e^{-rT}\Psi(S_T)]$. Moreover, the strategy π^* is minimal. This shows that the *rational price* given $S_t = \psi$ at time $t \in [0, T]$ is

$$\Upsilon(\Psi; t, \psi) = \tilde{E}[e^{-\lambda(T-t)}\Psi(S_T) \mid S_t = \psi]$$
$$= \tilde{E}[e^{-\lambda(T-t)}\Psi(S_{T-t}) \mid S_0 = \psi]. \quad □ \tag{6.21}$$

6.4.2 The American Contingent Claims

At this point, we will prove the following theorem for American contingent claims.

Theorem 6.4.7 *For $ACC(\Psi)$, we have the following:*

1. *The rational price of an American contingent claim given $S_t = \psi$ at time $t \in [0,T]$ is*

$$
\Upsilon(\Psi;t,\psi) = \sup_{\tau \in \mathcal{M}_t^T} \tilde{E}\left[e^{-\lambda(T-\tau)}\Psi(S_\tau) \mid S_t = \psi\right]
$$

$$
= \sup_{\tau \in \mathcal{M}_t^T} \tilde{E}\left[e^{-\lambda(T-\tau)}\Psi(S_{\tau-t}) \mid S_0 = \psi\right].
$$

2. *In the class $SF^+(x)$ there exists a minimal $ACC(\Psi)$-hedge $\pi^*(\cdot)$ whose capital $X^{\pi^*}(\cdot) = \{X^{\pi^*}(s), s \in [t,T]\}$ is given by*

$$
X^{\pi^*}(s) = \operatorname{ess\,sup}_{\tau \in \mathcal{M}_t^T} \tilde{E}\left[e^{-\lambda(\tau-t)}\Psi(S_\tau)|S_t = \psi\right]. \tag{6.22}
$$

3. *The minimal trading strategy $\pi^*(\cdot) = \{(\pi_1^*(s), \pi_2^*(s)), s \in [t,T]\}$ is given by*

$$
\pi_2^*(s) = \frac{\beta(s)B(s)}{S(s)g(S_s)} \tag{6.23}
$$

and, therefore,

$$
\pi_1^*(s) = \frac{X^{\pi^*}(s) - \pi_2^*(s)S(s)}{B(s)}, \tag{6.24}
$$

where $\beta(\cdot)$ is found from the martingale representation of the discounted wealth process, that is,

$$
\frac{X^{\pi^*}(t)}{B(t)} = Y^{\pi^*}(t) = Y^{\pi^*}(0) + \int_0^t \beta(s)\,d\tilde{W}(s). \tag{6.25}
$$

4. *A stopping time $\tau^* \in T_0^T$ is a rational exercise time of the American option if and only if*

$$
\Upsilon(\Psi;t,\psi) = \tilde{E}\left[e^{-\lambda(\tau^*-t)}\Psi(S_{\tau^*}^*) \mid S_t = \psi\right]
$$

$$
= \tilde{E}\left[e^{-\lambda(\tau^*-t)}\Psi(S_{\tau^*-t}) \mid S_0 t = \psi\right].
$$

6.5 Infinite-Dimensional Black-Scholes Equation

Suppose the (B, S)-market is described by (6.1) and (6.4). In this section, an infinitesimal equation for the pricing function $V : [0, T] \times \mathbf{C} \to \Re$ will be derived with a given convex payoff function $\Psi : \mathbf{C} \to \Re$, where

$$V(t, \psi) = \tilde{E}\left[e^{-\lambda(T-t)}\Psi(S_T) \mid S_t = \psi\right].$$

Note that we used the simpler notation $V(t, \psi)$ instead of $\Upsilon(\Psi; t, \psi)$, since the award function Ψ is assumed to be given and fixed. The same substitution of Υ by V remains valid throughout the end of the chapter.

This infinitesimal equation will be referred to as the infinite-dimensional Black-Scholes equation for the European contingent claim $ECC(\Psi)$ and it takes the form of a partial differential equation that involves Fréchet derivatives with respect to elements in $\mathbf{C}$.

The concept of Fréchet derivatives, $DV(t, \psi) \in \mathbf{C}^*$ and $D^2V(t, \psi) \in \mathbf{C}^\dagger$, and their extensions, $\overline{DV(t, \psi)} \in (\mathbf{C} \oplus \mathbf{B})^*$ and $\overline{D^2V(t, \psi)} \in (\mathbf{C} \oplus \mathbf{B})^\dagger$, as well as the infinitesimal shift operator $(\mathcal{S}V)(t, \psi)$ have been introduced in Chapter 2. They will be used in deriving the infinite-dimensional Black-Scholes equation in this section and the infinite-dimensional HJBVI for the American contingent claim $ACC(\Psi)$ in next section without repeating their definitions.

6.5.1 Equation Derivation

For the convex payoff function $\Psi : \mathbf{C} \to \Re$, we have the (rational) pricing function $V : [0, T] \times \mathbf{C} \to \Re$ from Section 6.4.2 that

$$\begin{aligned} V(t, \psi) &= e^{-\lambda(T-t)}\tilde{E}[\Psi(S_T)|S_t = \psi] \\ &= e^{-\lambda(T-t)}\tilde{E}[\Psi(S_{T-t})|S_0 = \psi]. \end{aligned} \tag{6.26}$$

In the following, we explore the applicability of the infinitesimal generator

$$\mathbf{A}V(t, \psi) = \lim_{\epsilon \downarrow 0} \frac{E[V(t + \epsilon, S_{t+\epsilon}) - V(t, \psi)]}{\epsilon},$$

where $V : [0, T] \times \mathbf{C} \to \Re$ is given in (6.26).

Let

$$P_t(\Phi)(\psi) = E\left[\Phi(S_t) \mid S_0 = \psi\right]$$

for $\Phi : \mathbf{C} \to \Re$ with $\psi \in \mathbf{C}$. Let $C_b \subset \mathbf{C}$ be the Banach space of all bounded uniformly continuous functions $\Phi : \mathbf{C} \to \Re$. Here, we are using weak convergence.

By Dynkin's formula (Theorem 2.4.1 in Chapter 2), if

$$dx(t) = H(x_t)\,dt + G(x_t)\,dW(t)$$

with $x_0 = \psi$, and $\Phi \in C_{lip}^2(\mathbf{C})\mathcal{D}(\mathcal{S})$, then $\Phi \in \mathcal{D}(\mathbf{A})$ and for each $\psi \in \mathbf{C}$,

$$\mathbf{A}(\Phi)(\psi) = \mathcal{S}(\Phi)(\psi) + \overline{D\Phi(\psi)}(H(\psi)\mathbf{1}_{\{0\}})$$
$$+ \frac{1}{2}\overline{D^2\Phi(\psi)}(G(\psi)\mathbf{1}_{\{0\}}, G(\psi)\mathbf{1}_{\{0\}}). \qquad (6.27)$$

It is known (see, e.g., Black and Scholes [BS73], SKKM94]) that the classical Black-Scholes equation is a deterministic parabolic partial differential equation (with a suitable auxiliary condition), the solution of which gives the value of the European option contract at a given time. Using the infinitesimal generator, a generalized version of the classical Black-Scholes equation can be derived when the (B, S)-market model uses (6.1) and (6.4). In the second part of the following theorem, we use Theorem 2.4.1 in Chapter 2, which states that for $\Lambda \in \mathcal{D}(\mathbf{A})$, $u_t = P_t(\Lambda) = E[\Lambda(S_t)]$ is the unique solution of

$$\frac{du_t}{dt} = \mathbf{A}u_t$$

that satisfies the following:
(i) u_t is weakly continuous, $\frac{du_t}{dt}$ is weakly continuous from the right, and $\|\frac{du_t}{dt}\|$ is bounded on every finite interval.
(ii) $\|u_t\| \le ce^{kt}$ for some constants c and k.
(iii)

$$u_0 = (w)\lim_{t\downarrow 0} u_t = \Lambda.$$

Theorem 6.5.1 *Let* $X^{\pi^*}(t) = V(t,\psi) = e^{-\lambda(T-t)}\tilde{E}[\Psi(S_{T-t})|S_0 = \psi]$ *be the wealth process for the minimal* (Ψ, x)*-hedge, where* $\psi \in \mathbf{C}_+$, $x = X^{\pi^*}(0)$, *and* $t \in [0,T]$. *Define* $\Phi_t(\psi) = \tilde{E}[\Psi(S_t)|S_0 = \psi]$ *and let* $\Phi_t \in C^2_{lip}(\mathbf{C}) \cap \mathcal{D}(\mathcal{S})$ *for* $t \in [0,T]$. *Finally, assume that* $\Psi \in C^2_{lip}(\mathbf{C}) \cap \mathcal{D}(\mathcal{S})$. *Then*

$$\lambda V(t,\psi) = (\partial_t + \mathbf{A})V(t,\psi)$$
$$= \mathcal{S}(V)(t,\psi) + \overline{DV(t,\psi)}(\lambda\psi(0)\mathbf{1}_{\{0\}})$$
$$+\frac{1}{2}\overline{D^2V(t,\psi)}(\psi(0)g(\psi)\mathbf{1}_{\{0\}}, \psi(0)g(\psi)\mathbf{1}_{\{0\}}),$$
$$\forall(t,\psi) \in [0,T) \times \mathbf{C}_+, \qquad (6.28)$$

where

$$V(T,\psi) = \Psi(\psi), \quad \forall\psi \in \mathbf{C}_+, \qquad (6.29)$$

and the trading strategy $(\pi_1^*(t), \pi_2^*(t))$ *is defined by*

$$\pi_2^*(t) = \frac{\beta(t)B(t)}{S(t)g(S_t)}$$

and

$$\pi_1^*(t) = \frac{1}{B(t)}\left[X^{\pi^*}(t) - S(t)\pi_2^*(t)\right].$$

$\beta(t)$ *is found from (6.17). Furthermore, if (6.28) and (6.29) hold, then* $V(t,\psi)$ *is the wealth process for the* (Ψ, x)*-hedge with* $\pi_2^*(t) = \frac{\beta(t)B(t)}{S(t)g(S_t)}$ *and* $\pi_1^*(t) = \frac{1}{B(t)}\left[X^{\pi^*}(t) - S(t)\pi_2^*(t)\right]$.

Note: Equations (6.28) and (6.29) are the infinite-dimensional Black-Scholes equations for the (B, S)-market with hereditary price structure as described by (6.1) and (6.4).

Proof. Letting

$$dS(t) = \lambda S(t)\, dt + S(t)g(S_t)\, d\tilde{W}(t)$$

and

$$P_t(\Psi)(\psi) = \tilde{E}\left[\Psi\left(S_t\right) | S_0 = \psi\right],$$

we have

$$e^{-\lambda(T-t)}P_{T-t}(\Psi)(\psi) = e^{-\lambda(T-t)}\tilde{E}\left[\Psi\left(S_{T-t}\right) | S_0 = \psi\right] = V(t,\psi).$$

From the above, we therefore have

$$\partial_t V(t,\psi) = \frac{\partial}{\partial t}\left\{e^{-\lambda(T-t)}P_{T-t}(\Psi)(\psi)\right\}$$

$$= \lambda e^{-\lambda(T-t)}P_{T-t}(\Psi)(\psi) - e^{-\lambda(T-t)}\mathbf{A}(P_{T-t}(\lambda)(\psi)).$$

Using the linearity of the $\mathbf{A}$ operator and $\Psi \in C^2_{lip}(\mathbf{C}) \cap \mathcal{D}(\mathcal{S})$, we have

$$\lambda V(t,\psi) = (\partial_t + \mathbf{A}V(t,\psi)$$

$$= \partial_t V(t,\psi) + \mathcal{S}V(t,\psi) + \overline{DV(t,\psi)}(\lambda\psi(0)\mathbf{1}_{\{0\}})$$

$$+ \frac{1}{2}\overline{D^2V(t,\psi)}(\psi(0)g(\psi)\mathbf{1}_{\{0\}}, \psi(0)(\psi)\mathbf{1}_{\{0\}}).$$

Therefore,

$$\lambda V(t,\psi) = \partial_t V(t,\psi) + \mathcal{S}V(t,\psi) + \overline{DV(t,\psi)}(\lambda\psi(0)\mathbf{1}_{\{0\}})$$

$$+ \frac{1}{2}\overline{D^2V(t,\psi)}(\psi(0)g(\psi))\mathbf{1}_{\{0\}}, \psi(0)g(\psi)\mathbf{1}_{\{0\}}).$$

Notice that

$$V(T,\psi) = e^{-\lambda(T-T)}\tilde{E}\left[\Psi\left(S_{T-T}\right) | S_0 = \psi\right] = \Psi(\psi).$$

Note that $\{S_t, t \geq 0\}$ is a **C**-valued Markov process (see [CY07]), so by the derivation of (6.28) and (6.29),

$$V(t,\psi) = e^{-\lambda(T-t)}\tilde{E}\left[\Psi\left(S_{T-t}\right) | S_0 = \psi\right].$$

However, $V(t,\psi) = X^{\pi^*}(t) = B(t)Y^{\pi^*}(t)$, so

$$dV(t,\psi) = B(t)\, dY^{\pi^*}(t) + Y^{\pi^*}(t)\, dB(t)$$

$$= B(t)\beta(t)\, d\tilde{W}(t) + \lambda Y^{\pi^*}(t)B(t)\, dt$$

$$= \pi_2^*(t)S(t)g(S_t)\, d\tilde{W}(t) + \lambda\Psi(t,\psi)\, dt.$$

Therefore,

$$
\begin{aligned}
V(t, S_t) - V(0, \psi) &= \lambda \int_0^t V(s, S_s)\, ds + \int_0^t \pi_2^*(s) S(s) g(S_s)\, d\tilde{W}(s) \\
&= \lambda \int_0^t \left[\pi_1^*(s) B(s) + \pi_2^*(s) S(s) \right] ds \\
&\quad + \int_0^t \pi_2^*(s) S(s) g(S_s)\, d\tilde{W}(s) \\
&= \lambda \int_0^t \pi_1^*(s)\, dB(s) \\
&\quad + \left[\int_0^t \pi_2^*(s) S(s)\, ds + \int_0^t \pi_2^*(s) S(s) g(S_s)\, d\tilde{W}(s) \right] \\
&= \lambda \int_0^t \pi_1^*(s)\, dB(s) + \int_0^t \pi_2^*(s)\, dS(s).
\end{aligned}
$$

Therefore, $V(t, \psi) = X^{\pi^*}(t)$ is self-financing. $\square$

Following Definition 2.4.4 in Chapter 2, a function $\Phi : \mathbf{C} \to \Re$ is said to be *quasi-tame* if there are C^∞-bounded maps $h : (\Re)^k \to \Re$ and $f_j : \Re \to \Re$ and piecewise C^1 functions $g_j : [-r, 0] \to \Re$, $1 \le j \le k-1$, such that

$$
\Phi(\eta) = r\left(\int_{-r}^0 f_1(\eta(s)) g_1(s)\, ds, \ldots, \int_{-r}^0 f_{k-1}(\eta(s)) g_{k-1}(s)\, ds, \eta(0) \right)
$$

for all $\eta \in \mathbf{C}$. For the quasi-tame function $\Phi : \mathbf{C} \to \Re$, the following Itô formula holds (see Section 2.5 in Chapter 2):

$$
\begin{aligned}
d\Phi(x_t) &= \mathcal{S}(\Phi)(x_t)\, dt + \overline{D\Phi(x_t)}(H(x_t)\mathbf{1}_{\{0\}})\, dt \\
&\quad + \frac{1}{2}\overline{D^2\Phi(x_t)}(G(x_t)\mathbf{1}_{\{0\}}, G(x_t)\mathbf{1}_{\{0\}})\, dt \\
&\quad + \overline{D\phi(x_t)}(H(x_t)\mathbf{1}_{\{0\}})\, dW(s), \quad (6.30)
\end{aligned}
$$

where

$$
dx(t) = H(x_t)\, dt + G(x_t)\, dW(t)
$$

and $x_0 = \psi \in \mathbf{C}$. Using this formula, we have the following corollary.

Corollary 6.5.2 *Let $\Psi(t, \psi)$ be a quasi-tame solution to (6.28) and (6.29), then $\pi_2^*(t) = \overline{D\Psi(t, \psi)}(\mathbf{1}_{\{0\}})$.*

Proof. To find the trading strategy $\{\pi(t) = (\pi_1(t), \pi_2(t)), t \in [0, T]\}$, we start by using Itô's formula for a quasi-tame function applied to $e^{-\lambda t}\Psi(S_t)$, which gives that

$$e^{-\lambda t}\Phi(t, S_t) - \Phi(0, \psi)$$

$$= -\lambda \int_0^t e^{-\lambda s}\Phi(s, S_s)\, ds + \int_0^t e^{-\lambda s}\partial_s\Phi(s, S_s)\, ds$$

$$+ \int_0^t e^{-\lambda s}\mathcal{S}(\Psi)(s, \tilde{\psi}_s)\, ds$$

$$+ \int_0^t e^{-\lambda s}\overline{D\Phi(s, S_s)}(\lambda S(s)\mathbf{1}_{\{0\}})\, ds$$

$$+ \int_0^t e^{-\lambda s}\overline{D\Phi(s, S_s)}(S(s)g(S_s)\mathbf{1}_{\{0\}})\, d\tilde{W}(s)$$

$$+ \frac{1}{2}\int_0^t e^{-\lambda s}\overline{D^2\Phi(s, S_s)}(S(s)g(S_s)\mathbf{1}_{\{0\}}, S(s)g(S_s)\mathbf{1}_{\{0\}})\, ds.$$

Since Φ satisfies the generalized Black-Scholes equation (6.28), the above equation gives,

$$e^{-\lambda t}\Phi(t, S_t) - \Phi(0, \psi) = \int_0^t e^{-\lambda s}\overline{D\Phi(s, S_s)}(S(s)g(S_s)\mathbf{1}_{\{0\}})\, d\tilde{W}(s). \quad (6.31)$$

Using the self-financing condition,

$$dY^{\pi^*}(t) = \frac{B(t)\, dX^{\pi^*}(t) - X^{\pi^*}(t)\, dB(t)}{B^2(t)}$$

$$= \frac{B(t)\left[\pi_1^*(t)\, dB(t) + \pi_2^*(t)\, dS(t)\right] - \left[\pi_1^*(t)B(t) + \pi_2^*(t)S(t)\right]\lambda B(t)\, dt}{B^2(t)}$$

$$= \frac{\left[\pi_1^*(t)\, dB(t) + \pi_2^*(t)\, dS(t)\right] - \left[\pi_1^*(t)B(t) + \pi_2^*(t)S(t)\right]\lambda\, dt}{B(t)}$$

$$= \frac{\pi_2^*(t)\left[\lambda S(t)\, dt + S(t)g(S_t)\, d\tilde{W}(t)\right] - \lambda\pi_2^*(t)S(t)\, dt}{B(t)}$$

$$= \frac{\pi_2^*(t)}{B(t)}S(t)g(S_t)\, d\tilde{W}(t).$$

Consequently,

$$Y^{\pi^*}(t) = Y^{\pi^*}(0) + \int_0^t \frac{\pi_2^*(s)}{B(s)}S(s)g(S_s)\, d\tilde{W}(s),$$

and by (6.2),

$$Y^{\pi^*}(t) = \frac{X^{\pi^*}(t)}{B(t)} = e^{-\lambda t}\frac{X^{\pi^*}(t)}{\phi(0)}.$$

Therefore,

$$e^{-\lambda t}X^{\pi^*}(t) - X^{\pi^*}(0) = \phi(0)\int_0^t \frac{\pi_2^*(s)S(s)g(S_s)}{B(s)}d\tilde{W}(s)$$

$$= \int_0^t e^{-\lambda s}\pi_2^*(s)S(s)g(S_s)d\tilde{W}(s). \quad (6.32)$$

Combining (6.31) and (6.32), we have

$$\int_0^t e^{-\lambda s}\overline{D\Phi(s,S_s)}(S(s)g(S_s)\mathbf{1}_{\{0\}})\,d\tilde{W}(s)$$

$$= \int_0^t e^{-\lambda s}\pi_2^*(s)S(s)g(S_s)\,d\tilde{W}(s), \quad \forall t \in [0,T],$$

and by the uniqueness of the martingale representation, we have

$$\pi_2^*(t) = \overline{D\Phi(t,S_t)}(\mathbf{1}_{\{0\}}) \quad \text{a.s.}$$

and the proof is complete.   □

The following corollary shows that when there is no time delay (i.e., $r = 0$) and the stock price $\{S(t), t \geq 0\}$ satisfies the nonlinear stochastic differential equation

$$\frac{dS(t)}{S(t)} = f(S(t))\,dt + g(S(t))\,dW(t), \quad t \geq 0, \quad (6.33)$$

and

$$dB(t) = \lambda B(t)\,dt, \quad t \geq 0, \quad (6.34)$$

then the generalized Black-Scholes equation (6.28) reduces to that of the classical one.

Corollary 6.5.3 *If $\Phi(t,S_t) = V(t,S(t))$ and there is no hereditary structure so that (6.1) and (6.4) reduce to (6.33) and (6.34), respectively, then the infinite-dimensional Black-Scholes equation (6.28)-(6.29) reduces to the classical Black-Scholes equation.*

Proof. If there is no delay, $r = 0$, so $\Phi(t,S_t) = V(t,S(t))$. Also, we have $\lambda\Phi(t,S_t) = \lambda V(t,S(t)) = \lambda V(t,x)$, where $S(t) = x$, and

$$\partial_t V(t,S_t) = \partial_t V(t,x),$$

$$\overline{DV(t,S_t)}(\lambda S(t)\mathbf{1}_{\{0\}}) = \partial_x V(t,x)\lambda x,$$

$$\overline{DV(t,S_t)}(S(s)g(S_t)\mathbf{1}_{\{0\}}) = \partial_x V(t,x)xg(x),$$

and

$$\overline{D^2V(t,S_t)}(S(s)g(S_t)\mathbf{1}_{\{0\}}, S(s)g(S_t)\mathbf{1}_{\{0\}}) = \frac{\partial^2}{\partial x^2}V(t,x)x^2g^2(x).$$

Also, $\mathcal{S}V(t, \tilde{\psi}_t) = 0$ if $r = 0$. Therefore,

$$\lambda V(t, x) = \partial_t V(t, x) + \partial_x V(t, x)\lambda x$$
$$+ \frac{1}{2}\frac{\partial^2}{\partial x^2}V(t, x)x^2 g^2(x), \quad \forall x \geq 0, \; \forall t \in [0, T].$$

For

$$g(S_t) = \sigma \geq 0,$$

$$\lambda V(t, x) = \frac{\partial}{\partial t}V(t, x) + \lambda x \frac{\partial}{\partial x}V(t, x)$$
$$+ \frac{1}{2}\sigma^2 x^2 \frac{\partial^2}{\partial x^2}V(t, x), \qquad (6.35)$$

which is the classical Black-Scholes equation.   □

6.5.2 Viscosity Solution

In this subsection, we will show that the option price $V : [0, T] \times \mathbf{C} \to \Re$ is actually a viscosity solution of the equation (6.28)-(6.29). The general theory for viscosity solutions can be found Chapters 3 and 4 for various settings.

First, let us define the viscosity solution of (6.28)-(6.29) as follows.

Definition 6.5.4 *Let $w \in C([0, T] \times \mathbf{C})$. We say that w is a viscosity sub-solution of (6.28)-(6.29) if, for every $\Gamma \in C^{1,2}_{lip}([0, T] \times \mathbf{C}) \cap \mathcal{D}(\mathcal{S})$ and for $(t, \psi) \in [0, T] \times \mathbf{C}$ satisfying $\Gamma \geq w$ on $[0, T] \times \mathbf{C}$ and $\Gamma(t, \psi) = w(t, \psi)$, we have*

$$\lambda\Gamma(t, \psi) - \partial_t\Gamma(t, \psi) - \mathcal{S}(\Gamma)(t, \psi) - \overline{D\Gamma(t, \psi)}(\lambda\psi(0)\mathbf{1}_{\{0\}})$$
$$-\frac{1}{2}\overline{D^2\Gamma(t, \psi)}(\psi(0)g(\psi)\mathbf{1}_{\{0\}}, \psi(0)g(\psi)\mathbf{1}_{\{0\}}) \leq 0.$$

We say that w is a viscosity super solution of (6.28)-(6.29) if, for every $\Gamma \in C^{1,2}_{lip}([0, T] \times \mathbf{C}) \cap \mathcal{D}(\mathcal{S})$, and for $(t, \psi) \in [0, T] \times \mathbf{C}$ satisfying $\Gamma \leq w$ on $[0, T] \times \mathbf{C}$ and $\Gamma(t, \psi) = w(t, \psi)$, we have

$$\lambda\Gamma(t, \psi) - \partial_t\Gamma(t, \psi) - \mathcal{S}(\Gamma)(t, \psi) - \overline{D\Gamma(t, \psi)}(\lambda\psi(0)\mathbf{1}_{\{0\}})$$
$$-\frac{1}{2}\overline{D^2\Gamma(t, \psi)}(\psi(0)g(\psi)\mathbf{1}_{\{0\}}, \psi(0)g(\psi)\mathbf{1}_{\{0\}}) \geq 0.$$

We say that w is a viscosity solution of (6.28)-(6.29) if it is a viscosity supersolution and a viscosity subsolution of (6.28)-(6.29).

Lemma 6.5.5 *Let $u, t \in [0, T]$ with $t \leq u$; we have*

$$V(t, \psi) = \tilde{E}[e^{-\lambda(u-t)}V(u, S_u)|S_t = \psi]. \qquad (6.36)$$

Proof. Let $u, t \in [0, T]$, such that $t \leq u$. Note that

$$V(u, S_u) = \tilde{E}[e^{-\lambda(T-u)}\Psi(S_T)|S_u]$$
$$= \tilde{E}[e^{-\lambda(T-u)}\Psi(S_T)|\mathcal{F}(u)]. \tag{6.37}$$

In view of this and using the fact the $S_t(\psi)$ is Markovian, we have

$$\tilde{E}[e^{-\lambda(u-t)}V(u, S_u)|S_t = \psi] = \tilde{E}[e^{-\lambda(u-t)}\tilde{E}[e^{-\lambda(T-u)}\Psi(S_T)|\mathcal{F}(u)]|S_t = \psi]$$
$$= \tilde{E}[e^{-\lambda(u-t)}e^{-\lambda(T-u)}\tilde{E}[\Psi(S_T)|\mathcal{F}(u)]|\mathcal{F}(t)]$$
$$= \tilde{E}[e^{-\lambda(T-t)}\Psi(S_T)|\mathcal{F}(t)]$$
$$= \tilde{E}[e^{-\lambda(T-t)}\Psi(S_T)|S_t = \psi]$$
$$= V(t, \psi). \tag{6.38}$$

This proves the lemma. □

Theorem 6.5.6 *The pricing function $V : [0, T] \times \mathbf{C} \to \Re$ is a viscosity solution of the infinite-dimensional Black-Scholes equation (6.28)-(6.29).*

Proof. Let $\Gamma \in C^{1,2}_{lip}([0, T] \times \mathbf{C}) \cap \mathcal{D}(\mathcal{S})$, for $(t, \psi) \in [0, T] \times \mathbf{C}$ such that $\Gamma \leq V$ on $[0, T] \times \mathbf{C}$ and $\Gamma(t, \psi) = V(t, \psi)$. We want to prove the viscosity supersolution inequality, that is,

$$\lambda V(t, \psi) - \partial_t V(t, \psi) - \mathcal{S}(V)(t, \psi) - \overline{D\Gamma(t, \psi)}(\lambda\psi(0)\mathbf{1}_{\{0\}})$$
$$-\frac{1}{2}\overline{D^2\Gamma(t, \psi)}(\psi(0)g(\psi)\mathbf{1}_{\{0\}}, \psi(0)g(\psi)\mathbf{1}_{\{0\}}) \geq 0.$$

Notation: Throughout the proof of the theorem we will use the following notation:

$$\tilde{E}^\psi[Y] = \tilde{E}[Y|S_t = \psi] \qquad \text{for any process } Y.$$

Let $\Gamma \in C^{1,2}_{lip}([0, T] \times \mathbf{C}) \cap \mathcal{D}(\mathcal{S})$. For $0 \leq t \leq t_1 \leq T$, by virtue of Theorem 2.4.1 in Chapter 2, we have

$$\tilde{E}^\psi[e^{-\lambda(s-t)}\Gamma(s, S_s)] - \Gamma(t, \psi)$$
$$= \tilde{E}^\psi\left[\int_t^s e^{-\lambda(u-t)}\left(\partial_t\Gamma(u, S_u) + \mathcal{S}(\Gamma)(u, S_u)\right.\right. \tag{6.39}$$
$$+ \overline{D\Gamma(u, S_u)}(\lambda S_u(0)\mathbf{1}_{\{0\}})$$
$$\left.\left.+ \frac{1}{2}\overline{D^2\Gamma(u, S_u)}(S_u(0)g(S_u)\mathbf{1}_{\{0\}}, S_u(0)g(S_u)\mathbf{1}_{\{0\}}) - \lambda\Gamma(u, S_u)\right)du\right].$$

For any $s \in [t, T]$, Lemma 6.5.5 gives

$$V(t, \psi) \geq e^{-\lambda(s-t)}\tilde{E}^\psi[V(s, S_s)].$$

Using the fact that $\Gamma \leq V$, we can get

$$0 \geq \tilde{E}^\psi\left[e^{-\lambda(s-t)}V(s,S_s)\right] - V(t,\psi)$$

$$\geq \tilde{E}^\psi\left[e^{-\lambda(s-t)}\Gamma(s,S_s)\right] - V(t,\psi)$$

$$\geq \tilde{E}^\psi\int_t^s e^{-\lambda(u-t)}\left[\left(\partial - t\Gamma(u,S_u) + \mathcal{S}(\Gamma)(u,S_u)\right.\right. \tag{6.40}$$

$$+\overline{D\Gamma(u,S_u)}(\lambda S_u(0)\mathbf{1}_{\{0\}})$$

$$\left.\left.+ \frac{1}{2}\overline{D^2\Gamma(u,S_u)}(S_u(0)g(S_u)\mathbf{1}_{\{0\}}, S_u(0)g(S_u)\mathbf{1}_{\{0\}}) - \lambda\Gamma(u,S_u)\right)\right]du.$$

Dividing by $(s-t)$ and letting $s \downarrow t$ in the previous inequality, we obtain

$$\lambda V(t,\psi) - \partial_t\Gamma(t,\psi) - \mathcal{S}(V)(t,\psi) - \overline{D\Gamma(t,\psi)}(\lambda\psi(0)\mathbf{1}_{\{0\}})$$

$$-\frac{1}{2}\overline{D^2\Gamma(t,\psi)}(\psi(0)g(\psi)\mathbf{1}_{\{0\}}, \psi(0)g(\psi)\mathbf{1}_{\{0\}}) \geq 0. \tag{6.41}$$

So we have proved the inequality.

Next, we want to prove that V is a viscosity subsolution. Let $\Gamma \in C_{lip}^{1,2}([0,T] \times \mathbf{C}) \cap \mathcal{D}(\mathcal{S})$. For $(t,\psi) \in [0,T] \times \mathbf{C}$ satisfying $\Gamma \geq V$ on $[0,T] \times \mathbf{C}$ and $\Gamma(t,\psi) = V(t,\psi)$, we want to prove that

$$\lambda V(t,\psi) - \partial_t\Gamma(t,\psi) - \mathcal{S}(V)(t,\psi) - \overline{D\Gamma(t,\psi)}(\lambda\psi(0)\mathbf{1}_{\{0\}})$$

$$-\frac{1}{2}\overline{D^2\Gamma(t,\psi)}(\psi(0)g(\psi)\mathbf{1}_{\{0\}}, \psi(0)g(\psi)\mathbf{1}_{\{0\}}) \leq 0. \tag{6.42}$$

For any $s \in [t,T]$, Lemma 6.5.5 gives

$$V(t,\psi) \leq \tilde{E}^\psi e^{-\lambda(s-t)}[V(s,S_s)],$$

so we can get

$$0 \leq \tilde{E}^\psi\left[e^{-\lambda(s-t)}V(s,S_s)\right] - V(t,\psi)$$

$$\leq \tilde{E}^\psi\left[e^{-\lambda(s-t)}\Gamma(s,S_s)\right] - \Gamma(t,\psi)$$

$$\leq \tilde{E}^\psi\int_t^s e^{-\lambda(u-t)}\left[\left(\partial_t\Gamma(u,S_u) + \mathcal{S}(\Gamma)(u,S_u) + \overline{D\Gamma(u,S_u)}(\lambda S_u(0)\mathbf{1}_{\{0\}})\right.\right.$$

$$\left.\left.+\frac{1}{2}\overline{D^2\Gamma(u,S_u)}(S_u(0)g(S_u)\mathbf{1}_{\{0\}}, S_u(0)g(S_u)\mathbf{1}_{\{0\}}) - \lambda\Gamma(u,S_u)\right)\right]du. \tag{6.43}$$

Dividing by $(s-t)$ and letting $s \downarrow t$ in the previous inequality, we obtain

$$\lambda V(t,\psi) - \partial_t\Gamma(t,\psi) - \mathcal{S}(V)(t,\psi) - \overline{D\Gamma(t,\psi)}(\lambda\psi(0)\mathbf{1}_{\{0\}})$$

$$-\frac{1}{2}\overline{D^2\Gamma(t,\psi)}(\psi(0)g(\psi)\mathbf{1}_{\{0\}}, \psi(0)g(\psi)\mathbf{1}_{\{0\}}) \leq 0. \tag{6.44}$$

So we have proved the inequality. This completes the proof of the theorem.
□

Proposition 6.5.7 *(Comparison Principle) If $V_1(t, \psi)$ and $V_2(t, \psi)$ are both continuous with respect to the argument (t, ψ) and are respectively the viscosity subsolution and supersolution of (6.28)-(6.29) with at most a polynomial growth, then*

$$V_1(t, \psi) \leq V_2(t, \psi), \quad \forall(t, \psi) \in [0, T] \times \mathbf{C}. \tag{6.45}$$

Proof. The proof follows the same argument as in Theorem 3.6.1 with some modifications and therefore is omitted here.   □

Given the above comparison principle, it is easy to prove that (6.28)−(6.29) has only a unique viscosity solution.

Theorem 6.5.8 *The rational pricing function $V : [0, T] \times \mathbf{C} \to \Re$ for $ECC(\Psi)$ is the unique viscosity solution of the infinite-dimensional Black-Scholes equation (6.28)-(6.29).*

6.6 HJB Variational Inequality

Since we are also interested in the secondary market of the American option that is sold at $t \in [0, T]$, characterizations of the rational price of the option at time t is also very desirable. In the following, we state the HJBVI that characterizes the pricing function $V : [0, T] \times \mathbf{C} \to [0, \infty)$ of the American option under the assumption that it is sufficiently smooth (i.e., $V \in C_{lip}^{1,2}([0, T] \times \mathbf{C}) \cap \mathcal{D}(\mathcal{S})$). The derivation follows easily from Chapter 4 as a consequence of optimal stopping problem.

From Theorem 6.4.7, it is known that the pricing function $V : [0, T] \times \mathbf{C} \to R$ is the value function of the following optimal stopping problem:

$$V(t, \psi) = \sup_{\tau \in \mathcal{M}_t^T} \tilde{E} \left[e^{-\lambda(\tau - t)} \Psi(S_\tau) | S_t = \psi \right].$$

It is clear that $V(T, \psi) = \Psi(\psi)$ for all $\psi \in \mathbf{C}$ and $V(0, \psi) = C^*(\Psi)$.

We have the following theorem.

Theorem 6.6.1 *Assume that the reward function $\Psi : \mathbf{C} \to [0, \infty)$ is globally convex. Then the pricing function $V \in C([0, T] \times \mathbf{C})$. Furthermore, $V(t, \cdot) : \mathbf{C} \to \Re$ is Lipschitz in $\mathbf{C}$ for each $t \in [0, T]$, that is,*

$$|V(t, \varphi) - V(t, \phi)| \leq K\|\varphi - \phi\|$$

for all $\varphi, \phi \in \mathbf{C}$ and for some $K > 0$.

Proof. Since the reward function Ψ satisfies Assumption 6.1.9, it is also Lipschitz on $\mathbf{C}$ (see Lemma 6.1.10). By following the proof of Lemma 4.5.2 in Chapter 4, one can show that the pricing function $V(t, \cdot) : \mathbf{C} \to [0, \infty)$ satisfies the Lipschitz condition for each $t \in [0, T]$.   □

It will be shown in the next section that the pricing function $V : [0, T] \times$ $\mathbf{C} \to \Re$ is a viscosity solution of the infinite-dimensional HJBVI (6.46)-(6.47). To derive (6.46)-(6.47), it is necessary to introduce some notations and concept as follows.

Consider the following infinite-dimensional HJBVI with the terminal condition

$$\max\{V - \Psi, \ \partial_t V + \mathbf{A}V - \lambda V\} = 0 \quad \text{on } [0, T] \times \mathbf{C} \tag{6.46}$$

and

$$V(T, \psi) = \Psi(\psi), \quad \forall \psi \in \mathbf{C}, \tag{6.47}$$

where

$$\mathbf{A}V(t, \psi) = \mathcal{S}V(t, \psi) + \overline{DV(t, \psi)}(\lambda\psi(0))\mathbf{1}_{\{0\}})$$
$$+ \frac{1}{2}\overline{D^2V(t, \psi)}(\psi(0)g(\psi)\mathbf{1}_{\{0\}}, \psi(0)\mathbf{g}(\psi)\mathbf{1}_{\{0\}}). \tag{6.48}$$

We can partition the domain of this function into the *continuation region* $\mathcal{C}$ and the *stopping region* $(\mathcal{C})^c \equiv \mathbf{C} - \mathcal{C}$ as follows:

$$\mathcal{C} \equiv \{(t, \psi) \in [0, T] \times \mathbf{C} : V(t, \psi) > \Psi(\psi)\}$$

and

$$(\mathcal{C})^c \equiv \{(t, \psi) \in [0, T] \times \mathbf{C} : V(t, \psi) = \Psi(\varphi)\},$$

since $V \geq \Psi$. It is clear from the continuity of $V : [0, T] \times \mathbf{C} \to \Re$ and $\Psi : \mathbf{C} \to \Re$ that $(\mathcal{C})^c$ is closed and $\mathcal{C}$ is open. Define

$$\mathcal{C}(t) \equiv \{\psi : (t, \psi) \in \mathcal{C}\}$$

and

$$(\mathcal{C})^c(t) \equiv \{\psi : (t, \psi) \in \mathcal{S}\}$$

are connected for every $t \in [0, T)$. Detailed characterizations of these two regions can be found in Section 4.4 in Chapter 4.

If it happens that the pricing function is sufficiently smooth, then it satisfies the infinite-dimensional HJBVI (6.46)-(6.47) as stated in the following result.

Theorem 6.6.2 *If the pricing function V is such that $V \in C^{1,2}_{lip}([0, T] \times \mathbf{C}) \cap \mathcal{D}(\mathcal{S})$, then the pricing function satisfies (6.46)-(6.47) in the classical sense.*

Proof. It is clear that $V(T, \psi) = \Psi(\psi)$ for all $\psi \in \mathbf{C}$. The rest of the theorem follows the derivation of HJBVI in Section 4.3 of Chapter 4. $\square$

Definition 6.6.3 *Let $\Phi \in C([0, T] \times \mathbf{C})$. We say that Φ is a viscosity subsolution of (6.46)-(6.47) if for every $(t, \psi) \in [0, T] \times \mathbf{C}$ and for every $\Gamma \in C^{1,2}_{lip}([0, T] \times \mathbf{C}) \cap \mathcal{D}(\mathcal{S})$ satisfying $\Gamma \geq \Phi$ on $[0, T] \times \mathbf{C}$ and $\Gamma(t, \psi) = \Phi(t, \psi)$, we have*

$$\max\{\Psi(\psi) - \Gamma(t, \psi), \partial_t \Gamma(t, \psi) + \mathbf{A}\Gamma(t, \psi)\} \leq 0.$$

We say that Φ is a viscosity supersolution of (6.46)-(6.47) if for every $(t, \psi) \in [0, T] \times \mathbf{C}$ and for every $\Gamma \in C_{lip}^{1,2}([0, T] \times \mathbf{C}) \cap \mathcal{D}(\mathcal{S})$ satisfying $\Gamma \leq \Phi$ on $[0, T] \times \mathbf{C}$ and $\Gamma(t, \psi) = \Phi(t, \psi)$, we have

$$\max \{\Psi(\phi) - \Gamma(t, \psi), \partial_t \Gamma(t, \psi) + \mathbf{A}\Gamma(t, \psi)\} \geq 0.$$

We say that Φ is a viscosity solution of (6.46)-(6.47) if it is a viscosity supersolution and a viscosity subsolution of (6.46)-(6.47).

We have the following theorem as a consequence of the results obtained in Section 4.5 of Chapter 4.

Theorem 6.6.4 *The pricing function $V : [0, T] \times \mathbf{C} \to \Re$ is the unique viscosity solution of (6.46)-(6.47).*

6.7 Series Solution

In this section, we consider a series solution for the (classical) solution Ψ : $[0, T] \times \mathbf{C} \to \Re$ of the infinite-dimensional Black-Scholes equation (6.28)-(6.29).

6.7.1 Derivations

We start by simplifying our infinite-dimensional Black-Scholes equation

$$\lambda V(t, \psi) = \partial_t V(t, \psi) + \mathcal{S}(V)(t, \psi) + \overline{DV(t, \psi)}(\lambda \psi(0)\mathbf{1}_{\{0\}})$$
$$+ \frac{1}{2}\overline{D^2 V(t, \psi)}(\psi(0)g(\psi)\mathbf{1}_{\{0\}}, \psi(0)g(\psi)\mathbf{1}_{\{0\}}),$$
$$\forall(t, \psi) \in [0, T) \times \mathbf{C}_+,$$

and

$$V(T, \psi) = \Psi(\psi) \quad \forall \psi \in \mathbf{C}_+.$$

Since we know that the solution is of the form $V(t, \psi) = e^{-\lambda(T-t)}\tilde{E}[\Psi(S_T)|S_t = \psi]$, we use this characterization of the solution with the differential equation and obtain

$$0 = \partial_t \tilde{\Phi}(t, \psi) + \mathcal{S}(\tilde{\Phi})(t, \psi) + \overline{D\tilde{\Phi}(t, \psi)}(\lambda \psi(0)\mathbf{1}_{\{0\}})$$
$$+ \frac{1}{2}\overline{D^2\tilde{\Phi}(t, \psi)}(\psi(0)g(\psi)\mathbf{1}_{\{0\}}, \psi(0)g(\psi)\mathbf{1}_{\{0\}})$$
$$\forall(t, \psi) \in [0, T) \times \mathbf{C}_+,$$

with

$$\tilde{\Phi}(T, \psi) = \Psi(\psi), \quad \forall \psi \in \mathbf{C}_+,$$

where $\tilde{\Phi}(t, \psi) = \tilde{E}[\Psi(S_T)|S_t]$. We complete the simplification by using a change of variables, replacing t with $\tau = T - t$, which gives the equation

$$\partial_\tau \Phi(\tau, \psi) = \mathcal{S}(\Phi)(\tau, \psi) + \overline{D\Phi(\tau, \psi)}(\lambda\psi(0)\mathbf{1}_{\{0\}})$$
$$+ \frac{1}{2}\overline{D^2\Phi(\tau, \psi)}(\psi(0)g(\psi)\mathbf{1}_{\{0\}}, \psi(0)g(\psi)\mathbf{1}_{\{0\}}),$$
$$\forall (\tau, \psi) \in [0, T) \times \mathbf{C}_+,$$

with

$$\Phi(0, \psi) = \Psi(\psi), \quad \forall \psi \in \mathbf{C}_+$$

We now seek a solution of the form

$$\Phi(\tau, \psi) = \sum_{i=0}^{\infty}\sum_{j=0}^{\infty} a_{ij}\tau^i\varphi^j(\psi),$$

where $\varphi : \mathbf{C} \to \Re$ is a bounded linear functional. Analogous to ordinary points for differential equations, we consider cases where

$$\lambda\psi(0) = \sum_{k=0}^{\infty} r_k\Theta^k(\psi)$$

and

$$\frac{1}{2}\psi^2(0)g^2(\psi) = \sum_{k=0}^{\infty} g_k\Theta^k(\psi).$$

The bounded linear functional Θ may be chosen so as to simplify the expressions for Ψ, $\lambda\psi(0)$, and $\frac{1}{2}\psi^2(0)g^2(\psi)$.

Using the series expression for Φ, we have

$$\partial_\tau\Phi(\tau, \psi) = \sum_{i=1}^{\infty}\sum_{j=0}^{\infty} a_{ij}i\tau^{i-1}\Theta^j(\psi),$$

$$\overline{D\Phi(\tau, \psi)}(\mathbf{1}_{\{0\}}) = \sum_{i=0}^{\infty}\sum_{j=1}^{\infty} a_{ij}\tau^i j\Theta^{j-1}(\psi)\Theta(\mathbf{1}_{\{0\}}),$$

$$\overline{D^2\Phi(\tau, \psi)}(\mathbf{1}_{\{0\}}, \mathbf{1}_{\{0\}}) = \sum_{i=0}^{\infty}\sum_{j=2}^{\infty} a_{ij}\tau^i j(j-1)\Theta^{j-2}(\psi)\Theta^2(\mathbf{1}_{\{0\}}),$$

and

$$\mathcal{S}(\Phi)(\tau, \psi) = \sum_{i=0}^{\infty}\sum_{j=1}^{\infty} a_{ij}\tau^i j\Theta^{j-1}(\psi)\mathcal{S}(\Theta)(\psi).$$

Therefore,

$$\sum_{i=1}^{\infty}\sum_{j=0}^{\infty} a_{ij} i \tau^{i-1} \Theta^j(\psi)$$

$$= \sum_{i=0}^{\infty}\sum_{j=1}^{\infty} a_{ij}\tau^i j \Theta^{j-1}(\psi)\mathcal{S}(\Theta)(\psi)$$

$$+ \left(\sum_{k=0}^{\infty} r_k \Theta^k(\psi)\right)\left(\sum_{i=0}^{\infty}\sum_{j=1}^{\infty} a_{ij}\tau^i j \Theta^{j-1}(\psi)\Theta(\mathbf{1}_{\{0\}})\right)$$

$$+ \left(\sum_{k=0}^{\infty} g_k \Theta^k(\psi)\right)\left(\sum_{i=0}^{\infty}\sum_{j=2}^{\infty} a_{ij}\tau^i j(j-1)\Theta^{j-2}(\psi))\Theta^2(\mathbf{1}_{\{0\}})\right),$$

$$\forall (\tau,\psi) \in [0,T) \times \mathbf{C}_+,$$

By reworking the indexes, we have

$$\sum_{i=0}^{\infty}\sum_{j=0}^{\infty} a_{i+1j}(i+1)\tau^i\Theta^j(\psi)$$

$$= \sum_{i=0}^{\infty}\sum_{j=0}^{\infty} a_{ij+1}\tau^i(j+1)\Theta^j(\psi)\mathcal{S}(\Theta)(\psi)$$

$$+ \left(\sum_{k=0}^{\infty} r_k \Theta^k(\psi)\right)\left(\sum_{i=0}^{\infty}\sum_{j=0}^{\infty} a_{ij+1}\tau^i(j+1)\Theta^j(\psi)\Theta(\mathbf{1}_{\{0\}})\right)$$

$$+ \left(\sum_{k=0}^{\infty} g_k \Theta^k(\psi)\right)\left(\sum_{i=0}^{\infty}\sum_{j=0}^{\infty} a_{ij+2}\tau^i(j+2)(j+1)\Theta^j(\psi))\Theta^2(\mathbf{1}_{\{0\}})\right)$$

$$\forall (\tau,\psi) \in [0,T) \times \mathbf{C}_+.$$

By collecting coefficients that go with $\tau^i\varphi^j(\psi)$, we have

$$(i+1)a_{i+1j} = (j+1)a_{ij+1}\Gamma(\varphi)(\psi)$$

$$+ \Big[(j+1)r_0 a_{ij+1} + jr_1 a_{ij} + (j-1)r_2 a_{ij-1} + \cdots$$

$$+ (j-k+1)r_k a_{ij-k+1} + \cdots + r_j a_{i1}\Big]\Theta(\mathbf{1}_{\{0\}})$$

$$+ \Big[(j+2)(j+1)g_0 a_{ij+2} + (j+1)jg_1 a_{ij+1}$$

$$+ j(j-1)g_2 a_{ij} + \cdots + (j-k+2)(j-k+1)g_k a_{ij-k+2}$$

$$+ \cdots + 2g_j a_{i2}\Big]\Theta^2(\mathbf{1}_{\{0\}}),$$

which when rearranged gives

$$(i+1)a_{i+1j} = \left[r_j\Theta(1_{\{0\}})\right]a_{i1} + \left[2r_{j-1}\Theta(1_{\{0\}}) + 2g_j\Theta^2(1_{\{0\}})\right]a_{i2}$$
$$+ \left[3r_{j-2}\Theta(1_0) + (3)(2)g_{j-1}\Theta^2(1_{\{0\}})\right]a_{i3} + \cdots$$
$$+ \left[kr_{j-k+1}\Theta(1_{\{0\}}) + k(k-1)g_{j-k+2}\Theta^2(1_{\{0\}})\right]a_{ik}$$
$$+ \cdots + \left[(j+1)r_0\Theta(1_{\{0\}}) + (j+1)(j)g_1\Theta^2(1_{\{0\}})\right.$$
$$\left. + (j+1)S(\Theta)(\psi)\right]a_{ij+1} + \left[(j+2)(j+1)g_0\Theta^2(1_{\{0\}})\right]a_{ij+2}$$

subject to

$$\Psi(\psi) = \sum_{j=0}^{\infty} a_{0j}\Theta^j(\psi). \qquad (6.49)$$

Our procedure for finding the a_{ij} is then to find $a_{0,j}$ using (6.49) and then to find the a_{1j} using

$$a_{1j} = b_{j1}a_{01} + b_{j2}a_{02} + \cdots + b_{jj+2}a_{0j+2},$$

and so on, with

$$a_{ij} = \frac{b_{j1}}{i}a_{i-1,1} + \frac{b_{j2}}{i}a_{i-1,2} + \cdots + \frac{b_{jj+2}}{i}a_{i-1,j+2}, \qquad (6.50)$$

where

$$b_{j1} = r_j\Theta(1_{\{0\}}),$$
$$b_{jk} = kr_{j-k+1}\Theta(1_{\{0\}}) + k(k-1)g_{j-k+2}\Theta^2(1_{\{0\}})$$

for $k \neq j+1, j+2$,

$$b_{jj+1} = (j+1)r_0\Theta(1_{\{0\}}) + (j+1)(j)g_1\Theta^2(1_{\{0\}}) + (j+1)S(\Theta)(\psi),$$

and

$$b_{jj+2} = (j+2)(j+1)g_0\Theta^2(1_{\{0\}}).$$

6.7.2 An Example

Consider

$$dS(t) = \lambda S(t)\,dt + \sigma S(t)\,dW(t).$$

Here, there is no hereditary structure (i.e., $r = 0$), so we get the classical Black-Scholes equation

$$\lambda V(t, S(t)) = \partial_t V(t, S(t)) + \lambda S(t)\partial_x V(t, S(t))$$
$$+ \frac{1}{2}\sigma^2 S^2(t)\partial_x^2 V(t, S(t)). \qquad (6.51)$$

Our modified problem becomes

$$\partial_\tau \Phi(\tau, S(\tau)) = \lambda S(\tau) \partial_x \Phi(\tau, S(\tau)) + \frac{1}{2} \sigma^2 S^2(\tau) \partial_x^2 \Phi(\tau, S(\tau)), \qquad (6.52)$$

and $\Phi(0, S(0)) = \Psi(S(0))$. We let $\Theta(\psi) = S(0)$. The series solution coefficients now take the form $a_{11} = b_{11} a_{01}$, $a_{12} = b_{22} a_{02}$, and so on, with $a_{1j} = b_{jj} a_{0j}$. In general,

$$a_{ij} = \frac{b_{jj}}{i} a_{i-1j}.$$

Here,

$$b_{jj} = j r_1 + j(j-1) g_2,$$

where $r_1 = \lambda$ and $g_2 = \frac{1}{2} \sigma^2$. Therefore,

$$a_{ij} = \frac{(b_{jj})^i}{i!} a_{0j},$$

with

$$V(t, S(t)) = e^{-\lambda(T-t)} \sum_{i=0}^{\infty} \sum_{j=0}^{\infty} a_{ij} (T-t)^i S(0)^j$$

$$= e^{-\lambda(T-t)} \sum_{i=0}^{\infty} \sum_{j=0}^{\infty} \frac{(b_{jj})^i}{i!} a_{0j} (T-t)^i S(0)^j.$$

6.7.3 Convergence of the Series

Another consideration is that of convergence of the series solution. Again, assume a solution of the form

$$\Phi(\tau, \psi) = \sum_{i=0}^{\infty} \sum_{j=0}^{\infty} a_{ij} \tau^i \Theta^j(\psi).$$

Since ψ is fixed for a given problem, we can simplify the notation by letting $x \equiv \Theta(\psi)$ and defining

$$u(t, x) = \sum_{i=0}^{\infty} \sum_{j=0}^{\infty} a_{ij} t^i x^j, \qquad (6.53)$$

with

$$u(0, x) = \Psi(x) = \sum_{j=0}^{\infty} a_{0j} x^j.$$

If we let

$$f_0^{(1)} = \mathcal{S}(\Theta)(\psi) + \Theta(1_{\{0\}}) r_0$$

and

$$f_k^{(1)} = \Theta(1_{\{0\}})r_k$$

for $k = 1, 2, \ldots$, and

$$f_k^{(2)} = \Theta^2(1_{\{0\}})g_k$$

for $k = 0, 1, 2, \ldots$, we can define

$$f^{(1)}(x) = \sum_{k=0}^{\infty} f_k^{(1)} x^k$$

and

$$f^{(2)}(x) = \sum_{k=0}^{\infty} f_k^{(2)} x^k$$

so that the equation to solve is

$$\frac{\partial u}{\partial t}(t, x) = f^{(1)}(x)\frac{\partial u}{\partial x}(t, x) + f^{(2)}(x)\frac{\partial^2 u}{\partial x^2}(t, x).$$

The functions $f^{(1)}$ and $f^{(2)}$ are clearly analytic, and so per the proof of the Cauchy-Kowalewski theorem found in Petrovsky [Pet91], there is a unique analytic solution of this partial differential equation.

A final consideration is the size of the region of convergence of the solution. The proof in [Pet91] uses the method of majorants, and so the region of convergence of (6.53) is at least as large as that of the solution of the majorizing problem. Let the series representing $f^{(1)}$ converge in $|x| \leq R_{f^{(1)}}$, the series representing $f^{(2)}$ converge for $|x| \leq R_{f^{(2)}}$, and the series representing Ψ converge for $|x| \leq R_{\Psi}$. Then

$$|f^{(1)}(x)| \leq \frac{M_{f^{(1)}}}{1 - \frac{x}{R_{f^{(1)}}}},$$

$$|f^{(2)}(x)| \leq \frac{M_{f^{(2)}}}{1 - \frac{x}{R_{f^{(2)}}}},$$

and

$$|\Psi(x)| \leq \frac{M_{\Psi}}{1 - \frac{x}{R_{\Psi}}}.$$

Let $M = \max\{M_{f^{(1)}}, M_{f^{(2)}}, M_{\Psi}\}$ and $a = \min\{R_{f^{(1)}}, R_{f^{(2)}}, R_{\Psi}\}$, then

$$\frac{M}{1 - \frac{x}{a}}$$

majorizes $f^{(1)}$, $f^{(2)}$, and λ. The proof in Petrovsky [Pet91] actually uses

$$A(z) = \frac{M}{1 - \frac{z}{a}},$$

where $z = \frac{t}{\alpha} + x$ and $0 < \alpha < 1$ is chosen so that

$$\alpha < \frac{1}{2A(z)}.$$

Therefore, the series solution (6.53) converges at least in the region where $|\frac{t}{\alpha} + x| \leq a$.

6.7.4 The Algorithm

Summarizing the above derivations, we have the following algorithm for computing a series solution $V : [0, T] \times \mathbf{C} \to \Re$, with the following terminal condition $V(T, \psi) = \Psi(\psi)$.

Step 1. Identify a bounded linear functional $\Theta \in \mathbf{C}^*$ such that the series expressions for Ψ, $\lambda\psi(0)$, and $\frac{1}{2}\psi^2(0)g^2(\psi)$ are straightforward to obtain.

Step 2. For $j = 0, 1, 2, \ldots$, obtain the coefficients a_{0j}, r_j, and g_j for Ψ, $\lambda\psi(0)$, and $\frac{1}{2}\psi^2(0)g^2(\psi)$.

Step 3. Compute $a_{i,j}$ according to (6.50).

Step 4. The series solution $V : [0, T] \times \mathbf{C}_+ \to \Re$ is then found from

$$V(t, \psi) = \sum_{i=0}^{\infty} \sum_{j=0}^{\infty} a_{ij}(T - t)^i \Theta^j(\psi).$$

6.8 Conclusions and Remarks

In this chapter, we have derived an infinite-dimensional Black-Scholes equation for the European option pricing problem and the infinite-dimensional HJBVI for the American option pricing problem under the (B, S)-market model given by (6.1) and (6.4). Both infinite-dimensional equations involved Fréchet derivatives as well as the infinitesimal generator of shift operators and, therefore, have characteristics that are distinctive from partial differential equations that appear in the literature. For both of the option types, it is concluded that the pricing function is the unique viscosity solution of the corresponding equations. A computational algorithm for the solution of the infinite-dimensional Black-Scholes equation is also obtained via a double sequence of polynomials of a certain bounded linear functional on a Banach space and the time variable.

7

Hereditary Portfolio Optimization

This chapter treats an infinite time horizon hereditary portfolio optimization problem in a financial market that consists of one *savings* account and one *stock* account. It is assumed that the *savings* account compounds continuously with a constant interest rate $\lambda > 0$ and the unit price process, $\{S(t), t \geq 0\}$, of the underlying *stock* follows a nonlinear stochastic hereditary differential equation (SHDE) (see (7.1)) with an infinite but fading memory. The main purpose of the *stock* account is to keep track of the inventories (i.e., the time instants and the base prices at which shares were purchased or short sold) of the underlying stock for the purpose of calculating the capital gains taxes and so forth. In the stock price dynamics, we assume that both $f(S_t)$ (the mean rate of return) and $g(S_t)$ (the volatility coefficient) depend on the entire history of stock prices S_t over the time interval $(-\infty, t]$ instead of just the current stock price $S(t)$ at time $t \geq 0$ alone. Within the solvency region $\mathbf{S}_\kappa$ (to be defined in (7.6)) and under the requirements of paying a fixed plus proportional transaction costs and capital gains taxes, the *investor* is allowed to consume from his *savings* account in accordance with a consumption rate process $C = \{C(t), t \geq 0\}$ and to make transactions between his *savings* and *stock* accounts according to a trading strategy $\mathcal{T} = \{(\tau(i), \zeta(i)), i = 1, 2, \ldots\}$, where $\tau(i), i = 0, 1, 2, \ldots$, denote the sequence of transaction times and $\xi(i)$ stands for quantities of transactions at time $\tau(i)$ (see Definition 7.1.7).

The *investor* will follow the following set of consumption, transaction, and taxation rules (Rules 7.1-7.6). Note that an action of the *investor* in the market is called a transaction if it involves trading of shares of the *stock*, such as buying and selling:

Rule 7.1. At the time of each transaction, the *investor* has to pay a transaction cost that consists of a fixed cost $\kappa > 0$ and a proportional transaction cost with the cost rate of $\mu \geq 0$ for both selling and buying shares of the *stock*. All of the purchases and sales of any number of stock shares will be considered one transaction if they are executed at the same time instant and therefore incurs only one fixed fee $\kappa > 0$ (and, of course, in addition to a proportional transaction cost).

Rule 7.2. Within the solvency region $\mathbf{S}_\kappa$, the *investor* is allowed to consume and to borrow money from his *savings* account for *stock* purchases. The investor can also sell and/or buyback at the current price shares of the *stock* he bought and/or short-sold at a previous time.

Rule 7.3. The proceeds for the sales of the *stock* minus the transaction costs and capital gains taxes will be deposited in his/her *savings* account and the purchases of stock shares together with the associated transaction costs and capital gains taxes (if short shares of the *stock* are bought back at a profit) will be financed from his *savings* account.

Rule 7.4. Without loss of generality it is assumed that the interest income in the *savings* account is tax-free by using the effective interest rate $\lambda > 0$, where the effective interest rate equals the interest rate paid by the bank minus the tax rate for the interest income.

Rule 7.5. At the time of a transaction (say $t \geq 0$), the *investor* is required to pay a capital gains tax (respectively, be paid a capital loss credit) in the amount that is proportional to the amount of profit (respectively, loss). A sale of stock shares is said to result in a profit if the current stock price $S(t)$ is higher than the base price $B(t)$ of the stock, and it is a loss otherwise. The base price $B(t)$ is defined to be the price at which the stock shares were previously bought or shortsold; that is, $B(t) = S(t - \tau(t))$, where $\tau(t) > 0$ is the time duration for which those shares (long or short) have been held at time t. The *investor* will also pay capital gains taxes (respectively, be paid capital loss credits) for the amount of profit (respectively, loss) by short-selling shares of the *stock* and then buying back the shares at a lower (respectively, higher) price at a later time. The tax will be paid (or the credit shall be given) at the buying-back time. Throughout, a negative amount of tax will be interpreted as a capital-loss credit. The capital gains tax and capital loss credit rates are assumed to be the same, as $\beta > 0$ for simplicity. Therefore, if $|m|$ ($m > 0$ stands for buying and $m < 0$ stands for selling) shares of the stock having the base price $B(t) = S(t - \tau(t))$ are traded at the current price $S(t)$, then the amount of tax due at the transaction time is given by

$$|m|\beta(S(t) - S(t - \tau(t))).$$

Rule 7.6. The tax and/or credit will not exceed all other gross proceeds and/or total costs of the stock shares, that is,

$$m(1 - \mu)S(t) \geq \beta m |S(t) - S(t - \tau(t))| \text{ if } m \geq 0$$

and

$$m(1 + \mu)S(t) \leq \beta m |S(t) - S(t - \tau(t))| \text{ if } m < 0,$$

where $m \in \Re$ denotes the number of shares of the stock traded, with $m \geq 0$ being the number of shares purchased and $m < 0$ being the number of shares of the stock sold.

Convention 7.7. Throughout, we assume that $\mu + \beta < 1$. This implies that the expenses (proportional transaction cost and tax) associated with a trade will not exceed the proceeds.

Under the above assumptions and Rules 7.1-7.6, the *investor's* objective is to seek an optimal consumption-trading strategy $(C^*, \mathcal{T}^*)$ in order to maximize the expected utility from the total discounted consumption over the infinite time horizon

$$E\left[\int_0^\infty e^{-\alpha t} \frac{C^\gamma(t)}{\gamma}\, dt\right],$$

where $\alpha > 0$ represents the discount rate and $0 < \gamma < 1$ represents the *investor's* risk aversion factor.

Due to the fixed plus proportional transaction costs and the hereditary nature of the stock dynamics and inventories, the problem will be formulated as a combination of a classical control (for consumptions) and an impulse control (for the transactions) problem in infinite dimensions. A combined classical-impulse control problem in finite dimensions is treated in Brekke and Øksendal [BØ98] for diffusion processes. In this chapter a Hamilton-Jocobi-Bellman quasi-variational inequality (HJBQVI) for the value function together with its boundary conditions are derived, and the verification theorem for the optimal investment-trading strategy is established. It is also shown that the value function is a viscosity solution of the HJBQVI (see HJBQVI (*) in Section 7.3.3 (D).

In recent years there has been an extensive amount of research on the optimal consumption-trading problems with proportional transaction costs (see, e.g., Akian et al. [AMS96], Akian et al. [AST01], Davis and Norman [DN90], Shreve and Soner [SS94], and references contained therein) and fixed plus proportional transaction costs (see, e.g., Øksendal and Sulem [ØS02]), all within the geometric Brownian motion financial market. In all these papers, the objective has been to maximize the expected utility from the total discounted or averaged consumption over the infinite time horizon without considering the issues of capital gains taxes (respectively, capital loss credits) when stock shares are sold at a profit (respectively, loss). In different contexts, the issues of capital gains taxes have been studied in Cadenillas and Pliska [CP99], Constantinides [Con83, Con84], Dammon and Spatt [DS96], Leland [Lel99], Garlappi et al. [GNS01], Tahar and Touzi [TT03], Demiguel and Uppal [DU04], and references contained therein. In particular, [Con83] and [Con84] considered the effect of capital gains taxes and capital loss credits on capital market equilibrium without consumption and transaction costs. These two papers illustrated that under some conditions, it may be more profitable to cut one's losses short and never to realize a gain because of capital loss credits and capital gains taxes as some conventional wisdom will suggest. In [CP99], the optimal transaction time problem with proportional transaction costs and capital gains taxes was considered in order to maximize the the long-run growth rate of the investment (or the so-called Kelley criterion), that is,

$$\lim_{t \to \infty} \frac{1}{t} E[\log V(t)],$$

where $V(t)$ is the value of the investment measured at time $t > 0$. This paper is quite different from what is presented in this chapter in that the unit price of the *stock* is described by a geometric Brownian motion, and all shares of the stock owned by the investor are to be sold at a chosen transaction time and all of its proceeds from the sale are to be used to purchase new shares of the *stock* immediately after the sale without consumption. Fortunately, due to the nature of the geometric Brownian motion market, the authors of that paper were able to obtain some explicit results.

In recent years, the interest in stock price dynamics described by stochastic delay equations has increased tremendously (see, e.g., Chang and Youree [CY99] and [CY07]). To the best of the author's knowledge, Chang [Cha07a, Cha07b] was the first to treat the optimal consumption-trading problem in which the hereditary nature of the stock price dynamics and the issue of capital gains taxes are taken into consideration. Due to drastically different nature of the problem and the techniques involved, the hereditary portfolio optimization problem with taxes and proportional transaction costs (i.e., $\kappa = 0$ and $\mu > 0$) remains to be solved. Much of the material presented in this chapter are taken from [Cha07a] and [Cha07b].

This chapter is organized as follows. The description of the stock price dynamics, the admissible consumption-trading strategies, and the formulation of the hereditary portfolio optimization problem are given in Section 7.1. In Section 7.2, the properties of the controlled state process are further explored and the corresponding infinite-dimensional Markovian solution of the price dynamics is investigated. Section 7.3 contains the derivations of the HJBQVI together with its boundary conditions (HJBQVI(*)) using a Bellman-type dynamic programming principle. The verification theorem for the optimal consumption-trading strategy and the proof that the value function is a viscosity solution of the HJBQVI(*) are contained in Sections 7.4 and 7.5, respectively.

Similar to the SHDE with bounded memory, we again use the following convention for systems with infinite but fading memory:

Convention 7.8. If $t \geq 0$ and $\phi : \Re \to \Re$ is a measurable function, define $\phi_t : (-\infty, 0] \to \Re$ by $\phi_t(\theta) = \phi(t + \theta)$, $\theta \in (-\infty, 0]$.

7.1 The Hereditary Portfolio Optimization Problem

This section is devoted to formulation of the hereditary portfolio optimization problem with capital gains taxes and a fixed plus proportional transaction costs.

7.1.1 Hereditary Price Structure with Unbounded Memory

Let $\rho : (-\infty, 0] \to [0, \infty)$ be the *influence function with relaxation property* that satisfies the following conditions.

Assumption 7.1.1 *The function* $\rho : (-\infty, 0] \to [0, \infty)$ *satisfies the following two conditions:*

1. *ρ is summable on $(-\infty, 0]$, that is,*

$$0 < \int_{-\infty}^{0} \rho(\theta) \, d\theta < \infty.$$

2. *For every $\lambda \leq 0$, one has*

$$\bar{K}(\lambda) = ess \sup_{\theta \in (-\infty, 0]} \frac{\rho(\theta + \lambda)}{\rho(\theta)} \leq \bar{K} < \infty,$$

$$\underline{K}(\lambda) = ess \sup_{\theta \in (-\infty, 0]} \frac{\rho(\theta)}{\rho(\theta + \lambda)} < \infty.$$

Under Assumption 7.1.1, it can be shown that ρ is essentially bounded and strictly positive on $(-\infty, 0]$. Furthermore,

$$\lim_{\theta \to -\infty} \theta \rho(\theta) = 0.$$

Examples of $\rho : (-\infty, 0] \to [0, \infty)$ that satisfy Assumption 7.1.1 are given in Section 2.4.1 of Chapter 2.

Let $\mathbf{M} = \Re \times L_\rho^2(-\infty, 0)$ (or simply $\Re \times L_\rho^2$ for short) be the history space of the stock price dynamics, where L_ρ^2 is the class of ρ-weighted Hilbert space of measurable functions $\phi : (-\infty, 0) \to \Re$ such that

$$\int_{-\infty}^{0} |\phi(\theta)|^2 \rho(\theta) \, d\theta < \infty.$$

Note that any constant function defined on $(-\infty, 0]$ is an element of $L_\rho^2(-\infty, 0)$.

For $t \in (-\infty, \infty)$, let $S(t)$ denote the unit price of the *stock* at time t. It is assumed that the unit *stock* price process $\{S(t), t \in (-\infty, \infty)\}$ satisfies the following SHDE with an unbounded but fading memory:

$$dS(t) = S(t)[f(S_t) \, dt + g(S_t) \, dW(t)], \quad t \geq 0. \tag{7.1}$$

In the above equation, the process $\{W(t), t \geq 0\}$ is a one-dimensional standard Brownian motion defined on a complete filtered probability space $(\Omega, \mathcal{F}, P; \mathbf{F})$, where $\mathbf{F} = \{\mathcal{F}(t), t \geq 0\}$ is the P-augmented natural filtration generated by the Brownian motion $\{W(t), t \geq 0\}$. Note that $f(S_t)$ and $g(S_t)$ in (7.1) represent respectively the *mean growth rate* and the *volatility rate* of the *stock* price at time $t \geq 0$. Note that the *stock* is said to have a hereditary price structure with infinite but fading memory because both the drift term $S(t)f(S_t)$ and the diffusion term $S(t)g(S_t)$ on the right-hand side of (7.1) explicitly depend on the entire past history prices $(S(t), S_t) \in [0, \infty) \times L^2_{\rho,+}$ in a weighted fashion by the function ρ satisfying Assumption 7.1.1. Note that we have used the following notation in the above:

$$L^2_{\rho,+} = \{\phi \in L^2_\rho \mid \phi(\theta) \geq 0, \quad \forall \theta \in (-\infty, 0)\}.$$

It is assumed for simplicity and to guarantee the existence and uniqueness of a strong solution $S(t), t \geq 0$ that the initial price function $(S(0), S_0) = (\psi(0), \psi) \in \Re_+ \times L^2_{\rho,+}$ is given and the functions $f, g : L^2_\rho \to [0, \infty)$ are continuous and satisfy the following Lipschitz and linear growth conditions (see, e.g., Sections 1.4 and 1.5 in Chapter 1 and Sections 2.4 and 2.5 in Chapter 2 for the theory of SHDEs with an unbounded but fading memory).

Assumption 7.1.2 *The functions f and g satisfy the following conditions:*

(A7.2.1). (Linear Growth Condition) There exists a constant $K_g > 0$ such that

$$0 \leq |\phi(0)f(\phi) + \phi(0)g(\phi)|$$
$$\leq K_{grow}\left(1 + \|(\phi(0), \phi)\|_{\mathbf{M}}\right), \quad \forall(\phi(0), \phi) \in \mathbf{M}.$$

(A7.2.2) (Lipschitz Condition) There exists a constant $K_{lip} > 0$ such that

$$|\phi(0)f(\phi) - \varphi(0)f(\varphi)| + |\phi(0)g(\phi) - \varphi(0)g(\varphi)|$$
$$\leq K_{lip}\|(\phi(0), \phi) - (\varphi(0), \varphi)\|_{\mathbf{M}}, \quad \forall(\phi(0), \phi), (\varphi(0), \varphi) \in \mathbf{M},$$

where $\|(\phi(0), \phi)\|_{\mathbf{M}}$ is the norm for the ρ-weighted Hilbert space $\mathbf{M}$ defined by

$$\|(\phi(0), \phi)\|_{\mathbf{M}} = \sqrt{|\phi(0)|^2 + \int_{-\infty}^0 |\phi(\theta)|^2 \rho(\theta)\, d\theta}.$$

(A7.2.3) (Upper and Lower Bounds) There exist positive constants α and σ such that

$$0 < \lambda < \underline{b} \leq f(\phi) \leq \overline{b} \quad and \quad 0 < \sigma \leq g(\phi); \; \forall \phi \in L^2_{\rho,+}.$$

Note that the lower bound of the *mean rate of return f* in Assumption (A7.2.3) is imposed to make sure that the *stock* account has a higher mean growth rate than the interest rate $\lambda > 0$ for the *savings* account. Otherwise, it will be more profitable and less risky for the *investor* to put all his money in the *savings* account for the purpose of optimizing the expected utility from the total consumption.

As a reminder, let us repeat the meaning of the *strong* solution process $\{S(t), t \in (-\infty, \infty)\}$ of (7.1) as follows.

Definition 7.1.3 *A process $\{S(t), t \in (-\infty, \infty)\}$ in $\Re$ is said to be a strong solution of (7.1) through the initial price function $(\psi(0), \psi) \in [0, \infty) \times L^2, \rho, +$ if it satisfies the following conditions:*

1. $S(t) = \psi(t), \quad \forall t \in (-\infty, 0]$.
2. $\{S(t), t \geq 0\}$ is **F**-adapted; that is, $S(t)$ is $\mathcal{F}(t)$-measurable for each $t \geq 0$.
3. The process $\{S(t), t \geq 0\}$ is continuous and satisfies the following stochastic integral equation P-a.s. for all $t \geq 0$:

$$S(t) = \psi(0) + \int_0^t S(s) f(S_s) \, ds + \int_0^t S(s) g(S_s) \, dW(s) \quad t \geq 0.$$

4. $\int_0^T |S(t) f(S_t)| \, dt < \infty$ P-a.s. for each $T > 0$.
5. $\int_0^T |S(t) g(S_t)|^2 \, dt$ P-a.s. for each $T > 0$.

Definition 7.1.4 *The strong solution $\{S(t), t \in (-\infty, \infty)\}$ of (7.1) through the initial datum $(\psi(0), \psi) \in [0, \infty) \times L^2_{\rho, +}$ is said to be strongly (or pathwise) unique if $\{\tilde{S}(t), t \in (-\infty, \infty)\}$ is another strong solution of (7.1) through the same initial function. Then*

$$P\{S(t) = \tilde{S}(t), \quad \forall t \geq 0\} = 1.$$

Under Assumption 7.1.1 and Assumptions (A7.2.1)-(A7.2.3), it can be shown that for each initial historical price function $(\psi(0), \psi) \in [0, \infty) \times L^2_{\rho, +}$, the price process $\{S(t), t \geq 0\}$ exists and is a positive, continuous, and **F**-adapted process defined on $(\Omega, \mathcal{F}, P; \mathbf{F})$.

Theorem 7.1.5 *Assume that Assumption 7.1.1 and Assumptions (A7.2.1)-(A7.2.3) hold. Then, for each $(\psi(0), \psi) \in [0, \infty) \times L^2_{\rho, +}$, there exists a unique non-negative strong solution process $\{S(t), t \in (-\infty, \infty)\}$ through the initial datum $(\psi(0), \psi)$. Furthermore, the $[0, \infty) \times L^2_{\rho, +}$-valued segment process $\{(S(t), S_t), t \geq 0\}$ is strong Markovian with respect to the filtration $\mathbf{G}$, where $\mathbf{G} = \{\mathcal{G}(t), t \geq 0\}$ is the filtration generated by $\{S(t), t \geq 0\}$, that is,*

$$\mathcal{G}(t) = \sigma(S(s), 0 \leq s \leq t)(= \sigma((S(s), S_s), 0 \leq s \leq t)), \forall t \geq 0.$$

We also note here that since security exchanges have only existed in a finite past, it is realistic but not technically required to assume that the initial historical price function $(\psi(0), \psi)$ has the property that

$$\psi(\theta) = 0 \ \forall \theta \leq \bar{\theta} < 0 \text{ for some } \bar{\theta} < 0.$$

Although the modeling of *stock* prices is still under intensive investigations, it is not the intention to address the validity of the model stock price dynamics described in (7.1) but to illustrate an hereditary optimization problem that is explicitly dependent on the entire past history of the stock prices for computing capital gains taxes or capital loss credits. The term "hereditary portfolio optimization" was coined by Chang [Cha07a, Cha07b] for the first time. We, however, mention here that stochastic hereditary equation similar to (7.1) was first used to model the behavior of elastic material with infinite memory and that some special form of stochastic functional differential equations with bounded memory have been used to model stock price dynamics in option pricing problems (see, e.g., Chang and Youree [CY99, CY07]).

7.1.2 The Stock Inventory Space

The space of stock inventories, $\mathbf{N}$, will be the space of bounded functions $\xi : (-\infty, 0] \to \Re$ of the following form:

$$\xi(\theta) = \sum_{k=0}^{\infty} n(-k)\mathbf{1}_{\{\tau(-k)\}}(\theta), \ \theta \in (-\infty, 0], \tag{7.2}$$

where $\{n(-k), k = 0, 1, 2, \ldots\}$ is a sequence in $\Re$ with $n(-k) = 0$ for all but finitely many k,

$$-\infty < \cdots < \tau(-k) < \cdots < \tau(-1) < \tau(0) = 0,$$

and $\mathbf{1}_{\{\tau(-k)\}}$ is the indicator function at $\tau(-k)$. Let $\| \cdot \|_N$ (the norm of the space $\mathbf{N}$) be defined by

$$\|\xi\|_N = \sup_{\theta \in (-\infty, 0]} |\xi(\theta)|, \ \forall \xi \in \mathbf{N}.$$

As illustrated in Sections 7.1.3 and 7.1.5, $\mathbf{N}$ is the space in which the *investor's* stock inventory lives. The assumption that $n(-k) = 0$ for all but finitely many k implies that the *investor* can only have finitely many open positions in his stock account. However, the number of open positions may increase from time to time. Note that the *investor* is said to have an open long (respectively, short) position at time τ if he still owns (respectively, owes) all or part of the stock shares that were originally purchased (respectively, shortsold) at a previous time τ. The only way to close a position is to sell what he owns and buy back what he owes.

Remark 7.1.6 *The inventory at time $t = 0$ described in (7.2) can also be equivalently expressed as the following double sequence:*

$$\xi = \{(n(-k), \tau(-k)), k = 0, 1, 2, \ldots\},$$

where $n(-k) > 0$ (respectively, $n(-k) < 0$) denotes the number of share of the stock purchased (respectively, shortsold) by the investor at time $\tau(-k)$ for $k = 0, 1, 2, \ldots$.

If $\eta : \Re \to \Re$ is a bounded function of the form

$$\eta(t) = \sum_{k=-\infty}^{\infty} n(k)\mathbf{1}_{\{\tau(k)\}}(t), \ -\infty < t < \infty,$$

where

$$-\infty < \cdots < \tau(-k) < \cdots < 0 = \tau(0) < \tau(1) < \cdots < \tau(k) < \cdots < \infty,$$

then for each $t \geq 0$, we define, using Convention 7.8, the function $\eta_t :$ $(-\infty, 0] \to \Re$ by

$$\eta_t(\theta) = \eta(t + \theta), \ \theta \in (-\infty, 0].$$

In this case,

$$\eta_t(\theta) = \sum_{k=-\infty}^{\infty} n(k) \mathbf{1}_{\{\tau(k)\}}(t + \theta)$$

$$= \sum_{k=-\infty}^{Q(t)} n(k) \mathbf{1}_{\{\tau(k)\}}(\theta), \ \theta \in (-\infty, 0],$$

where $Q(t) = \sup\{k \geq 0 \mid \tau(k) \leq t\}$. Note that if η_t represents the inventory of the investor's stock account, then η_t can also be expressed as the following double sequence (see Remark 7.1.6):

$$\eta_t = \{(n(k), \tau(k)), k = \ldots, -2, -1, 0, 1, 2, \ldots, Q(t)\}.$$

7.1.3 Consumption-Trading Strategies

Let $(X(0-), N_{0-}, S(0), S_0) = (x, \xi, \psi(0), \psi) \in \Re \times \mathbf{N} \times [0, \infty) \times L^2_{\rho,+}$ be the *investor's* initial portfolio immediately prior to $t = 0$; that is, the investor starts with $x \in \Re$ dollars in his *savings* account, the initial stock inventory,

$$\xi(\theta) = \sum_{k=0}^{\infty} n(-k) \mathbf{1}_{\{\tau(-k)\}}(\theta), \ \theta \in (-\infty, 0),$$

and the initial profile of historical stock prices $(\psi(0), \psi) \in [0, \infty) \times L^2_{\rho,+}$, where $n(-k) > 0$ (respectively, $n(-k) < 0$) represents an open long (respectively, short) position at $\tau(-k)$. Within the *solvency region* $\mathbf{S}_\kappa$ (see (7.6)), the *investor* is allowed to consume from his *savings* account and can make transactions between his *savings* and *stock* accounts under Rules 7.1-7.6 and according to a consumption-trading strategy $\pi = (C, T)$ defined below.

Definition 7.1.7 *The pair* $\pi = (C, T)$ *is said to be a consumption-trading strategy if the following hold:*
(i) The consumption rate process $C = \{C(t), t \geq 0\}$ *is a non-negative* **G**-*progressively measurable process such that*

$$\int_0^T C(t) \, dt < \infty, \ P\text{-}a.s. \ \forall T > 0.$$

(ii) $T = \{(\tau(i), \zeta(i)), i = 1, 2, \ldots\}$ *is a trading strategy with* $\tau(i), i = 1, 2, \ldots,$ *being a sequence of trading times that are* **G**-*stopping times such that*

$$0 = \tau(0) \leq \tau(1) < \cdots < \tau(i) < \cdots$$

and

$$\lim_{i \to \infty} \tau(i) = \infty \ P\text{-}a.s. \ ,$$

and for each $i = 0, 1, \ldots,$

$$\zeta(i) = (\ldots, m(i-k), \ldots, m(i-2), m(i-1), m(i))$$

is an $\mathbf{N}$-valued $\mathcal{G}(\tau(i))$-measurable random vector (instead of a random variable in $\Re$) that represents the trading quantities at the trading time $\tau(i)$.

In the above, $m(i) > 0$ (respectively, $m(i) < 0$) is the number of stock shares newly purchased (respectively, short-sold) at the current time $\tau(i)$ and at the current price of $S(\tau(i))$ and, for $k = 1, 2, \ldots$, $m(i-k) > 0$ (respectively, $m(i-k) < 0$) is the number of stock shares bought back (respectively, sold) at the current time $\tau(i)$ and the current price of $S(\tau(i))$ in his open short (respectively, long) position at the previous time $\tau(i-k)$ and the base price of $S(\tau(i-k))$.

For each stock inventory ξ of the form expressed (7.2), Rules 7.1-7.6 also dictate that the investor can purchase or short sell new shares (i.e., $-\infty < m(0) < \infty$), can sell all or part of what he/she owns, that is,

$$m(-k) \leq 0 \text{ and } n(-k) + m(-k) \geq 0 \text{ if } n(-k) > 0,$$

and/or can buy back all or part what he/she owes, that is,

$$m(-k) \geq 0 \text{ and } n(-k) + m(-k) \leq 0 \text{ if } n(-k) < 0,$$

all at the same time instant. Therefore, the trading quantity $\{m(-k), k = 0, 1, \ldots\}$ must satisfy the constraint set $\mathcal{R}(\xi) \subset \mathbf{N}$ defined by

$$\mathcal{R}(\xi) = \{\zeta \in \mathbf{N} \mid \zeta = \sum_{k=0}^{\infty} m(-k) \mathbf{1}_{\{\tau(-k)\}}, -\infty < m(0) < \infty, \text{ and}$$

$$\text{either } n(-k) > 0, m(-k) \leq 0, \text{ and } n(-k) + m(-k) \geq 0 \text{ or}$$
$$n(-k) < 0, m(-k) \geq 0, \text{ and } (-k) + m(-k) \leq 0 \text{ for } k \geq 1\}. \ (7.3)$$

7.1.4 Solvency Region

Throughout, the investor's state space $\mathbf{S}$ is taken to be $\mathbf{S} = \Re \times \mathbf{N} \times [0, \infty) \times L^2_{\rho,+}$. An element $(x, \xi, \psi(0), \psi) \in \mathbf{S}$ is called a portfolio, where $x \in \Re$ is investor's holding in his *savings* account, ξ is the investor's stock inventory, and $(\psi(0), \psi) \in [0, \infty) \times L^2_{\rho,+}$ is the profile of historical stock prices. Define the function $H_\kappa : \mathbf{S} \to \Re$ as follows:

$$H_\kappa(x, \xi, \psi(0), \psi) = \max \Big\{ G_\kappa(x, \xi, \psi(0), \psi),$$

$$\min\{x, n(-k), k = 0, 1, 2, \cdots\} \Big\}, \qquad (7.4)$$

where $G_\kappa : \mathbf{S} \to \Re$ is the liquidating function defined by

$$G_\kappa(x, \xi, \psi(0), \psi) = x - \kappa$$
$$+ \sum_{k=0}^{\infty} \Big[\min\{(1 - \mu)n(-k), (1 + \mu)n(-k)\}\psi(0)$$
$$- n(-k)\beta(\psi(0) - \psi(\tau(-k)))\Big]. \tag{7.5}$$

On the right-hand side of the above expression,

$$x - \kappa = \text{ the amount in his } \textit{savings} \text{ account after}$$
$$\text{deducting the fixed transaction cost } \kappa,$$

and for each $k = 0, 1, \ldots$

$$\min\{(1 - \mu)n(-k), (1 + \mu)n(-k)\}\psi(0)$$
$$= \text{ the proceeds for selling } n(-k) > 0 \text{ or buying back } n(-k) < 0$$
$$\text{shares of the stock net of proportional transactional cost;}$$

$$-n(-k)\beta(\psi(0) - \psi(\tau(-k)))$$
$$= \text{ the capital gains tax to be paid forselling the } n(-k) \text{ shares of the stock}$$
$$\text{with the current price of } \psi(0) \text{ and base price of } \psi(\tau(-k)).$$

Therefore, $G_\kappa(x, \xi, \psi(0), \psi)$ defined in (7.5) represents the cash value (if the assets can be liquidated at all) after closing all open positions and paying all transaction costs (fixed plus proportional transactional costs) and taxes.

The *solvency region* $\mathbf{S}_\kappa$ of the portfolio optimization problem is defined as

$$\mathbf{S}_\kappa = \Big\{(x, \xi, \psi(0), \psi) \in \mathbf{S} \mid H_\kappa(x, \xi, \psi(0), \psi) \geq 0\Big\}$$
$$= \{(x, \xi, \psi(0), \psi) \in \mathbf{S} \mid G_\kappa(x, \xi, \psi(0), \psi) \geq 0\} \cup \mathbf{S}_+, \tag{7.6}$$

where $\mathbf{S}_+ = [0, \infty) \times \mathbf{N}_+ \times [0, \infty) \times L^2_{\rho,+}$, and $\mathbf{N}_+ = \{\xi \in \mathbf{N} \mid \xi(\theta) \geq 0, \forall \theta \in (-\infty, 0]\}$.

Note that within the *solvency region* $\mathbf{S}_\kappa$, there are positions that cannot be closed at all, namely those $(x, \xi, \psi(0), \psi) \in \mathbf{S}_\kappa$ such that

$$(x, \xi, \psi(0), \psi) \in \mathbf{S}_+ \text{ and } G_\kappa(x, \xi, \psi(0), \psi) < 0.$$

This is due to the insufficiency of funds to pay for the transaction costs and/or taxes and so forth. Observe that the *solvency* region $\mathbf{S}_\kappa$ is an unbounded and nonconvex subset of the state space $\mathbf{S}$. The boundary $\partial \mathbf{S}_\kappa$ will be described in detail in Section 7.3.3.

7.1.5 Portfolio Dynamics and Admissible Strategies

At time $t \geq 0$, the investor's portfolio in the financial market will be denoted by the quadruplet $(X(t), N_t, S(t), S_t)$, where $X(t)$ denotes the *investor's* holdings in his *savings* account, $N_t \in \mathbf{N}$ is the *inventory* of his *stock* account, and $(S(t), S_t)$ describes the profile of the unit prices of the *stock* over the past history $(-\infty, t]$, as described in Section 7.1.1.

Given the initial portfolio

$$(X(0-), N_{0-}, S(0), S_0) = (x, \xi, \psi(0), \psi) \in \mathbf{S}$$

and applying a consumption-trading strategy $\pi = (C, \mathcal{T})$ (see Definition 7.1.7), the portfolio dynamics of $\{Z(t) = (X(t), N_t, S(t), S_t), t \geq 0\}$ can then be described as follows.

First, the *savings* account holdings $\{X(t), t \geq 0\}$ satisfies the following differential equation between the trading times:

$$dX(t) = [\lambda X(t) - C(t)]dt, \quad \tau(i) \leq t < \tau(i+1), \quad i = 0, 1, 2, \ldots, \quad (7.7)$$

and the following jumped quantity at the trading time $\tau(i)$:

$$
\begin{aligned}
X(\tau(i)) = X(\tau(i)-) &- \kappa - \sum_{k=0}^{\infty} m(i-k)\Big[(1-\mu)S(\tau(i)) \\
&- \beta(S(\tau(i)) - S(\tau(i-k)))\Big]\mathbf{1}_{\{n(i-k)>0, -n(i-k)\leq m(i-k)\leq 0\}} \\
&- \sum_{k=0}^{\infty} m(i-k)\Big[(1+\mu)S(\tau(i)) \\
&- \beta(S(\tau(i)) - S(\tau(i-k)))\Big]\mathbf{1}_{\{n(i-k)<0, 0\leq m(i-k)\leq -n(i-k)\}}.
\end{aligned}
$$
$$(7.8)$$

As a reminder, $m(i) > 0$ (respectively, $m(i) < 0$) means buying (respectively, selling) new stock shares at $\tau(i)$, and $m(i-k) > 0$ (respectively, $m(i-k) < 0$) means buying back (respectively, selling) some or all of what he owed (respectively, owned).

Second, the inventory of the *investor's* stock account at time $t \geq 0$, $N_t \in \mathbf{N}$, does not change between the trading times and can be expressed as

$$N_t = N_{\tau(i)} = \sum_{k=-\infty}^{Q(t)} n(k)\mathbf{1}_{\tau(k)} \quad \text{if } \tau(i) \leq t < \tau(i+1), \quad i = 0, 1, 2 \ldots, \quad (7.9)$$

where $Q(t) = \sup\{k \geq 0 \mid \tau(k) \leq t\}$. It has the following jumped quantity at the trading time $\tau(i)$:

$$N_{\tau(i)} = N_{\tau(i)-} \oplus \zeta(i), \quad (7.10)$$

where $N_{\tau(i)-} \oplus \zeta(i) : (-\infty, 0] \to \mathbf{N}$ is defined by

$$(N_{\tau(i)-} \oplus \zeta(i))(\theta) = \sum_{k=0}^{\infty} \hat{n}(i-k)\mathbf{1}_{\{\tau(i-k)\}}(\tau(i)+\theta) \tag{7.11}$$

$$= m(i)\mathbf{1}_{\{\tau(i)\}}(\tau(i)+\theta) + \sum_{k=1}^{\infty}\Big[n(i-k)$$

$$+ m(i-k)(\mathbf{1}_{\{n(i-k)<0,0\leq m(i-k)\leq -n(i-k)\}}$$

$$+ \mathbf{1}_{\{n(i-k)>0,-n(i-k)\leq m(i-k)\leq 0\}})\Big]\mathbf{1}_{\{\tau(i-k)\}}(\tau(i)+\theta),$$

for $\theta \in (-\infty, 0]$.

Third, since the *investor* is small, the unit stock price process $\{S(t), t \geq 0\}$ will not be in anyway affected by the *investor's* action in the market and is again described as in (7.1).

Definition 7.1.8 *If the investor starts with an initial portfolio*

$$(X(0-), N_{0-}, S(0), S_0) = (x, \xi, \psi(0), \psi) \in \mathbf{S}_\kappa,$$

the consumption-trading strategy $\pi = (C, \mathcal{T})$ *defined in Definition 7.1.7 is said to be admissible at* $(x, \xi, \psi(0), \psi)$ *if*

$$\zeta(i) \in \mathcal{R}(N_{\tau(i)-}), \quad \forall i = 1, 2, \ldots$$

and

$$(X(t), N_t, S(t), S_t) \in \mathbf{S}_\kappa, \quad \forall t \geq 0.$$

The class of consumption-investment strategies admissible at $(x, \xi, \psi(0), \psi) \in \mathbf{S}_\kappa$ *will be denoted by* $\mathcal{U}_\kappa(x, \xi, \psi(0), \psi)$.

7.1.6 The Problem Statement

Given the initial state $(X(0-), N_{0-}, S(0), S_0) = (x, \xi, \psi(0), \psi) \in \mathbf{S}_\kappa$, the *investor's* objective is to find an admissible consumption-trading strategy $\pi^* \in \mathcal{U}_\kappa(x, \xi, \psi(0), \psi)$ that maximizes the following expected utility from the total discounted consumption:

$$J_\kappa(x, \xi, \psi(0), \psi; \pi) = E^{x,\xi,\psi(0),\psi;\pi}\Big[\int_0^{\infty} e^{-\alpha t}\frac{C^\gamma(t)}{\gamma}dt\Big] \tag{7.12}$$

among the class of admissible consumption-trading strategies $\mathcal{U}_\kappa(x, \xi, \psi(0), \psi)$, where $E^{x,\xi,\psi(0),\psi;\pi}[\cdots]$ is the expectation with respect to $P^{x,\xi,\psi(0),\psi;\pi}\{\ldots\}$, the probability measure induced by the controlled (by π) state process $\{(X(t), N_t, S(t), S_t), t \geq 0\}$ and conditioned on the initial state

$$(X(0-), N_{0-}, S(0), S_0) = (x, \xi, \psi(0), \psi).$$

In the above, $\alpha > 0$ denotes the discount factor, and $0 < \gamma < 1$ indicates that the utility function $U(c) = \frac{c^\gamma}{\gamma}$, for $c > 0$, is a function of HARA (hyperbolic absolute risk aversion) type that were considered in most optimal consumption-trading literature (see, e.g., Davis and Norman [DN90], Akian et al. [AMS96], Akian et al. [AST01], Shreves and Soner [SS94], and Øksendal and Sulem [ØS02]) with or without a fixed transaction cost. The admissible (consumption-trading) strategy $\pi^* \in \mathcal{U}_\kappa(x, \xi, \psi(0), \psi)$ that maximizes $J_\kappa(x, \xi, \psi(0), \psi; \pi)$ is called an optimal (consumption-trading) strategy and the function $V_\kappa : \mathbf{S}_\kappa \to \Re_+$ defined by

$$V_\kappa(x, \xi, \psi(0), \psi) = \sup_{\pi \in \mathcal{U}_\kappa(x, \xi, \psi(0), \psi)} J_\kappa(x, \xi, \psi(0), \psi; \pi)$$
$$= J_\kappa(x, \xi, \psi(0), \psi; \pi^*) \tag{7.13}$$

is called the value function of the hereditary portfolio optimization problem.

The hereditary portfolio optimization problem considered in this chapter is then formalized as the following combined classical-impulse control problem.

Problem (HPOP). For each given initial state $(x, \xi, \psi(0), \psi) \in \mathbf{S}_\kappa$, identify the optimal strategy π^* and its corresponding value function $V_\kappa : \mathbf{S}_\kappa \to [0, \infty)$.

7.2 The Controlled State Process

Given an initial state $(x, \xi, \psi(0), \psi) \in \mathbf{S}_\kappa$ and an admissible consumption-investment strategy $\pi = (C, \mathcal{T}) \in \mathcal{U}_\kappa(x, \xi, \psi(0), \psi)$, the $\mathbf{S}_\kappa$-valued controlled state process will be denoted by $\{Z(t) = (X(t), N_t, S(t), S_t), t \geq 0\}$. Note that the dependence of the controlled state process on the initial state $(x, \xi, \psi(0), \psi)$ and the admissible consumption-trading strategy π will be suppressed for notational simplicity.

The main purpose of this section is to establish the Markovian and Dynkin's formula for the controlled state process $\{Z(t), t \geq 0\}$. Note that the $[0, \infty) \times L^2_{\rho,+}(= \mathbf{M}_+)$-valued process $\{S(t), S_t), t \geq 0\}$ described by (7.1) is uncontrollable by the *investor* and is therefore independent of the consumption-trading strategy $\pi \in \mathcal{U}_\kappa(x, \xi, \psi(0), \psi)$ but is dependent on the initial historical price function $(S(0), S_0) = (\psi(0), \psi) \in [0, \infty) \times L^2_{\rho,+}$.

7.2.1 The Properties of the Stock Prices

To study the Markovian properties of the $[0, \infty) \times L^2_{\rho,+}$-valued solution process $\{(S(t), S_t), t \geq 0\}$, where $S_t(\theta) = S(t + \theta)$, $\theta \in (-\infty, 0]$, and $(S(0), S_0) = (\psi(0), \psi)$, we need the following notation and ancillary results.

Let $\mathbf{M}^*$ be the space of bounded linear functionals (or the topological dual of the space $\mathbf{M}$) equipped with the operator norm $\| \cdot \|_{\mathbf{M}}^*$ defined by

$$\|\Phi\|_{\mathbf{M}}^* = \sup_{(\phi(0), \phi) \neq (0, 0)} \frac{|\Phi(\phi(0), \phi)|}{\|(\phi(0), \phi)\|_{\mathbf{M}}}, \quad \Phi \in \mathbf{M}^*.$$

Note that $\mathbf{M}^*$ can be identified with $\mathbf{M} = \Re \times L^2_\rho$ by the well-known Riesz representation theorem.

Let $\mathbf{M}^\dagger$ be the space of bounded bilinear functionals $\Phi : \mathbf{M} \times \mathbf{M} \to \Re$; that is, $\Phi((\phi(0),\phi),(\cdot,\cdot)), \Phi((\cdot,\cdot),(\phi(0),\phi)) \in \mathbf{M}^*$ for each $(\phi(0),\phi) \in \mathbf{M}$, equipped with the operator norm $\|\cdot\|_{\mathbf{M}}^\dagger$ defined by

$$\|\Phi\|_{\mathbf{M}}^\dagger = \sup_{(\phi(0),\phi)\neq(0,0)} \frac{\|\Phi((\cdot,\cdot),(\phi(0),\phi))\|_{\mathbf{M}}^*}{\|(\phi(0),\phi)\|_{\mathbf{M}}}$$
$$= \sup_{(\phi(0),\phi)\neq(0,0)} \frac{\|\Phi((\phi(0),\phi),(\cdot,\cdot))\|_{\mathbf{M}}^*}{\|(\phi(0),\phi)\|_{\mathbf{M}}}.$$

Let $\Phi : \mathbf{M} \to \Re$. The function Φ is said to be Fréchet differentiable at $(\phi(0),\phi) \in \mathbf{M}$ if for each $(\varphi(0),\varphi) \in \mathbf{M}$,

$$\Phi((\phi(0),\phi)+(\varphi(0),\varphi)) - \Phi(\phi(0),\phi) = D\Phi(\phi(0),\phi)(\varphi(0),\varphi)+o(\|(\varphi(0),\varphi)\|_{\mathbf{M}}),$$

where $D\Phi : \mathbf{M} \to \mathbf{M}^*$ and $o : \Re \to \Re$ is a function such that

$$\frac{o(\|(\varphi(0),\varphi)\|_{\mathbf{M}})}{\|(\varphi(0),\varphi)\|_{\mathbf{M}}} \to 0 \quad \text{as} \quad \|(\varphi(0),\varphi)\|_{\mathbf{M}} \to 0.$$

In this case, $D\Phi(\phi(0),\phi) \in \mathbf{M}^*$ is called the (first-order) Fréchet derivative of Φ at $(\phi(0),\phi) \in \mathbf{M}$. The function Φ is said to be continuously Fréchet differentiable if its Fréchet derivative $D\Phi : \mathbf{M} \to \mathbf{M}^*$ is continuous under the operator norm $\|\cdot\|_{\mathbf{M}}^*$. The function Φ is said to be twice Fréchet differentiable at $(\phi(0),\phi) \in \mathbf{M}$ if its Fréchet derivative $D\Phi(\phi(0),\phi) : \mathbf{M} \to \Re$ exists and there exists a bounded bilinear functional $D^2\Phi(\phi(0),\phi) : \mathbf{M} \times \mathbf{M} \to \Re$, where for each $(\varphi(0),\varphi),(\varsigma(0),\varsigma) \in \mathbf{M}$,

$$D^2\Phi(\phi(0),\phi)((\cdot,\cdot),(\varphi(0),\varphi)), D^2\Phi(\phi(0),\phi)((\varsigma(0),\varsigma),(\cdot,\cdot)) \in \mathbf{M}^*$$

and where

$$\Big(D\Phi((\phi(0),\phi)+(\varphi(0),\varphi)) - D\Phi(\phi(0),\phi)\Big)(\varsigma(0),\varsigma)$$
$$= D^2\Phi(\phi(0),\phi)((\varsigma(0),\varsigma),(\varphi(0),\varphi)) + o(\|(\varsigma(0),\varsigma)\|_{\mathbf{M}},\|(\varphi(0),\varphi)\|_{\mathbf{M}}).$$

Here, $o : \Re \times \Re \to \Re$ is such that

$$\frac{o(\cdot,\|(\varphi(0),\varphi)\|_{\mathbf{M}})}{\|(\varphi(0),\varphi)\|_{\mathbf{M}}} \to 0 \quad \text{as} \quad \|(\varphi(0),\varphi)\|_{\mathbf{M}} \to 0$$

and

$$\frac{o(\|(\varphi(0),\varphi)\|_{\mathbf{M}},\cdot)}{\|(\varphi(0),\varphi)\|_{\mathbf{M}}} \to 0 \quad \text{as} \quad \|(\varphi(0),\varphi)\|_{\mathbf{M}} \to 0.$$

In this case, the bounded bilinear functional $D^2\Phi(\phi(0),\phi) : \mathbf{M} \times \mathbf{M} \to \Re$ is the second-order Fréchet derivative of Φ at $(\phi(0),\phi) \in \mathbf{M}$.

The second-order Fréchet derivative $D^2\Phi$ is said to be globally Lipschitz on $\mathbf{M}$ if there exists a constant $K > 0$ such that

$$\|D^2\Phi(\phi(0),\phi) - D^2\Phi(\varphi(0),\varphi)\|_{\mathbf{M}}^{\dagger} \leq K\|(\phi(0),\phi) - (\varphi(0),\varphi)\|_{\mathbf{M}},$$

$$\forall(\phi(0),\phi),(\varphi(0),\varphi) \in \mathbf{M}.$$

Assuming all the partial and/or Fréchet derivatives of the following exist, the actions of the first-order Fréchet derivative $D\Phi(\phi(0),\phi)$ and the second-order Fréchet $D^2\Phi(\phi(0),\phi)$ can be expressed as

$$D\Phi(\phi(0),\phi)(\varphi(0),\varphi) = \varphi(0)\partial_{\phi(0)}\Phi(\phi(0),\phi) + D_\phi\Phi(\phi(0),\phi)\varphi$$

and

$$D^2\Phi(\phi(0),\phi)((\varphi(0),\varphi),(\varsigma(0),\varsigma))$$
$$= \varphi(0)\partial^2_{\phi(0)}\Phi(\phi(0),\phi)\varsigma(0) + \varsigma(0)\partial_{\phi(0)}D_\phi\Phi(\phi(0),\phi)\varphi$$
$$+ \varphi(0)D_\phi\partial_{\phi(0)}\Phi(\phi(0),\phi)(\varphi,\varsigma) + D^2_\phi\Phi(\phi(0),\phi)\varsigma,$$

where $\partial_{\phi(0)}\Phi$ and $\partial^2_{\phi(0)}\Phi$ are the first- and second-order partial derivatives of Φ with respect to its first variable $\phi(0) \in \Re$, $D_\phi\Phi$ and $D^2_\phi\Phi$ are the first- and second-order Fréchet derivatives with respect to its second variable $\phi \in L^2_\rho$, $\partial_{\phi(0)}D_\phi\Phi$ is the second-order derivative first with respect to ϕ in the Fréchet sense and then with respect to $\phi(0)$ and so forth.

Let $C^{2,2}(\Re \times L^2_\rho) \equiv C^2(\mathbf{M})$ be the space of functions $\Phi : \Re \times L^2_\rho(\equiv \mathbf{M}) \to \Re$ that are twice continuously differentiable with respect to both its first and second variable. The space of $\Phi \in C^{2,2}(\Re \times L^2_\rho)(= C^2(\mathbf{M}))$ with $D^2\Phi$ being globally Lipschitz will be denoted by $C^{2,2}_{lip}(\Re \times L^2_\rho)$ (or equivalently $C^2_{lip}(\mathbf{M})$).

The Weak Infinitesimal Generator $\mathcal{S}$

For each $(\phi(0),\phi) \in \Re \times L^2_\rho$, define $\tilde{\phi} : \Re \to \Re$ by

$$\tilde{\phi}(t) = \begin{cases} \phi(0) & \text{for } t \in [0,\infty) \\ \phi(t) & \text{for } t \in (-\infty,0). \end{cases}$$

Then for each $\theta \in (-\infty,0]$ and $t \in [0,\infty)$,

$$\tilde{\phi}_t(\theta) = \tilde{\phi}(t+\theta) = \begin{cases} \phi(0) & \text{for } t+\theta \geq 0 \\ \phi(t+\theta) & \text{for } t+\theta < 0. \end{cases}$$

A bounded measurable function $\Phi : \Re \times L^2_\rho \to \Re$ (i.e., $\Phi \in C_b(\Re \times L^2_\rho)$), is said to belong to $\mathcal{D}(\mathcal{S})$, the domain of the weak infinitesimal operator $\mathcal{S}$, if the following limit exists for each fixed $(\phi(0),\phi) \in \Re \times L^2_\rho$:

$$\mathcal{S}\Phi(\phi(0),\phi) \equiv \lim_{t\downarrow 0} \frac{\Phi(\phi(0),\tilde{\phi}_t) - \Phi(\phi(0),\phi)}{t}. \tag{7.14}$$

Remark 7.2.1 *Note that $\Phi \in C_{lip}^{2,2}(\Re \times L_\rho^2)$ does not guarantee that $\Phi \in \mathcal{D}(\mathcal{S})$. For example, let $\bar{\theta} > 0$ and define a simple tame function $\Phi : \Re \times L_\rho^2 \to \Re$ by*

$$\Phi(\phi(0), \phi) = \phi(-\bar{\theta}), \quad \forall (\phi(0), \phi) \in \Re \times L_\rho^2.$$

Then it can be shown that $\Phi \in C_{lip}^{2,2}(\Re \times L_\rho^2)$ and yet $\Phi \notin \mathcal{D}(\mathcal{S})$.

It is shown in Section 2.5 of Chapter 2, however, that any tame function of the above form can be approximated by a sequence of quasi-tame functions that are in $\mathcal{D}(\mathcal{S})$.

Again, consider the associated Markovian $\Re \times L_\rho^2$-valued process $\{(S(t), S_t), t \geq 0\}$ described by (7.1) with the initial historical price function $(S(0), S_0) = (\psi(0), \psi) \in \Re \times L_\rho^2$.

The following four theorems are repetitions of those in Section 2.5 of Chapter 2. Proofs can be found in that section and therefore are omitted here.

Theorem 7.2.2 *If $\Phi \in C_{lip}^{2,2}(\Re \times L_\rho^2) \cap \mathcal{D}(\mathcal{S})$, then*

$$\lim_{t \downarrow 0} \frac{E[\Phi(S(t), S_t) - \Phi(\psi(0), \psi)]}{t} = \mathbf{A}\Phi(\psi(0), \psi), \qquad (7.15)$$

where

$$\mathbf{A}\Phi(\psi(0), \psi) = \mathcal{S}\Phi(\psi(0), \psi) \frac{1}{2} \partial_x^2 \Phi(\psi(0), \psi) \psi^2(0) g^2(\psi)$$
$$+ \partial_x \Phi(\psi(0), \psi) \psi(0) f(\psi) \qquad (7.16)$$

and $\mathcal{S}(\Phi)(\psi(0), \psi)$ is as defined in (7.14).

It seems from a glance at (7.16) that $\mathbf{A}\Phi(\psi(0), \psi)$ requires only the existence of the first- and second-order partial derivatives $\partial_{\phi(0)}\Phi$ and $\partial_{\phi(0)}^2\Phi$ of $\Phi(\psi(0), \psi)$ with respect to its first variable $\psi(0) \in \Re$. However, detailed derivations of the formula reveal that a stronger condition than $\Phi \in C_{lip}^{2,2}(\Re \times L_\rho^2)$ is required.

We have the following Dynkin's formula (see Section 2.5 in Chapter 2):

Theorem 7.2.3 *Let $\Phi \in C_{lip}^{2,2}(\Re \times L_\rho^2) \cap \mathcal{D}(\mathcal{S})$. Then*

$$E[e^{-\alpha\tau}\Phi(S(\tau), S_\tau)] - \Phi(\psi(0), \psi) = E\left[\int_0^\tau e^{-\alpha t}(\mathbf{A} - \alpha I)\Phi(S(t), S_t)\, dt \right],$$
$$(7.17)$$

for all P-a.s. finite $\mathbf{G}$-stopping time τ.

The function $\Phi \in C_{lip}^{2,2}(\Re \times L_\rho^2) \cap \mathcal{D}(\mathcal{S})$ that has the following special form is referred to as a quasi-tame function:

$$\Phi(\phi(0), \phi) = \Psi(q(\phi(0), \phi)), \qquad (7.18)$$

where

$$q(\phi(0), \phi) = \left(\phi(0), \int_{-\infty}^0 \eta_1(\phi(\theta))\lambda_1(\theta)\, d\theta,\right.$$

$$\left. \ldots, \int_{-\infty}^0 \eta_n(\phi(\theta))\lambda_n(\theta)\, d\theta\right), \quad \forall(\phi(0), \phi) \in \Re \times L_\rho^2, \quad (7.19)$$

for some positive integer n and some functions $q \in C(\Re \times L_\rho^2; \Re^{n+1})$, $\eta_i \in C^\infty(\Re)$, $\lambda_i \in C^1((-\infty, 0])$ with

$$\lim_{\theta \to -\infty} \lambda_i(\theta) = \lambda_i(-\infty) = 0$$

for $i = 1, 2, \ldots, n$, and $h \in C^\infty(\Re^{n+1})$ of the form $h(x, y_1, y_2, \ldots, y_n)$.

We have the following Ito's formula (see Section 2.6 in Chapter 2) in case $\Phi \in \Re \times L_\rho^2$ is a quasi-tame function in the sense defined above.

Theorem 7.2.4 *Let $\{(S(t), S_t), t \geq 0\}$ be the $\Re \times L_\rho^2$-valued solution process corresponding to (7.1) with an initial historical price function $(\psi(0), \psi) \in \Re \times L_\rho^2$. If $\Phi \in C(\Re \times L_\rho^2)$ is a quasi-time function, then $\Phi \in \mathcal{D}(\mathbf{A})$ and*

$$e^{-\alpha\tau}\Phi(S(\tau), S_\tau) = \Phi(\psi(0), \psi) + \int_0^\tau e^{-\alpha t}(\mathbf{A} - \alpha I)\Phi(S(t), S_t)\, dt$$

$$+ \int_0^\tau e^{-\alpha t}\partial_{\phi(0)}\Phi(S(t), S_t)S(t)f(S_t)\, dW(t) \quad (7.20)$$

for all finite $\mathbf{G}$-stopping time τ, where I is the identity operator. Moreover, if $\Phi \in C(\Re \times L_\rho^2)$ is of the form described in (7.18) and (7.19), then

$$\mathbf{A}\Phi(\psi(0), \psi)$$

$$= \sum_{i=1}^n h_{y_i}(q(\psi(0), \psi)) \times \left(\eta_i(\psi(0))\lambda_i(0) - \int_{-\infty}^0 \eta_i(\psi(\theta))\dot{\lambda}_i(\theta)\, d\theta\right)$$

$$+ \partial_x h(q(\psi(0), \psi))\psi(0)f(\psi) + \frac{1}{2}\partial_x^2(q(\psi(0), \psi))\psi^2(0)g^2(\psi), \quad (7.21)$$

where Ψ_x, Ψ_{y_i}, and Ψ_{xx} denote the partial derivatives of $\Psi(x, y_1, \ldots, y_n)$ with respect to its appropriate variables.

In the following, we will state that the above Ito's formula also holds for any tame function $\Phi : \Re \times C(-\infty, 0] \to \Re$ of the following form:

$$\Phi(\phi(0), \phi) = h(q(\phi(0), \phi)) = h(\phi(0), \phi(-\theta_1), \ldots, \phi(-\theta_k)) \quad (7.22)$$

where $C(-\infty, 0]$ is the space continuous functions $\phi : (-\infty, 0] \to \Re$ equipped with uniform topology, $0 < \theta_1 < \theta_2 < \ldots < \theta_k < \infty$, and $h(x, y_1, \cdots, y_k)$ is such that $h \in C^\infty(\Re^{k+1})$.

Theorem 7.2.5 *Let $\{(S(t), S_t), t \geq 0\}$ be the $\Re \times L_\rho^2$-valued process corresponding to Equation (7.1) with an initial historical price function $(\psi(0), \psi) \in$*

$\Re \times L^2_\rho$. If $\Phi : \Re \times C(-\infty, 0] \to \Re$ is a tame function defined by (7.22), then $\Phi \in \mathcal{D}(\mathbf{A})$ and

$$
\begin{aligned}
&e^{-\alpha\tau}h(S(\tau), S(\tau - \theta_1), \dots, S(\tau - \theta_k)) \\
&= h(\psi(0), \psi(-\theta_1), \dots, \psi(-\theta_k)) \\
&\quad + \int_0^\tau e^{-\alpha t}(\mathbf{A} - \alpha I)h(S(t), S(t - \theta_1), \dots, S(t - \theta_k))\, dt \\
&\quad + \int_0^\tau e^{-\alpha t}\partial_x h(S(t), S(t - \theta_1), \dots, S(t - \theta_k))S(t)f(S_t)\, dW(t) \quad (7.23)
\end{aligned}
$$

for all finite **G**-stopping times τ, where

$$
\begin{aligned}
&\mathbf{A}h(\psi(0), \psi(-\theta_1), \dots, \psi(-\theta_k)) \\
&= \partial_x h(\psi(0), \psi(-\theta_1), \dots, \psi(-\theta_k))\psi(0)f(\psi) \\
&\quad + \frac{1}{2}\partial_x^2 h(\psi(0), \psi(-\theta_1), \dots, \psi(-\theta_k))\psi^2(0)g^2(\psi), \quad\quad (7.24)
\end{aligned}
$$

with $\partial_x h$ and $\partial_x^2 h$ being the first- and second-order derivatives with respect to x of $h(x, y_1, \cdots, y_k)$.

7.2.2 Dynkin's Formula for the Controlled State Process

Similar to the processes $\{X(t), t \geq 0\}$ and $\{(S(t), S_t), t \geq 0\}$, the **N**-valued controlled inventory process $\{N_t, t \geq 0\}$ of the *investor's* stock account described by (7.9) and (7.10) also satisfies the following change-of-variable formula:

$$
e^{-\alpha\tau}\Phi(N_\tau) = \Phi(\xi) + \sum_{0 \leq t \leq \tau} e^{-\alpha t}[\Phi(N_t) - \Phi(N_{t-})], \quad\quad (7.25)
$$

for all $\Phi \in C_b(\mathbf{N})$ (the space of bounded and continuous function from **N** to $\Re$), and finite **G**-stopping times τ, where $N_{\tau-} = \lim_{t\downarrow 0} N_{\tau-t}$. The above change-of-variable formula is rather self-explanatory.

Notation. In the following, we will use the convention that $C^{1,0,2,2}_{lip}(\mathcal{O})$ is the collection of continuous functions $\Phi : \mathcal{O} \to \Re$ ($\mathcal{S}_\kappa \subset \mathcal{O}$) such that $\Phi(\cdot, \xi, \psi(0), \psi) \in C^1(\Re)$ for each $(\xi, \psi(0), \psi)$, and $\Phi(x, \xi, \cdot, \psi) \in C^{2,2}_{lip}(\Re \times L^2_\rho) \cap \mathcal{D}(\mathcal{S})$ for each (x, ξ).

Combining the above results in this section, we have the following Dynkin's formula for the controlled (by the admissible strategy π) $\mathcal{S}_\kappa$-valued state process $\{Z(t) = (X(t), N_t, S(t), S_t), t \geq 0\}$:

$$
\begin{aligned}
E[e^{-\alpha\tau}\Phi(Z(\tau))] &= \Phi(Z(0-)) + E\left[\int_0^\tau e^{-\alpha t}\mathcal{L}^{C(t)}\Phi(Z(t))\, dt\right] \\
&\quad + E\left[\sum_{0 \leq t \leq \tau} e^{-\alpha t}\Big(\Phi(Z(t)) - \Phi(Z(t-))\Big)\right], \quad (7.26)
\end{aligned}
$$

for all $\Phi : \Re \times \mathbf{N} \times \Re \times L_\rho^2 \to \Re$ such that $\Phi(\cdot, \xi, \psi(0), \psi) \in C^1(\Re)$ for each $(\xi, \psi(0), \psi) \in \mathbf{N} \times \Re \times L_\rho^2$ and $\Phi(x, \xi, \cdot, \cdot) \in C_{lip}^{2,2}(\Re \times L_\rho^2) \cap \mathcal{D}(\mathcal{S})$ for each $(x, \xi) \in \Re \times \mathbf{N}$, where

$$\mathcal{L}^c \Phi(x, \xi, \psi(0), \psi) = (\mathbf{A} - \alpha I + (rx - c)\partial_x)\Phi(x, \xi, \psi(0), \psi) \qquad (7.27)$$

and $\mathbf{A}$ is as defined in (7.16). Note that $E[\cdots]$ in the above stands for $E^{x, \xi, \psi(0), \psi; \pi}[\cdots]$.

In the case that $\Phi \in C(\Re \times \mathbf{N} \times \Re \times L_\rho^2)$ is such that $\Phi(x, \xi, \cdot, \cdot) : \Re \times L_\rho^2 \to \Re$ is a quasi-tame (respectively, tame) function on $\Re \times L_\rho^2$ of the form described in (7.18) and (7.19) (respectively (7.22)), then the following Ito formula for the controlled state process $\{Z(t) = (X(t), N_t, S(t), S_t), t \geq 0\}$ also holds true.

Theorem 7.2.6 *If* $\Phi \in C(\Re \times \mathbf{N} \times \Re \times L_\rho^2)$ *is such that* $\Phi(x, \xi, \cdot, \cdot) : \Re \times L_\rho^2 \to \Re$ *is a quasi-tame function (respectively, tame) on* $\Re \times L_\rho^2$, *then*

$$e^{-\alpha \tau} \Phi(Z(\tau)) = \Phi(Z(0-)) + \int_0^\tau e^{-\alpha t} \mathcal{L}^{C(t)} \Phi(Z(t)) \, dt$$
$$+ \int_0^\tau e^{-\alpha t} \partial_{\psi(0)} \Phi(Z(t)) S(t) f(S_t) \, dW(t)$$
$$+ \Big[\sum_{0 \leq t \leq \tau} e^{-\alpha t} \Big(\Phi(Z(t)) - \Phi(Z(t-)) \Big) \Big], \qquad (7.28)$$

for every P-a.s. finite $\mathbf{G}$-*stopping time* τ. *Moreover, if* $\Phi(x, \xi, \psi(0), \psi) = h(x, \xi, q(\psi(0), \psi))$, *where* $h \in C(\Re \times \mathbf{N} \times \Re^{n+1})$ *and* $q(\psi(0), \psi)$ *is given by* *(7.18) and (7.19) (respectively, (7.22)), then*

$$\mathcal{L}^c \Phi(x, \xi, \psi(0), \psi) = (\mathbf{A} - \alpha I + (rx - c)\partial_x)h(x, \xi, q(\psi(0), \psi))$$

and $\mathbf{A}h(x, \xi, q(\psi(0), \psi))$ *is as given in (7.15)(respectively, (7.14)) for each fixed* $(x, \xi) \in \Re \times \mathbf{N}$.

7.3 The HJBQVI

The main objective of this section is to derive the dynamic programming equation for the value function in form of an infinite-dimensional Hamilton-Jacobi-Bellman quasi variational inequality (HJBQVI) (see HJBQVI (*) in Section 7.3.3(D)).

7.3.1 The Dynamic Programming Principle

The following Bellman-type dynamic programming principle (DPP) was established in Section 3.3.3 in Chapter 3 and still holds true in our problem by combining with that obtained in Theorem 3.3.9 in Section 3.3.3 (see also Kolmanovskii and Shaikhet [KS96]). For the sake of saving space, we take the following result as the starting point without proof for deriving our dynamic principle equation:

Proposition 7.3.1 *Let* $(x, \xi, \psi(0), \psi) \in \mathcal{S}_\kappa$ *be given and let* $\mathcal{O}$ *be an open subset of* $\mathcal{S}_\kappa$ *containing* $(x, \xi, \psi(0), \psi)$. *For* $\pi = (C, \mathcal{T}) \in \mathcal{U}_\kappa(x, \xi, \psi(0), \psi)$, *let* $\{(X(t), N_t, S(t), S_t), t \geq 0\}$ *be given by (7.7)-(7.11) and (7.1). Define*

$$\tau = \inf\{t \geq 0 \mid (X(t), N_t, S(t), S_t) \notin \bar{\mathcal{O}}\},$$

where $\bar{\mathcal{O}}$ *is the closure of* $\mathcal{O}$. *Then, for each* $t \in [0, \infty)$, *we have the following optimality equation:*

$$V_\kappa(x, \xi, \psi(0), \psi) = \sup_{\pi \in \mathcal{U}_\kappa(x, \xi, \psi(0), \psi)} E\Big[\int_0^{t \wedge \tau} e^{-\alpha s} \frac{C^\gamma(s)}{\gamma} ds \qquad (7.29)$$
$$+ 1_{\{t \wedge \tau < \infty\}} e^{-\alpha(t \wedge \tau)} V_\kappa(X(t \wedge \tau), N_{t \wedge \tau}, S(t \wedge \tau), S_{t \wedge \tau})\Big],$$

where we have used the notation $a \wedge b = \min\{a, b\}$ *for* $a, b \in \Re$.

7.3.2 Derivation of the HJBQVI

In this subsection, we will derive HJBQVI (see QVHJBI (*) in Subsection 7.3.3(D) based on the DPP described in Proposition 7.3.1). We emphasize here that it is not our intension to rigorously verify every step involved in the derivations since the rigorous verification is to be done in Sections 7.4, where we prove that the value function is a viscosity solution of HJBQVI (*). To derive QVHJBI (*), we consider the effects on the value function when there is consumption but no transaction and when there is transaction but no consumption.

(A) Consumptions Without Transaction
Assume first that there is no transaction then the corresponding state process $\{Z(t) = (X(t), N_t, S(t), S_t), t \geq 0\}$ satisfies the following set of equations:

$$dX(t) = [\lambda X(t) - C(t)] dt, \quad t \geq 0, \qquad (7.30)$$

$$\frac{dS(t)}{S(t)} = f(S_t) dt + g(S_t) dW(t), \quad t \geq 0, \qquad (7.31)$$

$$N_t = \xi, \ t \geq 0, \qquad (7.32)$$

with the initial state $(X(0-), N_{0-}, S(0), S_0) = (x, \xi, \psi(0), \psi) \in \mathbf{S}_\kappa$. In this case, $V_\kappa(X(t), N_t, S(t), S_t) = V_\kappa(X(t-), N_{t-}, S(t), S_t)$ for all $t \geq 0$, since there is no jump transaction. Assume that the value function $V_\kappa : \mathbf{S}_\kappa \to \Re_+$ is sufficiently smooth. From Proposition 7.3.1 and (7.15), we have

$$0 \geq \lim_{t \downarrow 0} \frac{E\left[e^{-\alpha t}V_\kappa(X(t), N_t, S(t), S_t) - V_\kappa(x, \xi, \psi(0), \psi)\right]}{t}$$

$$+ \lim_{t \downarrow 0} \frac{1}{t}E\left[\int_0^t e^{-\alpha s}\frac{C^\gamma(s)}{\gamma}ds\right].$$

Therefore,

$$0 \geq \lim_{t \downarrow 0} \frac{E\left[e^{-\alpha t}(V_\kappa(X(t), N_t, S(t), S_t) - V_\kappa(x, \xi, \psi(0), \psi))\right]}{t}$$

$$+ \lim_{t \downarrow 0} \frac{\left[(e^{-\alpha t} - 1)V_\kappa(x, \xi, \psi(0), \psi)\right]}{t}$$

$$+ \lim_{t \downarrow 0} \frac{1}{t}E\left[\int_0^t e^{-\alpha s}\frac{C^\gamma(s)}{\gamma}ds\right].$$

First, we note that

$$\lim_{t \downarrow 0} \frac{1}{t}E\left[\int_0^t e^{-\alpha s}\frac{C^\gamma(s)}{\gamma}ds\right] = \frac{c^\gamma}{\gamma},$$

since the consumption strategy $C(\cdot) = \{C(t), t \geq 0\}$ is right-continuous at $t = 0$ with $\lim_{t \downarrow 0} C(t) = c$.

Second, we note that

$$\lim_{t \downarrow 0} \frac{\left[(e^{-\alpha t} - 1)V_\kappa(x, \xi, \psi(0), \psi)\right]}{t} = -\alpha V_\kappa(x, \xi, \psi(0), \psi).$$

Third, we note that

$$\lim_{t \downarrow 0} \frac{E\left[e^{-\alpha t}(V_\kappa(X(t), N_t, S(t), S_t) - V_\kappa(x, \xi, \psi(0), \psi))\right]}{t}$$

$$= \lim_{t \downarrow 0} \frac{\left[e^{-\alpha t}(V_\kappa(X(t), N_t, S(t), S_t) - V_\kappa(x, N_t, S(t), S_t))\right]}{t}$$

$$+ \lim_{t \downarrow 0} \frac{E\left[e^{-\alpha t}(V_\kappa(x, N_t, S(t), S_t) - V_\kappa(x, \xi, S(t), S_t))\right]}{t}$$

$$+ \lim_{t \downarrow 0} \frac{E\left[e^{-\alpha t}(V_\kappa(x, \xi, S(t), S_t) - V_\kappa(x, \xi, \psi(0), \psi))\right]}{t}.$$

Using (7.30) and (7.32), we have

$$\lim_{t \downarrow 0} \frac{E\left[e^{-\alpha t}(V_\kappa(X(t), N_t, S(t), S_t) - V_\kappa(x, \xi, \psi(0), \psi))\right]}{t}$$

$$= (rx - c)\partial_x V_\kappa(x, \xi, \psi(0), \psi)$$

and

$$\lim_{t\downarrow 0} \frac{E\Big[e^{-\alpha t}(V_\kappa(x, N_t, S(t), S_t) - V_\kappa(x, \xi, S(t), S_t))\Big]}{t} = 0.$$

Also by (7.15), we have

$$\lim_{t\downarrow 0} \frac{E\Big[e^{-\alpha t}(V_\kappa(x, \xi, S(t), S_t) - V_\kappa(x, \xi, \psi(0), \psi))\Big]}{t} = \mathbf{A}V_\kappa(x, \xi, \psi(0), \psi)).$$

Combining the above terms, we have

$$0 \geq \Big(\mathbf{A} + (rx - c)\partial_x - \alpha I\Big)V_\kappa(x, \xi, \psi(0), \psi) + \frac{c^\gamma}{\gamma}, \quad \forall c \geq 0.$$

This shows that

$$\begin{aligned}
0 \geq \mathcal{A}V_\kappa(x, \xi, \psi(0), \psi) \\
\equiv \sup_{c \geq 0} \Big(\mathcal{L}^c V_\kappa(x, \xi, \psi(0), \psi) + \frac{c^\gamma}{\gamma}\Big) \\
= \Big(\mathbf{A} + rx\partial_x - \alpha I\Big)V_\kappa(x, \xi, \psi(0), \psi) \\
+ \sup_{c \geq 0}\Big(\frac{c^\gamma}{\gamma} - c\partial_x V_\kappa(x, \xi, \psi(0), \psi)\Big) \\
= \Big(\mathbf{A} + rx\partial_x - \alpha I\Big)V_\kappa(x, \xi, \psi(0), \psi) \\
+ \frac{1-\gamma}{\gamma}(\partial_x V_\kappa)^{\frac{\gamma}{\gamma-1}}(x, \xi, \psi(0), \psi),
\end{aligned} \tag{7.33}$$

since the maximum of the the above expression is achieved at

$$c^* = (\partial_x V_\kappa)^{\frac{1}{\gamma-1}}(x, \xi, \psi(0), \psi). \tag{7.34}$$

Note that the Fréchet differential operator $\mathbf{A}$ and $\mathcal{S}$ are defined in (7.15) and (7.14), respectively.

(B) Transactions Without Consumption
We next consider the case where there are transactions but no consumption. For each locally bounded $\Phi : \mathbf{S}_\kappa \to [0, \infty)$ and each $(x, \xi, \psi(0), \psi) \in \mathbf{S}_\kappa$ define the *intervention operator*

$$\begin{aligned}
\mathcal{M}_\kappa \Phi(x, \xi, \psi(0), \psi) \\
= \sup\{\Phi(\hat{x}, \hat{\xi}, \hat{\psi}(0), \hat{\psi}) \mid \zeta \in \mathcal{R}(\xi) - \{\mathbf{0}\}, (\hat{x}, \hat{\xi}, \hat{\psi}(0), \hat{\psi}) \in \mathcal{S}_\kappa\}, \tag{7.35}
\end{aligned}$$

where $(\hat{x}, \hat{\xi}, \hat{\psi}(0), \hat{\psi})$ are defined as follows.

$$\hat{x} = x - \kappa - (m(0) + \mu|m(0)|)\psi(0)$$
$$- \sum_{k=1}^{\infty} \Big[(1+\mu)m(-k)\psi(0)$$
$$- \beta m(-k)(\psi(0) - \psi(\tau(-k)))\Big] \mathbf{1}_{\{n(-k)<0, 0\le m(-k)\le -n(-k)\}}$$
$$- \sum_{k=1}^{\infty} \Big[(1-\mu)m(-k)\psi(0) - \beta m(-k)(\psi(0) - \psi(\tau(-k)))\Big]$$
$$\cdot \mathbf{1}_{\{n(-k)>0, -n(-k)\le m(-k)\le 0\}}, \tag{7.36}$$

and for all $\theta \in (-\infty, 0]$,

$$\hat{\xi}(\theta) = (\xi \oplus \zeta)(\theta)$$
$$= m(0)\mathbf{1}_{\{\tau(0)\}}(\theta)$$
$$+ \sum_{k=1}^{\infty} \Big(n(-k) + m(-k)[\mathbf{1}_{\{n(-k)<0, 0\le m(-k)\le -n(-k)\}}$$
$$+ \mathbf{1}_{\{n(-k)>0, -n(-k)\le m(-k)\le 0\}}]\Big) \mathbf{1}_{\{\tau(-k)\}}(\theta), \tag{7.37}$$

and, again,

$$(\hat{\psi}(0), \hat{\psi}) = (\psi(0), \psi). \tag{7.38}$$

If $(\hat{x}, \hat{\xi}, \hat{\psi}(0), \hat{\psi}) \notin \mathbf{S}_\kappa$ for all $\zeta \in \mathcal{R}(\xi) - \{\mathbf{0}\}$, we set $\mathcal{M}_\kappa \Phi(x, \xi, \psi(0), \psi) = 0$. If for all $(x, \xi, \psi(0), \psi) \in \mathbf{S}_\kappa$ there exists $(\hat{x}, \hat{\xi}, \hat{\psi}(0), \hat{\psi}) \in \mathbf{S}_\kappa$ such that

$$\mathcal{M}_\kappa \Phi(x, \xi, \psi(0), \psi) = \Phi(\hat{x}, \hat{\xi}, \hat{\psi}(0), \hat{\psi}),$$

then we set

$$\hat{\zeta}(x, \xi, \psi(0), \psi) = \hat{\zeta}_\Phi(x, \xi, \psi(0), \psi) = (\hat{x}, \hat{\xi}, \hat{\psi}(0), \hat{\psi}) \in \mathcal{R}(\xi). \tag{7.39}$$

Note that we let $\hat{\zeta}(x, \xi, \psi(0), \psi)$ denote a measurable selection of the map

$$(x, \xi, \psi(0), \psi) \mapsto (\hat{x}, \hat{\xi}, \hat{\psi}(0), \hat{\psi})$$

defined in (7.39).

We make the following technical assumption regarding the existence of a measurable selection

$$\hat{\zeta}(x, \xi, \psi(0), \psi) = \hat{\zeta}_{V_\kappa}(x, \xi, \psi(0), \psi)$$

for the value function $V_\kappa : \mathbf{S}_\kappa \to \Re$; that is, there exists a measurable function $\hat{\zeta}_{V_\kappa} : \mathbf{S}_\kappa \to \Re$ such that

$$V_\kappa(\hat{\zeta}(x, \xi, \psi(0), \psi)) = \mathcal{M}_\kappa V_\kappa(x, \xi, \psi(0), \psi), \quad \forall (x, \xi, \psi(0), \psi) \in \mathbf{S}_\kappa. \tag{7.40}$$

Assumption 7.3.2 *For each* $(x, \xi, \psi(0), \psi) \in \mathbf{S}_\kappa$, *there exists a measurable function* $\hat{\zeta}_{V_\kappa} : \mathbf{S}_\kappa \to \Re$ *such that (7.40) is satisfied for every* $(x, \xi, \psi(0), \psi) \in \mathbf{S}_\kappa$.

Assume without loss of generality that the *investor's* portfolio immediately prior to time t is $(X(t-), N_{t-}, S(t), S_t) = (x, \xi, \psi(0), \psi) \in \mathbf{S}_\kappa$. An immediate transaction of the amount $\zeta \in \mathcal{R} - \{0\}$ without consumption at time t (i.e., $C(t) = 0$) yields that $(X(t), N_t, S(t), S_t) = (\hat{x}, \hat{\xi}, \hat{\psi}(0), \hat{\psi})$, where $\hat{x}$, $\hat{\xi}$, and $(\hat{\psi}(0), \hat{\psi})$ are as given in (7.36)-(7.38). It is clear that

$$V_\kappa(x, \xi, \psi(0), \psi) \geq \mathcal{M}_\kappa V_\kappa(x, \xi, \psi(0), \psi), \quad \forall (x, \xi, \psi(0), \psi) \in \mathbf{S}_\kappa. \qquad (7.41)$$

Combining Sections 7.3.2(A) and 7.3.2(B), we have the following inequality:

$$\max\left\{ \mathcal{A}V_\kappa, \mathcal{M}_\kappa V_\kappa - V_\kappa \right\} \leq 0 \text{ on } \mathbf{S}_\kappa^\circ,$$

where $\mathbf{S}_\kappa^\circ$ denotes the interior of the *solvency region* $\mathbf{S}_\kappa$.

Using a standard technique in deriving the variational HJB inequality for stochastic classical-singular and classical-impulse control problems (see Bensoussan and Lions [BL81, BL84] and Cadenillas and Zapataro [CZ00] for stochastic impulse controls, Bekke and Øksendal [BØ98] and Øksendal and Sulem [ØS02] for stochastic classical-impulse controls, and Larssen [Lar02] and Larssen and Risebro [LR03] for classical and singular controls of stochastic delay equations), one can show that on the set

$$\{(x, \xi, \psi(0), \psi) \in \mathbf{S}_\kappa^\circ \mid \mathcal{M}_\kappa V_\kappa(x, \xi, \psi(0), \psi) < V_\kappa(x, \xi, \psi(0), \psi)\},$$

we have $\mathcal{A}V_\kappa = 0$, and on the set

$$\{(x, \xi, \psi(0), \psi) \in \mathbf{S}_\kappa^\circ \mid \mathcal{A}V_\kappa(x, \xi, \psi(0), \psi) < 0\},$$

we have $\mathcal{M}_\kappa V_\kappa = V_\kappa$. Therefore, we have the following HJBQVI on $\mathbf{S}_\kappa^\circ$:

$$\max\left\{ \mathcal{A}V_\kappa, \mathcal{M}_\kappa V_\kappa - V_\kappa \right\} = 0 \text{ on } \mathbf{S}_\kappa^\circ, \qquad (7.42)$$

where

$$\mathcal{A}\Phi = (\mathbf{A} + rx\partial_x - \alpha)\Phi + \sup_{c \geq 0} \left(\frac{c^\gamma}{\gamma} - c\partial_x\Phi \right), \qquad (7.43)$$

$\mathcal{M}_\kappa\Phi$ is as given in (7.35), and the operator $\mathbf{A}$ is given in (7.16).

7.3.3 Boundary Values of the HJBQVI

(A) The Solvency Region for $\kappa = 0$ and $\mu > 0$
When there is proportional but no fixed transaction cost (i.e., $\kappa = 0$ and $\mu > 0$), the solvency region can be written as

$$\mathbf{S}_0 = \{(x, \xi, \psi(0), \psi) \mid G_0(x, \xi, \psi(0), \psi) \geq 0\} \cup \mathbf{S}_+,$$

where G_0 is the liquidating function given in (7.5) with $\kappa = 0$, that is,

$$G_0(x, \xi, \psi(0), \psi) = x + \sum_{k=0}^{\infty} \Big[\min\{(1 - \mu)n(-k), (1 + \mu)n(-k)\}\psi(0)$$
$$- n(-k)\beta(\psi(0) - \psi(\tau(-k)))\Big]. \tag{7.44}$$

In the case $\kappa = 0$, we claim that

$$\mathbf{S}_+ \subset \{(x, \xi, \psi(0), \psi) \mid G_0(x, \xi, \psi(0), \psi) \geq 0\}.$$

This is because

$$x \geq 0 \text{ and } n(-i) \geq 0, \quad \forall i = 0, 1, 2, \dots \Rightarrow G_0(x, \xi, \psi(0), \psi) \geq 0.$$

In this case, all shares of the *stock* owned or owed can be liquidated because of the absence of a fixed transaction cost $\kappa = 0$. Therefore,

$$\mathbf{S}_0 = \{(x, \xi, \psi(0), \psi) \mid G_0(x, \xi, \psi(0), \psi) \geq 0\}.$$

We easily observe that $\mathbf{S}_0$ is an unbounded convex set.

(B) Decomposition of $\partial \mathbf{S}_\kappa$
For $I \subset \aleph_0 \equiv \{0, 1, 2, \dots\}$, the boundary $\partial \mathbf{S}_\kappa$ of $\mathbf{S}_\kappa$ can be decomposed as follows:

$$\partial \mathbf{S}_\kappa = \bigcup_{I \subset \aleph_0} (\partial_{-,I}\mathbf{S}_\kappa \cup \partial_{+,I}\mathbf{S}_\kappa), \tag{7.45}$$

where

$$\partial_{-,I}\mathbf{S}_\kappa = \partial_{-,I,1}\mathbf{S}_\kappa \cup \partial_{-,I,2}\mathbf{S}_\kappa, \tag{7.46}$$

$$\partial_{+,I}\mathbf{S}_\kappa = \partial_{+,I,1}\mathbf{S}_\kappa \cup \partial_{+,I,2}\mathbf{S}_\kappa, \tag{7.47}$$

$$\partial_{+,I,1}\mathbf{S}_\kappa = \{(x, \xi, \psi(0), \psi) \mid G_\kappa(x, \xi, \psi(0), \psi) = 0, x \geq 0, n(-i) < 0$$
$$\text{for all } i \in I \And n(-i) \geq 0 \text{ for all } i \notin I\}, \tag{7.48}$$

$$\partial_{+,I,2}\mathbf{S}_\kappa = \{(x, \xi, \psi(0), \psi) \mid G_\kappa(x, \xi, \psi(0), \psi) < 0, x \geq 0, n(-i) = 0$$
$$\text{for all } i \in I \And n(-i) \geq 0 \text{ for all } i \notin I\}, \tag{7.49}$$

$$\partial_{-,I,1}\mathbf{S}_\kappa = \{(x, \xi, \psi(0), \psi) \mid G_\kappa(x, \xi, \psi(0), \psi) = 0, x < 0, n(-i) < 0$$
$$\text{for all } i \in I \And n(-i) \geq 0 \text{ for all } i \notin I\}, \tag{7.50}$$

and

$$\partial_{-,I,2}\mathbf{S}_\kappa = \{(x,\xi,\psi(0),\psi) \mid G_\kappa(x,\xi,\psi(0),\psi) < 0, x = 0, n(-i) = 0$$
$$\text{for all } i \in I \ \& \ n(-i) \geq 0 \text{ for all } i \notin I\}. \tag{7.51}$$

The interface (intersection) between $\partial_{+,I,1}\mathbf{S}_\kappa$ and $\partial_{+,I,2}\mathbf{S}_\kappa$ is denoted by

$$Q_{+,I} = \{(x,\xi,\psi(0),\psi) \mid G_\kappa(x,\xi,\psi(0),\psi) = 0, x \geq 0, n(-i) = 0$$
$$\text{for all } i \in I \ \& \ n(-i) \geq 0 \text{ for all } i \notin I\}, \tag{7.52}$$

whereas the interface between $\partial_{-,I,1}\mathbf{S}_\kappa$ and $\partial_{-,I,2}\mathbf{S}_\kappa$ is denoted by

$$Q_{-,I} = \{(0,\xi,\psi(0),\psi) \mid G_\kappa(0,\xi,\psi(0),\psi) = 0, x = 0, n(-i) = 0$$
$$\text{for all } i \in I \ \& \ n(-1) \geq 0 \text{ for all } i \notin I\}. \tag{7.53}$$

For example, if $I = \aleph$, then $n(-i) < 0, \quad \forall i = 0, 1, 2, \ldots$, and

$$G_\kappa(x,\xi,\psi(0),\psi) \geq 0 \Rightarrow x \geq \kappa.$$

In this case, $\partial_{-,\aleph}\mathbf{S}_\kappa = \emptyset$ (the empty set),

$$\partial_{+,\aleph_0,1}\mathbf{S}_\kappa = \{(x,\xi,\psi(0),\psi) \mid G_\kappa(x,\xi,\psi(0),\psi) = 0, x \geq 0,$$
$$\text{and } n(-i) < 0 \text{ for all } i \in \aleph\},$$

and

$$\partial_{+,\aleph_0,2}\mathbf{S}_\kappa = \{(x,\xi,\psi(0),\psi) \mid G_\kappa(x,\xi,\psi(0),\psi) < 0, x \geq 0,$$
$$\text{and } n(-i) = 0 \text{ for all } i \in \aleph_0\}$$
$$= \{(x,\mathbf{0},\psi(0),\psi) \mid 0 \leq x \leq \kappa\}.$$

On the other hand, if $I = \emptyset$ (the empty set) (i.e., $n(-i) \geq 0$ for all $i \in \aleph_0$), then

$$\partial_{+,\emptyset,1}\mathbf{S}_\kappa = \{(x,\xi,\psi(0),\psi) \mid G_\kappa(x,\xi,\psi(0),\psi) = 0, x \geq 0,$$
$$\text{and } n(-i) \geq 0 \text{ for all } i \in \aleph_0\},$$

$$\partial_{+,\emptyset,2}\mathbf{S}_\kappa = \{(x,\xi,\psi(0),\psi) \mid G_\kappa(x,\xi,\psi(0),\psi) < 0, x \geq 0,$$
$$\text{and } n(-i) \geq 0 \text{ for all } i \in \aleph_0\},$$

$$\partial_{-,\emptyset,1}\mathbf{S}_\kappa = \{(x,\xi,\psi(0),\psi) \mid G_\kappa(x,\xi,\psi(0),\psi) = 0, x < 0,$$
$$\text{and } n(-i) \geq 0 \text{ for all } i \in \aleph\},$$

and

$$\partial_{-,\emptyset,2}\mathbf{S}_\kappa = \{(x,\xi,\psi(0),\psi) \mid G_\kappa(x,\xi,\psi(0),\psi) < 0, x = 0,$$
$$\text{and } n(-i) \geq 0 \text{ for all } i \in \aleph_0\}.$$

(C) Boundary Conditions for the Value Function

Let us now examine the conditions of the value function $V_\kappa : \mathbf{S}_\kappa \to \Re_+$ on the boundary $\partial \mathbf{S}_\kappa$ of the *solvency region* $\mathbf{S}_\kappa$ defined in (7.45)-(7.51).

We make the following observations regarding the behavior of the value function V_κ on the boundary $\partial \mathbf{S}_\kappa$.

Lemma 7.3.3 *Let* $(x, \xi, \psi(0), \psi) \in \mathbf{S}_\kappa$ *and let* $\hat{x}$, $\hat{\xi}$, *and* $(\hat{\psi}(0), \hat{\psi})$ *be as defined in (7.36)- (7.38). Then*

$$G_0(\hat{x}, \hat{\xi}, \hat{\psi}(0), \hat{\psi}) = G_0(x, \xi, \psi(0), \psi) - \kappa. \tag{7.54}$$

Proof. Suppose the *investor's* current portfolio is $(x, \xi, \psi(0), \psi) \in \mathbf{S}_\kappa$; then an instantaneous transaction of the quantity $\zeta = \{m(-k), k = 0, 1, 2, \ldots\} \in \mathcal{R}(\xi)$ will facilitate an instantaneous jump of the state from $(x, \xi, \psi(0), \psi)$ to the new state $(\hat{x}, \hat{\xi}, \hat{\psi}(0), \hat{\psi})$. The result follows immediately by substituting $(\hat{x}, \hat{\xi}, \hat{\psi}(0), \hat{\psi})$ into G_0 defined by (7.44). This proves the lemma. $\square$

Lemma 7.3.4 *If there is no fixed transaction cost (i.e., $\kappa = 0$ and $\mu > 0$) and if $(x, \xi, \psi(0), \psi) \in \partial_{I,1} \mathbf{S}_0$, that is,*

$$G_0(x, \xi, \psi(0), \psi) = 0,$$

then the only admissible strategy is to do no consumption but to close all open positions in order to bring his portfolio to $\{0\} \times \{\mathbf{0}\} \times [0, \infty) \times L^2_{\rho,+}$ after paying proportional transaction costs, capital gains taxes, and so forth.

Proof. For a fixed $(x, \xi, \psi(0), \psi) \in \mathbf{S}_0$, let $I \subset \aleph_0 \equiv \{0, 1, 2 \ldots\}$ be such that

$$i \in I \Rightarrow n(-i) < 0 \text{ and } i \notin I \Rightarrow n(-i) \geq 0.$$

To guarantee that $(X(t), N_t, S(t), S_t) \in \mathbf{S}_0$, we require that

$$G_0(X(t), N_t, S(t), S_t) \geq 0 \text{ for all } t \geq 0.$$

Applying Theorem 7.2.5 to the process

$$\{e^{-\lambda t} G_0(X(t), N_t, S(t), S_t), t \geq 0\},$$

we obtain

$$e^{-\lambda \tau} G_0(X(\tau), N_\tau, S(\tau), S_\tau)$$
$$= G_0(x, \xi, \psi(0), \psi) + \int_0^\tau (\partial_t + \mathbf{A})[e^{-\lambda t} G_0(X(t), N_t, S(t), S_t)]\, dt$$
$$+ \int_0^\tau \partial_{\psi(0)}[e^{-\lambda t} G_0(X(t), N_t, S(t), S_t)] S(t) f(S_t)\, dW(t)$$
$$+ \int_0^\tau \partial_x [e^{-\lambda t} G_0(X(t), N_t, S(t), S_t)](\lambda X(t) - C(t))\, dt$$
$$+ \sum_{0 \leq t \leq \tau} e^{-\lambda t}[G_0(X(t), N_t, S(t), S_t) - G_0(X(t-), N_{t-}, S(t), S_t)], \tag{7.55}$$

for all almost surely finite **G**-stopping time τ, where $X(t)$ and N_t are given in (7.8), and (7.9) and (7.10), respectively, with $\kappa = 0$.

Taking into the account of (7.8)-(7.10) and substituting into the function G_0, we have

$$G_0(X(t), N_t, S(t), S_t) = G_0(X(t-), N_{t-}, S(t), S_t).$$

Intuitively, this is also because of the invariance of the liquidated value of the assets without increase of stock value. Hence, (7.55) becomes the following by grouping the terms $n(Q(t) - i)$ according to $i \in I$ and $i \notin I$:

$$
\begin{aligned}
&d[e^{-\lambda t} G_0(X(t), N_t, S(t), S_t)] \\
&= e^{-\lambda t}\Big[-C(t) + \sum_{i \in I}(1 + \mu - \beta)n(Q(t) - i)S(t)(f(S_t) - \lambda) \\
&\quad + \sum_{i \notin I}(1 - \mu - \beta)n(Q(t) - i)S(t)(f(S_t) - \lambda) \\
&\quad - r\beta \sum_{i \in I} n(Q(t) - i)S(\tau(Q(t) - i)) \\
&\quad - \lambda\beta \sum_{i \notin I} n(Q(t) - i)S(\tau(Q(t) - i))\Big] dt \\
&\quad + e^{-\lambda t}\Big[\sum_{i \in I}(1 + \mu - \beta)n(Q(t) - i) \\
&\quad + \sum_{i \notin I}(1 - \mu - \beta)n(Q(t) - i)\Big] S(t)g(S_t)\, dW(t).
\end{aligned}
\tag{7.56}
$$

Now, the first exit time $\hat{\tau}$ ($\hat{\tau}$ is a **G**-stopping time) is defined by

$$
\begin{aligned}
\hat{\tau} \equiv 1 \wedge \inf \Big\{ t \geq 0 \mid & n(Q(t) - i)S(\tau(Q(t) - i)) \notin \\
& \text{the interval } (n(-i)\psi(\tau(-i)) - 1, 0) \text{ for } i \in I, \\
& \text{and } n(Q(t) - i)S(\tau(Q(t) - i)) \notin \\
& \text{the interval } (0, n(-i)\psi(\tau(-i)) + 1) \text{ for } i \notin I \Big\}.
\end{aligned}
$$

We can integrate (7.56) from 0 to $\hat{\tau}$, keeping in mind that $(x, \xi, \psi(0), \psi) \in \partial_{I,1}\mathbf{S}_0$ (or, equivalently, $G_0(x, \xi, \psi(0), \psi) = 0$), to obtain

$$0 \leq e^{-\lambda \hat{\tau}} G_0(X(\hat{\tau}), N_{\hat{\tau}}, S(\hat{\tau}), S_{\hat{\tau}})$$

$$= \int_0^{\hat{\tau}} e^{-\lambda s} \Big[-C(s) + \sum_{i \in I} (1 + \mu - \beta) n(Q(s) - i) S(s)(f(S_s) - \lambda)$$

$$+ \sum_{i \notin I} (1 - \mu - \beta) n(Q(s) - i) S(s)(f(S_s) - \lambda)$$

$$- \lambda \beta \sum_{i \in I} n(Q(s) - i) S(\tau(Q(s) - i))$$

$$- r\beta \sum_{i \notin I} n(Q(s) - i) S(\tau(Q(s) - i)) \Big] ds$$

$$+ \int_0^{\hat{\tau}} e^{-\lambda s} \Big[\sum_{i \in I} (1 + \mu - \beta) n(Q(s) - i)$$

$$+ \sum_{i \notin I} (1 - \mu - \beta) n(Q(s) - i) \Big] S(s) g(S_s) \, dW(s) \qquad (7.57)$$

Now, use the facts that $0 < \mu + \beta < 1$, $C(t) \geq 0$, $\alpha \geq f(S_t) > r > 0$, $n(-i) < 0$ for $i \in I$ and $n(-i) \geq 0$ for $i \notin I$, and Rule 7.6 to obtain the following inequality:

$$0 \leq e^{-\lambda \hat{\tau}} G_0(X(\hat{\tau}), N_{\hat{\tau}}, S(\hat{\tau}), S_{\hat{\tau}})$$

$$\leq \int_0^{\hat{\tau}} e^{-\lambda s} \Big[\sum_{i \notin I} (1 - \mu - \beta) n(Q(s) - i) S(t)(\alpha - \lambda) \, dt$$

$$+ \int_0^{\hat{\tau}} e^{-\lambda rs} \Big[\sum_{i \in I} (1 + \mu - \beta) n(Q(t) - i)$$

$$+ \sum_{i \notin I} (1 - \mu - \beta) n(Q(s) - i) \Big] S(s) g(S_s) \, dW(s) \qquad (7.58)$$

It is clear that

$$E \Big[\int_t^{\hat{\tau}} e^{-rs} \Big(\sum_{i \in I} (1 + \mu - \beta) n(Q(s) - i) \Big) S(s) g(S_s) \, dW(s) \Big] = 0.$$

Now, define the process

$$\tilde{W}(t) = \frac{\sigma - \lambda}{g(S_t)} t + W(t), \quad t \geq 0.$$

Then by the Girsanov transformation (see Theorem 1.2.16 in Chapter 1), $\{\tilde{W}(t), t \geq 0\}$ is a Brownian motion defined on a new probability space $(\Omega, \mathcal{F}, \tilde{P}; \mathbf{F})$, where $\tilde{P}$ and P are equivalent probability measures and, hence,

$$0 = E^{\tilde{P}}\Big[\int_0^{\hat{\tau}} e^{-\lambda s}\Big(\sum_{i \notin I}(1-\mu-\beta)n(Q(s)-i)\Big)S(s)g(S_s)d\tilde{W}(s)\Big]$$

$$= E\Big[\int_0^{\hat{\tau}} e^{-\lambda s}\Big(\sum_{i \notin I}(1-\mu-\beta)n(Q(s)-i)S(s)(\sigma-\lambda)\Big)ds$$

$$+ \int_0^{\hat{\tau}} e^{-\lambda s}\Big(\sum_{i \notin I}(1-\mu-\beta)n(Q(s)-i)S(s)g(S_s)dW(s)\Big)\Big].$$

Therefore,

$$\Big[\int_0^{\hat{\tau}} e^{-\lambda s}\Big(\sum_{k \notin I}(1-\nu-\beta)n(Q(s)-k)S(s)g(S_s)d\tilde{W}(s)\Big)\Big] = 0, \quad \tilde{P}\text{-a.s.}$$

Since $G_0(X(t), N_t, S(t), S_t) \geq 0$ for all $t \geq 0$, this implies that $\hat{\tau} = 0$ a.s., that is,

$$(X(\hat{\tau}), N_{\hat{\tau}}, S(\hat{\tau}), S_{\hat{\tau}}) = (x, \xi, \psi(0), \psi) \in \partial_{I,1}\mathcal{S}_0.$$

We need to determine the conditions under which the exit time occurred. Let k be the index of the shares of the stock where the state process violated the condition for the stopping time $\hat{\tau}$. In other words, if $k \in I$, then

$$\text{if } k \in I, \text{ then } n(Q(\hat{\tau})-k)S(\tau(Q(\hat{\tau})-k)) \notin (n(-k)\psi(-k))-1, 0)$$

or

$$\text{if } k \notin I, \text{ then } n(Q(\hat{\tau})-k)S(\tau(Q(\hat{\tau})-k) \notin (0, n(-k)\psi(-k))+1).$$

We will examine both cases separately.

Case 1. Suppose $k \in I$. Then

$$\text{either } n(Q(\hat{\tau})-k)S(\tau(Q(\hat{\tau})-k)) \leq n(-k)\psi(\tau(-k))-1$$
$$\text{or } n(Q(\hat{\tau})-k)S(\tau(Q(\hat{\tau})-k)) \geq 0.$$

We have established that

$$(X(\hat{\tau}), N_{\hat{\tau}}, S(\hat{\tau}), S_{\hat{\tau}}) \in \partial_{I,1}\mathbf{S}_0,$$

and this is inconsistent with

$$\text{both } n(Q(\hat{\tau})-k)S(\tau(Q(\hat{\tau})-k)) \leq n(-k)\psi(-k)-1$$
$$\text{and } n(Q(\hat{\tau})-k)S(\tau(Q(\hat{\tau})-k)) > 0.$$

Therefore, we know $n(Q(\hat{\tau}-k)) = 0$.

Case 2. Suppose $k \notin I$. Then

either $n(Q(\hat{\tau}) - k)S(\tau(Q(\hat{\tau}) - k)) \geq n(-k)\psi(-k) + 1$

or $n(Q(\hat{\tau}) - k)S(\tau(Q(\hat{\tau}) - k)) \leq 0$.

Again, since

$$(X(\hat{\tau}), N_{\hat{\tau}}, S(\hat{\tau}), S_{\hat{\tau}}) \in \partial_{I,1}\mathbf{S}_0,$$

we see that $n(Q(\hat{\tau}) - k) = 0$. We conclude from both cases that $(X(\hat{\tau}), N_{\hat{\tau}}) = (0, \{\mathbf{0}\})$. This means the only admissible strategy is to bring the portfolio from $(x, \xi, \psi(0), \psi)$ to $(0, \mathbf{0}, \psi(0), \psi)$ by an appropriate amount of the transaction specified in the lemma. This proves the lemma. $\square$

We have the following result.

Theorem 7.3.5 *Let* $\kappa > 0$ *and* $\mu > 0$. *On* $\partial_{I,1}\mathbf{S}_\kappa$ *for* $I \subset \aleph$; *then the investor should not consume but close all open positions in order to bring his portfolio to* $\{0\} \times \{\mathbf{0}\} \times \Re_+ \times L^2_{p,+}$. *In this case, the value function* $V_\kappa : \partial_{I,1}\mathbf{S}_\kappa \to \Re_+$ *satisfies the following equation:*

$$(\mathcal{M}_\kappa \Phi - \Phi)(x, \xi, \psi(0), \psi) = 0. \tag{7.59}$$

Proof. Suppose the *investor's* current portfolio is $(x, \xi, \psi(0), \psi) \in \partial_{I,1}\mathbf{S}_\kappa$ for some $I \subset \aleph_0$. A transaction of the quantity $\zeta = \{m(-k), k = 0, 1, 2, \ldots\} \in \mathcal{R}(\xi) - \{\mathbf{0}\}$ will facilitate an instantaneous jump of the state from $(x, \xi, \psi(0), \psi)$ to the new state $(\hat{x}, \hat{\xi}, \hat{\psi}(0), \hat{\psi})$ as given in (7.36)-(7.38).

We observe that since $\zeta = (m(-k), k = 0, 1, 2, \ldots) \in \mathcal{R}(\xi) - \{\mathbf{0}\}$, $n(-k) < 0$ implies $\hat{n}(-k) = n(-k) + m(-k) \leq 0$ and $n(-k) > 0$ implies $\hat{n}(-k) = n(-k) + m(-k) \geq 0$ for $k = 0, 1, 2, \ldots$. Taking into the account the new portfolio $(\hat{x}, \hat{\xi}, \hat{\psi}(0), \hat{\psi})$, we have from Lemma (4.3) that

$$G_0(\hat{x}, \hat{\xi}, \hat{\psi}(0), \hat{\psi}) = G_0(x, \xi, \psi(0), \psi) - \kappa. \tag{7.60}$$

Therefore, if $(x, \xi, \psi(0), \psi) \in \partial_{I,1}\mathbf{S}_\kappa$ for some $I \subset \aleph_0$, then

$$G_\kappa(x, \xi, \psi(0), \psi) = G_0(x, \xi, \psi(0), \psi) - \kappa = 0 = G_0(\hat{x}, \hat{\xi}, \hat{\psi}(0), \hat{\psi}).$$

This implies $(\hat{x}, \hat{\xi}, \hat{\psi}(0), \hat{\psi}) \in \partial_{I,1}\mathbf{S}_0$. From Lemma 7.3.3, we prove that the only admissible strategy is to make no consumption but to make another trading from the new state $(\hat{x}, \hat{\xi}, \hat{\psi}(0), \hat{\psi}) \in \partial_{I,1}\mathbf{S}_0$. Therefore, starting from $(x, \xi, \psi(0), \psi) \in \partial_{I,1}\mathbf{S}_\kappa$ we make two immediate instantaneous transactions (which will be counted as only one transaction), with the total amount specified by the following two equations:

$$0 = x - \kappa + \sum_{i \in I^c}[n(-i)\psi(0)(1 - \mu - \beta) + \beta n(-i)\psi(\tau(-i))]$$

$$+ \sum_{i \in I}[n(-i)\psi(0)(1 + \mu - \beta) + \beta n(-i)\psi(\tau(-i))], \tag{7.61}$$

$$0 = \xi \oplus \zeta \tag{7.62}$$

to reach the final destination $(0, \mathbf{0}, \psi(0), \psi)$. This proves the theorem. $\square$

We conclude the following boundary conditions from some simple observations and Theorem (4.5).

Boundary Condition (i). On the hyperplane

$$\partial_{-,\emptyset,2}\mathbf{S}_\kappa = \{(0, \xi, \psi(0), \psi) \in \mathbf{S}_\kappa \mid G_\kappa(0, \xi, \psi(0), \psi) < 0, n(-i) \geq 0 \; \forall i\},$$

the only strategy for the *investor* is to do no transaction and no consumption, since $x = 0$ and $G_\kappa(0, \xi, \psi(0), \psi) < 0$ (hence, there is no money to consume and not enough money to pay for the transaction costs, etc.), but to let the stock prices grow according to (7.1). Thus, the value function V_κ on $\partial_{-,\emptyset,2}\mathbf{S}_\kappa$ satisfies the equation

$$\mathcal{L}^0\Phi \equiv (\mathbf{A} - \alpha + rx\partial_x)\Phi = 0, \tag{7.63}$$

provided that it is smooth enough.

Boundary Condition (ii). On $\partial_{I,1}\mathbf{S}_\kappa$ for $I \subset \aleph_0$, the *investor* should not consume but buy back $n(-i)$ shares for $i \in I$ and sell $n(-i)$ shares for $i \in I^c$ of the stock in order to bring his portfolio to $\{0\} \times \{\mathbf{0}\} \times [0, \infty) \times L^2_{\rho,+}$ after paying transaction costs, capital gains taxes, and so forth. In other words, the investor brings his portfolio from the position $(x, \xi, \psi(0), \psi) \in \partial_{I,1}\mathbf{S}_\kappa$ to $(0, \mathbf{0}, \psi(0), \psi)$ by the quantity that satisfies (7.61) and (7.62). In this case, the value function $V_\kappa : \partial_{I,1}\mathbf{S}_\kappa \to \Re$ satisfies the following equation:

$$(\mathcal{M}_\kappa\Phi - \Phi)(x, \xi, \psi(0), \psi) = 0. \tag{7.64}$$

Note that this is a restatement of Theorem 7.3.5.

Boundary Condition (iii). On $\partial_{+,I,2}\mathbf{S}_\kappa$ for $I \subset \aleph_0$, the only optimal strategy is to make no transaction but to consume optimally according to the optimal consumption rate function $c^*(x, \xi, \psi(0), \psi,) = (\frac{\partial V_\kappa}{\partial x})^{\frac{1}{\gamma-1}}(x, \xi, \psi(0), \psi)$, which is obtained via

$$c^*(x, \xi, \psi(0), \psi) = \arg\max_{c\geq 0}\left\{\mathcal{L}^cV_\kappa(x, \xi, \psi(0), \psi) + \frac{c^\gamma}{\gamma}\right\},$$

where $\mathcal{L}^c$ is the differential operator defined by

$$\mathcal{L}^c\Phi(x, \xi, \psi(0), \psi) \equiv (\mathbf{A} - \alpha + (rx - c)\partial_x)\Phi. \tag{7.65}$$

This is because the cash in his *savings* account is not sufficient to buy back any shares of the *stock* but to consume optimally. In this case, the value function $V_\kappa : \partial_{+,I,2}\mathbf{S}_\kappa \to \Re_+$ satisfies the following equation provided that it is smooth enough:

$$\mathcal{A}\varPhi \equiv (\mathbf{A} - \alpha + rx\partial_x)\varPhi + \frac{1-\gamma}{\gamma}\left(\partial_x\varPhi\right)^{\frac{\gamma}{\gamma-1}} = 0. \qquad (7.66)$$

Boundary Condition (iv). On $\partial_{-,I,2}\mathbf{S}_\kappa$, the only admissible consumption-investment strategy is to do no consumption and no transaction but to let the stock price grows as in the Boundary Condition (i).

Boundary Condition (v). On $\partial_{+,\aleph_0,2}\mathbf{S}_\kappa = \{(x,\xi,\psi(0),\psi) \mid 0 \leq x \leq \kappa, n(-i) = 0, \forall i = 0,1,\dots\}$, the only admissible consumption-investment strategy is to do no transaction but to consume optimally as in Boundary Condition (iii).

Remark 7.3.6 *From Boundary Conditions (i)-(v), it is clear that the value function V_κ is discontinuous on the interfaces $Q_{+,I}$ and $Q_{-,I}$ for all $I \subset \aleph_0$.*

(D) The HJBQVI with Boundary Conditions
We conclude from the above subsections that the HJBQVI (together with the boundary conditions) can be expressed as

$$HJBQVI(*) = \begin{cases} \max\left\{\mathcal{A}\varPhi, \mathcal{M}_\kappa\varPhi - \varPhi\right\} = 0 & \text{on } \mathbf{S}_\kappa^\circ \\ \mathcal{A}\varPhi = 0, & \text{on } \bigcup_{I\subset\aleph_0}\partial_{+,I,2}\mathbf{S}_\kappa \\ \mathcal{L}^0\varPhi = 0, & \text{on } \bigcup_{I\subset\aleph_0}\partial_{-,I,2}\mathbf{S}_\kappa \\ \mathcal{M}_\kappa\varPhi - \varPhi = 0 & \text{on } \bigcup_{I\subset\aleph_0}\partial_{I,1}\mathbf{S}_\kappa, \end{cases}$$

where $\mathcal{A}\varPhi$, $\mathcal{L}^0\varPhi$ ($\mathcal{L}^c\varPhi$ with $c = 0$), and $\mathcal{M}_\kappa$ are as defined in (7.43), (7.63), and (7.35), respectively.

7.4 The Verification Theorem

Let

$$\tilde{\mathcal{A}}\varPhi = \begin{cases} \mathcal{A}\varPhi & \text{on } \mathbf{S}_\kappa^\circ \cup \bigcup_{I\subset\aleph}\partial_{+,I,2}\mathbf{S}_\kappa \\ \mathcal{L}^0\varPhi & \text{on } \bigcup_{I\subset\aleph}\partial_{-,I,2}\mathbf{S}_\kappa. \end{cases}$$

We have the following verification theorem for the value function $V_\kappa :$ $\mathcal{S}_\kappa \to \Re$ for our hereditary portfolio optimization problem.

Theorem 7.4.1 *(The Verification Theorem)*
(a) Let $U_\kappa = \mathbf{S}_\kappa - \bigcup_{I\subset\aleph_0}\partial_{I,1}\mathbf{S}_\kappa$. Suppose there exists a locally bounded non-negative-valued function $\varPhi \in C_{lip}^{1,0,2,2}(\mathbf{S}_\kappa) \cap \mathcal{D}(\mathcal{S})$ such that

$$\tilde{\mathcal{A}}\varPhi \leq 0 \text{ on } \mathcal{U}_\kappa \qquad (7.67)$$

and

$$\varPhi \geq \mathcal{M}_\kappa\varPhi \text{ on } \mathcal{U}_\kappa. \qquad (7.68)$$

Then $\varPhi \geq V_\kappa$ on $\mathcal{U}_\kappa$.

(b) Define $D \equiv \{(x, \xi, \psi(0), \psi) \in U_\kappa \mid \Phi(x, \xi, \psi(0), \psi) > \mathcal{M}_\kappa \Phi(x, \xi, \psi(0), \psi)\}$. Suppose

$$\tilde{\mathcal{A}}\Phi(x, \xi, \psi(0), \psi) = 0 \text{ on } D \qquad (7.69)$$

and that $\hat{\zeta}(x, \xi, \psi(0), \psi) = \hat{\zeta}_\Phi(x, \xi, \psi(0), \psi)$ exists for all $(x, \xi, \psi(0), \psi) \in \mathbf{S}_\kappa$ by Assumption 7.3.2. Let

$$c^* = \begin{cases} (\partial_x \Phi)^{\frac{1}{\gamma - 1}} & \text{on } \mathbf{S}_\kappa^\circ \cup \bigcup_{I \subset \aleph_0} \partial_{+, I, 2} \mathbf{S}_\kappa \\ 0 & \text{on } \bigcup_{I \subset \aleph_0} \partial_{-, I, 2} \mathbf{S}_\kappa. \end{cases} \qquad (7.70)$$

Define the impulse control $T^* = \{(\tau^*(i), \zeta^*(i)), i = 1, 2, \ldots\}$ inductively as follows:

First, put $\tau^*(0) = 0$ and inductively

$$\tau^*(i+1) = \inf\{t > \tau^*(i) \mid (X^{(i)}(t), N_t^{(i)}, S(t), S_t)) \notin D\}, \qquad (7.71)$$

$$\zeta^*(i+1) = \hat{\zeta}(X^{(i)}(\tau^*(i+1)-), N_{\tau^*(i+1)-}^{(i)}, S(\tau^*(i+1)), S_{\tau^*(i+1)}), \qquad (7.72)$$

$\{(X^{(i)}(t), N_t^{(i)}, S(t), S_t), t \geq 0\}$ is the controlled state process obtained by applying the combined control

$$\pi^*(i) = (c^*, (\tau^*(1), \tau^*(2), \ldots \tau^*(i); \zeta^*(1), \zeta^*(2), \cdots, \zeta^*(i))), \quad i = 1, 2, \ldots.$$

Suppose $\pi^* = (C^*, T^*) \in \mathcal{U}_\kappa(x, \xi, \psi(0), \psi)$,

$$e^{-\alpha t}\Phi(X^*(t), N_t^*, S(t), S_t) \to 0, \text{ as } t \to \infty \text{ a.s.}$$

and that the family

$$\{e^{-\alpha \tau}\Phi(X^*(\tau), N_\tau^*, S(\tau), S_\tau)) \mid \tau \text{ is a } \mathbf{G}\text{-stopping time}\} \qquad (7.73)$$

is uniformly integrable. Then $\Phi(x, \xi, \psi(0), \psi) = V_\kappa(x, \xi, \psi(0), \psi)$ and π^* obtained in (7.71) and (7.72) is optimal.

Proof.
(a) Suppose $\pi = (C, T) \in \mathcal{U}_\kappa(x, \xi, \psi(0), \psi)$, where $C = \{C(t), t \geq 0\}$ is a consumption rate process and $T = \{(\tau(i), \zeta(i)), i = 1, 2, \ldots\}$ is a trading strategy. Denote the controlled state processes (by π) with the initial state by $(x, \xi, \psi(0), \psi)$ by

$$\{Z(t) = (X(t), N_t, S(t), S_t), t \geq 0\}.$$

For $R > 0$, put

$$T(R) = R \wedge \inf\{t > 0 \mid \|Z(t)\| \geq R\}$$

and set
$$\theta(i+1) = \theta(i+1; R) = \tau(i) \vee (\tau(i+1) \wedge T(R)),$$

where $\|Z(t)\|$ is the norm of $Z(t)$ in $\Re \times \mathbf{N} \times \Re \times L_\rho^2$ in the product topology. Then by the generalized Dynkin formula (see (7.26)), we have

$$E[e^{-\alpha\theta(i+1)}\Phi(Z(\theta(i+1)-)] = E[e^{-\alpha\tau(i)}\Phi(Z(\tau(i)))]$$
$$+ \int_{\tau(i)}^{\theta(i+1)-} e^{-\alpha t} \mathcal{L}^{C(t)} \Phi(Z(t)) \, dt]$$
$$\leq E[e^{-\alpha\tau(i)}\Phi(Z(\tau(i)))]$$
$$- E\Big[\int_{\tau(i)}^{\theta(i+1)-} e^{-\delta t} \frac{C^\gamma(t)}{\gamma} \, dt\Big], \qquad (7.74)$$

since $\tilde{\mathcal{A}}\Phi \leq 0$.

Equivalently, we have

$$E[e^{-\alpha\tau(i)}\Phi(Z(\tau(i)))] - E[e^{-\alpha\theta(i+1)}\Phi(Z(\theta(i+1)-))]$$
$$\geq E\Big[\int_{\tau(i)}^{\theta(i+1)-} e^{-\alpha t} \frac{C^\gamma(t)}{\gamma} dt\Big].$$

Letting $R \to \infty$, using the Fatou lemma, and then summing from $i = 0$ to $i = k$ gives

$$\Phi(x, \xi, \psi(0), \psi) + \sum_{i=1}^{k} E\Big[e^{-\alpha\tau(i)}\Big(\Phi(Z(\tau(i))) - \Phi(Z(\tau(i-)))\Big)\Big]$$
$$- E[e^{-\alpha\tau(k+1)}\Phi(Z(\tau(k+1)-))]$$
$$\geq E\Big[\int_{0}^{\theta(k+1)} e^{-\delta t} \frac{C^\gamma(t)}{\gamma} dt\Big]. \qquad (7.75)$$

Now,
$$\Phi(Z(\tau(i))) \leq \mathcal{M}_\kappa \Phi(Z(\tau(i)-)) \text{ for } i = 1, 2, \ldots \qquad (7.76)$$

and, therefore,

$$\Phi(x, \xi, \psi(0), \psi) + \sum_{i=1}^{k} E\Big[e^{-\alpha\tau(i)}\Big(\mathcal{M}_\kappa \Phi(Z(\tau(i)-)) - \Phi(Z(\tau(i)-))\Big)\Big]$$
$$\geq E\Big[\int_{0}^{\theta(k+1)-} e^{-\delta t} \frac{C^\gamma(t)}{\gamma} dt + e^{-\alpha\tau(k+1)}\Phi(Z(\tau(k+1)-))\Big]. \qquad (7.77)$$

It is clear that
$$\mathcal{M}_\kappa \Phi(Z(\tau(i)-)) - \Phi(Z(\tau(i)-)) \leq 0 \qquad (7.78)$$

and, hence,

$$\Phi(x,\xi,\psi(0),\psi) \geq E\Big[\int_0^{\theta(k+1)-} e^{-\alpha t} \frac{C^\gamma(t)}{\gamma} dt$$

$$+ e^{-\alpha\tau(k+1)}\Phi(Z(\tau(k+i)-))\Big]. \qquad (7.79)$$

Letting $k \to \infty$, we get

$$\Phi(x,\xi,\psi(0),\psi) \geq E\Big[\int_0^\infty e^{-\alpha t} \frac{C^\gamma(t)}{\gamma} dt \Big], \qquad (7.80)$$

since Φ is a locally bounded non-negative function. Hence,

$$\Phi(x,\xi,\psi(0),\psi) \geq J_\kappa(x,\xi,\psi(0),\psi;\pi), \quad \forall \pi \in \mathcal{U}_\kappa(x,\xi,\psi(0),\psi). \qquad (7.81)$$

Therefore, $\Phi(x,\xi,\psi(0),\psi) \geq V_\kappa(x,\xi,\psi(0),\psi)$.

(b) Define $\pi^* = (C^*, \mathcal{T}^*)$, where $\mathcal{T}^* = \{(\tau^*(i), \zeta^*(i)), i = 1, 2, \ldots\}$ by (7.71) and (7.72). Then repeat the argument in part (a) for $\pi = \pi^*$. It is clear that the inequalities (7.79)-(7.81) become equalities. So we conclude that

$$\Phi(x,\xi,\psi(0),\psi) = E\Big[\int_0^{\tau^*(k+1)} e^{-\alpha t} \frac{C^\gamma(t)}{\gamma} dt$$

$$+ e^{-\alpha\tau^*(k+1)}\Phi(Z(\tau^*(k+1)-))\Big], \quad \forall k = 1, 2, \ldots. \qquad (7.82)$$

Letting $k \to \infty$ in (7.82), by (7.73) we get

$$\Phi(x,\xi,\psi(0),\psi) = J_\kappa(x,\xi,\psi(0),\psi;\pi^*). \qquad (7.83)$$

Combining this with (7.81), we obtain

$$\Phi(x,\xi,\psi(0),\psi) \geq \sup_{\pi \in \mathcal{U}_\kappa(x,\xi,\psi(0),\psi)} J_\kappa(x,\xi,\psi(0),\psi;\pi)$$

$$\geq J_\kappa(x,\xi,\psi(0),\psi;\pi^*)$$

$$= \Phi(x,\xi,\psi(0),\psi). \qquad (7.84)$$

Hence, $\Phi(x,\xi,\psi(0),\psi) = V_\kappa(x,\xi,\psi(0),\psi)$ and π^* is optimal. This proves the verification theorem. □

7.5 Properties of Value Function

7.5.1 Some Simple Properties

Some basic properties of the value function $V_\kappa : \mathcal{S}_\kappa \to \Re^+$ defined by (7.13) are investigated in this section.

Suppose the *investor's* portfolio is at $(x,\xi,\psi(0),\psi) \in \mathbf{S}_\kappa$ and an instantaneous transaction of the **N**-valued quantity

$$\zeta = m(0)\mathbf{1}_{\{\tau(0)\}} + \sum_{k=1}^{\infty} m(-k)\mathbf{1}_{\{\tau(-k)\}}(\chi_{\{n(-k)<0,0\leq m(-k)\leq -n(-k)\}}$$

$$+ \chi_{\{n(-k)>0,-n(-k)\leq m(-k)\leq 0\}}) \in \mathcal{R}(\xi)$$

leads to a new state of the portfolio $(\hat{x},\hat{\xi},\hat{\psi}(0),\hat{\psi})$, where $\hat{x}$ (the *investor's* new holdings in the *savings* account), $\hat{\xi}$ (the *investor's* new inventory in the *stock* account), and $(\hat{\psi}(0),\hat{\psi})$ (the new profile of stock prices) are given in (7.36)-(7.38) and are repeated below for the convenience of the readers.

$$\hat{x} = x - \kappa - (m(0) + \mu|m(0)|)\psi(0)$$

$$- \sum_{k=1}^{\infty} \Big[(1 - \mu - \beta)m(-k)\psi(0)$$

$$+ \beta m(-k)\psi(\tau(-k))\Big]\chi_{\{n(-k)>,-n(-k)\leq m(-k)\leq 0\}}$$

$$- \sum_{k=1}^{\infty} \Big[(1 + \mu - \beta)m(-k)\psi(0) + \beta m(-k)\psi(\tau(-k))\Big]$$

$$\times \chi_{\{n(-k)<0,0\leq m(-k)\leq -n(-k)\}}, \tag{7.85}$$

$$\hat{\xi} = \xi \oplus \zeta, \tag{7.86}$$

and

$$(\hat{\psi}(0),\hat{\psi}) = (\psi(0),\psi), \tag{7.87}$$

where $\chi_{\{\cdots\}}$ is the indicator function of the set (or event) $\{\cdots\}$, and $\xi \oplus \zeta : (-\infty,0] \to \Re$ is defined by

$$(\xi \oplus \zeta)(\theta) = m(0)\mathbf{1}_{\{\tau(0)\}}(\theta) \tag{7.88}$$

$$+ \sum_{k=1}^{\infty} \Big[n(-k) + m(-k)(\chi_{\{n(-k)<0,0\leq m(-k)\leq -n(-k)\}}$$

$$+ \chi_{\{n(-k)>0,-n(-k)\leq m(-k)\leq 0\}})\Big]\mathbf{1}_{\{\tau(-k)\}}(\theta), \ \forall \theta \in (-\infty,0],$$

or simply by the sequence

$$\xi \oplus \zeta = \{\hat{n}(-k), k = 0,1,2,\ldots\}, \tag{7.89}$$

with

$$\hat{n}(0) = m(0)$$

and

$$\hat{n}(-k) = n(-k) + m(-k)(\chi_{\{n(-k)<0,0\leq m(-k)\leq -n(-k)\}}$$

$$+\chi_{\{n(-k)>0,-n(-k)\leq m(-k)\leq 0\}}) \text{ for } k = 1,2,\cdots.$$

In this case, the portfolio $(\hat{x},\hat{\xi},\hat{\psi}(0),\hat{\psi})$ is said to be reachable from $(x,\xi,\psi(0),\psi)$ in one transaction.

We have the following trivial result.

Theorem 7.5.1 *If the portfolio* $(\hat{x}, \hat{\xi}, \hat{\psi}(0), \hat{\psi}) \in \mathbf{S}_\kappa$ *is reachable from* $(x, \xi, \psi(0), \psi) \in \mathbf{S}_\kappa$ *in one transaction, then*

$$V_\kappa(\hat{x}, \hat{\xi}, \hat{\psi}(0), \hat{\psi}) \leq V_\kappa(x, \xi, \psi(0), \psi).$$

Moreover, we have $\forall (x, \xi, \psi(0), \psi) \in \mathbf{S}_\kappa,$

$$\mathcal{M}_\kappa V_\kappa(x, \xi, \psi(0), \psi) \leq V_\kappa(x, \xi, \psi(0), \psi).$$

Proof. Suppose for contradiction purposes that there exists a $(x, \xi, \psi(0), \psi) \in \mathbf{S}_\kappa$ such that

$$V_\kappa(\hat{x}, \hat{\xi}, \hat{\psi}(0), \hat{\psi}) > V_\kappa(x, \xi, \psi(0), \psi).$$

If this were true, then one would start at such a position and then make one immediate transaction to $(\hat{x}, \hat{\xi}, \hat{\psi}(0), \hat{\psi})$ to achieve a higher value without making any consumption. This contradicts the definition of the value function V_κ. Therefore,

$$V_\kappa(\hat{x}, \hat{\xi}, \hat{\psi}(0), \hat{\psi}) \leq V_\kappa(x, \xi, \psi(0), \psi).$$

The second statement follows from the first immediately.   □

7.5.2 Upper Bounds of Value Function

In reality, the *solvency region* $\mathbf{S}_\kappa$, the liquidating function $G_\kappa : \mathbf{S} \to \Re$, and the value function V_κ defined in (7.6), (7.5), and (7.13), respectively, depend not only on the fixed transaction cost $\kappa \geq 0$ but also on the proportional transaction and tax rates $\mu \geq 0$ and $\beta \geq 0$. In this section, we will express $\mathbf{S}_\kappa$ as $\mathbf{S}_{\kappa,\mu,\beta}$, G_κ as $G_{\kappa,\mu,\beta}$, and V_κ as $V_{\kappa,\mu,\beta}$ to reflect such effects. Therefore, $\mathbf{S}_{0,\mu,\beta}$, $G_{0,\mu,\beta}$, and $V_{0,\mu,\beta}$ denote respectively the solvency region, the liquidating function, and the value function, when there is no fixed transaction cost ($\kappa = 0$ and $\mu, \beta > 0$) but there are positive proportional transaction costs and taxes. All other expressions will be interpreted similarly. For example, $\mathbf{S}_{0,\mu,0}$, $G_{0,\mu,0}$, and $V_{0,\mu,0}$ will be interpreted respectively as the solvency region, the liquidating function, and the value function, when there are no fixed transaction cost and tax but there are proportional transaction costs ($\kappa = \beta = 0$ and $\mu > 0$) and so forth.

When $\kappa = 0$, the solvency region $\mathbf{S}_{0,\mu,\beta}$ reduces from (7.6) to

$$\mathbf{S}_{0,\mu,\beta} = \Big\{ (x, \xi, \psi(0), \psi) \in \mathbf{S} \mid G_{0,\mu,\beta}(x, \xi, \psi(0), \psi) \geq 0 \Big\} \cup \mathbf{S}_+, \qquad (7.90)$$

where

$$G_{0,\mu,\beta}(x, \xi, \psi(0), \psi) = x + \sum_{k=0}^{\infty} \Big[\min\{(1-\mu)n(-k), (1+\mu)n(-k)\}\psi(0)$$

$$-n(-k)\beta(\psi(0) - \psi(\tau(-k))) \Big], \qquad (7.91)$$

$\mathbf{S}_+ = [0,\infty) \times \mathbf{N}_+ \times [0,\infty) \times \Re_+ \times L^2_{\rho,+}$, and $\mathbf{N}_+ = \{\xi \in \mathbf{N} \mid \xi(\theta) \geq 0, \forall \theta \in (-\infty, 0]\}$.

Similarly, when $\kappa = \beta = 0$, the solvency region $\mathbf{S}_{0,\mu,0}$ can be described by

$$\mathbf{S}_{0,\mu,0} = \left\{ (x, \xi, \psi(0), \psi) \in \mathbf{S} \mid G_{0,\mu,0}(x, \xi, \psi(0), \psi) \geq 0 \right\} \cup \mathbf{S}_+, \qquad (7.92)$$

where

$$G_{0,\mu,0}(x, \xi, \psi(0), \psi) = x + \sum_{k=0}^{\infty} \left[\min\{(1-\mu)n(-k), (1+\mu)n(-k)\} \psi(0) \right].$$
$$(7.93)$$

For each $(\psi(0), \psi) \in [0,\infty) \times L^2_{\rho,+}$, let $\mathbf{S}_{0,\mu,\beta}(\psi(0), \psi)$ be the projection of the *solvency region* $\mathbf{S}_{0,\mu,\beta}$ along $(\psi(0), \psi)$ defined by

$$\begin{aligned}\mathbf{S}_{0,\mu,\beta}(\psi(0), \psi) &\equiv \{(x, \xi) \in \Re \times \mathbf{N} \mid H_{0,\mu,\beta}(x, \xi, \psi(0), \psi) \geq 0\} \\ &= \{(x, \xi) \in \Re \times \mathbf{N} \mid G_{0,\mu,\beta}(x, \xi, \psi(0), \psi) \geq 0\}.\end{aligned}$$

We have the following results.

Proposition 7.5.2 *For each $(\psi(0), \psi) \in [0,\infty) \times L^2_{\rho,+}$, the projected solvency region $\mathbf{S}_{0,\mu,\beta}(\psi(0), \psi)$ along $(\psi(0), \psi)$ is a convex subset of the space $\Re \times \mathbf{N}$. Furthermore, if $V_{0,\mu,\beta} : \mathbf{S}_{0,\mu,\beta} \to \Re$ is the value function of the optimal consumption-trading problem when there is no fixed transaction cost, then for each $(\psi(0), \psi) \in [0,\infty) \times L^2_{\rho,+}$, $V_{0,\mu,\beta}(\cdot, \cdot, \psi(0), \psi)$ is a concave function on the projected solvency region $\mathbf{S}_{0,\mu,\beta}(\psi(0), \psi)$.*

Proof. If

$$x \geq 0 \text{ and } n(-i) \geq 0, \quad \forall i = 0, 1, 2, \cdots,$$

then

$$G_{0,\mu,\beta}(x, \xi, \psi(0), \psi) = x + \psi(0) \sum_{k=0}^{\infty} (1 - \mu - \beta) n(-k)$$

$$+ \beta \sum_{k=0}^{\infty} n(-k) \psi(\tau(-k))$$

$$\geq 0, \quad \text{since } 1 - \mu - \beta > 0.$$

Hence,

$$[0,\infty) \times \mathbf{N}_+ \times [0,\infty) \times L^2_{\rho,+} \subset \{(x, \xi, \psi(0), \psi) \mid G_{0,\mu,\beta}(x, \xi, \psi(0), \psi) \geq 0\}.$$

Therefore, all shares of the *stock* owned or owed can be liquidated because of the absence of a fixed transaction cost $\kappa = 0$.

For a fixed $(\psi(0), \psi) \in [0, \infty) \times L^2_{\rho,+}$, let

$$(x_1, \xi_1), (x_2, \xi_2) \in \mathbf{S}_{0,\mu,\beta}(\psi(0), \psi) \quad \text{and} \quad 0 \le c \le 1;$$

then

$$c(x_1, \xi_1) + (1 - c)(x_2, \xi_2) \in \mathbf{S}_{0,\mu,\beta}(\psi(0), \psi).$$

This is because

$$G_{0,\mu,\beta}(cx_1 + (1-c)x_2, c\xi_1 + (1-c)\xi_2, \psi(0), \psi)$$
$$= G_{0,\mu,\beta}(cx_1, c\xi_1, \psi(0), \psi) + G_{0,\mu,\beta}((1-c)x_2, (1-c)\xi_2, \psi(0), \psi)$$
$$= cG_{0,\mu,\beta}(x_1, \xi_1, \psi(0), \psi) + (1-c)G_0(x_2, \xi_2, \psi(0), \psi) \ge 0.$$

The value function $V_{0,\mu,\beta} : \mathbf{S}_{0,\mu,\beta} \to \Re_+$ for the case $\kappa = 0$ and $\mu, \beta > 0$ has the following concavity property: For each fixed $(\psi(0), \psi) \in [0, \infty) \times L^2_{\rho,+}$, $V_{0,\mu,\beta}(\cdot, \cdot, \psi(0), \psi) : \mathbf{S}_{0,\mu,\beta}(\psi(0), \psi) \to \Re_+$ is a concave function; that is, if $(x_1, \xi_1), (x_2, \xi_2) \in \mathbf{S}_{0,\mu,\beta}(\psi(0), \psi)$ and $0 \le c \le 1$, then

$$V_{0,\mu,\beta}(cx_1 + (1-c)x_2, c\xi_1 + (1-c)\xi_2, \psi(0), \psi)$$
$$\ge cV_{0,\mu,\beta}(x_1, \xi_1, \psi(0), \psi) + (1-c)V_{0,\mu,\beta}(x_2, \xi_2, \psi(0), \psi).$$

The detailed proof of this statement is omitted.   □

Using the verification theorem (Theorem 7.4.1), we obtain some upper bounds of the value function $V_{\kappa,\mu,\beta} : \mathbf{S}_{\kappa,\mu,\beta} \to \Re$ defined by (7.13) by comparing it with other value functions such as $V_{0,0,0}$, $V_{\kappa,\nu,\beta}$, and so forth.

(A) When $\kappa = \mu = \beta = 0$

We consider the scenario when $\kappa = \mu = \beta = 0$. In this case, there is no need to keep track of the time instants and the base prices when shares of the stock were purchased or shortsold in the past and we can lump all shares together, since there are no tax consequences. We therefore let $Y(t)$ be the total shares of the stock owned (if $Y(t) > 0$) or owed (if $Y(t) < 0$) by the investor at time $t \ge 0$, that is,

$$Y(t) = \sum_{k=0}^{Q(t)} n(k).$$

The stock price dynamics $\{S(t), t \in \Re\}$ remains unchanged, but the investor's savings and stock accounts should then be modified as follows:

$$dX(t) = (\lambda X(t) - C(t))dt, \quad \tau(i) \le t < \tau(i+1), \tag{7.94}$$

$$X(\tau(i+1)) = X(\tau(i+1)-) - m(i+1)S(\tau(i+1)), \tag{7.95}$$

and

$$Y(\tau(i+1)) = Y(\tau(i+1)-) + m(i+1), \quad i = 0, 1, 2, \ldots, \tag{7.96}$$

where $m(i+1) \in \Re$ is the number of shares of the stock purchased or sold at the transaction time $\tau(i+1)$.

Let $V_{0,0,0}(x, \xi, \psi(0), \psi)$ be the value function of the portfolio optimization problem when $\kappa = \mu = \beta = 0$. Assume that it takes the following form for some constant $C_1 > 0$:

$$V_{0,0,0}(x, \xi, \psi(0), \psi) = C_1 \left(x + \psi(0) \sum_{k=0}^{\infty} n(-k) \right)^{\gamma}. \tag{7.97}$$

We have the following proposition.

Proposition 7.5.3 *Assume that the value function*

$$V_{0,0,0} : \mathbf{S}_{\kappa,\mu,\beta} \subset \mathbf{S}_{0,0,0} \to \Re$$

takes the form in (7.97). Then

$$C_1 = \frac{1}{\gamma} C_0^{\gamma-1} \quad with \quad C_0 = \frac{1}{1-\gamma} \left[\alpha - \gamma\lambda - \frac{\gamma(\underline{b}-\lambda)^2}{2\sigma(1-\gamma)} \right], \tag{7.98}$$

provided that

$$\alpha > \gamma \left[\lambda + \frac{(\underline{b}-\lambda)^2}{2\sigma(1-\gamma)} \right]. \tag{7.99}$$

Moreover,

$$V_{\kappa,\mu,\beta}(x, \xi, \psi(0), \psi) \le V_{0,0,0}(x, \xi, \psi(0), \psi), \quad \forall(x, \xi, \psi(0), \psi) \in \mathbf{S}_{\kappa,\mu,\beta}. \tag{7.100}$$

Proof. For notational simplicity, let

$$\Lambda_0(x, \xi, \psi(0), \psi) = x + \psi(0) \left(\sum_{k=0}^{\infty} n(-k) \right).$$

Therefore,

$$V_{0,0,0}(x, \xi, \psi(0), \psi) = C_1 [\Lambda_0(x, \xi, \psi(0), \psi)]^{\gamma}.$$

We will prove that $V_{0,0,0}$ satisfies part (a) of the verification theorem (Theorem 7.4.1). Then

$$V_{0,0,0} \ge V_{\kappa,\mu,\beta} \quad \text{on } \mathbf{U}_{\kappa,\mu,\beta} \equiv \mathbf{S}_{\kappa,\mu,\beta} - \cup_{I \subset \aleph_0} \partial_{I,1} \mathbf{S}_{\kappa,\mu,\beta}.$$

First, we must show that

$$\mathcal{M}_{\kappa,\mu,\beta}(V_{0,0,0}(x, \xi, \psi(0), \psi)) \le V_{0,0,0}(x, \xi, \psi(0), \psi) \text{ on } \mathbf{S}_{\kappa,\mu,\beta}^{\circ}.$$

Observe that $V_{0,0,0} \in C_{lip}^{1,0,2,2}(\mathbf{S}_{\kappa,\mu,\beta}^{\circ}) \cap \mathcal{D}(\mathcal{S})$. A straightforward calculation yields

$$\Lambda_0(\hat{x},\hat{\xi},\hat{\psi}(0),\hat{\psi}) = \hat{x} + \psi(0)\left(\sum_{k=0}^{\infty}\hat{n}(-k)\right)$$

$$\leq x + \psi(0)\left[\sum_{k=0}^{\infty}(n(-k) + m(-k)(\chi_{\{n(-k)<0,0\leq m(-k)\leq -n(-k)\}}\right.$$

$$+\chi_{\{n(-k)>0,-n(-k)\leq m(-k)\leq 0\}})\Big]$$

$$\leq x + \psi(0)\left[\sum_{k=0}^{\infty}n(-k)(\chi_{\{n(-k)>0}} + \chi_{\{n(-k)<0\}})\right]$$

$$= \Lambda_0(x,\xi,\psi(0),\psi)$$

where $\hat{x}$, $\hat{\xi}$, and $(\hat{\psi}(0),\hat{\psi})$ are given in (7.85)-(7.88). Therefore,

$$\mathcal{M}_{\kappa,\mu,\beta}V_{0,0,0}(x,\xi,\psi(0),\psi) = C_1\mathcal{M}_{\kappa,\mu,\beta}[\Lambda_0(x,\xi,\psi(0),\psi)]^{\gamma}$$

$$= C_1\sup\left\{[\Lambda_0(\hat{x},\hat{\xi},\hat{\psi}(0),\hat{\psi})]^{\gamma} \mid \zeta \in \mathcal{R}(\xi) - \{\mathbf{0}\}\right\}$$

$$\leq C_1[\Lambda_0(x,\xi,\psi(0),\psi)]^{\gamma}$$

$$= V_{0,0,0}(x,\xi,\psi(0),\psi). \tag{7.101}$$

Second, let us show that (7.67) also holds.

$$\mathcal{A}(V_{0,0,0}(x,\xi,\psi(0),\psi))$$

$$= \sup_{c\geq 0}\left\{\mathcal{L}^c(V_{0,0,0})(x,\xi,\psi(0),\psi) + \frac{c^{\gamma}}{\gamma}\right\}$$

$$= (\mathbf{A} - \alpha + \lambda x\partial_x)(V_{0,0,0})(x,\xi,\psi(0),\psi)$$

$$+\frac{1-\gamma}{\gamma}\left(\partial_x V_{0,0,0}\right)^{\frac{\gamma}{\gamma-1}}(x,\xi,\psi(0),\psi). \tag{7.102}$$

Now, by (7.21),

$$\mathcal{A}(V_{0,0,0})(x,\xi,\psi(0),\psi) = (\mathbf{A} - \alpha + rx\partial_x)(V_{0,0,0})(x,\xi,\psi(0),\psi)$$

$$+\frac{1-\gamma}{\gamma}(\partial_x V_{0,0,0})^{\frac{\gamma}{\gamma-1}}(x,\xi,\psi(0),\psi)$$

$$= C_1(\mathbf{A} - \alpha + \lambda x\partial_x)[\Lambda_0(x,\xi,\psi(0),\psi)]^{\gamma})$$

$$+\frac{1-\gamma}{\gamma}(C_1\partial_x\Lambda_0)^{\frac{\gamma}{\gamma-1}}(x,\xi,\psi(0),\psi)$$

$$= C_1\gamma\Lambda_0^{\gamma-1}\psi(0)f(\psi)y + C_1\gamma(\gamma-1)\Lambda_0^{\gamma-2}\frac{1}{2}y^2\psi^2(0)g^2(\psi)$$

$$-\alpha C_1\Lambda_0^{\gamma} + C_1\lambda x\gamma\Lambda_0^{\gamma-1} + C_1^{\frac{\gamma}{\gamma-1}}(1-\gamma)\gamma^{\frac{1}{\gamma-1}}\Lambda_0^{\gamma}, \tag{7.103}$$

where $y = \sum_{k=0}^{\infty} n(-k)$ and $\Lambda_0 = \Lambda_0(x, \xi, \psi(0), \psi)$. Since

$$\lambda < \underline{b} \le f(\phi) \le \bar{b} \quad \text{and} \quad \underline{\sigma} \le g(\phi) \le \bar{\sigma} \; \forall \phi \in L_\rho^2(\Re_-),$$

$$\mathcal{A}(C_1 \Psi_0)(x, \xi, \psi(0), \psi) \le 0, \quad \forall(x, \xi, \psi(0), \psi) \in \mathbf{S}_{\kappa, \mu, \beta},$$

if and only if

$$C_1 = \frac{1}{\gamma} C_0^{\gamma - 1} \quad \text{with} \quad C_0 = \frac{1}{1 - \gamma} \left[\alpha - \gamma \lambda - \frac{\gamma(\underline{b} - \lambda)^2}{2\sigma(1 - \gamma)} \right],$$

provided that

$$\alpha > \gamma \left[\lambda + \frac{(\underline{b} - \lambda)^2}{2\sigma(1 - \gamma)} \right].$$

Third, we must also show that

$$\mathcal{L}^0(V_{0,0,0})(x, \xi, \psi(0), \psi) \le 0, \quad \forall(x, \xi, \psi(0), \psi) \in \cup_{I \subset \aleph_0} \partial_{-,I} \mathbf{S}_{\kappa, \mu, \beta},$$

where

$$\mathcal{L}^0(C_1 \Psi_0)(x, \xi, \psi(0), \psi) = (\mathbf{A} + \lambda x \partial_x - \alpha I)(C_1 \Psi_0)(x, \xi, \psi(0), \psi).$$

The proof involves the boundary conditions of the value function and it is up to the readers to provide the details. Concluding from the above, we have proved the proposition.   □

Note that the above result is also implied by the historical work of Merton (see [Mer90]) on the optimal consumption-investment problem in a perfect market when there is no transaction cost and no tax, and the stock price follows the geometric Brownian motion, with $f(\phi) \equiv \underline{b} \ge \lambda$ and $g(\phi) \equiv \sigma$. In that case, it is shown that the optimal portfolio satisfies the following Merton line:

$$\frac{Y(t)}{X(t)} = \frac{c^*}{1 - c^*}, \quad t \ge 0, \tag{7.104}$$

where

$$c^* = \frac{\underline{b} - \lambda}{(1 - \gamma)\sigma^2}. \tag{7.105}$$

(B) When $\beta = 0$, $\kappa > 0$, and $\mu > 0$

In this case, we have the following result.

Proposition 7.5.4 *Assume that $\beta = 0$ but $\kappa > 0$ and $\mu > 0$. Let ν be a constant such that*

$$1 - \mu \le \nu \le 1 + \mu. \tag{7.106}$$

Suppose

$$\alpha > \gamma \underline{b}. \tag{7.107}$$

Then there exists $K < \infty$ such that $\forall (x, \xi, \psi(0), \psi) \in \mathbf{S}_{\kappa,\mu,\beta}$,

$$V_{\kappa,\mu,\beta}(x, \xi, \psi(0), \psi) \le K \left(x + \nu\psi(0) \sum_{k=0}^{\infty} n(-k) \right)^{\gamma}. \tag{7.108}$$

Proof. We proceed as in Proposition 7.5.3, except that now we choose $K < \infty$ and define $\Lambda_\nu : \mathbf{S}_{\kappa,\mu,\beta} \to \Re$ and $\Psi_\nu : \mathbf{S}_\kappa \to \Re$ by

$$\Lambda_\nu(x, \xi, \psi(0), \psi) = x + \nu\psi(0) \left(\sum_{k=0}^{\infty} n(-k) \right) \tag{7.109}$$

and

$$\Psi_\nu(x, \xi, \psi(0), \psi) = [\Lambda_\nu(x, \xi, \psi(0), \psi)]^{\gamma} \equiv \Lambda_\nu^{\gamma}(x, \xi, \psi(0), \psi). \tag{7.110}$$

Then we have, from (7.36)-(7.38),

$$\hat{x} + \nu\psi(0) \left(\sum_{k=0}^{\infty} \hat{n}(-k) \right)$$

$$= \begin{cases} x - \kappa + \nu\psi(0) \left(\sum_{k=0}^{\infty} n(-k) \right) - m(1 + \mu - \nu) & \text{for } m > 0 \\ x - \kappa + \nu\psi(0) \left(\sum_{k=0}^{\infty} n(-k) \right) - m(1 - \mu - \nu) & \text{for } m < 0. \end{cases}$$

Thus, in any case, we have, by (7.106),

$$\hat{x} + \nu\psi(0) \left(\sum_{k=0}^{\infty} \hat{n}(-k) \right) \le x + \nu\psi(0) \left(\sum_{k=0}^{\infty} n(-k) \right),$$

and this proves that

$$\mathcal{M}_{\kappa,\mu,\beta}\Psi_\nu(x, \xi, \psi(0), \psi) \le \Psi_\nu(x, \xi, \psi(0), \psi), \quad \forall (x, \xi, \psi(0), \psi) \in \mathbf{S}_{\kappa,\mu,\beta}.$$

Using the verification theorem (Theorem 7.4.1), it remains to verify that

$$\mathcal{A}\Psi_\nu(x, \xi, \psi(0), \psi) \le 0, \quad \forall (x, \xi, \psi(0), \psi) \in \mathbf{S}_{\kappa,\mu,\beta}^{\circ}.$$

A straightforward calculation yields

$$\mathcal{A}(K\Psi_\nu)(x, \xi, \psi(0), \psi)$$

$$= \sup_{c \ge 0} \left\{ \mathcal{L}^c(K\Psi_\nu)(x, \xi, \psi(0), \psi) + \frac{c^{\gamma}}{\gamma} \right\}$$

$$= (\mathbf{A} - \alpha + \lambda x \partial_x)(K\Psi_\nu)(x, \xi, \psi(0), \psi)$$

$$+ \frac{1 - \gamma}{\gamma} \left(K \partial_x \Psi_\nu \right)^{\frac{\gamma}{\gamma-1}} (x, \xi, \psi(0), \psi)$$

$$= \frac{1}{2}K\gamma(\gamma-1)\Lambda_\nu^{\gamma-2}(x,\xi,\psi(0),\psi)\Big[\sum_{k=0}^{\infty}(1+\nu)n(-k)\Big]^2\psi^2(0)g^2(\psi)$$

$$+ K\gamma\Lambda_\nu^{\gamma-1}(x,\xi,\psi)\Big[\sum_{k=0}^{\infty}(1+\nu)n(-k)\Big]\psi(0)f(\psi)$$

$$- \alpha K\Lambda_\nu^\gamma(x,\xi,\psi(0),\psi) + K\gamma\lambda x\Lambda_\nu^{\gamma-1}(x,\xi,\psi(0),\psi)$$

$$+ \frac{1-\gamma}{\gamma}(K\gamma)^{\frac{\gamma}{\gamma-1}}\lambda_\nu^\gamma(x,\xi,\psi(0),\psi)$$

$$= \Lambda_\nu^{\gamma-2}(x,\xi,\psi(0),\psi)\Big\{\Big(\frac{1-\gamma}{\gamma}(K\gamma)^{\frac{\gamma}{\gamma-1}} - \alpha K\Big)\Lambda_\nu^2(x,\xi,\psi(0),\psi)$$

$$+ K\gamma\Big[\lambda x + \Big(\sum_{k=0}^{\infty}(1+\nu)n(-k)\Big)\psi(0)f(\psi)\Big]\Lambda_\nu(x,\xi,\psi(0),\psi)$$

$$- \frac{1}{2}K\gamma(1-\gamma)\Big[\sum_{k=0}^{\infty}(1+\nu-\beta)n(-k)\Big]^2\psi^2(0)g^2(\psi)\Big\}.$$

Hence,

$$\mathcal{A}(K\Psi_\nu)(x,\xi,\psi(0),\psi) \le 0, \quad \forall(x,\xi,\psi(0),\psi) \in \mathbf{S}^\circ_{\kappa,\mu,\beta},$$

if and only if

$$\Big[\frac{1-\gamma}{\gamma}(K\gamma)^{\frac{\gamma}{\gamma-1}} - \alpha K + K\gamma\bar{b}\Big]\Lambda_\nu^2$$

$$\le \frac{1}{2}\sigma^2K\gamma(1-\gamma)\nu^2\psi^2(0)\Big(\sum_{k=0}^{\infty}n(-k)\Big)^2, \quad \forall(x,\xi,\psi(0),\psi) \in \mathbf{S}^\circ_{\kappa,\mu,\beta}.$$

This holds if and only if

$$\alpha > \gamma\underline{b} + (1-\gamma)(K\gamma)^{\frac{1}{\gamma-1}}. \tag{7.111}$$

If (7.107) holds, then (7.111) holds for K large enough. This shows that $\forall(x,\xi, \psi(0),\psi) \in \mathbf{S}_{\kappa,\mu,\beta}$,

$$V_{\kappa,\mu,\beta}(x,\xi,\psi(0),\psi) \le K\Big(x + \nu\psi(0)\sum_{k=0}^{\infty}n(-k)\Big)^\gamma.$$

This proves the proposition.   □

Remark 7.5.5 *Proposition 7.5.4 shows that the value function* $V_{\kappa,\mu,\beta}$: $\mathbf{S}_{\kappa,\mu,\beta} \to \Re$ *is a finite function. Moreover, it is bounded on the set of the following form:*

$$\{(x,\xi,\psi(0),\psi) \in \mathbf{S}_{\kappa,\mu,\beta} \mid \Lambda_\nu(x,\xi,\psi(0),\psi) = constant\}$$

for every $\nu \in (1-\mu, 1+\mu)$.

7.6 The Viscosity Solution

It is clear that the value function $V_\kappa : \mathbf{S}_\kappa \to \Re_+$ has discontinuity on the interfaces $Q_{I,+}$ and $Q_{I,-}$ and, hence, it cannot be a solution of HJBQVI (*) in the classical sense. In addition, even though V_κ is continuous on the open set $\mathbf{S}_\kappa^\circ$, it is not necessarily in $C^{1,0,2,2}(\mathbf{S}_\kappa)$ to be a solution of the HJBQVI (*). The main purpose of this section is to show that V_κ is a viscosity solution of the HJBQVI (*). See Ishii [Ish93] and Øksendal and Sulem [ØS02] for the connection of viscosity solutions of second-order elliptic equations with stochastic classical control and classical-impulse control problems for diffusion processes.

To give a definition of a viscosity solution, we first define the upper and lower semicontinuity concept as follows. Let $\varXi$ be a metric space and let $\varPhi : \varXi \to \Re$ be a Borel measurable function. Then the upper semicontinuous (USC) envelope $\bar{\varPhi} : \varXi \to \Re$ and the lower semicontinuous (LSC) envelope $\underline{\varPhi} : \varXi \to \Re$ of $\varPhi$ are defined respectively by

$$\bar{\varPhi}(\mathbf{x}) = \limsup_{\mathbf{y} \to \mathbf{x}, \mathbf{y} \in \varXi} \varPhi(\mathbf{y}) \text{ and } \underline{\varPhi}(\mathbf{x}) = \liminf_{\mathbf{y} \to \mathbf{x}, \mathbf{y} \in \varXi} \varPhi(\mathbf{y}).$$

We let $USC(\varXi)$ and $LSC(\varXi)$ denote the set of USC and LSC functions on $\varXi$, respectively. Note that, in general, one has

$$\underline{\varPhi} \le \varPhi \le \bar{\varPhi}$$

and that $\varPhi$ is USC if and only if $\varPhi = \bar{\varPhi}$ and $\varPhi$ is LSC if and only if $\varPhi = \underline{\varPhi}$. In particular, $\varPhi$ is continuous if and only if

$$\underline{\varPhi} = \varPhi = \bar{\varPhi}.$$

Let $(\Re \times L_\rho^2)^*$ and $(\Re \times L_\rho^2)^\dagger$ be the space of bounded linear and bilinear functionals equipped with the usual operator norms $\|\cdot\|^*$ and $\|\cdot\|^\dagger$, respectively.

To define a viscosity solution, let us consider the following equation:

$$F(\mathbf{A}, \mathcal{S}, \partial_x, V_\kappa, (x, \xi, \psi(0), \psi)) = 0, \quad \forall (x, \xi, \psi(0), \psi) \in \mathbf{S}_\kappa, \qquad (7.112)$$

where

$$F : (\Re \times L_\rho^2)^\dagger \times \mathcal{L}(\Re \times L_\rho^2) \times \Re \times \Re^{\mathbf{S}_\kappa} \times \mathbf{S}_\kappa \to \Re$$

is defined by

$$F = \begin{cases} \max \Big\{ \varLambda(\mathbf{A}, \mathcal{S}, \partial_x, \varPhi, (x, \xi, \psi(0), \psi)), \\ \quad (\mathcal{M}_\kappa \varPhi - \varPhi)(x, \xi, \psi(0), \psi) \Big\}, & \text{on } \mathbf{S}_\kappa^\circ \\ \varLambda(\mathbf{A}, \mathcal{S}, \partial_x, \varPhi, (x, \xi, \psi(0), \psi)), & \text{on } \bigcup_{I \subset \aleph_0} \partial_{+,I,2}\mathbf{S}_\kappa \\ \varLambda^0(\mathbf{A}, \mathcal{S}, \partial_x, \varPhi, (x, \xi, \psi(0), \psi)), & \text{on } \bigcup_{I \subset \aleph_0} \partial_{-,I,2}\mathbf{S}_\kappa \\ (\mathcal{M}\varPhi - \varPhi)((x, \xi, \psi(0), \psi)), & \text{on } \bigcup_{I \subset \aleph_0} \partial_{I,1}\mathbf{S}_\kappa, \end{cases} \qquad (7.113)$$

$$\Lambda(\mathbf{A}, \mathcal{S}, \partial_x, \Phi, (x, \xi, \psi(0), \psi)) = \mathcal{A}\Phi(x, \xi, \psi(0), \psi),$$

and

$$\Lambda^0(\mathbf{A}, \mathcal{S}, \partial_x, \Phi, (x, \xi, \psi(0), \psi)) = \mathcal{L}^0\Phi(x, \xi, \psi(0), \psi).$$

Note that

$$F(\mathbf{A}, \mathcal{S}, \partial_x, \Phi, (x, \xi, \psi(0), \psi)) = HJBQVI(*),$$

$$\bar{F}(\mathbf{A}, \mathcal{S}, \partial_x, \Phi, (x, \xi, \psi(0), \psi))$$
$$= \max\Big\{\Lambda(\mathbf{A}, \mathcal{S}, \partial_x, \Phi, (x, \xi, \psi(0), \psi)),$$
$$(\mathcal{M}_\kappa\Phi - \Phi)(x, \xi, \psi(0), \psi)\Big\}, \quad \forall (x, \xi, \psi(0), \psi) \in \mathbf{S}_\kappa, \qquad (7.114)$$

and

$$\underline{F}(\mathbf{A}, \mathcal{S}, \partial_x, \Phi, (x, \xi, \psi(0), \psi)) = F(\mathbf{A}, \mathcal{S}, \partial_x, \Phi, (x, \xi, \psi(0), \psi)).$$

Definition 7.6.1 *(i) A function $\Phi \in USC(\mathbf{S}_\kappa)$ is said to be a viscosity subsolution of (7.112) if for every function $\Psi \in C_{lip}^{1,0,2,2}(\mathbf{S}_\kappa) \cap \mathcal{D}(\mathcal{S})$ and for every $(x, \xi, \psi(0), \psi) \in \mathbf{S}_\kappa$ such that $\Psi \geq \Phi$ on $\mathbf{S}_\kappa$ and $\Psi((x, \xi, \psi(0), \psi)) = \Phi(x, \xi, \psi(0), \psi)$, we have*

$$\bar{F}(\mathbf{A}, \mathcal{S}, \partial_x, \Psi, (x, \xi, \psi(0), \psi)) \geq 0. \qquad (7.115)$$

(ii) A function $\Phi \in LSC(\mathbf{S}_\kappa)$ is a viscosity supersolution of (7.112) if for every function $\Psi \in C_{lip}^{1,0,2,2}(\mathbf{S}_\kappa) \cap \mathcal{D}(\mathcal{S})$ and for every $(x, \xi, \psi(0), \psi) \in \mathbf{S}_\kappa$ such that $\Psi \leq \Phi$ on $\mathbf{S}_\kappa$ and $\Psi(x, \xi, \psi(0), \psi) = \Phi(x, \xi, \psi(0), \psi)$, we have

$$\underline{F}(\mathbf{A}, \mathcal{S}, \partial_x, \Psi, (x, \xi, \psi(0), \psi)) \leq 0. \qquad (7.116)$$

(iii) A locally bounded function $\Phi : \mathbf{S}_\kappa \to \Re$ is a viscosity solution of (7.112) if $\bar{\Phi}$ is a viscosity subsolution and $\underline{\Phi}$ is a viscosity supersolution of (7.112).

The following properties of the *intervention operator* $\mathcal{M}_\kappa$ can be established.

Lemma 7.6.2 *The following statements hold true regarding $\mathcal{M}_\kappa$ defined by (7.35).*
(i) If $\Phi : \mathbf{S}_\kappa \to \Re$ is USC, then $\mathcal{M}_\kappa\Phi$ is USC.
(ii) If $\Phi : \mathbf{S}_\kappa \to \Re$ is continuous, then $\mathcal{M}_\kappa\Phi$ is continuous.
(iii) Let $\Phi : \mathbf{S}_\kappa \to \Re$. Then $\overline{\mathcal{M}_\kappa\Phi} \leq \mathcal{M}_\kappa\bar{\Phi}$.
(iv) Let $\Phi : \mathbf{S}_\kappa \to \Re$ be such that $\Phi \geq \mathcal{M}_\kappa\Phi$. Then $\underline{\Phi} \geq \mathcal{M}_\kappa\underline{\Phi}$.
(v) Suppose $\Phi : \mathbf{S}_\kappa \to \Re$ is USC and $\Phi(x, \xi, \psi(0), \psi) > \mathcal{M}_\kappa\Phi(x, \xi, \psi(0), \psi) + \epsilon$ for some $(x, \xi, \psi(0), \psi) \in \mathbf{S}_\kappa$ and $\epsilon > 0$. Then

$$\Phi(x, \xi, \psi(0), \psi) > \overline{\mathcal{M}_\kappa\Phi}(x, \xi, \psi(0), \psi) + \epsilon.$$

Proof. We only provide a proof of part (i). Parts (ii)-(v) follow as a consequence of part (i).

Let $(x, \xi, \psi(0), \psi) \in \mathbf{S}_\kappa$ with $I = \{i \in \aleph_0 \mid n(-i) < 0\}$ and $I^c = \aleph_0 - I = \{i \in \aleph_0 \mid n(-i) \geq 0\}$. Define

$$\mathcal{P}(x, \xi, \psi(0), \psi) = \{(\hat{x}, \hat{\xi}, \hat{\psi}(0), \hat{\psi}) \in \mathbf{S}_\kappa \mid \zeta \in \mathcal{R}(\xi) - \{\mathbf{0}\}\}$$
$$= \mathcal{P}_+(x, \xi, \psi(0), \psi) \bigcup \mathcal{P}_-(x, \xi, \psi(0), \psi), \qquad (7.117)$$

where

$$\mathcal{P}_+(x, \xi, \psi(0), \psi) = \{(\hat{x}, \hat{\xi}, \hat{\psi}(0), \hat{\psi}) \in \mathbf{S}_\kappa \mid m(0) \geq 0,$$
$$\text{and } 0 \leq m(-i) \leq -n(-i) \text{ for } i \in I - \{0\};$$
$$\text{and } -n(-i) \leq m(-i) \leq 0 \text{ for } i \in I^c - \{0\}\},$$

$$\mathcal{P}_-(x, \xi, \psi(0), \psi) = \{(\hat{x}, \hat{\xi}, \hat{\psi}(0), \hat{\psi}) \in \mathbf{S}_\kappa \mid m(0) < 0,$$
$$\text{and } 0 \leq m(-i) \leq -n(-i) \text{ for } i \in I - \{0\};$$
$$\text{and } -n(-i) \leq m(-i) \leq 0 \text{ for } i \in I^c - \{0\}\},$$

$\hat{x}$ and $\hat{\xi}$ are as defined in (7.36) and (7.37), and $(\hat{\psi}(0), \hat{\psi}) = (\psi(0), \psi)$ due to the continuity and the uncontrollability (by the *investor*) of the stock prices.

We claim that for each $(x, \xi, \psi(0), \psi) \in \mathbf{S}_\kappa^\circ$, both $\mathcal{P}_+(x, \xi, \psi(0), \psi)$ and $\mathcal{P}_-(x, \xi, \psi(0), \psi)$ are compact subsets of $\mathbf{S}_\kappa$. To see this, we consider the following two cases.

Case (i). $G_\kappa(x, \xi, \psi(0), \psi) \geq 0$. In this case $\mathcal{P}_+(x, \xi, \psi(0), \psi)$ intersects with the hyperplane $\partial_{I-\{0\},1}\mathbf{S}_\kappa$ (since $m(0) > 0$). By the facts that $0 \leq m(-i) \leq -n(-i)$ for $i \in I - \{0\}$ and $n(-i) = 0$ for all but finitely many $i \in \aleph_0$ as required in (7.2), $\mathcal{P}_+(x, \xi, \psi(0), \psi)$ is compact.

Case (ii). $G_\kappa(x, \xi, \psi(0), \psi) < 0$. In this case, $\mathcal{P}_+(x, \xi, \psi(0), \psi)$ is bounded by the set

$$\{(x, \xi, \psi(0), \psi) \mid G_\kappa(x, \xi, \psi(0), \psi)\} = 0$$

and the boundary of $[0, \infty) \times \mathbf{N}_+ \times \Re \times L^2_{\rho,+}$.

From Case (i) and Case (ii), $\mathcal{P}_+(x, \xi, \psi(0), \psi)$ is a compact subset of $\mathbf{S}_\kappa$. We can also prove the compactness of $\mathcal{P}_-(x, \xi, \psi(0), \psi)$ in a similar manner.

Since Φ is USC on $\mathcal{P}(x, \xi, \psi(0), \psi)$, there exists

$$(x^*, \xi^*, \psi^*(0), \psi^*) \in \mathcal{P}(x, \xi, \psi(0), \psi)$$

such that

$$\mathcal{M}_\kappa \Phi(x, \xi, \psi(0), \psi) = \sup\{\Phi(\hat{x}, \hat{\xi}, \hat{\psi}(0), \hat{\psi}) \mid \zeta \in \mathcal{R}(\xi) - \{\mathbf{0}\}\}$$
$$= \Phi(x^*, \xi^*, \psi^*(0), \psi^*).$$

Fix $(x^{(0)}, \xi^{(0)}, \psi^{(0)}(0), \psi^{(0)}) \in \mathbf{S}_\kappa$ and let $\left\{(x^{(k)}, \xi^{(k)}, \psi^{(k)}(0), \psi^{(k)})\right\}_{k=1}^\infty$ be a sequence in $\mathbf{S}_\kappa$ such that

$$(x^{(k)}, \xi^{(k)}, \psi^{(k)}(0), \psi^{(k)}) \to (x^{(0)}, \xi^{(0)}, \psi^{(0)}(0), \psi^{(0)}) \text{ as } k \to \infty.$$

To show that $\mathcal{M}_\kappa \Phi$ is USC, we must show that

$$\mathcal{M}_\kappa \Phi(x^{(0)}, \xi^{(0)}, \psi^{(0)}(0), \psi^{(0)}) \geq \limsup_{k \to \infty} \mathcal{M}_\kappa \Phi(x^{(k)}, \xi^{(k)}, \psi^{(k)}(0), \psi^{(k)})$$

$$= \limsup_{k \to \infty} \Phi(x^{(k)*}, \xi^{(k)*}, \psi^{(k)*}(0), \psi^{(k)*}).$$

Let $(\tilde{x}, \tilde{\xi}, \tilde{\psi}(0), \tilde{\psi})$ be a cluster point of

$$\left\{ (x^{(k)*}, \xi^{(k)*}, \psi^{(k)*}(0), \psi^{(k)*}) \right\}_{k=1}^{\infty};$$

that is, $(\tilde{x}, \tilde{\xi}, \tilde{\psi}(0), \tilde{\psi})$ is the limit of some of convergent subsequence

$$\left\{ (x^{(k_j)*}, \xi^{(k_j)*}, \psi^{(k_j)*}(0), \psi^{(k_j)*}) \right\}_{j=1}^{\infty} \text{ of } \left\{ (x^{(k)*}, \xi^{(k)*}, \psi^{(k)*}(0), \psi^{(k)*}) \right\}_{k=1}^{\infty}.$$

Since

$$(x^{(k)}, \xi^{(k)}, \psi^{(k)}(0), \psi^{(k)}) \to (x^{(0)}, \xi^{(0)}, \psi^{(0)}(0), \psi^{(0)}),$$

we see that

$$\mathcal{P}(x^{(k)}, \xi^{(k)}, \psi^{(k)}(0), \psi^{(k)}) \to \mathcal{P}(x^{(0)}, \xi^{(0)}, \psi^{(0)}(0), \psi^{(0)})$$

in Hausdorff distance. Hence, since

$$(x^{(k_j)*}, \xi^{(k_j)*}, \psi^{(k_j)*}(0), \psi^{(k_j)*}) \in \mathcal{P}(x^{(k_j)}, \xi^{(k_j)}, \psi^{(k_j)}(0), \psi^{(k_j)})$$

for all j, we conclude that

$$(\tilde{x}, \tilde{\xi}, \tilde{\psi}(0), \tilde{\psi})$$
$$= \lim_{j \to \infty} (x^{(k_j)*}, \xi^{(k_j)*}, \psi^{(k_j)*}(0), \psi^{(k_j)*}) \in \mathcal{P}(x^{(0)}, \xi^{(0)}, \psi^{(0)}(0), \psi^{(0)}).$$

Therefore,

$$\mathcal{M}_\kappa \Phi(x^{(0)}, \xi^{(0)}, \psi^{(0)}(0), \psi^{(0)}) \geq \Phi(\tilde{x}, \tilde{\xi}, \tilde{\psi}(0), \tilde{\psi})$$

$$\geq \limsup_{j \to \infty} \Phi(x^{(k_j)*}, \xi^{(k_j)*}, \psi^{(k_j)*}(0), \psi^{(k_j)*})$$

$$= \limsup_{k \to \infty} \mathcal{M}_\kappa \Phi(x^{(k)}, \xi^{(k)}, \psi^{(k)}(0), \psi^{(k)}). \qquad \square$$

Theorem 7.6.3 *Suppose $\alpha > \lambda\gamma$. Then the value function $V_\kappa : \mathbf{S}_\kappa \to \Re$ defined by (7.13) is a viscosity solution of the HJBQVI (*).*

The theorem can be proved by verifying the following two propositions, the first of which shows that the value function is a viscosity supersolution of the HJBQVI (*) and the second shows that the value function is a viscosity subsolution of the HJBQVI (*).

Proposition 7.6.4 *The lower semicontinuous envelope* $\underline{V_\kappa} : \mathbf{S}_\kappa \to \Re$ *of the value function V_κ is a viscosity supersolution of the HJBQVI (*).*

Proof. Let $\Phi : \mathbf{S}_\kappa \to \Re$ be any smooth function with $\Phi \in C^{1,0,2,2}_{lip}(\mathcal{O}) \cap \mathcal{D}(\mathcal{S})$ on a neighborhood $\mathcal{O}$ of $\mathbf{S}_\kappa$ and let $(x, \xi, \psi(0), \psi) \in \mathbf{S}_\kappa$ be such that $\Phi \le \underline{V_\kappa}$ on $\mathbf{S}_\kappa$ and $\Phi(x, \xi, \psi(0), \psi) = \underline{V_\kappa}(x, \xi, \psi(0), \psi)$. We need to prove that

$$\underline{F}(\mathbf{A}, \mathcal{S}, \partial_x, \Phi, (x, \xi, \psi(0), \psi)) = F(\mathbf{A}, \mathcal{S}, \partial_x, \Phi, (x, \xi, \psi(0), \psi)) \le 0.$$

Note that by Lemma 7.6.2(iv),

$$\mathcal{M}_\kappa V_\kappa \le V_\kappa \Rightarrow \mathcal{M}_\kappa \underline{V_\kappa} = \mathcal{M}_\kappa \Phi \le \underline{V_\kappa} = \Phi \text{ on } \mathbf{S}_\kappa.$$

In particular, this inequality holds on $\bigcup_{I \subset \aleph_0} \partial_{1,1} \mathbf{S}_\kappa$. Therefore, we only need to show that

$$\mathcal{A}\Phi \le 0 \text{ on } \mathbf{S}_\kappa^\circ \cup \bigcup_{I \subset \aleph_0} \partial_{+,I,2} \mathbf{S}_\kappa$$

and

$$\mathcal{L}^0 \Phi \le 0 \text{ on } \bigcup_{I \subset \aleph_0} \partial_{-,I,2} \mathbf{S}_\kappa.$$

For $\epsilon > 0$, let $B(\epsilon) = B(\epsilon; (x, \xi, \psi(0), \psi))$ be the open ball in $\mathbf{S}_\kappa$ centered at $(x, \xi, \psi(0), \psi)$ and with radius $\epsilon > 0$. Let

$$\pi(\epsilon) = (C^\epsilon, T^\epsilon) \in \mathcal{U}_\kappa(x, \xi, \psi(0), \psi)$$

be the admissible strategy beginning with a constant consumption rate $C(t) = c \ge 0$ and no transactions up to the first time $\tau(\epsilon)$ at which the controlled state process $\{(X(t), N_t, S(t), S_t), t \ge 0\}$ exits from the set $B(\epsilon)$. Note that $\tau(\epsilon) > 0$, P-a.s. since there is no transaction and the controlled state process $\{(X(t), N_t, S(t), S_t), t \ge 0\}$ is continuous on $B(\epsilon)$.

Choose $(x^{(k)}, \xi^{(k)}, \psi^{(k)}(0), \psi^{(k)}) \in B(\epsilon)$ such that

$$(x^{(k)}, \xi^{(k)}, \psi^{(k)}(0), \psi^{(k)}) \to (x, \xi, \psi(0), \psi)$$

and

$$V_\kappa(x^{(k)}, \xi^{(k)}, \psi^{(k)}(0), \psi^{(k)}) \to \underline{V_\kappa}(x, \xi, \psi(0), \psi) \text{ as } k \to \infty.$$

Then by the Bellman DPP (see Proposition 7.3.1), we have for all k,

$$V_\kappa(x^{(k)}, \xi^{(k)}, \psi^{(k)}(0), \psi^{(k)})$$
$$\ge E^{(k)} \left[\int_0^{\tau(\epsilon)} e^{-\alpha t} \frac{c^\gamma}{\gamma} dt + e^{-\alpha\tau(\epsilon)} V_\kappa(X(\tau(\epsilon)), N_{\tau(\epsilon)}, S(\tau(\epsilon)), S_{\tau(\epsilon)}) \right]$$
$$\ge E^{(k)} \left[\int_0^{\tau(\epsilon)} e^{-\alpha t} \frac{c^\gamma}{\gamma} dt + e^{-\alpha\tau(\epsilon)} \Phi(X(\tau(\epsilon)), N_{\tau(\epsilon)}, S(\tau(\epsilon)), S_{\tau(\epsilon)}) \right],$$

where $E^{(k)}[\cdots]$ is notation for $E^{x^{(k)}, \xi^{(k)}, \psi^{(k)}(0), \psi^{(k)}; \pi(\epsilon)}[\cdots]$, the conditional expectation given the initial state $(x^{(k)}, \xi^{(k)}, \psi^{(k)}(0), \psi^{(k)})$ and the strategy $\pi(\epsilon)$. In particular, for all $0 \le t \le \tau(\epsilon)$,

$$0 = \underline{V_\kappa}(x, \xi, \psi(0), \psi) - \Phi(x, \xi, \psi(0), \psi)$$

$$= \lim_{k \to \infty} \left[V_\kappa(x^{(k)}, \xi^{(k)}, \psi^{(k)}(0), \psi^{(k)}) - \Phi(x^{(k)}, \xi^{(k)}, \psi^{(k)}(0), \psi^{(k)}) \right]$$

$$\geq \lim_{k \to \infty} E^{(k)} \left[\int_0^t e^{-\alpha s} \frac{c^\gamma}{\gamma} ds \right.$$

$$\left. + e^{-\alpha t} V_\kappa(X(t), N_t, S(t), S_t) - \Phi(x^{(k)}, \xi^{(k)}, \psi^{(k)}(0), \psi^{(k)}) \right]$$

$$= E \left[\int_0^t e^{-\alpha s} \frac{c^\gamma}{\gamma} ds \right.$$

$$\left. + e^{-\alpha t} \Phi(X(t), N_t, S(t), S_t) - \Phi(x, \xi, \psi(0), \psi) \right].$$

Dividing both sides of the inequality by t and letting $t \to 0$, we have from Dynkin's formula or Itô's formula (Theorem 7.2.6) that

$$0 \geq \lim_{t \to 0} \frac{1}{t} E \left[\int_0^t e^{-\alpha s} \frac{c^\gamma}{\gamma} ds \right.$$

$$\left. + e^{-\alpha t} \Phi(X(t), N_t, S(t), S_t) - \Phi(x, \xi, \psi(0), \psi) \right]$$

$$= \frac{c^\gamma}{\gamma} + \lim_{t \to 0} \frac{1}{t} E \left[e^{-\alpha t} \Phi(X(t), N_t, S(t), S_t) - \Phi(x, \xi, \psi(0), \psi) \right]$$

$$= \frac{c^\gamma}{\gamma} + \mathcal{L}^c \Phi(x, \xi, \psi(0), \psi),$$

where

$$\mathcal{L}^c \Phi(x, \xi, \psi(0), \psi) = (\mathbf{A} - \alpha + rx\partial_x - c)\Phi(x, \xi, \psi(0), \psi).$$

We conclude from the above that

$$\mathcal{L}^c \Phi(x, \xi, \psi(0), \psi) + \frac{c^\gamma}{\gamma} \leq 0 \qquad (7.118)$$

for all $c \geq 0$ such that $\pi(\epsilon) \in \mathcal{U}_\kappa(x, \xi, \psi(0), \psi)$ for $\epsilon > 0$ small enough. This implies that

$$\mathcal{A}\Phi(x, \xi, \psi(0), \psi) \equiv \sup_{c \geq 0} \left(\mathcal{L}^c \Phi(x, \xi, \psi(0), \psi) + \frac{c^\gamma}{\gamma} \right) \leq 0.$$

If $(x, \xi, \psi(0), \psi) \in \mathbf{S}_\kappa^\circ \cup \bigcup_{I \subset \aleph_0} \partial_{+,I,2}\mathbf{S}_\kappa$, then this is clearly the case for all $c \geq 0$, and, therefore, HJBQVI (*) implies that $\mathcal{A}\Phi(x, \xi, \psi(0), \psi) \leq 0$. If $(x, \xi, \psi(0), \psi) \in \bigcup_{I \subset \aleph_0} \partial_{-,I,2}\mathbf{S}_\kappa$, then the only such admissible c is $c = 0$. Therefore, we get $\mathcal{L}^0\Phi(x, \xi, \psi(0), \psi) \leq 0$ as required. This proves the proposition. □

Proposition 7.6.5 *Suppose* $\alpha > \lambda\gamma$. *Then the upper semicontinuous envelope* $\overline{V_\kappa} : \mathbf{S}_\kappa \to \Re$ *of the value function* V_κ *is a viscosity subsolution of the QVHJBI (*).*

Proof. Suppose $\pi = (C, \mathcal{T}) \in \mathcal{U}_\kappa(x, \xi, \psi(0), \psi)$. Since $\tau(1)$ is a **G**-stopping time, the event $\{\tau(1) = 0\}$ is $\mathcal{G}(0)$-measurable. By the well-known zero-one law (see [KS91, p.94]), one has

$$\text{either } \tau(1) = 0, \ P\text{-a.s. or } \tau(1) > 0 \ P\text{-a.s.} \tag{7.119}$$

We first assume $\tau(1) > 0$, P-a.s.. Then by the Markovian property, the cost functional $J_\kappa(x, \xi, \psi(0), \psi; \pi)$ satisfies the following relation: For P-a.s. ω,

$$e^{-\alpha\tau} J_\kappa(X(\tau), N_\tau, S(\tau), S_\tau; \pi) = E\left[\int_\tau^\infty e^{-\alpha(s+\tau)} \frac{C^\gamma(s)}{\gamma} ds \mid \mathcal{G}(\tau) \right](\omega).$$

Hence,

$$J_\kappa(x, \xi, \psi(0), \psi; \pi) = E^{x, \xi, \psi(0), \psi; \pi}\left[\int_0^\tau e^{-\alpha s} \frac{C^\gamma(s)}{\gamma} ds \right.$$
$$\left. + e^{-\alpha\tau} J_\kappa(X(\tau), N_\tau, S(\tau), S_\tau; \pi) \right] \tag{7.120}$$

for all **G**-stopping times $\tau \leq \tau(1)$. It is clear that

$$V_\kappa(x, \xi, \psi(0), \psi) \geq \mathcal{M}_\kappa V_\kappa(x, \xi, \psi(0), \psi), \quad \forall (x, \xi, \psi(0), \psi) \in \mathbf{S}_\kappa. \tag{7.121}$$

We will prove that $\overline{V_\kappa}$, the upper semicontinuous envelope of $V_\kappa : \mathbf{S}_\kappa \to \Re$, is a viscosity subsolution of HJBQVI (*). To this end, let $\Phi : \mathbf{S}_\kappa \to \Re$ be any smooth function with $\Phi \in C_{lip}^{1,0,2,2}(\mathcal{O}) \cap \mathcal{D}(\mathcal{S})$ on a neighborhood $\mathcal{O}$ of $\mathbf{S}_\kappa$ and let $(x, \xi, \psi(0), \psi) \in \mathbf{S}_\kappa$ be such that $\Phi \geq \overline{V_\kappa}$ on $\mathbf{S}_\kappa$ and $\Phi(x, \xi, \psi(0), \psi) = \overline{V_\kappa}(x, \xi, \psi(0), \psi)$. We need to prove that

$$\bar{F}(\mathbf{A}, \mathcal{S}, \partial_x, \Phi, (x, \xi, \psi(0), \psi)) \geq 0.$$

We consider the following two cases separately.

Case 1. $\overline{V_\kappa}(x, \xi, \psi(0), \psi) \leq \mathcal{M}_\kappa \overline{V_\kappa}(x, \xi, \psi(0), \psi)$. Then, by (7.115),

$$\bar{F}(\mathbf{A}, \mathcal{S}, \partial_x, \Phi, (x, \xi, \psi(0), \psi)) \geq 0,$$

and, hence, the above inequality holds at $(x, \xi, \psi(0), \psi)$ for $\Phi = \overline{V_\kappa}$ in this case.

Case 2. $\overline{V_\kappa}(x, \xi, \psi(0), \psi) > \mathcal{M}_\kappa \overline{V_\kappa}(x, \xi, \psi(0), \psi)$. In this case, it suffices to prove that $\mathcal{A}\Phi(x, \xi, \psi(0), \psi) \geq 0$. We argue by contradiction: Suppose $(x, \xi, \psi(0), \psi) \in \mathbf{S}_\kappa$ and $\mathcal{A}\Phi(x, \xi, \psi(0), \psi) < 0$. Then from the definition of $\mathcal{A}$, we deduce that $\partial_x \Phi(x, \xi, \psi(0), \psi) > 0$. Hence, by continuity, $\partial_x \Phi > 0$ on a neighborhood G of $(x, \xi, \psi(0), \psi)$. However, then

$$\mathcal{A}\Phi = (\mathbf{A} - \alpha)\Phi + (rx - \hat{c})\partial_x \Phi + \frac{\hat{c}^\gamma}{\gamma}$$

with $\hat{c} = \hat{c} = (\partial_x \Phi)^{\frac{1}{\gamma-1}}$ for all $(\tilde{x}, \tilde{\xi}, \tilde{\psi}(0), \tilde{\psi}) \in G \cap \mathbf{S}_\kappa$. Hence, $\mathcal{A}\Phi$ is continuous on $G \cap \mathbf{S}_\kappa$ and so there exists a (bounded) neighborhood $G(\bar{K})$ of $(x, \xi, \psi(0), \psi)$ such that

$$G(\bar{K}) = G(x, \xi, \psi(0), \psi; c)$$
$$= \left\{ (\tilde{x}, \tilde{\xi}, \tilde{\psi}(0), \tilde{\psi}) \mid |x - \tilde{x}| < \bar{K}, \right.$$
$$\left. \|\xi - \tilde{\xi}\|_N < \bar{K}, \|(\psi(0), \psi) - (\tilde{\psi}(0), \tilde{\psi})\|_{\mathbf{M}} < \bar{K} \right\}$$

for some $\bar{K} > 0$ and

$$\mathcal{A}\Phi(\tilde{x}, \tilde{\xi}, \tilde{\psi}(0), \tilde{\psi}) < \frac{1}{2}\mathcal{A}\Phi(x, \xi, \psi(0), \psi) < 0, \quad \forall (\tilde{x}, \tilde{\xi}, \tilde{\psi}(0), \tilde{\psi}) \in G(\bar{K}) \cap \mathbf{S}_\kappa. \tag{7.122}$$

Now, since $\overline{V_\kappa}(x, \xi, \psi(0), \psi) > \mathcal{M}_\kappa \overline{V_\kappa}(x, \xi, \psi(0), \psi)$, let η be any number such that

$$0 < \eta < (\overline{V_\kappa} - \mathcal{M}_\kappa \overline{V_\kappa})(x, \xi, \psi(0), \psi). \tag{7.123}$$

Since $\overline{V_\kappa}(x, \xi, \psi(0), \psi) > \mathcal{M}_\kappa \overline{V_\kappa}(x, \xi, \psi(0), \psi) + \eta$, we can, by Lemma 7.6.2(v), find a sequence $\{(x^{(k)}, \xi^{(k)}, \psi^{(k)}(0), \psi^{(k)}\}_{k=1}^\infty \subset G(\bar{K}) \cap \mathbf{S}_\kappa$ such that

$$(x^{(k)}, \xi^{(k)}, \psi^{(k)}(0), \psi^{(k)}) \to (x, \xi, \psi(0), \psi)$$

and

$$V_\kappa(x^{(k)}, \xi^{(k)}, \psi^{(k)}(0), \psi^{(k)}) \to \overline{V_\kappa}(x, \xi, \psi(0), \psi) \text{ as } k \to \infty$$

and for all $k \geq 1$,

$$\mathcal{M}_\kappa V_\kappa(x^{(k)}, \xi^{(k)}, \psi^{(k)}(0), \psi^{(k)}) < V_\kappa(x^{(k)}, \xi^{(k)}, \psi^{(k)}(0), \psi^{(k)}) - \eta. \tag{7.124}$$

Choose $\epsilon \in (0, \eta)$. Since $\overline{V_\kappa}(x, \xi, \psi(0), \psi) = \Phi(x, \xi, \psi(0), \psi)$, we can choose $K > 0$ (a positive integer) such that for all $n \geq K$,

$$|V_\kappa(x^{(k)}, \xi^{(k)}, \psi^{(k)}(0), \psi^{(k)}) - \Phi(x^{(k)}, \xi^{(k)}, \psi^{(k)}(0), \psi^{(k)})| < \epsilon. \tag{7.125}$$

In the following, we fix $n \geq K$ and put

$$(\tilde{x}, \tilde{\xi}, \tilde{\psi}(0), \tilde{\psi}) = (x^{(k)}, \xi^{(k)}, \psi^{(k)}(0), \psi^{(k)}).$$

Let $\tilde{\pi} = (\tilde{C}, \tilde{T})$ with $\tilde{T} = \{(\tilde{\tau}(i), \tilde{\zeta}(i)), i = 1, 2, \ldots\}$ be an ϵ-optimal control for $(\tilde{x}, \tilde{\xi}, \tilde{\psi}(0), \tilde{\psi})$ in the sense that

$$V_\kappa(\tilde{x}, \tilde{\xi}, \tilde{\psi}(0), \tilde{\psi}) \leq J_\kappa(\tilde{x}, \tilde{\xi}, \tilde{\psi}(0), \tilde{\psi}; \tilde{\pi}) + \epsilon.$$

We claim that $\tilde{\tau}(1) > 0$, P-a.s.. If this were false, then $\tilde{\tau}(1) = 0$, P-a.s. by the zero-one law (see (7.119)).

Then the state process $\{Z^{\tilde{\pi}}(t) \equiv (X^{\tilde{\pi}}(t), N_t^{\tilde{\pi}}, S^{\tilde{\pi}}(t), S_t^{\tilde{\pi}}), t \geq 0\}$ makes an immediate jump from $(\tilde{x}, \tilde{\xi}, \tilde{\psi}(0), \tilde{\psi})$ to some point $(\hat{x}, \hat{\xi}, \hat{\psi}(0), \hat{\psi}) \in \mathbf{S}_\kappa$ according to (7.36)-(7.38), and, hence, by its definition,

$$J_\kappa(\tilde{x}, \tilde{\xi}, \tilde{\psi}(0), \tilde{\psi}; \tilde{\pi}) = E^{x, \xi, \psi(0), \psi; \tilde{\pi}}[J_\kappa(\hat{x}, \hat{\xi}, \hat{\psi}(0), \hat{\psi}; \tilde{\pi})]$$

Denoting the conditional expectation given the initial state $(\tilde{x}, \tilde{\xi}, \tilde{\psi}(0), \tilde{\psi})$ and the strategy $\tilde{\pi}$ by $\tilde{E}[\cdots]$, we have

$$
\begin{aligned}
V_\kappa(\tilde{x}, \tilde{\xi}, \tilde{\psi}(0), \tilde{\psi}) &\le J_\kappa(\tilde{x}, \tilde{\xi}, \tilde{\psi}(0), \tilde{\psi}; \tilde{\pi}) + \epsilon \\
&= \tilde{E}[J_\kappa(\hat{x}, \hat{\xi}, \hat{\psi}(0), \hat{\psi}; \tilde{\pi})] + \epsilon \\
&\le \tilde{E}[V_\kappa(\hat{x}, \hat{\xi}, \hat{\psi}(0), \hat{\psi})] + \epsilon \\
&\le \mathcal{M}_\kappa V_\kappa(\tilde{x}, \tilde{\xi}, \tilde{\psi}(0), \tilde{\psi})] + \epsilon,
\end{aligned}
$$

which contradicts (7.119). We therefore conclude that $\tilde{\tau}(1) > 0$, P-a.s.. Fix $R > 0$ and define the **G**-stopping time τ by

$$\tau = \tau(\epsilon) = \tilde{\tau}(1) \wedge R \wedge \inf\{t > 0 \mid Z^{\tilde{\pi}}(t) \notin G(\bar{K})\}.$$

By the Dynkin formula (see Theorem 7.2.6), we have

$$
\begin{aligned}
\tilde{E}\left[e^{-\alpha\tau}\Phi(Z^{\tilde{\pi}}(\tau))\right] &= \Phi(\tilde{x}, \tilde{\xi}, \tilde{\psi}(0), \tilde{\psi}) + \tilde{E}\left[\int_0^\tau e^{-\alpha t}\mathcal{L}^{\tilde{c}}\Phi(Z^{\tilde{\pi}}(t)\, dt\right] \\
&\quad + \tilde{E}\left[e^{-\alpha\tau}\left(\Phi(Z^{\tilde{\pi}}(\tau)) - \Phi(Z^{\tilde{\pi}}(\tau-))\right)\right],
\end{aligned}
$$

or

$$\tilde{E}\left[e^{-\alpha\tau}\Phi(Z^{\tilde{\pi}}(\tau-))\right] = \Phi((\tilde{x}, \tilde{\xi}, \tilde{\psi}(0), \tilde{\psi}) + \tilde{E}\left[\int_0^\tau e^{-\alpha t}\mathcal{L}^{\tilde{c}}\Phi(Z^{\tilde{\pi}}(t))\, dt\right].$$

Since $V_\kappa \ge \mathcal{M}_\kappa V_\kappa$,

$$
\begin{aligned}
V_\kappa(\tilde{x}, \tilde{\xi}, \tilde{\psi}(0), \tilde{\psi}) &\le J_\kappa(\tilde{x}, \tilde{\xi}, \tilde{\psi}(0), \tilde{\psi}); \tilde{\pi}) + \epsilon \\
&= \tilde{E}\left[\int_0^\tau e^{-\alpha t}\frac{\tilde{C}^\gamma(t)}{\gamma}dt + J_\kappa(Z^{\tilde{\pi}}(\tau); \tilde{\pi})\right] + \epsilon \\
&\le \tilde{E}\left[\int_0^\tau e^{-\alpha t}\frac{\tilde{C}^\gamma(t)}{\gamma}dt + e^{-\alpha\tau}V_\kappa(Z^{\tilde{\pi}}(\tau))\right] + \epsilon \\
&= \tilde{E}\left[\int_0^\tau e^{-\alpha t}\frac{\tilde{C}^\gamma(t)}{\gamma}dt + e^{-\alpha\tau}\{V_\kappa(Z^{\tilde{\pi}}(\tau-))\chi_{\{\tau < \tilde{\tau}(1)\}}\right. \\
&\quad \left. + \mathcal{M}_\kappa V_\kappa(Z^{\tilde{\pi}}(\tau-))\chi_{\{\tau = \tilde{\tau}(1)\}}\}\right] + \epsilon \\
&\le \tilde{E}\left[\int_0^\tau e^{-\alpha t}\frac{\tilde{C}^\gamma(t)}{\gamma}dt + e^{-\alpha\tau}V_\kappa(Z^{\tilde{\pi}}(\tau-))\right] + \epsilon \\
&= \tilde{E}\left[\int_0^\tau e^{-\alpha t}\frac{\tilde{C}^\gamma(t)}{\gamma}dt + e^{-\alpha\tau}\Phi(Z^{\tilde{\pi}}(\tau-))\right] + \epsilon \\
&= \Phi(\tilde{x}, \tilde{\xi}, \tilde{\psi}(0), \tilde{\psi}) \\
&\quad + \tilde{E}\left[\int_0^\tau e^{-\alpha t}\left(\mathcal{L}^{\tilde{c}}\Phi(Z^{\tilde{\pi}}(t)) + \frac{\tilde{C}^\gamma(t)}{\gamma}\right)dt\right] \\
&\le V_\kappa(\tilde{x}, \tilde{\xi}, \tilde{\psi}(0), \tilde{\psi}) + \tilde{E}\left[\int_0^\tau e^{-\alpha t}\mathcal{A}\Phi(Z^{\tilde{\pi}}(t))dt\right] + 2\epsilon.
\end{aligned}
$$

We conclude from this that

$$\tilde{E}\left[\int_0^\tau e^{-\alpha t}\mathcal{A}\Phi(Z^{\tilde{\pi}}(t))\,dt\right] \geq -2\epsilon. \qquad (7.126)$$

Note that one can deduce from (7.122) that

$$\tilde{E}\left[\int_0^\tau e^{-\alpha t}\mathcal{A}\Phi(Z^{\tilde{\pi}}(t))\,dt\right] \leq \frac{1}{2\delta}\mathcal{A}\Phi(\tilde{x},\tilde{\xi},\tilde{\psi}(0),\tilde{\psi})(1-\tilde{E}[e^{-\alpha\tau}]). \qquad (7.127)$$

We claim the following.

Lemma 7.6.6

$$0 < E^{(k)}[e^{-\alpha\tau(\epsilon)}] < 1 \ \text{ when } k \to \infty \text{ and } \epsilon \to 0. \qquad (7.128)$$

Note that if Lemma 7.6.6 is true, then (7.126) contradicts (7.122) if ϵ is small enough. This contradiction proves that $\mathcal{A}\Phi(x,\xi,\psi(0),\psi) \geq 0$ and, hence,

$$\bar{F}(\mathbf{A},\mathcal{S},(\partial_x,\partial_{\psi(0)}),\Phi,(x,\xi,\psi(0),\psi)) \geq 0.$$

Therefore, to complete the proof of Proposition 7.6.5, we must verify Lemma 7.6.6.

Proof of Lemma 7.6.6. First, note that for $t < \tau$, we have that

$$dX(t) = (\lambda X(t) - \tilde{C}(t))dt, \ \text{ for } 0 \leq t < \tau \leq \tilde{\tau}(1).$$

Consequently, for $t < \tau$, by (7.7) we have

$$X(t) = X(0)e^{\lambda t} - e^{\lambda t}\int_0^t e^{-\lambda s}\tilde{C}(s)\,ds \geq X(0) - \bar{\lambda},$$

and, hence, with some constant $K < \infty$,

$$\int_0^\tau e^{-\alpha t}\frac{\tilde{C}^\gamma(t)}{\gamma}\,dt \leq \frac{1}{\gamma}\left[\int_0^\tau e^{-\alpha t}\tilde{C}^\gamma(t)\,dt\right]^\gamma\left[\int_0^\tau e^{\frac{\lambda\gamma-\alpha}{1-\gamma}}\,dt\right]^{1-\gamma}$$

$$\leq K(X(0)(1-e^{-\alpha\tau})+\bar{\lambda}e^{-\alpha\tau})^\gamma,$$

since $\lambda\gamma - \alpha < 0$.

Combining this with (7.120), we get

$$V_\kappa(\tilde{x},\tilde{\xi},\tilde{\psi}(0),\tilde{\psi}) - \epsilon \leq J_\kappa(\tilde{x},\tilde{\xi},\tilde{\psi}(0),\tilde{\psi};\tilde{\pi})$$

$$\leq E\left[\int_0^\tau e^{-\alpha t}\frac{\tilde{C}^\gamma(t)}{\gamma}\,dt + e^{-\alpha\tau}V_\kappa(Z^{\tilde{\pi}}(\tau))\right]$$

$$\leq E\Big[K(x-(x-\bar{\lambda})e^{-\lambda\tau})^{\gamma}\Big]$$
$$+E\Big[e^{-\alpha\tau}V_{\kappa}(Z^{\tilde{\pi}}(\tau-))\chi_{\{\tilde{\tau}(1)>\tau\}}\Big]$$
$$+E\Big[e^{-\alpha\tau}\Big(V_{\kappa}(Z^{\tilde{\pi}}(\tau))-V_{\kappa}(Z^{\tilde{\pi}}(\tau-))\Big)\chi_{\{\tilde{\tau}(1)\leq\tau\}}\Big]$$
$$\leq E\Big[K(x-(x-\bar{\lambda})e^{-\lambda\tau})^{\gamma}\Big]$$
$$+E\Big[e^{-\alpha\tau}V_{\kappa}(Z^{\tilde{\pi}}(\tau-))\chi_{\{\tilde{\tau}(1)>\tau\}}\Big]$$
$$+E\Big[e^{-\alpha\tau}\mathcal{M}_{\kappa}V_{\kappa}(Z^{\tilde{\pi}}(\tau-))\chi_{\{\tilde{\tau}(1)\leq\tau\}}\Big]$$
$$\leq E\Big[K(x-(x-\bar{\lambda})e^{-\lambda\tau})^{\gamma}\Big]$$
$$+E\Big[e^{-\alpha\tau}\chi_{\{\tilde{\tau}(1)>\tau\}}\Big]\times\sup_{(\tilde{x},\tilde{\xi},\tilde{\psi}(0),\tilde{\psi})\in G(\bar{K})}V_{\kappa}(\tilde{x},\tilde{\xi},\tilde{\psi}(0),\tilde{\psi})$$
$$+E\Big[e^{-\alpha\tau}\chi_{\{\tilde{\tau}(1)\leq\tau\}}\Big]\times\sup_{(\tilde{x},\tilde{\xi},\tilde{\psi}(0),\tilde{\psi})\in G(\bar{K})}\mathcal{M}_{\kappa}V_{\kappa}(\tilde{x},\tilde{\xi},\tilde{\psi}(0),\tilde{\psi}). \quad (7.129)$$

Now, if there exists a sequence $\epsilon_j \to 0$ and a subsequence

$$\{(x^{(k_j)},\xi^{(k_j)},\psi^{(k_j)}(0),\psi^{(k_j)})\} \text{ of } \{(x^{(k)},\xi^{(k)},\psi^{(k)}(0),\psi^{(k)})\}$$

such that

$$E^{(k_j)}[e^{-\alpha\tau(\epsilon_j)}] \to 1 \text{ when } j \to \infty,$$

then

$$E\Big[e^{-\alpha\tau}\chi_{\{\tilde{\tau}(1)>\tau\}}\Big] \to 0 \text{ when } j \to \infty;$$

so by choosing $(x,\xi,\psi(0),\psi) = (x^{(k_j)},\xi^{(k_j)},\psi^{(k_j)}(0),\psi^{(k_j)})$, $\tau = \tau(\epsilon_j)$, and letting $k \to \infty$, we obtain

$$\bar{V}_{\kappa}(x,\xi,\psi(0),\psi) \leq K\bar{\lambda}^{\gamma} + \sup_{(x,\xi,\psi(0),\psi)\in G(\bar{K})}\mathcal{M}_{\kappa}V_{\kappa}(x,\xi,\psi(0),\psi).$$

Hence, by Lemma 7.6.2,

$$\overline{V_{\kappa}}(x,\xi,\psi(0),\psi) \leq \lim_{\bar{\lambda}\to 0}(K\bar{\lambda}^{\gamma} + \sup_{(x,\xi,\psi(0),\psi)\in G(\bar{K})}\mathcal{M}_{\kappa}V_{\kappa}(x,\xi,\psi(0),\psi))$$
$$= \overline{\mathcal{M}_{\kappa}V_{\kappa}}(x,\xi,\psi(0),\psi)$$
$$\leq \mathcal{M}_{\kappa}\overline{V_{\kappa}}(x,\xi,\psi(0),\psi)$$
$$< V_{\kappa}(x,\xi,\psi(0),\psi) - \eta.$$

This contradicts (7.123). This contradiction proves the proposition and completes the proof that $\overline{V_{\kappa}}$ is a viscosity subsolution. $\square$

7.7 Uniqueness

The proof that the value function $V_\kappa : \mathbf{S}_\kappa \to \Re$ is the unique viscosity solution of the HJBQVI (*) has not been established at this point. However, we make a conjecture on the comparison principle for viscosity super- and sub-solutions as follows.

Conjecture on Comparison Principle.

(i) Suppose Φ is a viscosity subsolution and Ψ is a viscosity supersolution of the HJBQVI (*) and that Φ and Ψ satisfy the following estimates:

$$-C\Psi_0(x, \xi, \psi(0), \psi) \leq \Phi(x, \xi, \psi(0), \psi), \quad \forall (x, \xi, \psi(0), \psi) \in \mathbf{S}_\kappa, \qquad (7.130)$$

$$\Phi(x, \xi, \psi(0), \psi) \leq C\Psi_0(x, \xi, \psi(0), \psi), \quad \forall (x, \xi, \psi(0), \psi) \in \mathbf{S}_\kappa, \qquad (7.131)$$

for some $C < \infty$. Then

$$\Phi \leq \Psi \text{ on } \mathbf{S}_\kappa^\circ.$$

(ii) Moreover, if, in addition,

$$\Psi(x, \xi, \psi(0), \psi) = \liminf_{(\tilde{x}, \tilde{\xi}, \tilde{\psi}(0), \tilde{\psi}) \to (x, \xi, \psi(0), \psi), (\tilde{x}, \tilde{\xi}, \tilde{\psi}(0), \tilde{\psi}) \in \mathbf{S}_\kappa^\circ} \Psi(\tilde{x}, \tilde{\xi}, \tilde{\psi}(0), \tilde{\psi})$$

$$\forall (x, \xi, \psi(0), \psi) \in \partial \mathbf{S}_\kappa, \qquad (7.132)$$

then

$$\Phi \leq \Psi \text{ on } \mathbf{S}_\kappa.$$

The following uniqueness result follows immediately from the Conjecture on Comparison Principle described above.

Theorem 7.7.1 *Suppose $\alpha > \gamma \underline{b}$. Then the value function $V_\kappa : \mathbf{S}_\kappa \to \Re$ defined in (7.13) is continuous on $\mathbf{S}_\kappa^\circ$, and is the unique viscosity solution of HJBQVI (*) with the property that there exists $C < \infty$ such that*

$$V_\kappa(x, \xi, \psi(0), \psi) \leq C\Psi_0(x, \xi, \psi(0), \psi), \quad \forall (x, \xi, \psi(0), \psi) \in \mathbf{S}_\kappa. \qquad (7.133)$$

Proof. It has been shown in Section 7.6 that the value function V_κ is continuous on $\mathbf{S}_\kappa^\circ$ and is a viscosity solution of HJBQVI (*) provided that $\alpha > \gamma \underline{b}$. Suppose $\Phi : \mathbf{S}_\kappa \to \Re$ is another viscosity solution of HJBQVI (*), it follows that $\bar{\Phi}$ is a viscosity subsolution and $\underline{\Phi}$ is a viscosity supersolution, and similarly for V_κ. Hence, by the Conjecture on Comparison Principle (provided it can be proved), we have

$$\bar{\Phi} \leq \underline{V}_\kappa \leq \bar{V}_\kappa \leq \underline{\Phi} \leq \bar{\Phi} \text{ on } \mathbf{S}_\kappa^\circ.$$

Therefore, $V_\kappa = \Phi$ on $\mathbf{S}_\kappa$. This proves the theorem. $\square$

7.8 Conclusions and Remarks

Instead of developing a general theory for optimal classical-impulse control problems, we focus on a specific problem that arises from a hereditary port-folio optimization problem in which a small investor's portfolio consists of a savings account that grows continuously with a constant interest rate and an account of one underlying stock whose prices follow a nonlinear stochas-tic hereditary differential equation with infinite but fading memory. The in-vestor is allowed to consume from the savings and make transactions between the savings and the stock accounts subject to Rules 7.1-7.6 that govern the transactions, transaction cost, and capital gains tax. The capital gains tax is proportional to the amount of profit in selling or buying back shares of the stock. A profit or a loss is defined to the difference between the sale and the base price of the stock, where the base price is the price at which shares of the stock was bought or short sold in the past. Within the solvency region, the investor's objective is to maximize a certain expected discounted util-ity from consumption over an infinite time horizon, by using an admissible consumption-trading strategy.

The results obtained in this chapter include (1) the derivation of an infinite dimensional Hamilton-Jacobi-Bellman quasi-variational inequality (HJBQVI) together with its boundary conditions; (2) the establishment of the verification theorem for the optimal consumption-trading strategy; (3) the properties and upper bounds for the value function; and (4) the introduction of a viscosity so-lution and a proof of the value function as a viscosity solution of the HJBQVI. In addition, it is also conjectured that the value function is the unique vis-cosity solution of the HJBQVI. Issues that have not been addressed in this chapter include the proof for the conjecture and computational methods for the optimal strategy and the value functions.

The material presented in this chapter is based on the two pioneer papers by the author (see [Cha07a, Cha07b]) in which the term "hereditary port-folio optimization" appeared for the first time in literature. The hereditary nature of the problem comes from the facts that the stock price dynamics is described by a nonlinear stochastic hereditary differential equation with infi-nite but fading memory and that the stock base price for calculating capital gains taxes depend on the historical stock prices. Based on the stock price dynamics and Rules 7.1-7.6, there are many open problems, including pricing of European and American options, optimal terminal wealth problem, optimal ergodic control problem, and so forth. Due to the presence of transaction costs (fixed and proportional) and capital gains taxation, it is anticipated that the market is not complete and the open problems mentioned above are expected to be extremely challenging ones.

References

[Ada75]   Adams, R.A.: Sobolev Spaces, Academic Press, New Yor (1975)

[AMS96]   Akian, M., Menaldi, J.L., Sulem, A.: On an investment-consumption
          model with transaction costs. SIAM J. Control & Optimization, **34**,
          329–364 (1996)

[AST01]   M. Akian, M., Sulem, A., M. I. Taksar, M.I.: Dynamic optimization of
          long term growth rate for a portfolio with transaction costs and logrith-
          mic utility. Mathematical Finance. **11**, 153-188 (2001)

[AP93]   Ambrosetti, A., Prodi, G.: A Primer of Nonlinear Analysis. Cambridge
          University Press (1993)

[Arr97]   Arriojas, M.: A Stochastic Calculus for Functional Differential Equa-
          tions, Doctoral Dissertation, Department of Mathematics, Southern Illi-
          nois University at Carbondale (1997)

[AHM07] Arriojas, M., Hu, Y., S.-E. Mohammed, S.-E., Pap, G.: A delayed Balck
          and Scholes formula. Stochastic Analysis and Applications, **25**, 471-492
          (2007)

[BM06]   Barbu, V., Marinelli, C.: Variational inequlities in Hilbert spaces with
          measures and optimal stopping. preprint (1996)

[BS91]   Barles, G., Souganidis, P.E.: Convergence of approximative schemes for
          fully nolinear second order equations. J. Asymptotic Analysis., **4**, 557–
          579 (1991)

[Bat96]   Bates, D.S.: Testing option pricing models. Stochastic Models in Fi-
          nance, Handbook of Statistics, **14**, 567-611 (1996)

[BR05]   Bauer, H., Rieder, U.: Stochastic control problems with delay. Math.
          Meth. Oper. Res., **62**(3), 411-427 (2005)

[BL81]   Bensousan, A., Lions, J.L.: Applications of Variational Inequalities.,
          North-Holland, Amsterdam New-York (1981)

[BL84]   Bensoussan, A., Lions, J.L.: Impulse Control and Quasi-Variational In-
          equalities. Gauthier-Villars, Paris (1984)

[BS05]   Bensoussan, A., Sulem, A.: Controlled Jumped Diffusion., Springer-
          Verlag, New York (2005)

[BBM03] Beta, C., Bertram, M., Mikhailov, A.S., Rotermund, H.H., Ertl, G.: Con-
          trolling turbulence in a surface chemical reaction by time-delay autosyn-
          chronization. Phys. Rev. E., **67**(**2**), 046224.1–046224.10 (2003)

[Bil99]   Billingsley, P.: Convergence of Probability Measures. 2nd edition, Wiley,
          New York (1999)

[BS73]   Black, F., Scholes, M.: The pricing of options and corporate liabilities.
           J. Political Economy., **81**, 637–659 (1973)
[Bou79]   Bourgain, J.: La propriete de Radon-Nikodym. Cours polycopie, **36**,
           universite P.et M. Curie, Paris (1979)
[BO98]   Brekke, K.A., Oksendal, B.: A verification theorem for combined sto-
           chastic control and impulse control. In: Stochastic Analysis and Related
           Topics., Vol. 6, Progress in Probability **42**, pp. 211–220, Birkhauser
           Boston, Cambridge, MA (1998)
[CP99]   Cadenillas, A., Pliska, S.R.: Optimal trading of a security when there
           are taxes and transaction costs. Finance and Stochastics., **3**, 137–165
           (1999)
[CZ00]   Cadenillas, A., Zapataro, P.: Classical and impulse stochastic control
           of the exchange rate using interest rates and reserves. Mathematical
           Finance., **10**, 141-156 (2000)
[CFN06]   Calzolari, A., Florchinger, P., Nappo, G.: Approximation of nonlinear
           filtering for Markov systems with delayed observations. SIAM J. Control
           Optim., **45**, 599–633 (2006)
[CFN07]   Calzolari, A., Florchinger, P., Nappo, G.: Convergence in nonlinear fil-
           tering for stochastic delay systems. Preprint (2007)
[Cha84]   Chang, M.H.: On Razumikhin-type stability conditions for stochastic
           functional differential equations. Mathematical Modelling, **5**, 299–307
           (1984)
[Cha87]   Chang, M.H.: Discrete approximation of nonlinear filtering for stochastic
           delay equations. Stochastic Anal. Appl., **5**, 267–298 (1987)
[Cha95]   Chang, M.H.: Weak infinitesimal generator for a stochastic partial dif-
           ferential equation with time delay. J. of Applied Math. & Stochastic
           Analysis, **8**,115–138 (1995)
[Cha07a]   Chang, M.H.: Hereditary Portfolio Optimization with Taxes and Fixed
           Plus Proportional Transaction I. J. Applied Math & Stochastic Analysis.
           (2007)
[Cha07b]   Chang, M.H.: Hereditary Portfolio Optimization with Taxes and Fixed
           Plus Proportional Transaction II. J. Applied Math & Stochastic Analy-
           sis. (2007)
[CPP07a]   Chang, M.H., Pang, T., Pemy, M.: Optimal stopping of stochastic func-
           tional differential equations. preprint (2007).
[CPP07b]   Chang, M.H., Pang, T., Pemy, M: Optimal control of stochastic func-
           tional differential equations with a bounded memory. preprint (2007)
[CPP07c]   Chang, M.H., Pang, T., Pemy, M.: Finite difference approximation of
           stochastic control systems with delay., preprint (2007)
[CPP07d]   Chang, M.H., Pang, T., Pemy, M.: Viscosity solution of infinite dimen-
           sional Black-Scholes equation and numerical approximations. preprint
           (2007)
[CPP07e]   Chang, M.H., Pang, T., Pemy, M.: Numerical methods for stochastic
           optimal stopping problems with delays. preprint (2007)
[CY99]   Chang, M.H., Youree, R.K.: The European option with hereditary price
           structures: Basic theory. Applied Mathematics and Computation., **102**,
           279–296 (1999)
[CY07]   Chang, M.H., Youree, R.K.: Infinite dimensional Black-Scholes equation
           with hereditary price structure. preprint (2007)

[CCP04]   Chilarescu, C., Ciorca, O., Preda, C.: Stochastic differential delay dqua-
          tions applied to portfolio selection problem. preprint (2004)
[CM66]   Coleman, B.D., Mizel, V.J.: Norms and semigroups in the theory of
          fading memory. Arch. Rational Mech. Anal., **23**, 87–123 (1966)
[Con83]   Constantinides, G.M.: Capital Market Equilibrium with Personal Tax.,
          Econometrica., **51**, 611–636 (1983)
[Con84]   Constantinides, G.M.: Optimal stock trading with personal taxes: im-
          plications for prices and the abnormal January returns. J. Financial
          Economics., **13**, 65–89 (1984)
[CIL92]   Crandall, M.g., Ishii, H., Lions, P.L.: User's guide to viscosity solutions
          of second order partial differential equations. Bull. Amer. Math. Soc.,
          27, 1–67 (1992)
[DZ92]   Da Prato, G., Zabczyk, J.: Stochastic Equations in Infinite Dimensions.
          Cambridge University Press, Cambridge (1992)
[DS96]   Dammon, R., Spatt, C.: The optimal trading and pricing of securities
          with asymmetric capital gains taxes and transaction costs. Reviews of
          Financial Studies., **9**, 921–952 (1996)
[DN90]   Davis, M.H.A., Norman, A.: Portfolio selection with transaction costs.
          Math. Operations Research, **15**, 676–713 (1990)
[DZ94]   Davis, M.H.A., Zariphopoulou, T.: American options and transaction
          fees. In: *Proceedings IMA Workshop on Mathematical Finance*, Springer,
          New York (1994)
[DU04]   Demiguel, A.V., Uppal, R.: Portfolio investment with the exact tax basis
          via nonlinear programming. preprint (2004)
[DU77]   Diestel, J., Uhl, J.J.: Vector Measures. Mathematical Surveys Mono-
          graph Series **15**, AMS, Providence (1977)
[Doo94]   Doob, H.: Measure Theory. Springer-Verlag, New York (1994)
[DS58]   Dunford, N., Schwartz, J.T.: Linear Operators, Part I: General Theory.
          Interscience Publishers, New York (1958)
[Dyn65]   Dynkin, E.B.: Markov Processes. Volumes I & II, Springer-Verlag,
          Berlin-Göttingen-Heidelberg (1965)
[DY79]   Dynkin, E.B., Yushkevich, A.A.: Controlled Markov Processes, Springer,
          Berlin-Heidelberg-New York (1979)
[EL76]   Ekeland, I. and Lebourg, G.: Generic differentiability and pertubed op-
          timization in Banach spaces. Trans. Amer. Math. Soc., **224**, 193-216
          (1976)
[Els00]   Elsanousi, I.: Stochastic Control for Systems with Memory. Dr. Scient.
          thesis, University of Oslo (2002)
[EL01]   Elsanousi, I., Larssen, B.: Optimal consumption under partial observa-
          tions for a stochastic system with delay. preprint (2001)
[EOS00]   Elsanousi, I., Oksendal, B., Sulem, A.: Some solvable stochastic con-
          trol problems with delay. Stochastics and Stochastic Reports, **71**, 69–89
          (2000)
[EK86]   Ethier, S.N., Kurtz, T.G.: Markov Processes: Characterization and Con-
          vergence. Wiley, New York (1986)
[Fab06]   Fabbi, G.: Viscosity solutions approach to economic models governed by
          DDEs., preprint (2006)
[FFG06]   Fabbri, G., Faggian, S., Gozzi, F.: On the dynamic programming ap-
          proach to economic models governed by DDE's. preprint (2006)

[FG06]   Fabbi, G., Gozzi, F.: Vintage capital in the AK growth model: a dynamics programming approach. preprint (2006)

[FR05]   Ferrante, M., Rovina, C.: Stochastic delay differential equations driven by fractional Brownian motion with Hurst parameter $H > \frac{1}{2}$. preprint (2005)

[FN06]   Fischer, M., Nappo, G.: Time discretisation and rate of convergence for the optimal control of continuous-time stochastic systems with delay. Preprint (2006)

[FR06]   Fischer, M., Reiss, M.: Discretization of stochastic control problems for continuous time dynamics with delay. J. Comput. Appl. Math., to appear (2006)

[FR75]   Fleming, W.H., Rishel, R.W.: Deterministic and Stochastic Optimal Control. Springer-Verlag, New York (1975)

[FS93]   Fleming, W.H., Soner, H.M.: Controlled Markov Processes and Viscosity Solutions. Springer-Verlag, New York (1993)

[FPS00]   Fouque, J.-P., Papanicolaou, G., Sircar, K.P.: Derivatives in Finanical Markets with Stochastic Volatility. Cambridge University Press (2000)

[Fra02]   Frank, T.D.: Multiviariate Markov processes for stochastic systmes with delays: application to the stochastic Gompertz model with delay. Phys. Rev. E, **66**, 011914–011918 (2002)

[FB01]   Frank, T.D., Beek, P.J.: Stationary solutions of linear stochastic delay differential equations: applications to biological system. Phys. Rev. E, **64**, 021917–021921 (2001)

[GR05a]   Gapeev, P. V., Reiss, M.: An optimal stopping problem in a diffusion-type model with delay. preprint (2005)

[GR05b]   Gapeev, P. V., Reiss, M.: A note on optimal stopping in models with delay. preprint (2005)

[GNS01]   Garlappi, L., Naik, V., Slive, J.: Portfolio selection with multiple risky assets and capital gains taxes., preprint (2001)

[GS99]   Gatarek, D., Świech, A.: Optimal stopping in Hilbert spaces and pricing of American options. Math. Methods Oper. Res., **50**, 135-147 (1999)

[GMM71]   Goel, N.S., Maitra, S.C., Montroll, E.W.: On the Volterra and other nonlinear models of interacting populations. Rev. Modern Phys., **43**, 231–276 (1971)

[GM04]   Gozzi, F., Marinelli, C.: Stochastic optimal control of delay equations arising in advertising models. preprint (2004)

[GMS06]   Gozzi, F., Marinelli, C., Saving, S: Optimal dynamic advertising under uncertainty with carryover effects. preprint (2006)

[GSZ05]   Gozzi, F., Swiech, A., Zhou, X.Y.: A corrected proof of the stochastic verification theorem within the framework of viscosity solutions. SIAM J. Control & Optimization, **43**, 2009–2019 (2005)

[Hal77]   Hale, J.: Theory of Functional Differential Equations. Springer, New York, Heidelberg, Berlin (1977)

[HL93]   Hale, J., Lunel, S. V.: Introduction to Functional Differential Equations. Applied Mathematical Sciences Series, Springer, New York, Heidelberg, Berlin (1993)

[HK79]   Harrison, J.M., Kreps, D.M.: Martingales and arbitrage in multi-period security markets. J. Economic Theory, **20**, 381-408 (1979)

[HP81]   Harrison, J.M., Pliska, S.R.: Martingales and stochastic integrals in theory of continuous trading. Stochastic Processess Appl., **11**, 215-260 (1981)

[HK90]   Horowitz, J., Karandikar, R.L.: Martingale problems associated with
            the Boltzman equation. In: (Ed: E. Cinlar et al) Seminar on Stochastic
            Processes, Birkhaüser, Boston, pp. 75-122, (1990)
[Hul00]   Hull, J.C.: Options, Futures and Other Derivatives. 4th edition, Prentice
            Hall (2000)
[HMY04]   Hu, Y., Mohammed, S.A., Yan, F.: Discrete-time approximations of sto-
            chastic delay equations: the Milstein scheme. Ann. Prob., **32**, 265-314,
            (2004)
[IW81]   Ikeda, N., Watanabe, S.: Stochastic Differential Equations and Diffusion
            Processes. North-Holland, Amsterdam (Kodansha Ltd., Tokyo) (1981)
[Ish93]   Ishii, K.: Viscosity solutions of nonlinear second order elliptic PDEs
            associated with impulse control problems. Funkcial. Ekvac., **36**, 123–
            141 (1993)
[Itô44]   Itô, K.: Stochastic integral. Proceedings of Imperial Academy., Tokyo,
            20, 519–524 (1944)
[Itô84]   Itô, K.: Foundations of Stochastic Differential Equations in Infinite Di-
            mensional Spaces. SIAM, Philadelphia, Pennsylvania (1984)
[IN64]   Itô, K., Nisio, M.: On stationary solutions of a stochastic differential
            equation. Kyoto J. Mathematics, **4**, 1–75 (1964)
[IS04]   Ivanov, A.F., Swishchuk, A.V.: Optimal control of stochastic differential
            delay equations with applications in economics. preprint (2004)
[JK95]   Jouini, E., Kallah, H.: Martingale and arbitrage in securities markets
            with transaction costs. J. Economic Theory, **66**, 178–197 (1995)
[JKT99]   Jouini, E., Koehl, P.F., Touzi, N.: Optimal investment with taxes: an
            optimal control problem with endogeneous delay. Nonlinear Analysis,
            Theory, Methods and Applications, **37**, 31–56 (1999)
[JKT00]   Jouini, E., Koehl, P.F., Touzi, N.: Optimal investment with taxes: an
            existence result. J. Mathematical Economics, **33**, 373–388 (2000)
[Kal80]   Kallianpur, G.: Stochastic Filtering Theory. Volume 13 of Applications
            of Mathematics, Springer-Verlag, New York (1980)
[KX95]   Kallianpur, G., Xiong, J.: Stochastic Differential Equations in Infinite
            Dimensional Spaces. Lecture Notes-Monograph Series Volume 26, Insti-
            tute of Mathematical Statistics (1995)
[KM02]   Kallianpur, G. Mandal, P.K.: Nonlinear filtering with stochastic de-
            lay equations. In: N. Balakrishnan (ed), Advance on Theoretical and
            Methodological Aspects of Probability and Statistics. IISA Volume 2,
            Gordon and Breach Science Publishers, 3–36 (2002)
[Kar96]   Karatzas, I.: Lectures on Mathematical Finance. CMR Series, American
            Mathematical Society (1996)
[KS91]   Karatzas, I., Shreve, S.E.: Brownian Motion and Stochastic Calculus.
            2nd edition, Springer-Verlag, New York (1991)
[KSW04a]   Kazmerchuk, Y.I., Swishchuk, A.V., Wu, J.H.: The pricing of options
            for securities markets with delayed response. preprint (2004)
[KSW04b]   Kazmerchuk, Y.I., Swishchuk, A.V., Wu, J.H.: Black-Scholes formula
            revisited: securities markets with delayed response. preprint (2004)
[KSW04c]   Kazmerchuk, Y.I., Swishchuk, A.V., Wu, J.H.: A continuous-time
            GARCH model for stochastic volatility with delay. preprint (2004)
[KS03]   Kelome, D., Swiech, A.: Viscosity solutions of infinite dimensional Black-
            Scholes-Barenbaltt equation. Applied Math. Optimization, **47**, 253-278
            (2003)

[KLR91]   Kind, P., Liptser, R., Runggaldier, W.: Diffusion approximation in past-dependent models and applications to option pricing. Annals of Probability, **1**, 379-405 (1991)

[KLV04]   Kleptsyna, M.L., Le Breton, A., Viot, M.: On the infinite time horizon linear-quadratic regulator problem under fractional Brownian perturbation. preprint (2004)

[Kni81]   Knight, F.B.: Essentials of Brownian Motion and Diffusion. Mahtematical Surveys, **18**, American Mathematical Society, Providence (1981)

[KS96]   Kolmanovskii, V.B., Shaikhet, L.E.: Control of Systems with Aftereffect. Translations of Mathematical Monographs Vol. 157, American Mathematical Society (1996)

[Kry80]   Krylov, N.V.: Controlled Diffusion Processes. Volume 14 of Applications of Mathematics Series, Springer-Verlag, New York (1980)

[Kus77]   Kushner, H.J.: Probability Methods for Approximations in Stochastic Control and for Elliptic Equations. Academic Press, New York (1977)

[Kus05]   Kushner, H.J.: Numerical approximations for nonlinear stochastic systems with delays. Stochastics, **77**, 211-240 (2005)

[Kus06]   Kushner, H.J.: Numerical approximations for stochastic systems with delays in the state and control. preprint (2006)

[KD01]   Kushner, H.J., Dupuis, P.: Numerical Methods for Stochastic Control Problems in Continuous Time. second edition, Springer-Verlag, Berlin and New York (2001)

[Kusu95] Kusuoka, S.: Limit theorem on option replication with transaction costs. Annals of Applied Probability, **5**, 198-221 (1995)

[Lan83]   Lang, S.: Real Analysis. 2nd ed., Addison-Wesley Publishing Company, Advanced Book Program/World Science Division, Reading, Massachusetts (1983)

[Lar02]   Larssen, B.: Dynamic programming in stochastic control of systems with delay. Stochastics & Stochastic Reports, **74**, 651-673 (2002)

[LR03]   Larssen, B., Risebro, N.H.: When are HJB-equations for control problems with stochastic delay equations finite dimensional. Stochastic Analysis and Applications, **21**, 643-661 (2003)

[Lel99]   Leland, H.E.: Optimal portfolio management with transaction costs and capital gains taxes. preprint (1999)

[Lio82]   Lions, P. L.: Generalized Solutions of Hamilton-Jacobi Equations. Research Notes in Mathematics, Research Note No. 69, Pitman Publishing, Boston (1982)

[Lio89]   Lions, P. L.: Viscosity solutions of fully nonlinear second-order equations and optimal stochastic control in infinite dimensions. III. Uniqueness of Viscosity solutions for general second-order equations. J. Functional anlysis, **62**, 379-396 (1989)

[Lio92]   Lions, P. L.: Optimal control of diffusion processes and Hamilton-Jacobi-Bellman equations. Part 1: The dynamic programming principle and applications and part 2:Viscosity solutions and uniqueness. Comm. P.D.E., **8** 1101-1174, 1229-1276 (1992)

[LMB90]   Longtin, A., Milton, J., Bos, J., Mackey, M.: Noise and critical behavior of the pupil light reflex at oscillation onset. Phys. Rev. A, **41**, 6992-7005 (1990)

[Mal97]   Malliavin, P.: Stochastic Analysis. Springer-Verlag, Berlin (1997)

[Mao97]   Mao, X.: Stochastic Differential Equations and Their Applications. Hor-
           wood Publishing, Chichester (1997)
[MM88]   Marcus, M., Mizel, V.J.: Stochastic functional differential equations
           modelling materials with selective recall. Stochastics, **25**, 195–232 (1988)
[MN06]   Mazliak, L., Nourdin, I.: Optimal control for rough differential equations.
           preprint (2006)
[MT84]   Mizel, V.J., Trutzer, V.: Stochastic hereditary equations: existence and
           asymptotic stability. J. of Integral Equations, **7**, 1–72 (1984)
[Mer73]   Merton, R.C.: Theory of rational option pricing. Bell J. Economics and
           Management Science, **4**, 141-183 (1973)
[Mer90]   Merton, R.C.: Continuous-Time Finance. Basil Blackwell, Oxford (1990)
[Moh84]   Mohammed, S.E.: Stochastic Functional Differential Equations. Re-
           search Notes in Mathematics **99**, Pitman Advanced Publishing Program,
           Boston, London, Melbourne (1984)
[Moh98]   Mohammed, S.E.: Stochastic differential systems with memory: theory,
           examples, and applications. In: Decreusefond, L., Gjerde, J., Oksendal,
           B., Ustunel, A.S. (eds), Stochastic Analysis and Related Topics VI, The
           Geilo Workshop 1996, Progress in Probability, Birkhauser (1998)
[NSK91]   Niebur, E., Schuster, H. G., Kammen, D.: Collective frequencies and
           metastability in networks of limit cycle oscillators with time delay. Phys.
           Rev. Lett., **20**, 2753–2756 (1991)
[Nua95]   Nualart, D.: The Malliavin Calculus and Related Topics. Springer-
           Verlag, Berlin-Heidelberg, New York (1995)
[NP88]   Nualart, D., Pardoux, E.: Stochastic calculus with anticipating inte-
           grands. Probability Theory and Related Fields, **78**, 535-581 (1988)
[Øks98]   Øksendal, B.: Stochastic Differential Equations. 5th edition, Springer-
           Verlag, Berlin, New York (2000)
[Øks04]   Øksendal, B.: Optimal stopping with delayed information. preprint
           (2004)
[OS00]   Oksendal, B., Sulem, A.: A maximum principle for optimal control of
           stochastic systems with delay with applications to finance. In: Mendaldi,
           J.M., Rofman, E., Sulem, A. (eds), Optimal Control and Partial Differ-
           ential Equations-Innovations and Applications, IOS Press, Amsterdam
           (2000)
[OS02]   Oksendal, B., Sulem, A.: Optimal consumption and portfolio with both
           fixed and proportional transaction costs. SIAM J. Control & Optimiza-
           tion, **40**, 1765–1790 (2002)
[Pet00]   Peterka, R.J.: Postual control model interpretation of stabilogram dif-
           fusion analysis. Biol. Cybernet., **82**, 335–343 (2000)
[Pet91]   Petrovsky, I.G.: Lectures on Partial Differential Equations. Dover Pub-
           lication, Inc., New York (1991)
[Pra03]   Prakasa-Rao, B.L.S.: Parametric estimation for linear stochastic delay
           differential equations driven by fractional Brownian motion. preprint
           (2003)
[Pro95]   Protter, P.: Stochastic Integration and Differential Equations. Springer-
           Verlag (1995)
[RBCY06]   Ramezani, V., Buche, R., Chang, M.H., Yang, Y.: Power control for a
           wireless queueing system with delayed state information: heavy traffic
           modeling and numerical analysis. In Proceedings of MILCOM, Wash-
           ington, DC (2006)

[RY99]   Revuz, D., Yor, M.: Continuous Martingales and Brownian Motion. 3rd
            edition, Springer-Verlag, New York (1999)
[Rud71]   Rudin, W.: Real and Complex Analysis. McGraw Hill, New York (1971)
[Shi78]   Shirayev, A.N.: Optimal Stopping Rules. Springer-Verlag, Berlin (1978)
[SKKM94]   Shiryaev, A.N., Kabanov, Y.M., Kramkov, D.O., Melnikov, A.V.: To-
            ward the theory of pricing of options of both European and American
            types II: continuous time. Theory of Probability and Appl., **39**, 61-102
            (1994)
[SS94]   Shreve, S.E., Soner, H.M.: Optimal investment and consumption with
            transaction costs. Ann. Appl. Probab., 4, 609–692 (1994)
[Sid04]   Siddiqi, A. H.: Applied Functional Analysis. Marcel Dekker, New York
            (2004)
[SF98]   Singh, P., Fogler, H.S.: Fused chemical reactions: the use of dispersion
            to delay reaction time in tubular reactors. Ind. Eng. Chem. Res., **37**,
            No. 6, pp. 2203–2207 (1998)
[Ste78]   Stegall, C.: Optimization of functions on certains subsets of Banach
            spaces. Math. Ann., **236**, 171-176 (1978)
[TT03]   Tahar, I.B., Soner, H.M., Touzi, N.: Modeling continuous-time financial
            markets with capital gains taxes. preprint (2003)
[VB93]   Vasilakos, K., Beuter, A.: Effects of noise on delayed visual feedback
            system. J. Theor. Bio., **165**, 389–407 (1993) **19**, 139–153 (1981)
[WZ77]   Wheeden, R.L., Zygmund, A.: Measure and Integral: An Introduction
            to Real Analysis. Pure and Applied Mathematics Series, Marcel Dekker,
            Inc., New York-Basel (1977)
[YW71]   Yamada, T., Watanabe, S.: On the uniqueness of solutions of stochastic
            differential equations. J. Math. Kyoto Univ., **11**, 155–167 (1971)
[Yan99]   Yan, F.: Topics on Stochastic Delay Equations. Ph.D. Dissertation,
            Southern Illinois University at Carbondale (1999)
[YM05]   Yan, F., Mohammed, S.E.: A stochastic calculus for systems with mem-
            ory. Stochastic Analy. and Appl., **23**, 613–657 (2005)
[YBCR06]   Yang, Y., Buche, R., Chang, M.H., Ramezani, V.: Power control for
            mobile communications with delayed state information in heavy traffic.
            In Proceedings of IEEE Conference on Decision and Control, San Diego,
            CA (2006)
[YZ99]   Yong, J., Zhou, X.Y.: Stochastic Control: Hamiltonian Systems and HJB
            Equations. No. 43 in Applications of Mathematics, Springer-Verlag, New
            York (1999)
[Zho93]   Zhou, X.Y.: Verification theorems within the framework of viscosity so-
            lutions. J. Math. Anal. Appl., **177**, 208–225 (1993).
[ZYL97]   Zhou, X.Y., Yong, J., Li, X.: Stochastic verification theorems within the
            framework of viscosity solutions. SIAM J. Control & Optimization, **35**,
            243–253 (1997)
[Zhu91]   Zhu, H.: Dynamic Programming and Variational Inequalities in Singular
            Stochastic Control. Ph.D. dissertation, Brown University (1991)
[ZK81]   Zvonkin, A.K., Krylov, N.V.: On strong solutions of stochastic differen-
            tial equations. Sel. Math. Sov., **1**, 19–61 (1981)

Index

Stochastic Modelling and Applied Probability
formerly: Applications of Mathematics

Stochastic Modelling and Applied Probability
formerly: Applications of Mathematics

51 Asmussen, **Applied Probability and Queues** (2nd. ed. 2003)
52 Robert, **Stochastic Networks and Queues** (2003)
53 Glasserman, **Monte Carlo Methods in Financial Engineering** (2004)
54 Sethi/Zhang/Zhang, **Average-Cost Control of Stochastic Manufacturing Systems** (2005)
55 Yin/Zhang, **Discrete-Time Markov Chains** (2005)
56 Fouque/Garnier/Papanicolaou/Sølna, **Wave Propagation and Time Reversal in Random Layered Media** (2007)
57 Asmussen/Glynn, **Stochastic Simulation: Algorithms and Analysis** (2007)
58 Kotelenez, **Stochastic Ordinary and Stochastic Partial Differential Equations: Transition from Microscopic to Macroscopic Equations** (2008)
59 Chang, **Stochastic Control of Hereditary Systems and Applications** (2008)